INTERNATIONAL SEMINAR ON
NUCLEAR WAR AND PLANETARY EMERGENCIES
25th Session:

WATER — POLLUTION, BIOTECHNOLOGY — TRANSGENIC PLANT VACCINE, ENERGY, BLACK SEA POLLUTION, AIDS — MOTHER–INFANT HIV TRANSMISSION, TRANSMISSIBLE SPONGIFORM ENCEPHALOPATHY, LIMITS OF DEVELOPMENT — MEGACITIES, MISSILE PROLIFERATION AND DEFENSE, INFORMATION SECURITY, COSMIC OBJECTS, DESERTIFICATION, CARBON SEQUESTRATION AND SUSTAINABILITY, CLIMATIC CHANGES, GLOBAL MONITORING OF PLANET, MATHEMATICS AND DEMOCRACY, SCIENCE AND JOURNALISM, PERMANENT MONITORING PANEL REPORTS, WATER FOR MEGACITIES WORKSHOP, BLACK SEA WORKSHOP, TRANSGENIC PLANTS WORKSHOP, RESEARCH RESOURCES WORKSHOP, MOTHER–INFANT HIV TRANSMISSION WORKSHOP, SEQUESTRATION AND DESERTIFICATION WORKSHOP, FOCUS AFRICA WORKSHOP

THE SCIENCE AND CULTURE SERIES

Nuclear Strategy and Peace Technology

Series Editor: Antonino Zichichi

1981 — International Seminar on Nuclear War — 1st Session: The World-wide Implications of Nuclear War

1982 — International Seminar on Nuclear War — 2nd Session: How to Avoid a Nuclear War

1983 — International Seminar on Nuclear War — 3rd Session: The Technical Basis for Peace

1984 — International Seminar on Nuclear War — 4th Session: The Nuclear Winter and the New Defence Systems: Problems and Perspectives

1985 — International Seminar on Nuclear War — 5th Session: SDI, Computer Simulation, New Proposals to Stop the Arms Race

1986 — International Seminar on Nuclear War — 6th Session: International Cooperation: The Alternatives

1987 — International Seminar on Nuclear War — 7th Session: The Great Projects for Scientific Collaboration East-West-North-South

1988 — International Seminar on Nuclear War — 8th Session: The New Threats: Space and Chemical Weapons — What Can be Done with the Retired I.N.F. Missiles-Laser Technology

1989 — International Seminar on Nuclear War — 9th Session: The New Emergencies

1990 — International Seminar on Nuclear War — 10th Session: The New Role of Science

1991 — International Seminar on Nuclear War — 11th Session: Planetary Emergencies

1991 — International Seminar on Nuclear War — 12th Session: Science Confronted with War (unpublished)

1991 — International Seminar on Nuclear War and Planetary Emergencies — 13th Session: Satellite Monitoring of the Global Environment (unpublished)

1992 — International Seminar on Nuclear War and Planetary Emergencies — 14th Session: Innovative Technologies for Cleaning the Environment

1992 — International Seminar on Nuclear War and Planetary Emergencies — 15th Session (1st Seminar after Rio): Science and Technology to Save the Earth (unpublished)

1992 — International Seminar on Nuclear War and Planetary Emergencies — 16th Session (2nd Seminar after Rio): Proliferation of Weapons for Mass Destruction and Cooperation on Defence Systems

1993 — International Seminar on Planetary Emergencies — 17th Workshop: The Collision of an Asteroid or Comet with the Earth (unpublished)

1993 — International Seminar on Nuclear War and Planetary Emergencies — 18th Session (4th Seminar after Rio): Global Stability Through Disarmament

1994 — International Seminar on Nuclear War and Planetary Emergencies — 19th Session (5th Seminar after Rio): Science after the Cold War

1995 — International Seminar on Nuclear War and Planetary Emergencies — 20th Session (6th Seminar after Rio): The Role of Science in the Third Millennium

1996 — International Seminar on Nuclear War and Planetary Emergencies — 21st Session (7th Seminar after Rio): New Epidemics, Second Cold War, Decommissioning, Terrorism and Proliferation

1997 — International Seminar on Nuclear War and Planetary Emergencies — 22nd Session (8th Seminar after Rio): Nuclear Submarine Decontamination, Chemical Stockpiled Weapons, New Epidemics, Cloning of Genes, New Military Threats, Global Planetary Changes, Cosmic Objects & Energy

1998 — International Seminar on Nuclear War and Planetary Emergencies — 23rd Session (9th Seminar after Rio): Medicine & Biotechnologies, Proliferation & Weapons of Mass Destruction, Climatology & El Nino, Desertification, Defence Against Cosmic Objects, Water & Pollution, Food, Energy, Limits of Development, The Role of Permanent Monitoring Panels

1999 — International Seminar on Nuclear War and Planetary Emergencies — 24th Session HIV/AIDS Vaccine Needs, Biotechnology, Neuropathologies, Development Sustainability — Focus Africa, Climate and Weather Predictions, Energy, Water, Weapons of Mass Destruction, The Role of Permanent Monitoring Panels, HIV Think Tank Workshop, Fertility Problems Workshop

2000 — International Seminar on Nuclear War and Planetary Emergencies — 25th Session Water — Pollution, Biotechnology — Transgenic Plant Vaccine, Energy, Black Sea Pollution, Aids — Mother–Infant HIV Transmission, Transmissible Spongiform Encephalopathy, Limits of Development — Megacities, Missile Proliferation and Defense, Information Security, Cosmic Objects, Desertification, Carbon Sequestration and Sustainability, Climatic Changes, Global Monitoring of Planet, Mathematics and Democracy, Science and Journalism, Permanent Monitoring Panel Reports, Water for Megacities Workshop, Black Sea Workshop, Transgenic Plants Workshop, Research Resources Workshop, Mother–Infant HIV Transmission Workshop, Sequestration and Desertification Workshop, Focus Africa Workshop

THE SCIENCE AND CULTURE SERIES

Nuclear Strategy and Peace Technology

INTERNATIONAL SEMINAR ON NUCLEAR WAR AND PLANETARY EMERGENCIES

25th Session:

WATER — POLLUTION, BIOTECHNOLOGY — TRANSGENIC PLANT VACCINE, ENERGY, BLACK SEA POLLUTION, AIDS — MOTHER-INFANT HIV TRANSMISSION, TRANSMISSIBLE SPONGIFORM ENCEPHALOPATHY, LIMITS OF DEVELOPMENT — EGACITIES, MISSILE PROLIFERATION AND DEFENSE, INFORMATION SECURITY, COSMIC OBJECTS, DESERTIFICATION, CARBON SEQUESTRATION AND SUSTAINABILITY, CLIMATIC CHANGES, GLOBAL MONITORING OF PLANET, MATHEMATICS AND EMOCRACY, SCIENCE AND JOURNALISM, PERMANENT MONITORING PANEL REPORTS, WATER FOR MEGACITIES WORKSHOP, BLACK SEA WORKSHOP, TRANSGENIC PLANTS WORKSHOP, RESEARCH RESOURCES WORKSHOP, MOTHER-INFANT HIV TRANSMISSION WORKSHOP, SEQUESTRATION AND DESERTIFICATION WORKSHOP, FOCUS AFRICA WORKSHOP

"E. Majorana" Centre for Scientific Culture
Erice, Italy, 19–24 August 2000

Series editor and Chairman: A. Zichichi

edited by R. Ragaini

World Scientific
Singapore • New Jersey • London • Hong Kong

Published by

World Scientific Publishing Co. Pte. Ltd.

P O Box 128, Farrer Road, Singapore 912805

USA office: Suite 1B, 1060 Main Street, River Edge, NJ 07661

UK office: 57 Shelton Street, Covent Garden, London WC2H 9HE

INTERNATIONAL SEMINAR ON NUCLEAR WAR AND PLANETARY EMERGENCIES 25TH SESSION: WATER — POLLUTION, BIOTECHNOLOGY — TRANSGENIC PLANT VACCINE, ENERGY, BLACK SEA POLLUTION, AIDS — MOTHER–INFANT HIV TRANSMISSION, TRANSMISSIBLE SPONGIFORM ENCEPHALOPATHY, LIMITS OF DEVELOPMENT — MEGACITIES, MISSILE PROLIFERATION AND DEFENSE, INFORMATION SECURITY, COSMIC OBJECTS, DESERTIFICATION, CARBON SEQUESTRATION AND SUSTAINABILITY, CLIMATIC CHANGES, GLOBAL MONITORING OF PLANET, MATHEMATICS AND DEMOCRACY, SCIENCE AND JOURNALISM, PERMANENT MONITORING PANEL REPORTS, WATER FOR MEGACITIES WORKSHOP, BLACK SEA WORKSHOP, TRANSGENIC PLANTS WORKSHOP, RESEARCH RESOURCES WORKSHOP, MOTHER–INFANT HIV TRANSMISSION WORKSHOP, SEQUESTRATION AND DESERTIFICATION WORKSHOP, FOCUS AFRICA WORKSHOP

ISBN 981-02-4669-2

Printed in Singapore by World Scientific Printers

CONTENTS

8. Limits of Development — Megacities

9. Missile Proliferation and Defense — Information Security

10. Cosmic Objects

11. Desertification, Carbon Sequestration and Sustainability

12. Climatic Changes — Cosmic Objects, Global Monitoring of Planet, Mathematics and Democracy, Science and Journalism

1. OPENING SESSION

PLANETARY EMERGENCIES–THE SCIENTISTS' JUBILEE

T.D. LEE, K.M.B. SIEGBAHN, ANTONINO ZICHICHI
Presented by Antonino Zichichi

Dear Colleagues, Ladies and Gentlemen,

I welcome you to the 25th Session of the Planetary Emergencies Seminars and declare the Seminar to be open.

This Seminar is conducted under the patronage of His Holiness John Paul II, as one of the World Federation of Scientists' contributions to the Scientists' Jubilee. The Programme of this Seminar and its associated workshops will include the following topics:

- Black Sea Pollution.
- Potable Water and Pollution.
- HIV transmission from Mother to Infant.
- Transgenic Plants Vaccine.
- Desertification, Carbon Sequestration in Soils and Sustainability.
- Sustainability of Development in Megacities.
- Missiles and Proliferation.
- Energy, Food, Cosmic Objects and Transmissible Spongiform Encephalopathy.

I would like to draw your attention to the workshop, held during the last two days in Erice, on *Mother-Infant HIV Transmission*. The World Federation of Scientists and the World Laboratory have a long tradition of catering to infants' and childrens' needs. Three of our largest and successful pilot projects dealt with heart disease, deafness and neonatology. I would like to encourage our PMP members to pay particular attention to the solving of infants' and childrens' needs.

You all know of course, that this year is a Jubilee Year. The World Federation of Scientists has been at the heart of an ongoing dialogue between Church and Science, for the last twenty-five years.

Twenty years ago, the meeting in the Vatican between H.H. John Paul II and a World Federation of Scientists' delegation was the start of an unprecedented collaboration between Science and Church. Differences over Galileo Galilei's motivations were reconciled and John Paul II has, ever since, given his constant support to our organisation. His visits to Erice, where he gave his blessings to the WFS community, and his ensuing seven statements (see annex) have been a constant reminder of his belief in our ideals.

Three years ago, I proposed to His Holiness to have a special celebration for a Scientists' Jubilee–the first ever in mankind's history. His Holiness readily agreed and included the Scientists' Jubilee in the official list of celebrations. The year 2000 Jubilee therefore marks the closing of the chapter of dissension between Church and Science, and promises an exemplary co-operation for the third millennium.

In commemoration, on Science Day of the Jubilee, 25 May 2000, the World Federation of Scientists, the Ettore Majorana Centre and the World Laboratory have dedicated all the Seminars, Courses and Workshops held in Erice in year 2000 to the Scientists' Jubilee.

Now I would like to remind you of what I said during my closing statement last year. We have entered a period where decision-makers have taken a growing interest in scientific activities. They take important decisions on the basis of what they hear from interdisciplinary experts, most of whom know very little in many fields but are capable of expressing their superficial thoughts in terms that are understood by everybody. Our answer to this is the constitution of strongly specialised Permanent Monitoring Panels, but which include experts from other fields of science.

POTENTIAL FOR TRANSGENIC PLANTS IN VACCINE PRODUCTION

DR. JULIAN K-C. MA
Dept. of Oral Medicine and Pathology, Unit of Immunology, Guy's Hospital, London UK

In this presentation I shall take the opportunity to describe work in a new and extremely exciting area of biotechnology, the development of transgenic plants as an expression system for recombinant vaccine production. This has real potential to benefit the health of mankind, not only in the West, but also, and most importantly, in the developing world.

Most people are aware by now that it is possible to genetically modify plants. This is a relatively recent technology that began in the early 80's. Such is the potential however, that the area has developed very rapidly with many applications. One can divide the uses and applications of genetically modified plants broadly into two areas—those that are designed to benefit plants and agricultural properties and those that are targeted towards improving health of both humans and animals. Those that you will be most familiar with are shown in the top half of this slide. These include the development of plants that are resistant to pests, those that are made resistant to herbicides in order to simplify farming practices, and those that give rise to so-called 'desirable traits'. In terms of medical applications, many of you will have heard of the 'Golden Rice Project' led by Dr. Potrykus. Here new genes have been introduced to encode an iron binding protein and to engineer a metabolic pathway in rice; these are designed to address vitamin A and iron deficiency, important forms of malnutrition in the Indian sub-continent.

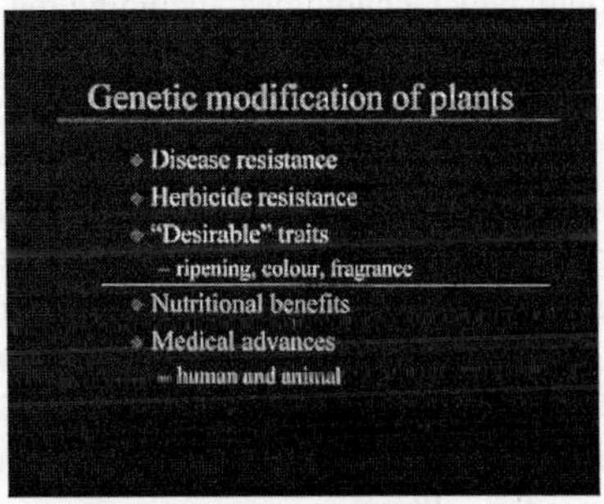

In the last two days our workshop has focused on the final topic—using plants to make vaccines and immunotherapeutic agents. Infectious disease is one of the most important global problems and of course children, who are a focus area for this symposium for Planetary Emergencies, are the main beneficiaries of vaccines.

So why are we interested in using plants? There are many potential advantages, but I consider the most important to be the following: firstly plants are higher eukaryotes. This means that as an expression system for recombinant proteins there are many benefits. They make proteins in a similar manner to mammalian cells, they have cellular machinery and enzymes that are homologous to mammalian counterparts - in short they are eminently suitable for the production of both simple and complex proteins of all kinds. Secondly, plants are the most efficient producers of protein on the planet. They have simple nutritional requirements: soil, sunshine and water. We also have thousands of years of expertise in agriculture. In terms of vaccine production there is a potential to scale up production to agricultural proportions and this would have the benefit of driving down the cost of production. In the Western world, a number of vaccines are available to us and we more or less take these for granted. The sad fact is that in developing countries, the vast majority of vaccines are far too expensive, so although existing technology is effective, it is not delivering products on a global scale. Even in the UK, the cost of the highly effective Hepatitis B vaccine was too high for a vaccination policy that included the entire population. Unfortunately, targeting high-risk groups only has seriously compromised the overall vaccination strategy against this disease. The major health organisations have placed a figure on the affordable cost of vaccines for developing countries. This is U.S. $1 per dose. We firmly believe this target can only be achieved through new technologies, including the use of plants.

In terms of the technical development of this system there are further benefits. We have a lot of experience in processing plants and purification of plant derived compounds. Of course, the extraction and purification of medicinal compounds from plants formed the basis for the science of pharmacology. Thus an enormous number of our best-known drugs from the Pharmacopoeia were originally isolated from plants. Plants are not of course, host to any animal viruses or prions, that might complicate purification methodology. I have already mentioned scale up. Plants are also easily stored and transported as seeds that are highly stable in adverse environments without the need for special facilities. All these factors contribute to low production costs. Furthermore for companies wanting to invest in this technology, the initial capital investment for a production facility is low, compared to alternative technologies.

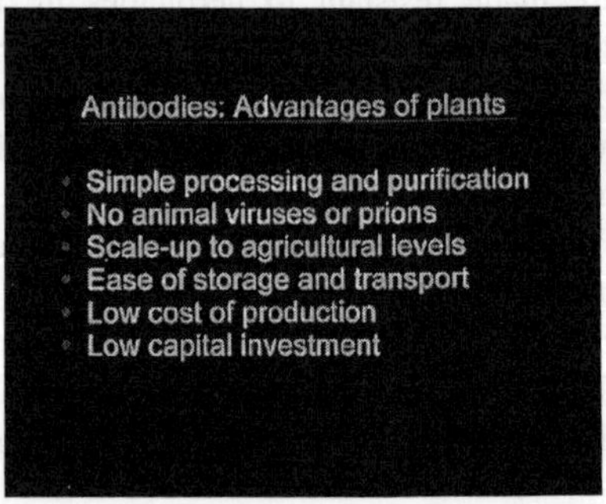

I am going to tell about work in plants that relate to the two approaches to vaccination, active and passive. In active immunisation one takes a virus or bacterial protein, the antigen, and this is usually administered by injection. The body is stimulated to mount an immune response that provides protection against infection by the organism to which the vaccine was made. In passive immunisation, pre-formed protective antibodies are administered directly to the patient, which gives immediate protection. However, this is usually short-lived unless the antibodies are administered repeatedly. Antibodies are proteins that are produced naturally by the white blood cells as part of the immune response against infection. Both active and passive immunisation have their respective advantages and the choice is largely dependent on the disease in question.

I am grateful to Dr. Charles Arntzen for allowing me to illustrate his pioneering work in active immunisation using plant-derived antigens. One of the diseases he has been working on is Hepatitis B. Immunisation with the surface antigen of this virus illicits a protective immune response, indeed this antigen is currently used as a commercial vaccine and is produced in yeast. The gene encoding this antigen was cloned into Agrobacterium, a natural pathogen of plants. This bacteria is used to transfect plant cells which can then be regenerated by *in vitro* techniques into whole plants (for details see Drake et al., Antibody production in plants. In P. Shepherd and Dean (eds). **Monoclonal Antibodies - A practical approach.** Oxford University Press). Many plants can be manipulated in this way, tobacco is a standard choice, but in this case the plant that has been used is potato. This brings us to the important consideration of oral vaccines.

Nobody is fond of injections, particularly children, furthermore in developing countries, the cost of a needle and syringe is an important consideration. Plants can certainly be used to produce vaccines for injection, but the use of edible plants also brings the possibility of immunisation by the oral route. This can be very effective, as demonstrated by the current oral polio vaccine.

The technical hurdle was to express antigens in plants at sufficient levels, but this has now been achieved. Indeed Dr. Arntzen has gone on to demonstrate proof of principle by a feeding study in humans. Volunteers fed transgenic potatoes expressing Hepatitis B surface antigen developed specific antibody responses, which is an important step towards the commercialisation of this plant vaccine.

Hepatitis B will probably be the first target for active immunisation using transgenic plant derived proteins. This slide shows some of the other antigens that are being developed in plants. It includes the gastro-intestinal diseases caused by *E. coli*, cholera toxin and Norwalk virus. This list grows, almost, on a daily basis. A concern however, is that the development of this technology is being undertaken either by academic groups or small biotechnology companies and not by the large pharmaceutical industries, which have the know-how and financial backing to market these products efficiently. One reason for this might be the undermining of the considerable investment that they have already put into conventional methods of vaccine production. Furthermore it is quite clear that vaccines developed particularly for the Third World are unlikely to generate significant profits, despite the large number of people they will benefit.

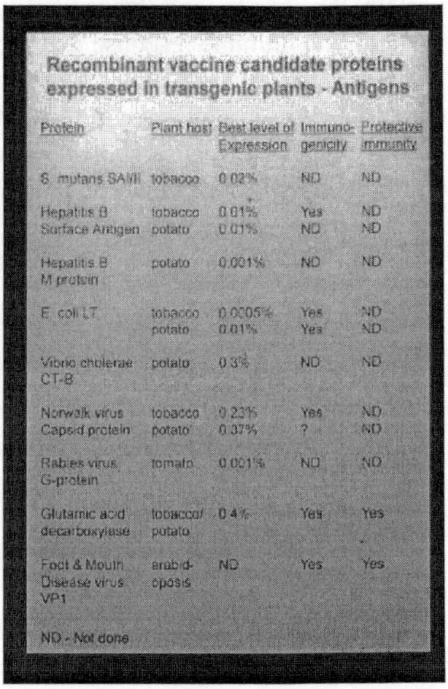

Recombinant vaccine candidate proteins expressed in transgenic plants - Antigens

Protein	Plant host	Best level of Expression	Immuno-genicity	Protective immunity
S. mutans SAI/II	tobacco	0.02%	ND	ND
Hepatitis B Surface Antigen	tobacco potato	0.01% 0.01%	Yes ND	ND ND
Hepatitis B M protein	potato	0.001%	ND	ND
E. coli LT	tobacco potato	0.0005% 0.01%	Yes Yes	ND ND
Vibrio cholerae CT-B	potato	0.3%	ND	ND
Norwalk virus Capsid protein	tobacco potato	0.23% 0.37%	Yes ?	ND ND
Rabies virus G-protein	tomato	0.001%	ND	ND
Glutamic acid decarboxylase	tobacco/ potato	0.4%	Yes	Yes
Foot & Mouth Disease virus VP1	arabid-opsis	ND	Yes	Yes

ND - Not done

I would like now to move onto passive immunisation which, as I have mentioned earlier, involves the administration of pre-formed antibodies to protect against an infectious disease. By using monoclonal antibodies, this can be an extremely safe approach as the antibodies can be safety tested, and there is no risk of generating unwanted immunological side effects. Antibodies are normal proteins, which are produced by white blood cells in response to infection, and they generally bring about the resolution of infection and help to protect against subsequent encounters against the same organism. They are complicated molecules consisting of four proteins: two each of a heavy and a light chain. What makes them difficult to produce in anything other than mammalian cells is the assembly and conformational requirements of the four constituent molecules in order to produce a functional molecule.

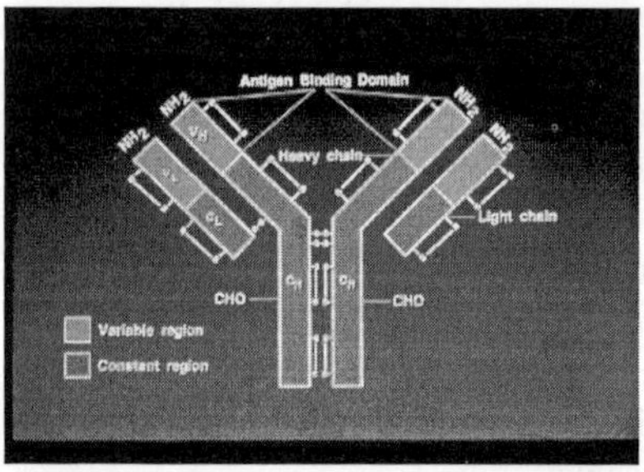

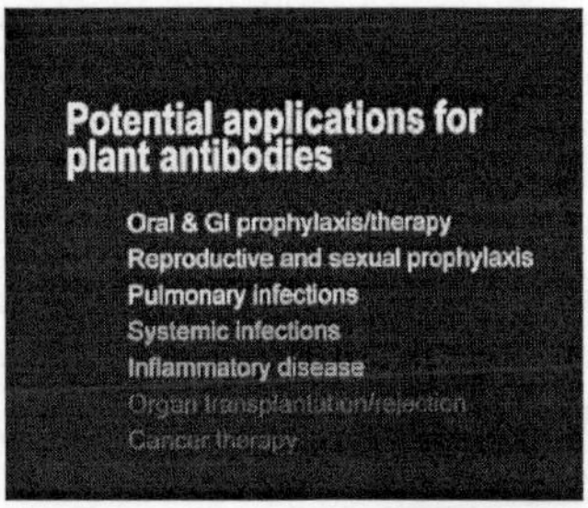

There are several clinical applications for passively administered antibodies. The vast majority of infections occur at the mucosal sites of the body, including the eyes, the respiratory passage, gastro-intestinal tract and the uro-genital tract. These sites are normally protected by mucosal antibodies, so it is an attractive possibility to produce large quantities of pure antibodies to apply directly at these sites to enhance this natural protection. In this slide, the first three applications relate to enhancing the protective immune system at mucosal sites. There are also a number of other applications in which one could use antibodies, either systemically delivered or in organ transplantation, prevention of rejection or possibly cancer therapy. One of the key issues in these applications is the quantity of antibodies that would be required.

On this slide is an estimate of the global requirement for antibodies, based on each type of application. Some are quite modest, for example the treatment of graft rejection for cancer therapy would probably require less than 10 kg of a particular antibody per year. For the treatment of a chronic disease, rather more antibody is required. But for the area of prophylaxis in which we are most interested, much greater quantities of antibody would be needed—in excess of thousands of kilograms per clinical indication.

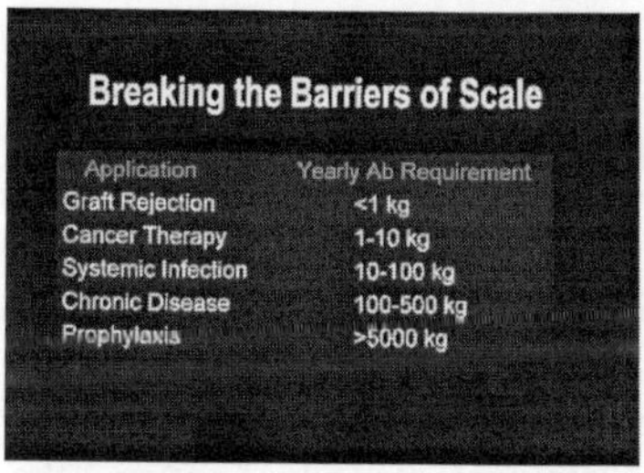

Let me give you one example—dental caries, the disease which we have been studying at Guy's Hospital, for several years now.

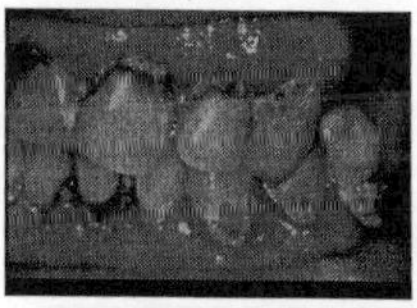

Tooth decay is not a fatal disease, although it is very prevalent affecting up to 96% of the population and it is clearly an infectious disease for which a vaccine would be desirable. The disease is caused by a bacterium, *Streptococcus mutans* which colonises the teeth during childhood.

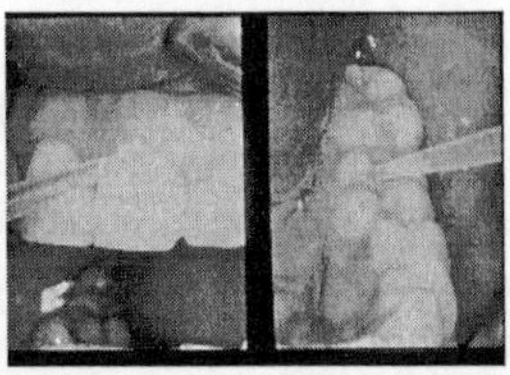

We have developed a monoclonal antibody which can be applied directly onto the teeth to prevent colonisation by this bacterium and therefore disease. The antibody is applied topically, in this instance by a pipette, but we envisage other applications, using either mouthwash or toothpaste or other similar vehicles. A single course of treatment gives long lasting protection and we envisage that for children, a single course of treatment per year might be adequate to prevent colonisation by

Streptococcus mutans. Of course there is also a large adult population who have suffered from tooth decay in the past and who would also benefit from this passive protection.

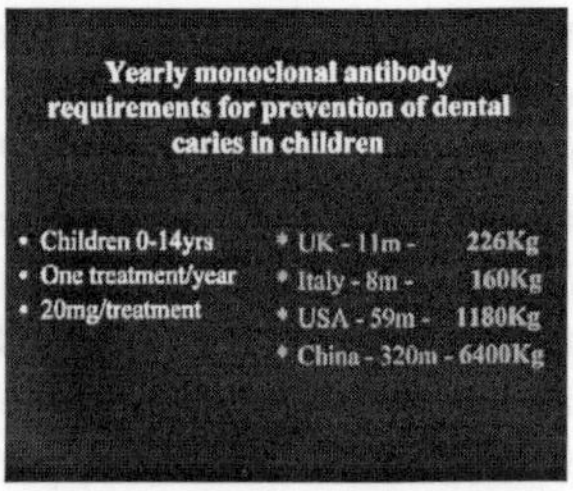

However, just based on the treatment of children up to the age of 14 years, we have calculated the antibody requirements for a commercial product. In most countries in Europe, such as the UK and Italy, hundreds of kilograms per year of this particular antibody would be required. However, in the United States, and China, it rapidly becomes clear that this requirement moves into the thousands of kilogram range.

To put these figures into perspective, current global facilities for production of monoclonal antibody amount to approximately 500 kilograms of a single antibody per year. Clearly then, prophylactic immunotherapy with monoclonal antibodies is beyond our capabilities at present and requires new technologies. Theoretically the production capacity using transgenic plants is limitless, but practically, yields of antibody between thousands and hundreds of thousand kilograms per year are realistic. This not only brings mucosal preventive immunotherapy into range, but also allows us to contemplate other applications of antibodies. These would include penetration of developing countries with potent immunotherapeutic agents, the possibility of pollution control and environmental clean up using pollutant specific antibodies and possibly the replacement of some industrial enzymes with cheaper plant derived antibodies.

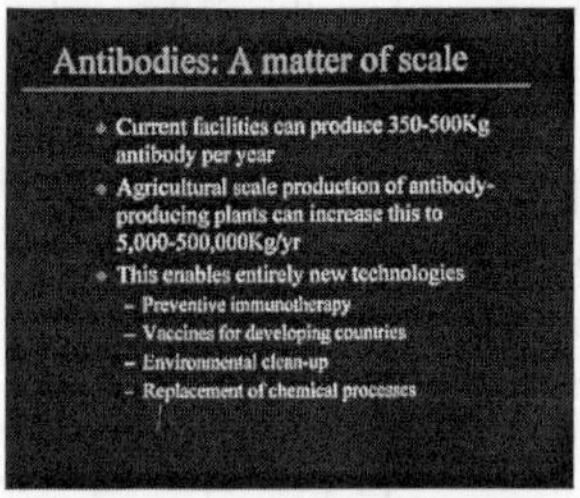

A number of groups, including ourselves have demonstrated the ability of plants to produce antibodies. The monoclonal antibody that I have spoken about earlier that prevents against tooth decay was the first example of this. This antibody has been effective in a human clinical trial and is currently going through phase II clinical trials. There are a number of biotechnology companies that are working to develop this technology and various projections have been made with regard to the cost benefits. One company using corn as the production plant, has proposed the cost of antibodies shown in this slide. Using a production estimate of 250 kilograms a year, the cost per gram of unpurified antibody would be around 20 cents. As the purification increases so does the cost to the GMP standard for systemic administration at a cost of 20

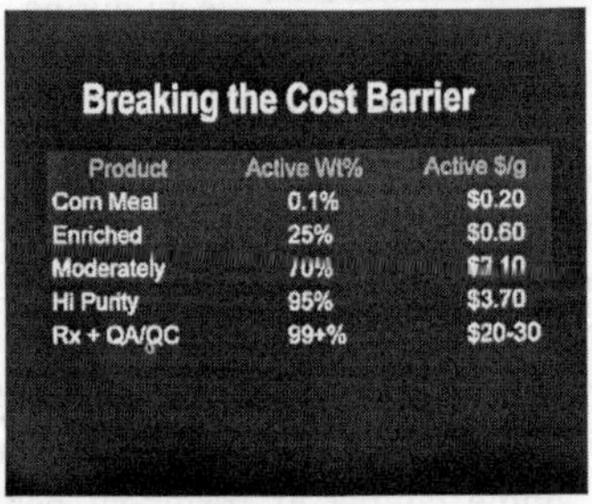

to 30 dollars per gram. This can be compared with the current cost of a monoclonal antibody produced in mammalian cell culture for which the price this year, was approximately $20,000 per gram. Even if the projected figures for plant derived antibodies are inaccurate by one or two orders of magnitude, it is still clear that plants can deliver considerable cost benefits.

There are still a number of obstacles to overcome. Not least of these is the current concern over genetic modification of plants. This has been prominent both in Europe and increasingly so in the United States. It should be recognised that the fundamental technology is shared between those who modify plants for food and those, such as ourselves, who are attempting to produce pharmaceutical products.

The problem is compounded when the perception of public feeling (that is often driven by the media) is acted upon by organisations such as food retailers, whose concerns are unrelated to scientific evidence, but more towards adopting a distinctive stance within a competitive business. These acts tend to aggravate concerns, by persuading the public that there may well be good medical reasons for supermarkets to act in this way.

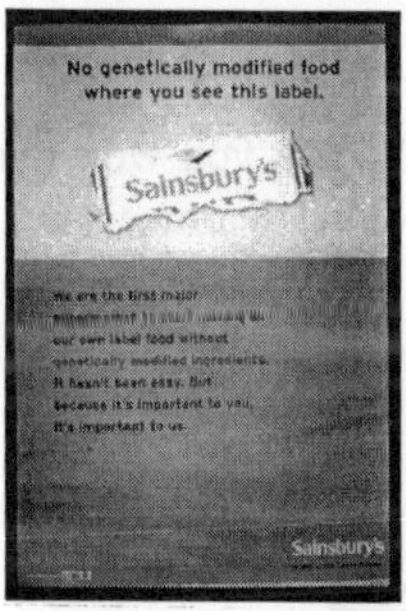

However, although there does seem to be a significant move against genetically modified foods, the public does appear to be in favour of medical advances and benefits to be gained from genetically modified plants. Indeed the anti-GM food lobby are often supportive of the medical applications of the technology, however these views are not widely publicised, lest they detract from the main thrust of these peoples' arguments.

The requirement now is for the scientific and medical community to maintain a clear distinction between GM foods and GM pharmaceuticals. Although we employ the same technology and even though some of the GM pharmaceuticals might be delivered in foods, it should be made perfectly clear that in terms of regulation and development, medical products are treated entirely as pharmaceuticals and go through the same rigorous safety testing and clinical trials required of all new pharmaceutical products.

A number of developmental challenges still face us, but the absolute priority must be to get the first product onto the market, so that the benefits of this technology can be clearly seen. Once this has happened, the development of further products will be facilitated. In this regard one of the most important immediate actions is for there to be harmonisation of regulatory issues regarding this technology on a world-wide basis. The USA has already started to define regulations through the USDA and the FDA. Our concern is that there is no equivalent move within Europe, where these issues are currently dealt with on a national basis. Clearly Europe urgently needs to form it's own guidelines in order that it can work in concert with those of the US. The bottom line is that the technology has been developed and the enormous potential is starting to be tapped. The regulatory bodies now need to assist us in the further development of products.

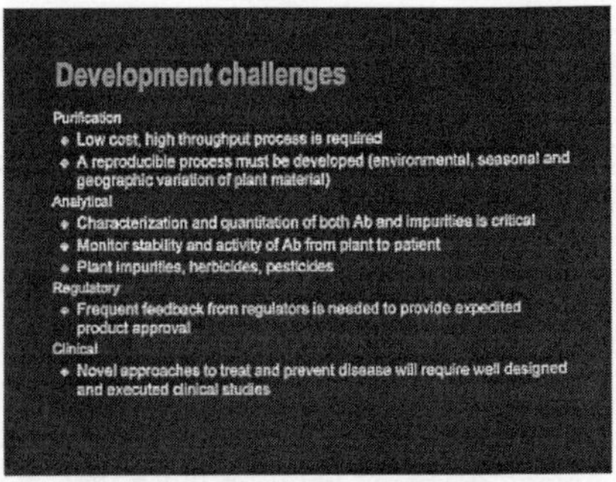

On that note I would like to thank you for your attention. I leave you with this vision of the future in which the landscape is entirely dominated by pharmaceutical factories.

GLOBAL ENERGY PROBLEMS AND PROSPECTS

DAVID BODANSKY
Department of Physics, University of Washington, Seattle, WA 98195, USA

OVERVIEW OF ENERGY-RELATED PROBLEMS

The availability and utilization of energy technologies has had a profound effect in determining the development of human societies. When the resources appeared to be stable—in the slowly changing era of wood, wind, water, and animals—there was relatively little attention paid to energy. Its role was accepted uncritically and even the concept of energy was not formulated.

All this has changed in the past two centuries, as energy from coal, oil, natural gas, uranium, and large-scale hydroelectric sources has transformed the way we live and work. Energy is now a matter of major interest, and concern has grown during the past several decades that the world is too dependent on energy sources that may not prove sufficient for the future and whose use may harm the environment.

This concern is related to a number of fundamental energy problems. The impact of these problems has still not been felt with much direct force, but their collective effect could become critical as this century progresses. They include:

- *The rising demand for energy.* Rising world population and the low present use of energy in developing countries creates the need for additional energy sources.
- *The depletion of fossil fuels.* Fossil fuels now provide about 85% of world primary energy and these resources are limited.
- *Possible global climate change.* The use of fossil fuels, and especially of coal, threatens substantial climate change with possibly serious adverse effects.
- *Difficulties in finding replacements for fossil fuels.* No replacements exist that are accepted, beyond any serious dissent or doubt, as being abundant, practical, and clean.
- *Possible social disruption or war.* If competition for energy sources and for water supplies becomes sufficiently severe, the result could be social disruption or war.

This listing makes clear the central role of fossil fuels. We are highly dependent on them, but their use creates environmental problems and their loss may create severe social and economic problems.

PATTERNS OF ENERGY USE

Distribution of Energy Among Countries

The overall energy problems are exacerbated by the very uneven distribution of energy consumption among the countries of the world, reflecting the wide disparities in economic and technological development. The United States and Western Europe, with 13% of the world's population, accounted in 1998 for 43% of the world's primary energy consumption and 47% of the world's electricity generation[1].

Illustrative data on per capita electricity and energy consumption are presented in Table 1. It is seen, for example, that Japan consumes about nine times as much electricity per capita and six times as much primary energy as China. In a wider extreme, the per capita consumption of electricity and primary energy in the United States is more than 100 times that in Bangladesh.

Table 1. Per capita consumption of electric power (average kilowatts) and total primary energy (gigajoules) in selected countries, 1998).

Country	**Electricity (average kWe)**	**Primary Energy (GJ)**
United States	1.42	370
Japan	0.84	178
Western Europe	0.62	155
Eastern Europe & FSU	0.40	132
World	0.24	67
China	0.09	29
India	0.05	14
Bangladesh	0.01	3

Given the finite nature of the world's energy resources and the environmental impacts of energy use, it is important for all countries to produce and use energy efficiently and particularly for the largest per capita users to moderate their use rates. But the key difficulty is not so much that the industrialized countries are using more than "their share," but that the developing countries are using too little energy. This reflects the fact that they have not yet reached a desirable level of industrialization, where "desirable" is defined not only in terms of the standards of the industrialized countries but also by the goals of the developing ones.

Although there are still very great disparities, it should be noted that some of the developing countries have been closing the gap at a substantial rate. Thus, total electrical generation in China was only 13% of the U.S. total in 1980 but has risen to 30% in 1998. But for China and the rest of the developing world there is still a long way to go.

Sources of Energy in the World Today

World consumption of primary energy in 1998 was about 400 exajoules (EJ)[2,3]. The breakdown by individual energy sources is given in Table 2. About 85% of this energy comes from fossil fuels. Petroleum products head the list, because they are crucial in transportation and are used also for heating, electricity generation, and as a chemical feedstock. Renewable energy provides 8% of the reported primary energy with the overwhelming share coming from hydroelectric power. Nuclear energy provides the remaining 7% of primary energy.

Electricity generation totaled 13,600 billion kWh in 1998 or, in alternative units, 1550 gigawatt-years (GWyr). Of this 63% was generated using fossil fuels (mostly coal), 20% with renewable sources, and 17% with nuclear reactors. The 1550 GWyr corresponds to a primary energy of about 150 EJ, roughly 38% of all primary energy[4].

One of the major trends in energy use is the increase in the relative importance of electricity, as the world's energy economy is gradually becoming electrified. The average annual growth in world electricity consumption from 1980 to 1998 (3.1%) was roughly twice the growth rate for primary energy consumption (1.5%)[5]. In the United States, electricity generation in 1998 was roughly 10 times that of 1950, although total energy consumption did not even triple[6].

Table 2. Scenarios for energy consumption and CO_2 emissions in 2050, compared to actual in 1998.

		Scenarios for 2050			
	Actual	IIASA-WEC Scenarios			Sailor
	1998	A3	B	C2	et al.
Population (billion)	5.9	10.1	10.1	10.1	9.0
Primary Energy (EJ)	396	1033	830	597	900
petroleum	158	181	169	110	
coal	92	94	173	62	
natural gas	89	331	188	140	
total fossil fuel	339	606	531	311	300
renewables	31	308	185	211	300
nuclear	26	118	115	74	300
CO_2 emissions (GtC)	6.1	9.3	9.6	5.1	5.5

Data from International Energy Annual[1], Nakicenovic et al.,[7] and Sailor et al.[8] In the latter scenario, the CO_2 production rate is for the same relative mix of fossil fuels as in 1997.

ENERGY SCENARIOS FOR THE FUTURE

Anticipated Growth in Energy Consumption

One can expect world energy consumption to rise substantially, partly due to rising population and partly due to an increase in the per capita consumption in the developing countries. If the entire world now used energy at the per capita rate of Western Europe, world primary energy consumption would be over 900 EJ, not the actual 400 EJ. At one-half the U.S. rate, it would be about 1100 EJ.

A joint study of the International Institute for Applied System Analyses (IIASA) and the World Energy Council (WEC) developed a number of scenarios that describe possible future energy patterns. These are reported in *Global Energy Perspectives*, written by a number of the study participants[7]. Three of the IIASA-WEC scenarios for 2050 are summarized in Table 2: Scenario A3, a high growth, "technology driven," scenario in which heavy use is made of renewable energy (largely biomass) and nuclear energy; Scenario B, which is a "middle course" case; and Scenario C2, an "ecologically driven" scenario. Also listed is a scenario from Sailor *et al.* for a case where extensive use is made of nuclear energy[8]. The latter two scenarios are designed to hold down CO_2 production, but they differ greatly in the postulated use of nuclear energy and therefore in the total available primary energy.

Given the past failures of long-term energy predictions, there is no reason to expect that any scenario developed today will accurately depict the actual course of events. Aside from this generalized skepticism, specific difficulties can be seen in the scenarios of Table 2. Scenario A3 assumes the consumption of more oil and gas than may prove to be available. Scenario C2 couples a decrease in the world's per capita energy production with a large increase in the gross world product, requiring an increase in the efficiency of energy use that may not be achievable. The Sailor *et al.* scenario allows per capita energy consumption to rise through the large-scale use of nuclear energy, but the public support necessary for such a large expansion may be lacking. Scenario B may be closest to a "realistic" picture, but in this scenario carbon dioxide (CO_2) emissions rise substantially.

The value of such scenarios is that they suggest some of the possible directions in which an energy policy can try to move. Some of the constraints and opportunities are discussed in succeeding sections.

FOSSIL FUEL RESOURCES

Oil resources

Warnings of imminent oil shortages have been frequent over the past thirty years, but to date the shortages have not materialized. During this period additional countries have become important producers and estimates of world oil resources have risen. At the same time, oil consumption has not grown at the anticipated rate. Thus, although there have been abrupt increases in oil prices in response to policy-driven cutbacks in oil production, these have not been fundamentally due to global resource limitations.

It is not clear if the predicted world oil crisis is a realistic threat during the next several decades. Recent estimates from the IIASA/WEC study and from a United States Geological Survey assessment[9] reach similar totals for the remaining resource of conventional crude oil. The average of their results is about 2200 billion barrels (bbo). This corresponds to an energy resource of approximately 13,000 EJ.

At the 1999 rate of world oil production — 24 bbo per year[10] — this oil would suffice for about 90 years. But it is highly unlikely that oil production will be flat. Instead it is expected to rise. The profile of oil production is commonly taken to be that of the "Hubbert curve." The Hubbert curve is roughly bell-shaped and its hypothesized future evolution is extrapolated from the production history to date[11]. The date of concern is the year when production peaks, when one-half of the original resource – commonly termed the "estimated ultimate resource" (EUR) – has been extracted. This date is quite insensitive to the amount of the original resource because resource consumption rises to a higher peak if the EUR is greater[12,13,14]. Thus, for an EUR of 2000 bbo the calculated peak is reached in 2004 while for an EUR of 4000 bbo the peak is reached in 2030[14]. For a remaining resource of 2200 bbo, discussed above, the EUR is about 3000 bbo, the peak is reached in 2019, and by 2090 production would be only about one-fifth of the 1999 rate.

It should be noted that an EUR of 3000 bbo is considerably higher than the values estimated in most other studies made over the past two decades. These ranged, as reported in a 1996 summary, from about 1700 bbo to 2600 bbo, with a median near 2000 bbo[12]. In a 1998 estimate by two experienced oil analysts the remaining resource is given as only about 1000 bbo, implying an EUR of under 2000 bbo[13]. The differences between the low and high estimates are obviously more important in a model where consumption remains relatively flat than in the Hubbert model.

Although the Hubbert picture is widely cited, and at some level appears inescapable, predictions made on the basis of it should be taken as suggestive but not necessarily precise in quantitative detail. The Hubbert model achieved a notable triumph in anticipating the peak in U.S. oil production (excluding Alaska) that occurred in about 1970. However, world production has been rising more slowly in recent years than would be expected from this model. It was only 5% higher in 1999 than in 1979, and was actually less in 1999 than in 1997[10].

Beyond conventional oil, resources of unconventional oil (including tar sands, and heavy crude oil, and shale oil) are estimated in the WEC-IIASA studies to somewhat exceed the original conventional oil resource. If these can be economically extracted, the onset of oil shortages would be postponed. [The term "economically extracted" is a highly flexible one. For example, if the efficiency of automobiles (in kilometers per liter of oil) were doubled, then oil that is twice as expensive would remain "affordable."]

This may suggest that there is no reason to be concerned about future oil supplies. However, complacency is not justified. While an oil crisis may not be imminent, oil remains a limited resource and it would be unfortunate if it is further squandered. Our children may or may not pay the penalty but our grandchildren are likely to. Oil demand has been restrained in recent years by a combination of improved efficiencies in use, switching to other fuels, and economic difficulties in some parts of the world. Assuming that the world economy grows, the demand for oil will also grow unless its use is limited to applications where it is uniquely valuable (in particular, transportation).

Finally, it should be noted that oil resources are to a large extent concentrated in limited geographical areas. Oil, and especially oil that is extractable at low cost, is disproportionately found in the Middle East. This gives countries such as Saudi Arabia great power in influencing the global economy and gives countries like the United States a strong incentive to intervene politically or militarily. The greater the world's dependence on oil, the greater the risks of crises.

Natural gas resources

Estimated resources of conventional gas, expressed in energy terms, are somewhat greater than those of oil, and the consumption rate is less. The average of the IIASA/WEC and USGS estimates (which differ by about 20%) is in the neighborhood of 15,000 trillion cubic feet (Tcf), corresponding to roughly 16,000 EJ. Again, in the IIASA/WEC estimate, unconventional resources somewhat exceed conventional ones.

World production of natural gas in 1998 amounted to 83 Tcf[1]. Were the use rate to remain constant, conventional natural gas resources would therefore suffice for almost 200 years. However, there is an incentive to switch to natural gas as the preferred fossil fuel because less CO_2 is produced with natural gas than with coal, for the same energy output. If all the electricity generated in 1998 using fossil fuels were generated entirely by gas-fired plants at a 50% thermal efficiency, about 56 Tcf of gas would be required. If world electricity use doubles in the next two or three decades, as appears quite likely, that alone would mean a very large increase in natural gas use for an expansion based largely on gas. At the same time, gas is also a potential replacement for oil in the heating of buildings and perhaps even in transportation, further hastening the day when supply shortages might occur.

A very large additional natural gas resource may exist in the form of methane hydrates. It is not established that this methane can be extracted from the oceans in an economical and environmentally benign manner. If that can be done, the estimated resource is enormous. In the IIASA/WEC summary, it is the equivalent of about 800,000 EJ. However, this possibility must be viewed as speculative because there has been no significant exploitation of the methane hydrates to date, and it is not known whether this will prove to be a practical resource.

Coal resources

Coal resources are so ample that relatively little effort has gone into determining their actual magnitude. The IIASA/WEC study places these resources at an equivalent of about 140,000 EJ — roughly ten times the resources of conventional oil or natural gas. Were there no concern about the emission of CO_2 and other pollutants, this would suffice for well over a century even at greatly expanded levels of energy use. But the prospect of global climate change, as well as other pollutants from coal, makes the use of this coal — at least by present methods — an unattractive option.

FOSSIL FUELS AND GLOBAL CLIMATE CHANGE

A counterpart to the problems of limited supplies of fossil fuels is the problem of the environmental effects of the use of fossil fuels. In particular their combustion leads to the production of CO_2 with the consequent prospect of substantial climate changes. The expected effects are being studied by scientists in many countries, and some of this work is captured in continual studies by the Intergovernmental Panel on Climate Change (IPCC). The IPCC Second Assessment was published in 1995[15] and the Third Assessment is now in the final review stage.

While the details of the potential climate changes are not firmly established, the results of the analyses by the IPCC and by most atmospheric scientists point to a rise in global temperature, increasing sea level, changing rainfall patterns, and a possible increase in the frequency and severity of violent climate events, such as hurricanes.

The response of the world community to the perceived dangers is reflected in the 1997 Kyoto Protocol. The industrialized countries (a defined group of so-called Annex I countries) are responsible for a disproportionate share of the current emissions, and the protocol calls upon the original Annex I countries to reduce, on average, their greenhouse gas emissions to 95% of their 1990 totals by about 2010. This refers to all greenhouse gases collectively, but carbon dioxide is the dominant component. It was responsible, for example, for 83% of the 1996 greenhouse gas emissions by OECD countries (converted to CO_2 equivalents)[16]. As of early 2000, there were 84 signatories to the Protocol but only 22 of these had ratified it. Of the Annex I countries, all had signed the Protocol but none had ratified it[17].

Lack of ratification does not prevent countries from trying to implement at least the spirit of the Kyoto protocol, but so far there has been little progress in the right direction. Overall emissions by OECD countries rose by about 5.6% from 1990 to 1996 (excluding recently admitted European countries with economies in transition — Poland, Hungary, and the Czech Republic).[16] For the United States, which is slated to be 7% below the 1990 level, greenhouse gas emissions were 8% above this level in 1996 and 10% above in 1998, due almost entirely to a rise in CO_2 emissions[18].

The failure to date of most of the industrialized countries to curb their CO_2 emissions, coupled with the needs of the developing countries to increase their energy production, means that it will be difficult to avoid a large increase in the emission of CO_2 into the atmosphere over the next several decades. What will happen beyond that depends on the success of efforts to increase energy efficiency and to change the mix of energy sources.

FOSSIL FUEL SOLUTIONS TO REDUCING CARBON DIOXIDE EMISSIONS

Substitution of natural gas for coal

Natural gas, which is mostly methane (CH_4), is a relatively hydrogen-rich fossil fuel. In consequence, the CO_2 emission per unit energy production is only about 56% as great for natural gas as for coal. Therefore, switching from coal to natural gas gives an immediate payoff in reduced CO_2 production. The United Kingdom and to some extent Germany – the two main coal users in Western Europe – have decreased their CO_2 emissions by replacing some of their coal with gas. In particular, in the United Kingdom coal consumption was cut almost in half from 1990 to 1998 and the energy loss was made up almost entirely by an increase in natural gas consumption[1]. Of course, this partial solution is only possible to the extent that natural gas is available to a country, which is the case for United Kingdom from North Sea sources and for Germany through imports.

Natural gas has a further advantage over coal in electricity generation in that the recently developed combined-cycle combustion-turbine gas-fired plants run at considerably higher thermal efficiency than traditional coal-fired plants. Thus, the substitution of natural gas for coal in electricity generation provides a relatively quick fix to the CO_2 problem. However, it is only a partial measure and its long-term practicality is tied up with the uncertain status of natural gas resources.

Fossil fuels without carbon dioxide

Although in the combustion of any fossil fuel, carbon and oxygen are combined to produce CO_2, it is not inevitable that this CO_2 be emitted to the atmosphere. In recent years there has been increasing consideration of the option of capturing the CO_2 after combustion and storing it in the oceans or in geological sites. An alternative approach is to process natural gas or coal with steam, producing hydrogen gas and CO_2 as the main outputs. The hydrogen can be used as a clean fuel, perhaps in a fuel cell, and the CO_2 is collected and sequestered.

In one of the earliest implementation of sequestration, about one megaton of CO_2 (0.3 MtC) from natural gas combustion is being pumped annually into the seabed of the North Sea, off the coast of Norway. This is being done in response to a Norwegian tax of \$170 per ton of C on CO_2 emissions into the atmosphere[19]. Other suggested sites for CO_2 storage are the oceans themselves or various formations in the ground such as oil and gas reservoirs and coal beds.

Fossil fuel combustion produces a world total of over 6 gigatons of C (GtC) per year (over 20 Gt of CO_2). No one expects that the CO_2 from small, distributed sources such as automobiles will be captured, but the amount from power plants and other concentrated sources is large, and successful sequestration of this component would be a major advance.

However, this approach is still not proven on a large scale. There remain questions as to the permanence of the storage, the environmental effects of injecting large amounts of CO_2 into the oceans or ground, and on the possibility of accidents. Even the legal status of disposal of CO_2 in the oceans is in doubt, given the Law of the Sea. With the newness of the field in mind, a 1999 U.S. government report called for an expanded Research and Development program to be "oriented toward understanding more fully the fate of the sequestered CO_2 and the impacts it will have on the environment and on human safety, and toward developing options to ensure a flexible response.[20]"

REPLACEMENTS FOR FOSSIL FUELS: RENEWABLE ENERGY

General characteristics of renewable energy:

Renewable energy sources have a number of favorable features:

- *The potential resource is very large.* The most important renewable energy sources, including hydroelectric power and wind, derive their energy from the sun[21]. The annual solar flux at the surface of the Earth is about 3 million EJ and is even higher at the top of the atmosphere. Capturing 0.1% of this energy would provide more than enough energy for any plausible future scenario.
- *The source will be constant for millions of years.* The constancy of the sun's output makes solar energy truly "renewable."
- *The sources are clean.* None of the forms of solar-derived energy produce CO_2 or other important pollutants.

These characteristics make solar-derived sources attractive candidates as replacements for fossil fuels. However, there are also difficulties:

- *Solar energy sources are land intensive.* The solar source is very diffuse. The typical average flux in mid-latitude regions (e.g. the United States) is about 200 MW/km^2, which corresponds to an annual solar input of 0.006 EJ/km^2. Large areas are therefore required if substantial amounts of energy are to be obtained — especially for biomass energy where the efficiency for capturing solar energy is low.
- *Many renewable sources are intermittent.* The renewable energy forms with the greatest unlimited potential – wind, photovoltaic, and thermal solar – vary with weather conditions and time of day.
- *The base of experience is limited.* There is in an inverse correlation between ultimate expandability and present rates of use. Thus, we may be vulnerable to surprises if these sources are used on a large scale.

This last point is brought out in the data of Table 3. Hydroelectric power provided about 19% of world electricity in 1998 and biomass about 1%, together accounting for 98% of the electricity generation from renewable sources. The two renewable sources that have the most open-ended prospects for expansion — wind and direct solar (including photovoltaic) — together provided only about 0.1%. It is important to have a broader base of experience before concluding that these will be very major contributors to future energy supply.

Table 3. World use of renewable energy sources for the generation of electricity, 1998.

Source	Generation GW yr	Percent of Tool of all electricity	of renewables
Hydroelectric	293	19	93
Biomass	16	1	5
Geothermal	4.7	0.3	1.5
Wind	1.4	0.09	0.1
Solar and Photovoltaic	0.3	0.02	0.1
ALL	315	20	100

Note: Data from U.S. DOE, Energy Information Administration[22].

We now turn to a more detailed discussion of some of the renewable sources that have received particular attention as potential major future contributors to world energy supply.

Direct use of solar thermal energy

The most important impact of solar energy does not appear in any energy budgets and is often forgotten. This is the sun's warming and lighting of the Earth. Conscious efforts to design houses to capture and retain additional solar energy are often considered in the category of energy conservation, rather than as exploitation of solar energy. Whatever viewpoint is taken, however, important gains can be achieved with energy-conscious building designs that exploit the solar input and thereby reduce the demand for other sources of energy. It is also possible to use direct solar energy for the heating of water and this has been done on a large scale in relatively sunny regions.

It is also possible to use solar thermal energy directly for electricity generation. In such arrangements, solar energy is concentrated by mirror systems and used to heat a fluid that drives a heat engine coupled to an electric generator. A variety of mirror configurations have been employed for this purpose, but to date this has not become a major contributor.

Biomass

Extensive future expansion of the use of biomass is projected in many scenarios. Already biomass is an important contributor to the world's primary energy supply. It provides a great deal of "non-commercial" energy in the developing world through the intensive use of wastes and, unfortunately, the utilization and depletion of forests. In some industrialized countries, particularly the United States, forest wastes are used on a large scale to provide energy to the wood and paper industries.

A large expansion of the biomass contribution will require dedicated plantations. However, the efficiency of photosynthesis is low, the growing season is limited, and the land coverage is not complete. Roughly speaking, providing 1 GWyr of electrical energy would require about 3000 to 4000 km^2 in a dedicated biomass plantation. In principle, large amounts of land are available in the world for expanded biomass use[7]. However, the desirability of establishing such plantations has to be judged in the context of the need of land for food production, the value to be attached to keeping some land free of human use, and the requirements for water and fertilizers in biomass production.

On balance, assuming other alternatives exist, it appears that large-scale biomass plantations for energy production may be an extravagant use of land. The intensive use of waste products from agriculture and the forest industries could, on the other hand, provide a more modest but less intrusive source of additional biomass energy.

It may also be noted that biomass and agricultural soil can provide a sink for CO_2, assuming that the biomass is not later used as a fuel.

Wind

Large amounts of energy are potentially available from wind, and efforts have intensified over the past two decades to capture this energy using modern wind turbines. The United States, Germany, and Denmark have made the most extensive use of wind power, but together their 1998 generation amounted to only 1.0 Gwyr[22]. On a fractional basis, Denmark is the leader, with 8% of its 1998 electricity provided by wind power.

Taking into account the availability of both wind and appropriate locations for wind turbines, one estimate places the potential worldwide wind resource at about 12,000 TWh per year, i.e. an annual output of about 1400 Gwyr[23]. This is almost equal to total world electricity generation in 1998. With variations dependent upon the local wind, it requires in the neighborhood of 500 km^2 of land area per GWyr of annual output. Most of this land is empty space between the wind turbines and could be used for grazing or some types of agriculture.

A very large number of wind turbines would be required. For example, if turbines with a rated capacity of 1 MWe are used, over 3000 turbines would be required (allowing for a capacity factor of about 30%). The acceptability of so many large units – on economic, environmental, and aesthetic grounds – remains to be seen. That is a question that perhaps can be better judged after there is a further expansion of what to date has been a relatively small program. Were wind to provide a large fraction of a region's electricity, it would be necessary to address the electricity storage problem caused by the intermittency of the winds.

REPLACEMENTS FOR FOSSIL FUELS: NUCLEAR ENERGY

Nuclear energy status

Nuclear energy now provides about 17% of the world's electricity. Two extremes can be contemplated for nuclear energy's future:

- *Gradual phaseout.* In this scenario, few new nuclear plants are ordered in the next several decades and most existing plants are shut down when they reach the end of their normal lifetime – or perhaps sooner.
- *Revival and expansion.* In this scenario, nuclear power is embraced by the public and by the government in a large number of countries, with large-scale new construction beginning by about 2010. In such a scenario nuclear capacity

might rise from about 350 GWe today to the neighborhood of 4000 GWe by 2050, corresponding to roughly 300 EJ of primary energy.

It is unlikely that either of these extreme scenarios will be fulfilled. There are sufficient differences in national attitudes towards nuclear energy and in dependence on it that no path is likely to soon be adopted on a worldwide basis. France obtains close to 75% of its electricity from nuclear power and is essentially saturated, while South Korea obtained 43% in 1999 and is expanding rapidly[24]. On the other hand, Germany and Sweden are planning to phase out nuclear power and Italy shut down the last reactors in its small program in 1990.

A major present obstacle to nuclear power is economic. In many countries, for example the United States, it is cheaper to build a high-efficiency combined-cycle gas turbine generator, fueled by natural gas. As long as the price of natural gas remains low, neither nuclear power nor renewable sources can compete. This situation is likely to change only if the price of natural gas rises substantially or concern about greenhouse gas emissions leads to the imposition of penalties or rewards that alter the relative costs of different sources — for example, a carbon tax to reflect some of the external costs of gas-fired and (much worse) coal-fired power plants.

However, there are important obstacles to nuclear power, quite apart from economics. These are concerns over nuclear reactor safety, nuclear waste disposal, and nuclear weapons proliferation.

<u>Nuclear Reactor Safety</u>

Excluding the Chernobyl-type reactors (the RBMK reactors) there have by now been about 9000 reactor-years of operation with one accident that resulted in damage to the reactor core — the 1979 reactor accident at Three Mile Island (TMI) in Pennsylvania — and no accident which resulted in an appreciable exposure of the public outside the site. This corresponds to a historical rate of about 10^{-4} core damage accidents per reactor year. Since the TMI accident, the nuclear industry and government regulators have worked to improve reactor safety. As a result, in an illustration of the success of these efforts for existing U.S. reactors, the rate of small failures which might be precursors to full-fledged accidents, dropped by about a factor of 100 from 1974-78 to 1994-1998[25,26].

The next generation of nuclear reactors can be anticipated to be still safer than the present one. One option is to rely on "evolutionary" reactors that are fundamentally similar to existing models but whose design takes advantage of past experience to improve safety and economy. Two such reactors are already in operation in Japan. Additional options are the more fundamentally changed "advanced" reactors that place greater reliance on passive safety features. For any of these, it is not unreasonable to expect that the calculated probability of a TMI-scale accident will be under 10^{-6} per reactor-year and the probability of a Chernobyl-scale accident (i.e., with a large external release of radioactivity) will be a factor of 10 or 100 less.

In a world with a massive nuclear complement of, say, 4000 reactors, this would imply only a 4% chance of a TMI-scale accident per decade and a much smaller chance of another Chernobyl. If this record could actually be achieved, nuclear reactor accidents would not be a great problem. Of course to accomplish and maintain such performance will require sustained care in the design, construction, and operation of reactors.

Nuclear Waste Disposal

Most countries are now planning to put the wastes from reactor operation deep underground — so-called "deep geological disposal." The wastes are in the form either of spent fuel or reprocessed wastes and are to be contained in rugged protective canisters. Alternatives are to imbed waste canisters below the seabed or to transmute the wastes before disposal to reduce the amounts of selected long-lived radionuclides. Neither of these alternatives has yet been adopted by any country.

In geological disposal, the combined "defenses" of the canisters and their immediate surroundings and of the geological formations in which they are placed should prevent the escape of radionuclides into the accessible environment for thousands of years. Beyond 10,000 years, and especially beyond 100,000 years, it is possible that there will be some escape of radionuclides into the environment but by then there will be a large diminution in the overall activity and the releases involved are small. Overall, although much of the public is quite concerned about nuclear wastes, most of the scientists who have looked into the matter believe that on a technical level it is a solvable problem — in the sense that disposal can be accomplished without creating major dangers for future populations.

This sort of optimism is reflected in the planned regulatory standards. Thus, the U.S. Environmental Protection Agency put forth in 1999 a proposal calling for a radiation dose limit of 0.15 millisieverts per year for the "reasonably maximally exposed individual" living near the waste repository site[27]. This limit is to apply for 10,000 years. This dose is 1/20 of the radiation dose (roughly 3 millisieverts per year) received by the average person in the United States from natural sources. This standard has not yet been finally adopted, but its proposal suggests that the anticipated risks are small.

Nuclear Weapons Proliferation

The third concern, over nuclear weapons proliferation, is a very serious one. It involves high stakes and quite uncertain projections of the intentions and capabilities of different countries. The world will always face the danger that some new country will try to develop nuclear weapons. A country can undertake this even if it starts with no base of commercial nuclear energy — e.g., the United States, the Soviet Union, the United Kingdom, France, and China in their acknowledged programs and probably Israel, Iraq and North Korea in their suspected programs. On the other hand, its path will probably be easier and its chances of success greater if a country has a commercial nuclear program, because it will then have more people with relevant technical training as well as some pertinent equipment and materials. There is also the possibility that terrorist or sub-national groups will divert fissionable material from the civilian fuel cycle.

However, given the existence of nuclear weapons arsenals in many countries, the heavy dependence of some countries on nuclear power, and the importance of nuclear materials in medical diagnosis and treatment, it is unrealistic to imagine a truly "nuclear-free" world. There will always remain the potential for the development of nuclear weapons, based either on plutonium-239 or uranium-235. It is therefore desirable to have vigorous monitoring programs under the aegis of a strong and effective International Atomic Energy Agency. This may be more likely to be accomplished if commercial nuclear power is being widely pursued. Thus, while nuclear power may lessen some of the technical barriers against proliferation, strong commercial nuclear programs could lead to a better institutional base for safeguards against proliferation.

Having ample energy sources may also help to reduce some of the conflicts that lead to war. Concern over oil resources was an important factor in Japan's decision to enter into World War II, and the war with Iraq following its invasion of Kuwait was to a considerable extent a war over oil. More generally, scarcity of resources creates dangerous conflicts and one can readily imagine future wars over oil. To the extent that nuclear energy, or any other source, reduces the world's dependence on oil, the less the chance that oil will be an item of critical contention.

The emphasis here is on oil. But water is an even more vital resource, and demand is beginning to exceed the supply in some parts of the world as populations grow. Further, underground reservoirs are gradually being depleted. Water shortages, and their potential for causing conflict, can be alleviated through the desalination of seawater and pipeline transport of water, if there is sufficient affordable energy. Again, an energy-rich world is more likely to be peaceful than an energy-poor one.

Nuclear resources

Estimated uranium resources of 20 million tons suffice for about 100,000 GWyr of operation of present-day light water reactors, for uranium at an equivalent cost of under 0.006 U.S. dollars per kWh[28]. Roughly speaking, this is the equivalent of 10,000 EJ of primary energy — a resource of the same order of magnitude as the resources of conventional oil and gas.

There are also "unconventional" resources of fissionable material in uranium from seawater and in thorium. Seawater contains about 4 billion tons of uranium, which is 200 times the land-based uranium resource indicated above. If extractable at acceptable cost and effort, this could provide nuclear fission energy for over 1000 years. If the practicality of using this resource can be established, there will be no need to face up to the complications of breeder reactors at any time in the near future.

Nuclear Fusion

Nuclear fusion offers the prospect of essentially unlimited resources, but with a very uncertain timetable. An overall energy gain has not yet been achieved in the fusion devices under development, and a date for the practical generation of electricity from fusion power cannot be reliably estimated. Continued large research programs are justified by the payoff that will be realized if the efforts are successful. However, it would be imprudent to base any energy planning on the success of the fusion efforts at any time within, say, the next fifty years.

SUMMARY AND CONCLUSIONS

The energy problem is essentially that of increasing the world's energy supply without greatly increasing the world's use of fossil fuels. Eventually, in fact, this use must decrease in response to limited supplies and to the dangers of global climate change[29]. On the scale of a decade or two, the world is unlikely to feel any severe new consequences either from shortages of fossil fuels — politically-motivated cutbacks excepted — or from changes in global climate. However, on the scale of a century there is a very great problem. If the world continues on its present path, our descendants are likely to face severe shortages of oil and natural gas as well as a climate that has been substantially altered.

It is relatively easy to discuss the situation for these time extremes. It is less clear what will happen over a more workable time frame for "long-term" planning, say, 50-years. However, continuation of our present patterns of producing and using energy hold the potential for the following severe consequences:

- Energy shortages and high prices may hinder the economic progress of developing countries and intensify their material privation.
- The tensions produced by energy shortages and the scramble for the remaining resources of oil or gas may lead to political instability or war.
- Changes in global climate may begin to cause substantial difficulties in some parts of the world, with worse to come from the already committed input of CO_2 and other greenhouse gases into the atmosphere.

Potential approaches to addressing the problem fall into several categories:

The market vision. This is essentially a laissez-faire approach in which it is assumed that adequate energy supplies will be found in response to increased demand, and that environmental problems will be solved when they become apparent. The danger of this approach is that alternatives to fossil fuels, and particularly to coal, may not be developed in a short enough time and on a large enough scale to avoid environmental damage that cannot be satisfactorily ameliorated.

The green vision. This is essentially a vigorous interventionist approach in which energy problems are to be solved by a combination of energy conservation, the use of renewable resources, and the use of much cleaner fossil fuel energy (by some combination of natural gas and carbon sequestration). Again, however, there is a danger of energy shortages as well as the environmental burdens from the preemption of large land areas.

The eclectic approach. This approach is based on the premise that the problems are sufficiently difficult to require the pursuit of a wide variety of technologies. These include continued efforts to use energy more efficiently, the exploration of the practicality of sequestering CO_2 from fossil fuels, the development of the practical forms of renewable energy, the substantial expansion of nuclear fission energy, and continued efforts to develop nuclear fusion. The danger of this approach is that resources will be wasted on impractical or undesirable options.

Ignoring questions of emphasis, the key difference between the "green vision" and the "eclectic approach" is the latter's utilization of nuclear energy. Both approaches require explicit government intervention (for example, subsidies or taxes) to counteract short-term market influences. But they differ in the weight they give to the risks of using nuclear energy and to the risks of trying to get by without it. They also differ in their optimism about the prospects and costs of renewable energy.

It is unlikely that there will be a uniform international approach to energy problems. Different countries are likely to choose different paths, depending in part upon their objective situation with respect to resources and technical capabilities and in part on their attitudes on policy issues. Overall, however, it appears that the costs of pursuing all options are less than the costs of a failure to maintain an adequate energy supply. Thus, until there is a firmer basis for rejecting individual options, prudence favors pursuing a broad array of them, in the spirit of the eclectic approach.

REFERENCES

1. *International Energy Annual 1998*, Energy Information Administration report DOE/EIA-0219(98) (U.S. Department of Energy, Washington DC, 2000).
2. Energy equivalents: 1 gigaton oil equivalent (Gtoe) = 41.87 EJ; 1 quad = 1.055 EJ.
3. These figures reflect energy data reported by governments and other official bodies. There is also a large additional contribution from the use of biomass – wood and wastes – that does not go through commercial channels and is lost to most statistical tabulations.
4. There is an arbitrariness in relating primary energy to electricity generation, for example for hydroelectric power. The average efficiency of fossil fuel sources (about 33%) is here used to provide a conversion factor from kWh to nominal prinary energy (in EJ or quad).
5. From the *International Energy Annual* for 1998 (Ref. 1) and for earlier years.
6. *Annual Energy Review 1998*, Energy Information Administration report DOE/EIA-0384(98) (U.S. Department of Energy, Washington DC, 1999).
7. *Global Energy Perspectives*, Nebojsa Nakicenovic, Arnulf Grübler, and Alan McDonald, eds., (Cambridge University Press, Cambridge, 1998).
8. William C. Sailor, David Bodansky, Chaim Braun, Steve Fetter, and Bob van der Zwaan, "A Nuclear Solution to Climate Change?" *Science* **288**, 1177-8 (2000).
9. *U.S. Geological Survey World Petroleum Assessment 2000 – Description and Results*; at http://greenwood.cr.usgs.gov/energy/WorldEnergy/DDS-60/
10. *Monthly Energy Review, May 2000*, Energy Information Administration Report DOE/EIA-0035(2000/05) (U.S. Department of Energy, Washington DC, 2000).

11. The Hubbert curve is usually presented as a logistic curve, which can be approximated by a Gaussian curve (see Bartlett, Ref. 14).
12 James J. MacKenzie, "Heading Off the Permanent Oil Crisis," *Issues in Science and Technology* **XII**, no. 4, 48-54 (Summer 1996).
13. Colin J. Campbell and Jean H. Laherrère, "The End of Cheap Oil," *Scientific American* **278**, no. 3, 78-83 (March 1998).
14. Albert A. Bartlett, "An Analysis of U.S. and World Oil Production Patterns Using Hubbert-Style Curves," *Mathematical Geology* **32**, no. 1, 1-17 (January 2000).
15. *IPCC Second Assessment, Climate Change 1995*, A Report of the Intergovernmental Panel on Climate Change (World Meteorological Organization and United Nations Environmental Programme, 1995).
16. *National Climate Policies and the Kyoto Protocol* (Organisation for Economic Co-operation and Development, Paris, 1999).
17. From web site of the United Nations Framework Convention on Climate Change, at:
http://www.unfccc.de/resource/kpstats.pdf
18. *Emissions of Greenhouse Gases in the United States 1998*, Energy Information Administration Report DOE/EIA-0573(98) (U.S. Department of Energy, Washington DC, 1999).
19. E. A. Parson and D. W. Keith, "Fossil Fuels Without CO_2 Emissions," *Science* **282**, 1053-4 (1998).
20. *Carbon Sequestration, State of the Science*, Office of Fossil Science working paper, Draft, February 1999 (U.S. Department of Energy, 1999), p. 9-1.
21. We here omit geothermal and tidal energy, which are renewable sources that do not derive their energy from the sun. They are unlikely to play major parts in providing future energy supply.
22. DOE/EIA data printouts (private communication from Patricia Smith, Energy Information Administration).
23. Steve Fetter, *Climate Change and the Transformation of World Energy Supply* (Center for International Security and Cooperation, Stanford, 1999), p. 57.
24. "Nuclear Share of Electricity Generation," *IAEA Bulletin* **42**, no. 1, 58 (2000).
25. T. E. Murley, *Nuclear Safety* **31**, 1, 1990; T. E. Murley, MIT Safety Course (July 1999).
26. W. D. Travers, SECY-99-289 (Nuclear Regulatory Commission, 1999); available at: http://www.nrc.gov/NRC/COMMISSION/SECYS/secy1999-289/1999-289scy.html
27. "Environmental Protection Agency, 40 CFR 197. Environmental Radiation Protection Standards for Yucca Mountain, Nevada, Proposed Rule," *Federal Register* **64**, no. 166 (August 27, 1999), pp. 46976-47016.
28. David Bodansky, *Nuclear Energy: Principles, Practices, and Prospects* (American Institute of Physics Press/Springer-Verlag, Woodbury, NY, 1996).
29. As discussed above, this decrease could be delayed if coal could be used without substantial CO_2 emissions to the atmosphere.

GLOBAL MALNUTRITION

W. PHILIP T. JAMES
Chairman, International Obesity TaskForce/Director, Public Health Policy Group, 321 North Gower Street, London NW1 2NS, UK.

INTRODUCTION

The term global malnutrition conjures up issues of concern for the future of mankind and whether, with the population explosion, environmental deterioration and ever more limited water supplies the world can continue to feed itself. This is a topic considered elsewhere in this meeting. In nutritional terms, however, we are now moving to a new perspective on the world's food problems. We have come to think of the improvements in food supplies and health over the last 100 years as part of steady progress from a time when life expectancy was much lower and the health problems of *homo sapiens* were appalling.

In fact this perspective is probably wrong. If we take the height of adults as an index of the interaction between the availability of high quality foods and environmental factors which have an impact on infection rates, we can see that the height of affluent people in advanced economies is substantially greater than that of the poorer sections of the world. These height differences also apply to the different socio-economic groups within affluent countries such as those in Northern Europe or North America. It is accepted that the greater height of the affluent groups reflects the benefits of a high quality of childcare and healthy lifestyles including their good diet. We still, however, tend to think of national differences in height being the result of major genetic differences between races or regional subgroups. This concept of genetic differences in post-war years was taken to explain the failure of children to grow in Central America, Africa and Asia, but it slowly emerged that if children from these communities were brought up on an excellent diet with clean water supplies, good sanitation, appropriate immunisation programmes then even with only a simple health service they grew as well as their Dutch or American counterparts. We know that African and Indian babies, if born of adequate weight and exclusively breastfed, grow just as fast as the Scandinavian or American child. Therefore we have moved away from assuming that racial or other genetic subgroups have an intrinsically lower capacity for growth although accepting that within any population there are clear genetically-based differences in growth potential. So average adult heights can be taken as a simple index of the final outcome of children's wellbeing.

If with this perspective we look at the heights of adults over the last 11 millennia, it is apparent that the hunter-gatherers of 9000 B.C. were on average nearly as tall as those who achieved a maximum height at around 5000 BC. Subsequently the heights of adult males in the Eastern Mediterranean region where domestic agriculture first emerged 10,000 years ago began to fall. The ancient and modern Greeks grew less well than their ancestors at the beginning of agriculture. More detailed studies of the skeletons show that hunter-gatherers commonly displayed fewer signs of anaemia than the farmers that succeeded them and showed far fewer episodes of bone growth arrest associated with infection. Some specific diseases which are discernible in the skeleton, such as tuberculosis, only emerge relatively recently in prehistory amongst the more densely populated, settled groups[1]. Thus, although we congratulate ourselves on all the benefits that have accrued from the public health measures introduced in the 20th century such as good water supplies, improved sanitation, better food and the advent of modern immunisation and antibiotics, in practice we are now only reverting in stature to the heights of our ancestors seven millennia ago. The dramatic increases in the height of the Japanese over the last 50 years can mostly be attributed to dietary change, but we still have extraordinarily short populations as in Central America and Mexico, with many in the medical profession considering them as genetically short. New studies, however, are showing that, for example, Mexican children on typical maize tortilla and bean diets show growth spurts when they are given supplements of zinc or additional meat with its high zinc and other nutrient content. So it seems likely that we can expect a progressive, intergenerational increase in stature of people in developing countries until they reach an average height of about 1.8 metres.

This analysis may seem somewhat theoretical since for years it was customary to interpret a reduced stature as a useful response to limited food supplies. It was therefore "smart to be small". Unfortunately in the last 20 years the evidence for the hazards of poor growth in children have become overwhelming. We have at present about 180 million underweight children below the age of 5 years and I was privileged to lead a Commission for the inter-UN group, the ACC/SCN[2] to enquire as to why we still had so many "malnourished" children when it would seem that economic developments should be reducing the number at a steady rate. We documented the well-known fact that most of these children are underweight because they are actually too short for their age, but not necessarily wasted. Nevertheless, they all fall below the lowest limits for normal growth in a well-fed population and are therefore classified as showing severe growth retardation. The International Food Policy Institute (IFPRI) estimated that even with the current trends in economic growth and their associated changes in food availability and wellbeing there would still, by 2020, be 120 million stunted and underweight pre-school children. So one billion children will have gone through their early development by 2020 having been handicapped during this period by poor growth.

Of far greater importance than a low stature, however, is the recognition of a clear relationship between a slowing in growth and a slowing in mental development. Stunted children are stunted not only physically, but also mentally. There have been studies in many different parts of the world showing that stunting in children at the age of 2 is associated with lower school achievement, later development and even poor school attendance. Thus, for example, in Pakistan an improvement of only a quarter of a standard deviation score for the height-for-age of 2 year old children will lead, on average, to an increase in subsequent school enrolment rates of 2% for boys and 10% for girls. Jamaican studies also show that stunted children if left without any special attention remain well below the normal mental development track of well-fed children. If groups of stunted children are either fed or simply played with by the mother to stimulate them in verbal and motor skills, then both interventions will lead to very significant improvements in mental development. The combination of maternal training to encourage mental stimulation and food supplements will not only lead to marked improvements in height, but also return their mental development to that of well-nurtured children.

The link between poor children's growth and impaired mental development is remarkably consistent in terms of human experimental studies, both from general observational data and from economic analyses of the consistent associations between poor childhood growth and academic achievement. These findings have remarkable implications which are often underestimated. In a modern society it is increasingly evident that the economic development of a country depends not only on the innovation, flexibility and organisational skills of an élite, but particularly on the intellectual capacity and education of the mass of workers. This demand becomes every more obvious as information technology and modern science accelerate in their applications. We therefore have to confront the fact that the human capacity of many nations has been handicapped by generations of children who have been inadequately fed and brought up in infection-prone environments. Our UN Commission called for a dramatic rethink of social and political priorities because national leaders must confront the implications of these findings and what they signify for the economic development of their country. It has been estimated that 75% of the economic development of Britain over the last 300 years was linked to the improvement in nutrition and health of the population so that the industrial revolution could progress and be sustained by a healthier workforce. Certainly the remarkable improvements in infant mortality rates and increased life-expectancy in the UK which became dramatically apparent at the beginning of the 20th century occurred without any of the major medical advances such as the introduction of antibiotics etc. which came much later. There were, however, remarkable advances in public investment in clean water supplies, sewerage systems and housing, together with, by the 1930s, a dramatic realisation that the so-called genetically stunted children of the poor were in practice eating extremely poor diets, low in both animal protein, micronutrients and often energy. Boyd Orr's studies and that of others between the two world wars led to a dramatic change in government policies throughout the world, with food security being emphasised as fundamental and a new emphasis being placed on the essential requirements for enough food to meet the energy needs of even the most active workers and sufficient animal protein to promote the growth of children.

MICRONUTRIENT DEFICIENCIES

There are currently three micronutrient deficiencies which each threaten more than a billion people, i.e. iodine, vitamin A and iron deficiencies. To these can be added an increasing recognition that zinc deficiency may be a markedly underestimated problem. Modern nutritional science is also highlighting the widespread evidence of biochemical folic acid deficiency, even in affluent societies and there is increasing concern that we have underestimated the importance of adequate zinc intakes. Selenium deficiency is also being assessed since the role of selenium is becoming recognised as far more important than originally thought. Most work, however, has been done on the three prominent deficiencies of iodine, vitamin A and iron.

IODINE DEFICIENCY DISORDERS (IDD)

Iodine deficiency arises because over the millennia iodine has been leached out of the earth's rocks by glaciation, rain and snow; the older exposed mountain ranges are particularly affected, e.g. the Himalayas, the large mountainous areas of China, the Andes and the European Alps. Water derived from these areas is therefore grossly deficient in iodine, but iodine continues to be present in deep soils, river estuaries and is in abundance in the sea. Currently over 1 billion children, adults and their livestock are exposed in 130 countries to the risks of iodine deficiency unless special measures are taken to supplement the food, feed or water supply. For humans supplements are usually organised by iodization of the salt being produced from major salt sources in the affected regions.

Iodine is essential for animal but not plant growth because it serves as the key constituent of the thyroid hormones which control the metabolism of every cell in the body including brain cells. As iodine deficiency threatens, the thyroid enlarges (producing a goitre) in an attempt to capture more iodine. As production of the thyroid hormone, thyroxin, falls, so the metabolism and function of the body slows. Thyroxin's vital role during fetal development means that moderately deficient mothers are likely to produce babies with frank cretinism or children with mental deficiency. There are currently about 50 million iodine-related mentally deficient people and about 16 million cretins in the world[3].

Given the range of iodine deficiency disorders and the permanent effects on brain function, enormous efforts have been made in the last ten years by Hetzel and his colleagues in the International Council for the Control of IDD[4]. In 1990 only 5-10% of the third of the world at risk from IDD had access to iodised salt, but by 1999 the coverage had risen to 68% (with 90% of the people in the Americas being covered). In Europe, however, only 27% of the populations at risk have access to iodised salt because of the collapse of previous arrangements for iodization in some Eastern European countries and because there is still no national commitment in several Western European countries, e.g. Spain, France, Italy and Belgium where mild to moderate IDD persists. A

European initiative is therefore needed with continuing effort elsewhere to ensure the systematic retention of the iodization of all salt being consumed.

VITAMIN A

Progress with vitamin A deficiency has not been so impressive. Vitamin A, derived as such from animal and fish products, particularly liver and fish oils, can also be produced by the body's conversion of some of the carotenoids which colour vegetables and fruit. About 60 of the 600 complex carotenoid molecules can be converted to vitamin A and this is stored in the liver in man. Vitamin A has complex roles in the development of cells and vitamin A deficiency therefore impairs growth and the replication of tissues such as the skin and the mucus membranes of the lungs, intestine and urinary tract. Vitamin A is also vital for the production of the light-responsive pigment in the eye with vitamin A deficiency first becoming apparent as night blindness. The cellular integrity of the skin and membranes is so important that this in addition to a probable effect on the immune system means that vitamin A exerts an important role in preventing and combating bacterial and viral infections. Vitamin A deficiency is a public health problem in 96 countries and especially those in Africa, S.E. Asia and the Middle East. Up to 3 million children die each year from infections exacerbated by vitamin A deficiency and between 250,000 and 500,000 children go blind each year because of vitamin A deficiency. With between 140 and 250 million children under five years being known to be deficient with their greater risk of death and severe infection, vitamin A deficiency is clearly a public health problem for children in particular but also for adults.

Breastfeeding is a highly protective means of ensuring adequate vitamin A intakes provided the mother has vitamin A stores in her liver which, when replete, can last for up to seven years without further vitamin A intake. Vitamin A from animal and fish products are a major source of this vitamin in western countries, but in Africa and Asia children and adults are much more dependent on their intake of fresh vegetables and fruit. Paradoxically vitamin A deficiency occurs in the very countries where green leafy vegetables and fruit usually occur in abundance but are not eaten in adequate amounts either because these crops are now grown as cash crops for export or because children in particular are not encouraged during weaning to start on fruit and vegetables as now advocated in the West. It has recently become apparent that the carotenoids from green leafy vegetables are less available for conversion to vitamin A than those carotenoids found in fruit, but a number of intervention studies in Asia and Africa have demonstrated that it is possible to change substantially the eating practices of the people and limit the occurrence of signs of vitamin A deficiency. Over the last decade major efforts have been made to promote fruit and vegetable consumption but given the huge numbers of children affected, there has also been a systematic attempt to provide a large dose of vitamin A by capsules or even injections so that the child's liver stores can rapidly be repleted and the tragedy of blindness from xeropthalmia eliminated. There has been some controversy between those who emphasise the importance of capsule provision and supplements and those who emphasise that it is essential long-term to transform the eating practices of the susceptible populations. Nevertheless, in practice, a combination of both approaches is important so that we do not wait for the impact of health education and changing cultural habits to alleviate this problem. Since 1990 the number of countries that have implemented a supplementation programme has risen substantially, and since 1998 the Micronutrient Initiative has allowed about 40% of the 96 countries where vitamin A deficiency is a public health problem to include vitamin A supplements in their national immunisation programmes. An increasing number of countries are also implementing food fortification programmes so that again vitamin A as such is given to children and adults. Unfortunately about 30% of all countries with a public health problem still have not identified the magnitude of their problem nor yet developed strategies for action against this problem. Clearly efforts need to be increased and it may be that novel strategies such as the new initiative to develop transgenic rice, rich in both beta-carotene and iron will in due course provide a new mechanism for routinely helping to prevent these micronutrient deficiencies. Much will depend, however, on testing the biological availability of these micronutrients and the access of the poorest farmers in Africa and Asia to this genetically manipulated rice.

IRON DEFICIENCY ANAEMIA

The problem of iron deficiency affects both industrialised and developing countries and is the world's most widespread nutritional problem. Iron deficiency leads to anaemia and this deficiency is the commonest cause of anaemia in industrialised societies. In the developing world other deficiencies, e.g. of folic acid and vitamin B_{12}, also contribute as do infections such as malaria and common genetic disorders such as sickle cell anaemia and thalassaemia. Iron deficiency not only arises because the dietary availability of iron is limited when derived from plant sources and in particularly from cereals, but iron is also lost when there is parasitic infestation of the intestine and urinary tract, e.g. by hookworm, schistosomiasis and amaebiosis. Chronic infections also shut down iron absorption even when it is readily available within the intestine and any other intestinal damage, e.g. from recurrent low-grade tropical enteropathies, also limit iron's uptake. It is therefore not surprising that 2 billion children and adults are known to suffer from anaemia and a billion or more also have iron deficiency signifying their lack of reserves should there be sudden iron losses. The prevalence of anaemia is pandemic in some countries with, for example, 87% of pregnant women having iron deficiency anaemia in India. The S.E. Asian region has a pandemic problem with more than 50% of all pregnant women being frankly anaemic; only Thailand shows what can be done since only 13.4% of Thai pregnant women have anaemia; this is a startling contrast.

The impact of anaemia is huge since it increases maternal and newborn mortality, reduces the health and development of children and probably causes permanent mental damage if children are anaemic in infancy. In adults it has been demonstrated that it impairs immune function and reduces working and productive capacity so that the earning power of anaemic men and women is less. This has clearly been shown to be reversible, but the measures for improving iron status, whilst well-recognised, have not been promoted in an effective way. With about half the world's population living without appropriate sanitation and with at least a third of the world without clean drinking water, it is a mistake to think of dietary measures alone as being sufficient. Iron supplements, now typically given twice a week to take account of the intestine's temporary shut-down of iron uptake when suddenly exposed to high concentrations, can substantially improve iron uptake but the child and adult status does depend on the balance between absorption and iron losses. Not only must iron losses be counteracted by improving sanitary and water conditions as a major priority, but it has to be recognised that women are particularly vulnerable because they lose twice as much iron during the reproductive phase of their lives through menstrual losses and in addition transfer substantial amounts of iron to the baby during pregnancy. Unfortunately, in Asia in particular, women's status is poor and they are less likely to have access to a high-quality diet. If dependent totally on cereals their chance of obtaining enough iron are small. In Asia the problems are amplified because of the dominant vegetarian cultures of so many societies. Modest amounts of meat have clearly been shown to improve substantially iron deficiency and anaemia. This is an issue which has not been properly confronted. Lean meat not only provides a rich source of very readily available iron, but also an excellent source of zinc and selenium as well as the constituents of animal protein and their associated complex molecules which seem to be particularly effective in promoting the growth of children. Recently the UN reviewed progress in combating iron deficiency and it was recognised that far too little had been done to deal with this scourge.

THE LONG-TERM IMPACT OF EARLY MALNUTRITION

Note has already been made of the permanent handicap of stunted and anaemic infants and children, but in the last ten years there has been a renewed recognition that the animal experimental data demonstrating the profound long-term impact of early poor nutrition on long-term growth and function can also be seen in humans. A wealth of evidence is now highlighting the fact that children born small are not only more likely to remain stunted in infancy and childhood, but are handicapped even in adult life. Gambian data has shown that children born during the hungry season of the year, with modest reductions in birth weight, begin to die early in adult life from both a susceptibility to the prevailing infections and because of reproductive problems in women. There is increasing emphasis therefore on the long-term impact of early events on the capacity of the body to build a sustaining and responsive immune system.

Data now emerging reveal that children born small are particularly liable to show signs of metabolic and physiological changes which are conducive to the development of later diabetes, high blood pressure and heart disease. Data from Jamaica, India and South Africa reveal that insulin resistance and a relatively elevated blood pressure become apparent by the age of 4 and persist into adolescence. Now as adults begin to gain more weight, populations in the developing world seem to have a remarkable propensity for depositing any excess fat in the abdominal area which is now recognised to be associated with a very much greater risk of diabetes, high blood pressure, heart disease and stroke.

The worst combination for precipitating these changes seems to be poor growth *in utero* and during the first year or so of life followed by a rapid growth in height and weight with subsequent weight gain in early adult life. It is as though early fetal and childhood experiences had programmed the babe metabolically to expect continued semi-starvation with a limited capacity to adapt to a sudden plentiful supply of food. What is emerging therefore is that our failure to optimise the growth and well-being of girls and young women before as well as during pregnancy, is condemning the next generation to an increased susceptibility to disease. This is of enormous public health significance given the rapid emergence of obesity on a global basis (see later). The determinants of this programming are still uncertain, but a high priority will have to be given to resolving this issue so that coherent public health initiatives can be developed.

ADULT MALNUTRITION

About ten years ago we developed a classification for adult under-nutrition which we termed at that stage chronic energy deficiency (CED)[5]. We then documented to our surprise that there was remarkably consistent evidence from South America, Sub-Saharan Africa and Asia indicating that weights below a BMI of 18.5 were associated with a reduced work capacity and below a BMI of 17 with a clearly documented increased susceptibility to ill-health and an inability to sustain attendance at work. Below a BMI of 16 there was indeed evidence of increased mortality. We then discovered that half the adult Indian population was underweight, this figure rising to 70% when studies were conducted in poor rural communities. In Africa about one third of the adult population at that stage was underweight but evidence was difficult to collate because the issue of malnutrition had been seen as only relevant to children under the age of five. Therefore, the UN, international charities and national governments had rarely bothered to measure adult weights and heights.

Ferro-Luzzi then went on to document the surprising extent of seasonal changes in weight, particularly in Africa and western parts of India. Although the weight reductions were modest it is precisely this degree of weight loss which is associated in the Gambia with mothers having smaller babes suffering long-term handicaps. In women it is clear that the likelihood of their producing low birthweight babies could be directly linked to their low BMI before pregnancy and their poor weight gain during pregnancy. About 20% of the world's babies are born with a low birth weight, i.e. below 2.5 kilos and these low birth weight babes constitute a major susceptible group.

Recent analyses in Africa show that adult weights can drop to BMIs as low as 13 or less and the problems of famine and war have not been properly documented in terms of their impact on adults. Indeed in refugee camps adult weights have, until recently, rarely been measured. It is now becoming clear that the traditional practice of providing modest amounts of food in the expectation that refugees would scavenge to supplement their meagre supplies, is wrong. Therefore we persuaded the UN to increase the minimum energy (calorie) *per caput* allowance for the 20 million or so living in refugee camps. It is however clear that western societies, despite their food abundancies, are not providing either enough food to satisfy the energy needs of the refugees nor food of the right quality to ensure that they do not succumb to multiple mineral and vitamin deficiencies. Some of the most florid episodes of pellagra, scurvy and anaemia have occurred in refugee camps controlled and organised by the international agencies. The whole problem of adult malnutrition therefore needs to be rethought with appropriate documentation. Ferro-Luzzi and I, in association with Indian colleagues, have demonstrated in rural Indian villages where multiple vitamin and mineral deficiencies with impaired immune functions are found, that these deficiencies and impairments are improved by systematic micronutrient supplementation. It seems therefore that we have become cavalier in our approach to the management of refugees, and the problem of adult undernutrition in general needs to be highlighted.

ADULT CHRONIC DISEASES

In the last five years there has been a transformation in our understanding of global health problems because it has become evident that those adult diseases which were considered the consequence of affluence, i.e. obesity, diabetes, coronary heart disease, high blood pressure and stroke, are such common causes of ill-health in the developing world that nearly twice as many people die of coronary heart disease in the developing world as succumb in the industrialised nations.

In 1997 the International Obesity Task Force helped WHO to review the problem of obesity, to standardise its measurement, to specify its multiple complications and to consider its underlying causes. It became apparent that an epidemic was underway in most parts of the world. Central and Eastern Europe had obesity rates in adults which affected 20-30% of the total adult population. Obesity is classified as a BMI of 30 or more, whereas the generous upper limit of normality is taken to be 25. The normal range of BMI has been taken by WHO as 18.5 to 24.9 where BMI equals weight in kilos divided by height in metres squared. Above a BMI of 25 there is an increased risk of diabetes, high blood pressure and abnormal blood lipids. By the time BMIs have risen to 30, then the burden of morbidity and mortality escalates. There is increasing concern that Asian people in particular, but perhaps those of Indian descent in the Americas also, are particularly susceptible to the consequences of excess weight gain with diabetes and high blood pressure emerging particularly in Asia even when BMIs are 23 or 24. It is estimated that two thirds of the world's diabetics have developed their disease because their BMIs have exceeded the designated upper limit of a BMI of 25. Obviously individuals vary in their susceptibility to diabetes, coronary heart disease and high blood pressure and familial genetic inheritance contributes to this susceptibility. What is of concern, however, is that the epidemics of diabetes, heart disease and stroke so evident in developing countries may relate not only to their current inappropriate diet and physical inactivity, but to their enhanced susceptibility programmed in early life. This, to me, seems to be a more likely explanation than an ethnically based, genetic sensitivity.

Most individuals who develop diabetes develop so-called maturity onset diabetes where the body's resistance to the pancreatic production of insulin is so great that it overwhelms the pancreas' capacity to compensate. With age this capacity declines and it would appear that in some poor communities there is a lower capacity than that found in affluent societies. Experimentally, long-term pancreatic insufficiency can be induced by poor fetal and post-natal nutrition, but the dominant feature of diabetes in the Third World seems to be substantial insulin resistance relating to an unusual propensity to abdominal obesity. Numerous claims are emerging that adult Indians living even in the slums of the major Indian cities, although appearing to be normal in weight, have a selective small accumulation of abdominal fat which in both men and women is associated with a surprisingly high proportion of diabetes and its precursor, glucose intolerance. It is estimated that about 12% of Indian slum dwellers have undiagnosed diabetes mellitus with a further 18% with glucose intolerance. Thus an epidemic of diabetes can be expected particularly in Asia and this is already overwhelming the health budgets of Third World countries as they seek to import insulin to deal with the demands. There are already 35 million adults with diabetes in China and if the Chinese lifestyle becomes like that of Chinese living in Mauritius then we might expect up to 150 million Chinese to suffer from diabetes within the next 20 years.

Excess weight gain, with all its complications, seems to arise because of two principal features. First the rise in dietary fat intake and second, the remarkable reductions in physical activity evident in most parts of the world. It must be remembered that recent data suggest that primitive man was not particularly physically active, but the evidence suggests that they were not obese. Fat and sugar were rare components of the diet until the last century or two. New evidence suggests that fat is particularly conducive to weight gain, but there is a synergistic interaction with physical inactivity so that one needs to be on a low fat diet, e.g. 20%, if one seeks to maintain normal body weights whilst remaining sedentary. It is very easy to overeat calories as fat and adults in most parts of the world are steadily gaining weight during adult life whereas traditionally this did not occur. Unfortunately the power of the media and the image of affluent western people eating high fat westernised foods has come to pervade the world and industrial interests obviously benefit if they can sell energy dense foods rich in fat and sugars. These are attractive, easily consumed, loaded with calories and convenient to store and transport. In the developing world the fat, sugar and salt intake of people moving from the countryside to towns increases and within weeks it is evident that adult weights and blood pressures begin to rise.

Salt is now clearly linked to the development of high blood pressure, although there are still some scientists and of course industrial groups who dispute the evidence. Babies brought up on infant formulae with the usual amounts of salt have much greater blood pressure levels at the age of 15 than infants fed a low salt formula and a clear relationship has been established across societies between the salt intake and the prevailing blood pressure levels in a population. New detailed studies form the United States document that the best way of reducing blood pressure in both normal adults and those with frank high blood pressure is to consume a low fat diet rich in vegetables and fruit with a salt intake of less than the 6 gram goal specified by WHO. Indeed, the latest evidence suggest that the ideal intake should be even lower and we have known for years that the hunter gatherers of old survived well on only half a gram, salt being a very rare commodity at that time.

The reason why people develop coronary heart disease is much clearer. The three major risk factors of high blood pressure, smoking and a high blood cholesterol have been repeatedly documented across the globe. CHD is one of the major effects of tobacco use, but it is interesting that the smokers of Japan and China, with their high blood pressure associated with high salt intakes, still do not suffer from CHD to nearly the same extent as other countries. Indeed, Japan has one of the lowest heart disease rates in the world. It would appear that the type of fat eaten is crucially important in the development of CHD. Hundreds of studies have shown that saturated fat, particularly with those fatty acids derived from dairy milk and coconut oil are particular stimulants of the body's synthesis and retention of cholesterol. There is a direct link between the prevailing cholesterol levels of individuals and societies and their risk of CHD. The n-3 fatty acids found in fish oils and vegetables limit the alterations in blood lipids and are protective against CHD in part because they also seem to limit aberrations in the electrical control of the heart. It is for these reasons that scientific expert groups throughout the world have highlighted the importance of reducing the saturated fat intake and having the appropriate balance of polyunsaturated fatty acids with a particular emphasis now on an adequate intake of the fish derived n-3 fatty acids. It is curious that plant breeding has for many years had as one of its objectives, a desire to reduce the n-3 fatty acid content of foods because these fats are much more liable to oxidisation and thereby make foods rancid. The need to preserve food, to lengthen the shelf-life, to ease the transport of foods around the world has meant that the n-3 fatty acid component of the diet has dropped substantially and we have only now begun to realise its consequences.

NUTRITIONAL CHALLENGES TO COME

Nutrition is entering a new and exciting phase. Not only are we beginning to realise how important nutrition is for so many of the prevailing public health problems of the world including the resistance to infections such as HIV, but the interaction of nutrients with the genome means that modern functional genomics is coming to take on nutrition as one of the principal determinants of the complex processes governing gene expression. From detailed laboratory studies we will have to move to clinical and population analyses, but it is clear that even on the basis of our current knowledge, we have an immense amount to do. Most countries only pay lip service to dietary improvements and assume, inappropriately, that all that is needed in a global society is to encourage people to eat a nutritionally balanced mixed diet. In fact the poor in both Western societies and in the developing world do not have access to all the elements that are required to make "informed choices". Those involved in health promotion have known for some years now that health education alone has proved to be a remarkably ineffective way of reducing the burden of disease within a community. In Scandinavia they have shown that sustained, multi-sectoral change with alterations in medical advice to mothers and their children, the galvanising of breastfeeding initiatives, the provision of set high quality menus for nursery and school children, the specification of nutritional quality for public catering and many other measures are needed to transform a nation's diet. The rate of CHD and stroke has been reduced to only a quarter of its previous level by these sustained societal changes, backed by governmental and local authority initiatives as well as changes in agriculture and food processing.

We now need to learn how best to help the world as a whole to cope with the welter of new information, the increasing globalisation of the food chain and the inappropriate perception that it is simply an individual and not a governmental responsibility to improve health. We need to take on board the new challenges coming from a recognition that we may have inter-generational handicaps unless we place special emphasis on the health and welfare of girls and women. In many parts of the world we need to see a transformation in policy and practice because without these changes we can predict that the current pandemic of diet-related diseases is going to escalate. This is, indeed a planetary emergency.

REFERENCES

1. Foster, P. and Leathers, H.D. 1999. The World Food Problem. Tackling the Causes of Undernutrition in the Third World. Second Edition. Lynn Rienner Publishers Inc, Colorado, USA. and London, England, UK.

2. James et al. 2000. Ending Malnutrition by 2020: an Agenda for Change in the Millennium. Final Report to the ACC/SCN by the Commission on the Nutrition Challenges of the 21st Century. Supplement to the Food and Nutrition Bulletin, September/October 2000. UNU International Nutrition Foundation, USA.

3. WHO, 1999. Nutrition for Health and Development. Progress and Prospects on the Eve of the 21st Century. WHO/NHD/99.9. WHO, Geneva.

4. Hetzel, Basil S. 1989. The Story of Iodine Deficiency. An International Challenge in Nutrition. Oxford Medical Publications, Oxford University Press, UK.

5. James, W.P.T., Ferro-Luzzi, A. & Waterlow, J.C. (1988). Definition of chronic energy deficiency in adults. Report of a Working Party of the International Dietary Energy Consultative Group. European Journal of Clinical Nutrition, 42(12), 969-981.

MOTHER TO INFANT TRANSMISSION OF HIV: SUCCESSFUL INTERVENTIONS AND IMPLEMENTATION

CATHERINE M. WILFERT, M.D.
Professor Emerita, Duke University Medical Center
Scientific Director, Elizabeth Glaser Pediatric AIDS Foundation, Chapel Hill, North Carolina, USA

More than 90% of the estimated 4-8 million children who have acquired HIV infection have been infected by transmission of the virus from their mothers.[1] It has been unequivocally proven that interventions can successfully prevent mother to child transmission of HIV. In 1994 the results of a randomized clinical trial (PACTG 076) were announced with the information that an antiretroviral drug (AZT) diminished transmission of HIV from mothers to their infants.[2] The magnitude of the effect was unexpected with a 67% reduction occurring in treated mothers.[2] Over the subsequent 5 years the number of infants reported with AIDS has decreased by an estimated 75% in the United States.[3] Similar decreases have occurred in other nations with the resources to institute antiretroviral treatment of pregnant women. The rapidity of the nationwide decline in pediatric HIV infection and the efficacy of the interventions which appear comparable to that of the clinical trials have been gratifying. This too was not predicted as problems of access, acceptance, and adherence to the regimen could have diminished the effectiveness in the "real world" without the resources and monitoring afforded by a randomized clinical trial.

There have been 8 clinical trials reported as of July, 2000, which demonstrate that mother to child transmission of HIV can be diminished. These trials are summarized briefly in Table 1. The first reported randomized, controlled, double blinded study (076) was conducted in the U.S. and France.[3] The nucleoside reverse transcriptase inhibitor AZT was assessed in a population of women with CD4 counts greater than 200 who did not breastfeed. AZT diminished transmission by 67% when initiated as early as the 14th week of pregnancy, continued through labor via an intravenous infusion, and administered orally to the infant for the first 6 weeks of life.[3] This clinical trial provided the information to change the course of the perinatal epidemic in the developed world. Subsequent surveillance has substantiated that a dramatic decrease in perinatal transmission has occurred in those nations with access to antiretroviral medications.

In 1999 the CDC trial in Bangkok reported the results of a study which initiated AZT at 36 weeks gestation, continued the drug orally during labor, and did not give the drug to the infant.[4] This regimen reduced transmission with a relative efficacy of 51% in a population of women who did not breastfeed. Thailand then began the process of implementation of the study results.

A second trial in Thailand conducted with Harvard and National Institute of Child Health and Development was initiated with a preliminary analysis presented first in Montreal in September 1999.[5] Table 1 and the final analysis was available and presented in Durban in July 2000.[6] The study was a four-armed factorial design combining different durations of drug administration to the mother and baby. The arm with the shortest treatment to mothers (4 weeks) and to babies (3 days) had results comparable to the first CDC Thailand trial (transmission rate = 10.5%). (4) The preliminary analysis showed a significant difference between this arm and that with the longest duration of treatment so the "inferior" arm of the trial was stopped in March 1999.[5] The final analysis showed that all three of the remaining arms were more efficient than the discontinued arm (4.7, 6.5, 8.6% transmission rates respectively) but not significantly different from each other in reducing transmission. (Table 1)[6]

Two trials were conducted in Western Africa in breastfeeding populations.[7-8] AZT was administered antepartum, intrapartum, and in the ANRS trial[(7)] also administered post partum to the mother for one week. These two trials reduced transmission by 37-38% when infants were assessed at 3-6 months. There was no discernible advantage conferred by the post partum week of AZT to the mother in the ANRS study. The efficacy in breastfeeding populations was somewhat less than in Thailand.

The PETRA Trial in Southern Africa was conducted in 5 sites in South Africa, Uganda and Tanzania.[9] AZT/3TC was administered in three different regimens to a population of women who could elect to formula feed in two of the sites. The results demonstrated a 50% reduction in the mothers who received antepartum drugs initiated at 36 weeks gestation. A 37% reduction in transmission occurred with the initiation of drug intrapartum and continued for one week to the infant. AZT/3TC administered only intrapartum failed to reduce transmission. AZT/3TC administered intrapartum and postpartum reduced transmission comparably to AZT alone administered antepartum for 4 weeks in other trials. The administration of drug to the infant was important if mother's medication was started during labor.[9]

The results of the study known as HIVNET 012 conducted in Uganda in a breastfeeding population were greeted with enthusiasm and excitement because of the feasibility of a simple regimen of nevirapine for the developing world.[10] A single dose of nevirapine administered at the onset of labor and a single dose administered to the infant decreased transmission by 47% in comparison to AZT administered during labor and for a week post partum to the newborn infant. The intervention could be given to women in an antenatal clinic and then initiated at the onset of labor by the pregnant woman. Nevirapine is stable, has rapid effects on infectivity and viral replication and is safe.

At the International AIDS meeting in Durban, South Africa the results of additional studies were presented. The SAINT trial compared nevirapine to AZT/3TC and equivalent efficacy was shown at 6 weeks post partum.[11] A trial comparing ddI, ddI/D4T, AZT, and D4T, showed equivalent transmission rates in small numbers of mother infant pairs.[12] The latter observations are important as they extend the number of antiretroviral agents which have been tested and appear to be safe in the perinatal setting.

Therefore, there is no doubt that antiretroviral regimens can diminish the transmission of HIV from mothers to their babies. The data suggest that transmission may occur prior to labor, intrapartum and post partum when the infant is breastfed. Transmission in proximity to delivery is estimated to account for approximately two thirds of the infected infants in a non-breastfeeding population. Breastfeeding may increase transmission by an estimated 14%.[13] It may be responsible for almost 50% of the total infections in populations where breastfeeding is the established and essential means of feeding infants.[14] The successful interventions have primarily diminished transmission which occurs in proximity to delivery. In the developed world where it is possible for women to receive combination antiretroviral treatment the intrauterine transmission of HIV is also being diminished. There are data to indicate that virus burden does correlate with the risk of transmission[15-16] and circumstantial evidence has begun to indicate that optimal suppression of virus burden during pregnancy will further diminish the probability of transmission to the fetus/infant. In the developing world optimal suppression of maternal virus is not yet possible nor is a proven alternative to breastfeeding readily available.

The assessment of overall benefit of antiretroviral interventions in the setting where breastfeeding is essential is a necessary prerequisite to consideration of plans or policies. Published data[17] and information recently presented in Durban document an overall benefit in the reduction of transmission by administration of antepartum/ intrapartum/post partum antiretroviral drugs.[18-19] The longitudinal follow up of available studies in breastfeeding populations is summarized in Table 2.

Table 2. Durability of Perinatal Prevention of HIV.

Study	Transmission rate		reduction	p
Western Africa				
(CDC + ANSRS) (17,18)	5 mos	24 mo		
Placebo	30.6%	30.3%		
AZT	21.5%	21.9%	26%	
PETRA*(HIV or death) (20)	18 mo			
AP,IP,PP	21%			p = .11
IP,PP	25%			p = .60
IP	28%			p = .77
Placebo	27%			
HIVNET 012 (19)	12 mo			
AZT	24.1%			
Nevirapine	15.7%			p = .003
*HIV or death				
AZT	28.8%			p = .004
Nevirapine	19.5%			

The reported results from PETRA do not show an overall benefit of reducing perinatal transmission of HIV in breastfeeding populations but it is important to note the results were reported as mortality or HIV transmission and all other studies report results as a reduction in transmission.[20] Since this study is the only one failing to show an overall benefit it is reasonable to look carefully at the analyses. First, it was done in 5 sites and they differed as to rates of breastfeeding, C-section, and overall mortality. In South Africa where the mortality was lowest the overall benefit from the intervention was sustained.[2] The data for 12 months to two years follow up substantiate a benefit from efficacious antiretroviral interventions administered around the time of delivery.

HIV transmission does occur via breastfeeding but it seems to occur at the same rate in infants who have received an intervention such as AZT or nevirapine. It is important to acknowledge that there does not appear to be an increase in the number of infants infected post partum after a successful reduction in transmission. The interventions do significantly decrease transmission of HIV. Observations on potential ways to further decrease transmission are currently being evaluated and reported. Coutsaudis has reported that exclusive breastfeeding transmits infection at a lower rate than "mixed" feeding[21-22] in an observational study. In some settings the practice of offering water, tea, or porridge in addition of breast milk is common, so the term "mixed feeding" refers to offering anything in addition to breast milk. If exclusion of these other feeds can diminish transmission this could be a practical approach to further reduction in perinatal transmission. There are also studies considering whether the administration of an antiretroviral drug to the infant during the period of breastfeeding can diminish transmission.

Implementation of efficacious interventions should be initiated as quickly as possible. Thailand has conducted pilot studies and the data are convincing that this nation can implement AZT for all HIV infected pregnant women within a year. The annual birth cohort is approximately 1,000,000 babies, with an estimated HIV seroprevalence of approximately 1%. There are an estimated 10,000-15,000 HIV+ deliveries annually. The country has committed itself to providing AZT for mothers, infants, and formula for infants born to HIV infected mothers. To date the acceptance of VCT and the intervention has been very high. (~ 80%) Counselors are being trained for the entire country. It is projected that the implementation will be in place for the entire country by the end of 2000.

Implementation of successful interventions is not easy, requires resources and a commitment to the essential infrastructure. The crisis of HIV/AIDS demands that interventions be made available and accessible as soon as possible. UNAIDS and UNICEF have initiated pilot projects with VCT and a course of AZT similar to that in the Western African Trials. The projects have been underway for approximately 2 years and an estimated 10,000 women have been offered counseling and testing but uptake is variable and the numbers of women receiving AZT is in the hundreds.

The Elizabeth Glaser Pediatric AIDS Foundation and Global Strategies to Prevent HIV Infection initiated the Call to Action in September of 1999. In February 2000, 8 sites were funded. These sites are Ministry of Public Health, Thailand; Central Hospital of Kigali, Gitega, Biryogo and Gikondo Health Center:Kigali, Rwanda, Mulago Hospital in Kampala, Uganda; City Health Clinic/Cato Manor in Durban, South Africa; Chris Hani Baragwanath and Lilian Nygoyi Clinic, Soweto/South Africa; Nyanza Provincial Hospital, Kisumu, Kenya; Kijabe Hospital/Africa Inland Church Medical Missionaries, Kijabe Kenya; and Cameroon Baptist Convention Health Board, Banso Baptist Hospital and Mbingo Baptist Hospital: NW Province Cameroon. The sites in sub-Saharan Africa have a seroprevalence of HIV ranging from 15.5-38%. The number of deliveries range from 1800-35,000 annually. Five of the 8 sites have completed training of counselors and are offering counseling and testing to antenatal attendees. Each of these sites have begun providing the intervention to seropositive women. Reported acceptance of testing ranges from 48-98% with all sites except one achieving greater than 80% acceptance of testing. The initial 5 months of these projects have resulted in initiation of successful implementation projects.

The questions attendant upon the study of perinatal transmission, successful interventions, and implementation of the technology are not all answered. There are continuing ethical dilemmas which need appropriate consideration. This area of investigation was extensively publicized and created open controversy over the issue of the use of placebo controlled trials. The controversies have not ended. The data generated by locations attempting to provide voluntary counseling and testing have lessons for the future of implementation programs. Some of these experiences have demonstrably poor uptake of the testing. That is, fewer than half of the women elect to be tested and even fewer return for results.

When these data are scrutinized, questions are raised concerning the optimal means to proceed with interventions to diminish mother to child transmission. For example:

1. Is it appropriate to withhold the intervention in this setting? Should the intervention be made available to seropositive women as well as to those who refuse testing?
2. Alternatively, in settings where antenatal care is accessed poorly, does one wait for the capacity to offer VCT to be developed?
3. Alternatively, does one offer a successful intervention in a high seroprevalence area without knowing the serostatus of women?

It is provocative to look at the theoretical cost benefit analyses when considering these questions. VCT costs more than nevirapine and is difficult to establish. Routine offering of nevirapine is "cost effective" for areas with a seroprevalence greater than 5%. Recognizing that this is an incomplete means of assessment, we must at least be willing to consider various approaches to implementation of successful interventions. The evaluation of these interventions must occur in order to provide documentation of their success or failure.

HIV infection is unique by virtue of the almost universally fatal nature of this infection, and the rapidity with which it has become a global pandemic. There are multiple interdigitations of cultural beliefs with behaviors which transmit infection and fears creating stigma and discrimination for those who are infected. This epidemic places exorbitant economic demands on infected persons, affected communities and nations where the seroprevalence of this infection has decreased life expectancy, and increased infant and childhood mortality. The HIV pandemic is increasing health and economic disparities on a daily basis. The reduction in perinatal transmission of HIV by existing feasible interventions is the most effective prevention which is available to date. It is essential that resources be mobilized to actively implement these interventions as efficiently as possible.

REFERENCES

1. UNAIDS. AIDS epidemic update: December 1999. Geneva: UNAIDS, 1999 (available online at www.unaids.org/publications/documents/epidemiology/surveillance/wad1999/embacc.pdf).
2. Connor EM, Sperling RS, Gelber R, et al. Reduction of maternal-infant transmission of human immunodeficiency virus type 1 with zidovudine treatment. N Engl J Med 1994; 331: 1173-80.
3. Lindegren ML, Steinberg S, Byers RH "Epidemiology of HIV/AIDS in Children" Chapter 1 in HIV/AIDS in Infants, Children and Adolescents. Pediatric Clinics of N Amer. vol. 47: 1, Feb 2000, p12. Ed. Martha Rogers WB Saunders Co. Phil.
4. Shaffer N, Chuschoowong R, Mock PA, et al. Short-course zidovudine for perinatal HIV-1 transmission in Bangkok, Thailand: a randomised controlled trial. Lancet 1999; 353: 773-80.
5. Lallemont M. Jourdain G, Soyeon K, et al. Perinatal HIV Prevention Trial (PHPT), Thailand: DSMB recommends termination of short-term arm after first interim analysis. 2nd Conference on Global Strategies for the Prevention of HIV Transmission from Mothers to Infants. Montreal, September, 1999 (abstr 016).
6. LeCoeur S. XIII AIDS Conference, July 2000, Durban, South Africa (abstr LbOr03).
7. Wiktor SZ, Ekpini E, Karon JM, et al. Short-course zidovudine for prevention of mother-to-child transmission of HIV-1 in Abidjan, Cote d'Ivoire: a randomised trial. Lancet 1999; 353: 781-85.
8. Dabis F, Msellati P, Meda N, et al. 6-month efficacy, tolerance and acceptability of a short regimen of oral zidovudine to reduce vertical transmission of HIV in breastfed children in Cote d'Ivoire and Burkina Faso: a double-blind placebo-controlled multicentre trial. Lancet 1999; 353: 786-92.
9. Saba J, on behalf of the PETRA Trial Study Team. Interim analysis of early efficacy of three short ZDV/3TC combination regimens to prevent mother-to-child transmission of HIV-1: the PETRA Trial. 6th Conference on Retroviruses and Opportunistic Infections. Chicago, January-February, 1999 (abstr 57).
10. Guay LA, Musoke P, Fleming T, et al. Intrapartum and neonatal single-dose nevirapine compared with zidovudine for prevention of mother-to-child

transmission of HIV-1 in Kampala, Uganda: HIVNET 012 randomised trial. Lancet 1999; 354: 795-802.

11. Moodley D. XIII AIDS Conference, July 2000, Durban, South Africa (abstr LbOr2).
12. Gray G. XII AIDS Conference, July 2000, Durban, South Africa (abstr TuOrB355).
13. Dunn DT, Newell ML, Ades AE, et al. Risk of HIV-1 transmission through breastfeeding. Lancet 1992; 340: 585-588, 1992.
14. Nduati R, John G, Mbori-Ngacha D, et al. Effect of breastfeeding and formula feeding on transmission of HIV-1: a randomized clinical trial. JAMA 2000; 283: 1167-74.
15. Katzenstein DA, Mbizvo M, Zijenah L, et al. Serum level of maternal human immunodeficiency virus (HIV) RNA, mortality, and vertical transmission of HIV to Zimbabwe. J Infect Dis 1999; 179: 1382-87.
16. Mofenson LM, Lambert JS, Stiehm ER, et al. Risk factors for perinatal transmission of human immunodeficiency virus type 1 in women treated with zidovudine. N Engl J Med 1999; 341: 385-93.
17. Ditrame ANRS 049 Study Group. 15-month efficacy of maternal oral zidovudine to decrease vertical transmission of HIV-1 in breastfed African children. Lancet 1999; 354: 2050-51.
18. Wiktor S. XIII AIDS Conference Durban, South Africa, July 2000 (abstr TuOrB354).
19. Owen M. XIII AIDS Conference Durban, South Africa, July 2000 (abstr LbOr01).
20. Gray G. XII AIDS Conference Durban, South Africa, July 2000 (abstr LbOr 05).
21. Coutsoudia A, Pillay K, Spooner E, et al. Influence of infant-feeding patterns on early mother-to-child transmission of HIV-1 in Durban, South Africa: a prospective cohort study. Lancet 1999; 354: 471-76.
22. Coutsoudia A. XIII AIDS Conference, Durban, South Africa, July 2000 (abstr LbOr6).

THE GLOBAL BURDEN OF DISEASE 1990-2020

ALAN D. LOPEZ
World Health Organization, Geneva, Switzerland

Reliable information on the causes of disease and injury in populations and how these patterns of ill-health are changing, is a critical input into the formulation and evaluation of health policies and programs and for the determination of priorities for health research and action. Such assessments must take into account, not only causes of death, but the impact of non-fatal outcomes and the comparative importance of major health hazards or risk factors. The Global Burden of Disease Study, which commenced in 1992, is perhaps the first comprehensive assessment of global health conditions, providing quantitative estimates of premature death and disability from over 100 diseases and injuries and 10 major risk factors, for eight geographical regions of the world, by age and sex. Contributions from death, disability and risk factors have been assessed using a time-based metric of future potential years of life lost, or lived, with a disability, namely Disability-Adjusted Life Years, or DALYs. In 1990, about 1.3 billion DALYs were lost as a result of new cases of disease and injury in that year, almost 90% of which occurred in developing regions. Of the global total, about 52% of DALYs lost in 1990 arose from male mortality and morbidity, compared with 48% among females. The pattern of DALYs lost varied quite markedly between the sexes. For example, at ages 15-44 years, the leading causes of DALYs lost for women (worldwide) were depression, tuberculosis, anaemia, suicide, bipolar disorder and obstructed labor, whereas for men the leading causes were road traffic accidents, depression, alcohol use, homicide, tuberculosis and war.

Of the 10 major risk factors evaluated, malnutrition was by far the leading cause of DALYs worldwide, causing an estimated 16% of the global burden of disease in 1990 (18% in developing regions), with the contributions to disease burden being particularly evident in Sub-Saharan Africa (33%) and India (22%). This was substantially more than for other exposures assessed, including safe water and sanitation (7%), unsafe sex, tobacco, alcohol and occupation (3-4% each).

Projections of the burden of disease were made based on scenarios according to the degree of optimism or pessimism about changes in the variables used to project health status. The baseline assumptions suggest that by 2020, ischemic heart disease will be the leading cause of DALYs worldwide (rising from 5th place in 1990), followed by depression (4th), road traffic accident (9th), stroke (6th), COPD (12th) and lower respiratory infections (1st). On current trends, tobacco is expected to be the leading underlying cause of death and disability worldwide in 2020, causing more deaths (8-9 million) than AIDS, tuberculosis and complications of childbirth combined.

MTBE—THE MEGA CITY PUBLIC HEALTH DEBACLE

LORNE G. EVERETT, PH.D., D.SC.
Chief Scientist, The IT Group, 3700 State Street, Suite 350, Santa Barbara, CA 93105. Tel: (805) 569-9825: Fax: (805) 569-6496: lorne.everett@theitgroup.com

Methyl Tertiary Bethyl Ether (MTBE) was selected primarily by the major oil companies in America to satisfy mandates associated with the Clean Air Act amendments. MTBE has been referred to as the "green additive" and as such it's purpose was to improve air quality conditions in America by causing combustion engines to run cleaner. MTBE replaced lead in gasoline and rather than improving air quality has resulted in enormous surface and subsurface water damages. Many countries internationally are using or considering using MTBE and as such this green additive creates international concern. MTBE is the fourth largest organic produced in America and the number one contaminant identified in America's waters. MTBE is a known animal carcinogen and a potential human carcinogen. This paper describes the various characterization and remediation technologies considered for MTBE cleanup. Examples of MTBE remediation are described relative to the United States Department of Defense National MTBE test site at Port Hueneme and the US EPA National MTBE test site. In addition to exhibiting persistence in the subsurface MTBE has a very low threshold for taste and odor, i.e. in the 5 parts per billion range. Because of the inefficiencies of two stroke gasoline engines many of California's recreation lakes are in serious need of environmental attention. Lake Tahoe, for example, has MTBE concentrations in excess of 30 parts per billion. Based upon recommendations from the University of California the Governor of California has chosen to eliminate the use of MTBE in gasoline in California. Considerable apprehension exists relative to MTBE in new systems. MTBE has been found in groundwater with state of the art tanks and state of the art monitoring systems, which indicate that leakage is not occurring. After a ten year upgrade of underground tanks in America it is unnerving to consider that MTBE is being detected at sites where all indications are, as expressed by the monitoring systems, that the systems are tight. MTBE is found at greater than 85% of the sites where it is considered for analysis. Currently the cleanup costs, as projected by the University of California for MTBE is one to three billion dollars per year.

2. WATER — POLLUTION

COST BENEFIT ANALYSIS FOR THE USE OF MTBE AND ALTERNATIVES

ARTURO A. KELLER
Bren School of Environmental Science and Management, University of California, Santa Barbara, CA 93106, keller@bren.ucsb.edu
LINDA FERNANDEZ
Depts. of Environmental Science and Economics, University of California, Riverside, CA 92501, linda.fernandez@ucr.edu

ABSTRACT

The recent experience in the U.S., and in particular in California, indicates that there are significant costs associated with gasoline additives to address air pollution, when the entire health and environmental impact is assessed. Methyl tert-Butyl Ether (MTBE) is highly soluble, and thus transfers easily to groundwater and surface water bodies, either as a result of gasoline leaks or spills. It presents possible health concerns and definitely affects the taste and odor quality of the water. Thus, the air quality benefits achieved by better combustion of the improved gasoline formulation may be superceded by water treatment costs. In addition, there are several other cost categories that have to taken into consideration in such policy decisions, such as monitoring costs, ecological damages, and restrictions on recreational activities. Complicating the policy making process, one has to take into account the fact that air quality benefits decrease with time, since vehicle technologies are improving, such that the reduction in emissions is not necessarily only due to the gasoline additive, but also to other factors. The current work presents an analysis of the situation in California, as well as a discussion on the aspects of the cost-benefit analysis which may differ for situations such as Mexico City, Beijing, Athens or other cities with rather different air pollution levels and vehicle technologies.

INTRODUCTION

The search for solutions to air quality problems has led to the development of gasoline additives which can have a positive impact on combustion efficiency, significantly reducing emissions of carbon monoxide, ozone precursors and hazardous air pollutants, such as benzene. However, a careful consideration of the impact of these gasoline additives on other media must be made for each circumstance. The case of Methyl tert-

Butyl Ether (MTBE) has served to highlight the potential for cross-media contamination at large scales, when the environmental impact assessment is incomplete.

A recent study[1], as part of a wider evaluation of the health and environmental impacts of MTBE[2], concluded that the air quality benefits for California, derived from the use of MTBE as a gasoline additive at 11 to 15% by volume, were relatively small, and are decreasing with time. The decrease is based on the fact that other policies and technologies implemented to reduce air emissions are becoming more important as the vehicle fleet modernizes. Older vehicles (pre-1990) do not have many of the emissions control devices (e.g. advanced catalytic converters, oxygen sensor feedback, fuel injection), and thus may emit large amounts of carbon monoxide and other air pollutants. The addition of MTBE (or other oxygenated compounds) can reduce the emissions of carbon monoxide from older vehicles. In addition, MTBE is used to replace some of the high octane rating that benzene and other aromatics provide; the substitution reduces the aromatic fraction of the gasoline and thus lowers the emission of air toxics.

The cost-benefit analysis conducted for California examined the human health benefits derived from controlling air pollution, and then systematically analyzed the costs associated with the use of MTBE across the following categories:

- Human health costs due to air pollution from MTBE and its combustion byproducts
- Human health costs due to water pollution derived from MTBE
- Water treatment or alternative water supply costs
- Fuel price increase costs
- Costs due to increased fuel consumption
- Monitoring costs
- Recreational costs
- Ecosystem damages

Although cost-benefit analysis has been used extensively to evaluate alternative policies, there have been to date no studies that examine the cross-media implications of different gasoline (or other fuels) formulations. For example, a recent study focused on the costs and benefits of the Clean Air Act of 1970[3], but considered only the impact of reducing criteria air pollutants on human health, without addressing specific policies to achieve the reductions, such as modifications to fuel formulations, new vehicle technologies, or emission control technologies on stationary sources. Schwing et al.[4] used cost benefit analysis to compare leaded and unleaded fuels, but did not consider the impact of leaded fuels on water supplies. In developing countries where MTBE is being evaluated as a substitute for tetraethyl lead, the water quality damages associated with the use of leaded fuels must also be considered. Krupnick and Walls[5] compared methanol to conventional gasoline in terms of reducing motor vehicle emissions and urban ozone, yet avoided discussing health effects or potential impacts on water quality.

For policy makers charged with deciding between gasoline formulations to achieve improved air quality, a cost benefit analysis can provide answers to the following questions:

- Do the costs of using MTBE outweigh the benefits it produces?
- What are the policy options that are available to reduce the costs of using MTBE and what are the trade-offs?
- How do alternative formulations compare in terms of costs and benefits?

METHODOLOGY

The first step in the analysis is to identify the benefits and cost categories. In addition to the categories indicated in the introduction, it is possible to add other costs, such as litigation, or replacement land for sites contaminated with MTBE, or the cost of drilling new wells for public drinking water supply. The number of categories considered depends to a large extent on the availability of adequate data; some cost categories have too much uncertainty associated with them, or may be deemed not as significant as the principal cost categories. Although there may be cultural differences that place a different value on some of the cost categories, we believe that for most studies the categories we have identified will serve to make adequate policy decisions.

The next step is to identify alternative policies, or in our case, alternative formulations. In the case of MTBE, there are gasoline formulations which use ethanol, toluene or iso-octane, which can provide essentially the same air quality benefits[6,7], with different costs. In addition, since we were interested in determining the cost of adding MTBE, we used as our baseline conventional unleaded gasoline, which was the gasoline formulation sold prior to the introduction of the reformulated oxygenated gasolines in California and other parts of the U.S. Table 1 presents the typical composition of these gasoline formulations.

Table 1. Composition of Gasoline Formulations.

Property	Typical Conventional Gasoline	Oxygenated CaRFG2	Typical CaRFG2 with MTBE	Non-oxygenated CaRFG2[1]
Aromatics, (vol. %)	32.0	max. 25.0	25	22.7
Olefins, (vol. %)	9.2	max. 10.0	4.1	4.6
Benzene, (vol. %)	1.53	max. 1.0	0.93	0.94
Oxygen content (%)	0	1.8-2.7	2.1	0
Sulfur (ppm by weight)	339	max. 40	31	38
Reid Vapor Pressure (psi)	8.7	max. 7.0	6.8	6.9
T90, °F	330	max. 300	293	297
T50, °F	218	max. 210	202	208

[1]based on AQIRP (1997)

Complete details of the methods used to value each benefit or cost category are presented in Fernandez and Keller[8]. Briefly, health benefits were valued using cost of illness[9] or the value of a statistical life[10], depending on whether the pollutant causes morbidity, mortality, or both. The health effects associated with MTBE and other pollutants were derived from the study by Froines et al.[11]. Water treatment costs were based on experimental and field studies[12,13], plus data from Reuter et al.[14] on the number of water reservoirs that are contaminated, and Fogg et al.[15] on the number of groundwater sites requiring treatment. With this information, we were able to integrate an estimate of the cost of water treatment for the State of California.

Market prices were used to estimate the direct cost paid by consumers at the pump due to the mandated use of oxygenated fuels. The increase in gasoline consumption was based on engineering estimates of the decrease in fuel energy content[16], which result in an increased cost of operation. Market prices were also used to calculate monitoring costs incurred to track the extent of contamination, in surface and ground waters, or in ambient air concentrations. We use the travel cost method to value recreational costs from possible restrictions of boats and jet-skis on bodies of water which also serve as drinking water sources. The factor income and restoration cost methods are used to value environmental health costs which account for damage to important environmental goods, such as fish and other sensitive fauna and flora[17,18,19].

Conceptually, the valuation of environmental impacts of each alternative is straightforward. In practice, there are significant difficulties since important elements of the valuation process are not measured or have large uncertainties associated with them. For example, although MTBE may be associated with asthma, the epidemiological studies have not been conducted. Similarly, there is a large uncertainty in the valuation of the effects of reducing air pollutant levels on human health once they are below the air quality standards, or even when they are slightly above, since the toxicological data is at much higher concentrations.

RESULTS

Details of the assumptions used for constructing the figures can be found in Keller, et al.[1] We present here the main features of our results. Table 2 presents the cost and benefits for the three formulations studied (gasoline with MTBE, gasoline with ethanol, and non-oxygenated gasoline, with either toluene or iso-octane). The health benefits are essentially the same across all three formulations, since studies have shown that all these formulations achieve essentially the same carbon monoxide and ozone precursors emissions reductions, within statistical significance. An additional benefit achieved in the California Phase II Reformulated Gasoline formulations[20], is the reduction of benzene content. Benzene is a known human carcinogen[21,22], and the reduced content (less than 1%) in reformulated gasoline results in decreased volatile emissions. We have estimated that this alone reduces 33 to 920 cancer cases per year in California, which are valued at \$165 million to \$4.6 billion dollars over a lifetime (70 years) using the value of a statistical life of \$5 million; on an annual basis, this represents \$2 to \$66 million in savings. The uncertainty in this estimate is derived largely from the uncertainty in the cancer potency factor. In addition, we estimate that the reduction in carbon monoxide may have decreased the number of hospital admissions due to congestive heart failure by up to 840 cases per year.

Table 2. Annualized Cost Benefit Analysis of Fuel Alternatives.

	CaRFG2-MTBE	CaRFG2-Ethanol	Non-oxy CaRFG2
Air Quality Benefits	\$2 to \$84 million	\$2 to \$84 million	\$2 to \$84 million
Health Costs			
air quality damages	\$0 to \$27 million	\$3 to \$200 million	N.S.[1]
water treatment	\$340 to \$1480 million	N.S.[1]	\$1 to\$10 million
alternate water supplies	\$1 to 30 million	N.S.[1]	N.S.[1]
Direct Costs			
fuel price increase	\$135 to \$675 million	\$290 to \$991 million	\$141 to \$1300 million
fuel efficiency decrease	\$310 to \$400 million	\$290 to \$580 million	(\$150) to (\$230) million
Other Costs			
water monitoring costs	\$2 to \$4 million	N.S.[1]	N.S.[1]
recreational costs	\$160 to \$200 million	N.S.[1]	N.S.[1]
ecosystem damages	N.S.[1]	N.S.[1]	N.S.[1]
Costs Subtotal	\$0.9 to 2.8 billion	\$0.6 to \$1.8 billion	(\$0.09) to \$1.2 billion
Net Benefit or (Cost)	(\$0.9) to (\$2.7) billion	(\$0.5) to (\$1.8) billion	\$0 to (\$1.2) billion

[1]Not significant

Similarly, the reductions in ozone concentrations may lead to a reduction in hospital admissions due to acute respiratory illnesses of up to 300 cases per year. However, it is difficult to determine the fraction of the air pollutant's reduction due strictly to the fuel formulation, or due to other factors such as vehicle technology, meteorology, driving patterns, economic activity, etc. Thus, these estimates should be seen as an upper bound. Significant as they are with respect to human health, we estimate that these benefits only represent around \$14 to \$78 million per year.

There are several reasons why these benefits are small in California:

There have been many programs successfully implemented by the California Air Resources Board to reduce the emissions of air pollutants, and thus the overall concentrations of carbon monoxide and ozone are much lower than in previous decades. For example, carbon monoxide levels prior to the introduction of the current version of oxygenated, reformulated gasoline were less than 12 ppm in practically all air basins in California. The National Ambient Air Quality Standard (NAAQS) for carbon monoxide is 9 ppm for an 8-hour average. Only 3 monitoring stations in California exceeded the NAAQS in 1996, and the downward trend in concentrations since the late 1970s continues[23]. At present, there is only one non-attainment area in California for carbon monoxide, namely the South Coast air basin. The maximum CO concentrations registered in this region since 1994 are around 11 ppm; these levels are only observed a few days during each month, and are concentrated in episodes when adverse meteorological conditions intersect with high levels of emissions. Thus, the benefit of oxygenating the gasoline is small, since it is generally considered that the NAAQS are set such that if they are attained there is no health impact even to sensitive populations. A decrease of 2 ppm in peak carbon monoxide concentrations would only benefit people with ischemic heart disease, which represent around 3 % of the population. In addition, only a fraction of this population would be affected enough to require hospitalization[24,25].

Similarly, ozone concentrations have been declining even in the South Coast region over the last two decades. However, in this case, although the number of days the California standard is exceeded has decreased noticeably, from a high of 261 days of exceedance in 1981 to around 135 to 180 days of exceedance in recent years[26], the levels are significantly above the California standard of 0.09 ppm (1-hour average) during the summer months, peaking at 0.18 to 0.20 ppm. The decrease in ozone concentrations due to the oxygenate could be up to 0.015 ppm, which would benefit sensitive populations with reduced respiratory function[27].

The average age of the vehicle fleet has decreased in recent years, due to the economic recovery in southern California since 1993. This means that newer vehicles with improved emissions controls are replacing older vehicles. In addition, programs implemented to remove older vehicles are reducing the number of high emitters. The emissions from stationary sources and other mobile sources (trucks, airplanes, ships, trains, etc.) are becoming more important relative to emissions from vehicles that use gasoline with MTBE. Programs to control emissions from these other sources lag the successful control of emissions from vehicles.

These conditions may be quite different in other cities and air quality basins, in particular in developing countries. For example, Mexico City has seen increasing ozone and carbon monoxide concentrations in the past few years. A program targeted at these air pollutants would have a major impact on human health. In addition, the vehicle fleet is much older, and instead of a somewhat steady state vehicular population, it is increasing rapidly. Transportation bottlenecks and adverse meteorological conditions compound the problem. Thus, a full evaluation of the health benefits for these situations should be conducted prior to making any decision to use or ban MTBE. And alternative formulations (e.g. non-oxygenated) should also be considered.

There are air quality costs associated with the new gasoline formulations. For example, the use of MTBE results in volatilization of MTBE (from fuel lines, fueling stations, pipelines, etc.). In addition, the combustion of MTBE increases the emissions of formaldehyde[6,7] in controlled combustion studies. Formaldehyde is a known human carcinogen[22,28,29]. Although the observed MTBE concentrations are below the cancer risk level[11], the potential increase in formaldehyde concentrations could result in up to 380 cancer cases per year, negating some of the benefits of reducing benzene emissions. It should be noted that to date, the concentration of formaldehyde in various Californian cities has shown no upward (or downward) trend since the introduction of MTBE

Similarly, adding ethanol to gasoline (as a replacement of MTBE) increases the emissions of acetaldehyde[6,7,30]. Acetaldehyde is also a probable human carcinogen[22,31-33].

If the concentrations were to increase as they did in New Mexico during the introduction of ethanol-gasoline mixtures, by 1-2 parts per billion (ppb), the number of cancer cases could increase to 2,800 per year. However, it should be mentioned that ethanol has been used in the Midwestern U.S. with no noticeable increase in acetaldehyde concentrations. A more complete study is required to determine whether there is really a concern with the use of ethanol.

If either toluene or iso-octane are used to replace MTBE in non-oxygenated formulations, the concentrations of either chemical would increase in ambient air. However, the levels of toluene would probably be below the Reference Concentration (RfC) in air of 0.4 mg/m^3 or 400 $\mu g/m^3$ [34,35]. In California, the mean concentration in air is 8.5 $\mu g/m^3$. This concentration could increase significantly and still not be close to the RfC, where adverse effects would be measurable. None of the data suggest that toluene is carcinogenic. Iso-octane is not classified as a hazardous air pollutant by U.S. EPA, and there is no toxicological information from the Agency for Toxic Substances and Disease Registry (ATSDR). It is a normal component of gasoline, and thus can produce acute effects on the central nervous system when inhaled at high concentrations, but the risk is similar to conventional gasoline. There have been limited studies of the combustion byproducts of these formulations, so it is highly recommendable to make a full assessment before proceeding to their wide-scale introduction.

The introduction of reformulated gasoline with MTBE resulted in an estimated price increase of 1 to 5 cents per gallon. This translates into a cost to the economy of \$135 million to \$675 million. In addition, due to the lower energy content of MTBE[16], there is an additional cost to the California economy of \$300 to 380 million due to the increase in fuel consumption to maintain the same driving pattern. A recent study by the California Energy Commission[36] estimated that the cost of using ethanol as a substitute for MTBE would be around 1.9 to 6.7 cents per gallon, or an annual cost of \$260 to 900 million. The increase in fuel consumption due to the use of ethanol would cost an additional \$560 million (a 3% increase in consumption). For non-oxygenated gasoline, CEC[36] estimates a price increase from 0.9 to 8.8 cents per gallon, or \$121 million to \$1.3 billion per year. The CEC estimates that in the short term (1-3 years), the price increase would be at the high end of the range (4.3 to 8.8 cents per gallon), whereas once refiners have been able to install the necessary equipment or long-term import contracts are established (3-6 years), the price increase should be only around 0.9 to 3.7 cents per gallon[36]. However, in this case, the use of either toluene or isooctane would result in decreased fuel consumption, due to the higher energy content of these chemicals, saving the economy from \$150 to 220 million.

Reuter et al.[14] and Fogg et al.[15] estimated the number of groundwater supplies, leaking fuel tanks and surface water reservoirs that are currently contaminated with MTBE. Based on their information, and the study by Keller et al.[12], we made an estimate of the aggregate cost of water treatment in California, of around \$340 to 1,480 million per year (Table 2). These costs are based on the premise that contaminated water must be treated to a concentration below the 5 $\mu g/L$ level set by California's EPA as the Secondary Water Quality Standard, based on taste and odor considerations.

A literature review indicates that the cost of using ethanol, in terms of risk to the water supplies, is low. Ethanol plumes biodegrade fairly rapidly. In the event that water supplies become contaminated with ethanol, the available toxicological information does not support treating the water to the low levels required by MTBE, and filtration in biologically active GAC would probably be a cost-effective option. We consider the incremental costs of water treatment to be negligible relative to conventional gasoline, since BTEX compounds in the gasoline fraction would determine the treatment design, rather than ethanol.

For non-oxygenated gasoline, the differential cost of remediation and/or water treatment relative to conventional gasoline is small. The increased volumetric fraction of toluene in non-oxygenated CaRFG2 will result in higher initial toluene concentrations, but toluene is easily biodegraded by the intrinsic microbial communities. If isooctane is used instead of MTBE, it has a very low solubility in water, and it is readily biodegraded along with other components of conventional gasoline. It is likely that natural attenuation will be applicable at the same rates as for conventional gasoline. Above ground treatment costs may increase at most 10% relative to treating water contaminated by conventional gasoline.

Some utilities may be forced to purchase water from other supplies, at least in the short to intermediate term. For example, the city of Santa Monica has been purchasing water from the Metropolitan Water District due to the contamination of most of their drinking water wells with MTBE. The cost per year for alternate water supply, assuming that 20% of the contaminated water has to be replaced at a cost of $1.65/1000 gallons that Santa Monica pays for water from the Metropolitan Water District[37], is around $1 million to $30 million. These costs would not be significant for other gasoline formulations, relative to conventional gasoline.

There are some incremental monitoring costs, since water utilities are required to sample more frequently, in particular in surface water reservoirs where boating is allowed. Statewide, this cost is expected to amount to $1 million to $4 million. For groundwater sources, the current costs could increase to $1 and $2 million annually. Monitoring air quality is done by collecting samples on a regular basis and running a standardized analysis, which provides information on a number of air toxics. We do not consider any additional costs will be incurred to monitor ambient air concentrations of MTBE, formaldehyde, acetaldehyde, benzene or combustion by-products. We consider that this cost would not be significant for ethanol-based gasoline formulations or non-oxygenated gasoline, relative to conventional gasoline.

One alternative that can be considered for minimizing the impact of MTBE is banning motorcraft from surface water reservoirs. If boating was completely eliminated from all the reservoirs in California, we estimate that the cost, in terms of recreational value lost, would be on the order of $160 to $200 million. It is likely that only a partial ban would be implemented, and probably not a year-round ban. MTBE is volatile enough to quickly escape to the atmosphere from a contaminated reservoir, in the order of weeks. We consider that this cost would not be significant for ethanol-based gasoline formulations or non-oxygenated gasoline, relative to conventional gasoline.

Ecological risk assessment studies[38] indicate that the concentrations of MTBE that have been detected in lakes and water reservoirs should not result in significant damages to biota in aquatic ecosystem. Localized spills may have an impact, but there is insufficient data to estimate the ecosystem damages, and they are likely to be small relative to other MTBE costs. Note that all damages and costs are estimated relative to the use of conventional gasoline. For example, local ecosystem damages due to a pipeline rupture would be very similar whether the gasoline contained MTBE or not. We consider that this cost would not be significant for ethanol-based gasoline formulations or non-oxygenated gasoline, relative to conventional gasoline.

DISCUSSION

Comparing the bottom line for each formulation, it is clear that using MTBE in reformulated gasoline is a very expensive option. The costs far outweigh the benefits. More significantly, the benefits are small and decreasing with time, as the vehicle fleet modernizes, incorporating emissions control technologies. The main cost driver is water treatment, based on the very tight standards set for MTBE in California. Ethanol-based formulations don't fare much better, although the mid-point of the costs range is smaller than the mid-point of the MTBE formulation. The biggest uncertainty is the impact of a much larger demand for ethanol, which could drive prices up significantly, at least in the short term. One hidden advantage of ethanol is the fact that it can be produced from agricultural wastes such as rice straw, reducing greenhouse emissions. Many developing countries have the potential of producing ethanol from these sources, thus reducing their dependence on imported oil, while improving their air quality.

Non-oxygenated gasoline formulations are apparently the best option for California. There is currently one refiner commercializing a non-oxygenated formulation that meets the strict California Phase II Reformulated Gasoline specifications, except for the oxygen content, presumably at a profitable price. However, the technologies may not be available to the entire industry. Capital expenditures are needed to convert to these formulations. In the mean time, imported toluene or isooctane may drive the costs towards the high end of the range. However, if air quality must be maintained, this appears to be the lowest cost strategy for California. An important consideration is the need to evaluate the toxicology (human and ecological) and fate and transport of any gasoline additives before making drastic changes. The mistakes made with MTBE should be avoided at all costs.

For developing countries, there are a number of factors that must be considered when choosing gasoline formulations. First, an assessment of the air quality benefits must be made, to evaluate whether the change in formulations is warranted. This requires information on trends of ambient air concentrations of pollutants, average vehicle age and technology, as well as number of hospitalizations from congestive heart failure and acute respiratory illnesses, and their cost. Next, an assessment of the vulnerability of water sources must be made. The use of Geographical Information Systems, which can overlay well locations with the location of underground storage tanks, can serve to make an assessment of vulnerability. An inventory of leaking tanks is also needed. California also was able to reduce the impact on water resources due to the decade-long program to upgrade underground storage tanks. If the tanks do not have double-containment and leak detectors, the probability of failure is around 2% per year[39], and the water treatment costs will certainly overwhelm the air quality benefits.

REFERENCES

1. Keller, A.A., Fernandez, L.F., Hitz, S., Kun, H., Peterson, A., Smith, B. and Yoshioka, M. 1998. An Integral Cost-Benefit Analysis of Gasoline Formulations Meeting California Phase II Reformulated Gasoline Requirements. in *Health and environmental assessment of MTBE, vol. 5.* UC Toxics Research and Teaching Program, UC Davis.
2. Keller, A.A., Froines, J., Koshland, C., Reuter, J., Suffet, I. and Last, J. 1998. Health & Environmental Assessment of MTBE, Summary and Recommendations. *UC TSR&TP Report to the Governor of California.*
3. USEPA. 1997. *The Benefits and Costs of the Clean Air Act, 1970 to 1990.* Environmental Protection Agency, Washington, D.C.
4. Schwing, R., B. Southworth, C. Von Buseck, and C. Jackson. 1980. Benefit-Cost Analysis of Automotive Emission Reductions. *Journal of Environmental Economics and Management*, Vol. 7, No.1.
5. Krupnick, A., Walls, M. 1992. The Cost Effectiveness of Methanol for Reducing Motor Vehicle Emissions and Urban Ozone. *Journal of Policy Analysis and Management*, Vol. 11, No. 3, 1992.
6. AQIRP. 1997. Program Final Report, *Auto/Oil Air Quality Improvement Research Program*. January 1997.
7. Koshland, C.P., Sawyer, R.F., Lucas, D., Franklin, P. 1998. Evaluation of Automotive MTBE Combustion Byproducts, in *Health and environmental assessment of MTBE, vol. 2.* UC Toxics Research and Teaching Program, UC Davis.
8. Fernandez, L.F. and Keller, A.A., 2000. Cost Benefit Analysis of MTBE and Alternative Gasoline Formulations. Submitted to *Environmental Science and Policy.*
9. Abdalla, C., Roach, B., Epp, D. 1992. Valuing Environmental Quality Changes Using Averting Expenditures: An Application to Groundwater Contamination, *Land Economics*, Vol. 68, No. 2.
10. Fisher, A., L. Chestnut, Violette, D. 1989. The Value of Reducing Risks to Death: A Note on New Evidence. *Journal of Policy Analysis and Management*, Vol. 8, No. 1.
11. Froines et al. 1998. In *Health and environmental assessment of MTBE, vol. 2.* UC Toxics Research and Teaching Program, UC Davis.
12. Keller, A.A., Sandall, O.C., Rinker, R.G., Mitani, M.M., Bierwagen, B.G., Michael, M.J. 1998. Cost and Performance Evaluation of Treatment Technologies

for MTBE-Contaminated Water, in *Health and environmental assessment of MTBE, vol. 4.* UC Toxics Research and Teaching Program, UC Davis.

13. Keller, A.A.; Sandall, O.C.; Rinker, R.G.; Mitani, M.M.; Bierwagen, B.; Snodgrass, M. J. *Ground Water Monitoring and Remediation* **2000**, in Press.
14. Reuter, J.E., Allen, B.C., Goldman, C.R. 1998. Methyl tert-butyl ether in surface drinking water supplies, in *Health and environmental assessment of MTBE, vol. 3.* UC Toxics Research and Teaching Program, UC Davis.
15. Fogg, G.E., Meays, M.E., Trask, J.C., Green, C.T., LaBolle, E.M., Shenk, T.W., Rolston, D.E. 1998. Impacts of MTBE on California groundwater, in *Health and environmental assessment of MTBE, vol. 3.* UC Toxics Research and Teaching Program, UC Davis.
16. NTSC. 1997. *Interagency assessment of oxygenated fuels.* Executive Office of the President of the United States, National Science and Technology Council, Committee on Environment and Natural Resources.
17. Anderson, R., Rockel, M. 1991. Economic Valuation of Wetlands, Discussion Paper #065, *American Petroleum Institute*, Washington, D.C.
18. Bell, F. 1989. Application of Wetland Valuation Theory to Florida Fisheries. *SGR-95, Sea Grant Publication*, Florida State University, Tallahassee, FL, 1989.
19. Shabman, L. and S. Batie. 1987. Mitigating Damages from Coastal Wetlands Development: Policy, Economics, and Financing. *Marine Resource Economics*, Vol. 4, No. 3.
20. CARB. 1991. *Proposed Regulations for California Phase 2 Reformulated Gasoline*, California Air Resources Board, Staff Report, Stationary Source Division.
21. ATSDR. 1991. Toxicological Profile for Benzene. *Agency for Toxic Substances and Disease Registry*, U.S. Public Health Service, U.S. Department of Health and Human Services, Atlanta, GA.
22. IARC. 1985. *IARC Monographs on the Evaluation of the Carcinogenic Risk of Chemicals to Humans: Allyl Compounds, Aldehydes, Epoxides and Peroxides. Volume 36.* International Agency for Research on Cancer, World Health Organization, Lyon, France.
23. USEPA. 1999. Air Quality Criteria for Carbon Monoxide. EPA Report EPA-600/P-99/001. Office of Research and Development, U.S. Environmental Protection Agency, Washington, D.C.
24. Morris, R.D., Naumova, E.N., Munasinghe, R.L. 1995. Ambient air pollution and hospitalization for congestive heart failure among elderly people in seven large U.S. cities. *American Journal of Public Health*, vol. 85, no. 10, pp.1361-1365
25. Graves, E.J., Owings, M.F. 1998. 1996 Summary: National Hospital Discharge Survey. Advance data from vital and health statistics. National Center for Health Statistics, Hyattsville, MD
26. CARB. 1999. *California Air Quality Data.* California Air Resources Board (http://arbis.arb.ca.gov/aqd/aqd.htm)

27. Burnett, R.T., Brook, J.R., Yung, W.T., Dales, R.E. and Krewski, D. 1997. Association between ozone and hospitalization for respiratory diseases in 16 Canadian cities. *Environmental Research* **72**(1): 24-31
28. USEPA. 1993. *Integrated Risk Information System (IRIS) on Formaldehyde*. U.S. Environmental Protection Agency, Environmental Criteria and Assessment Office, Office of Health and Environmental Assessment, Office of Research and Development, U.S. Environmental Protection Agency, Cincinnati, OH.
29. USEPA. 1988. *Health and Environmental Effects Profile for Formaldehyde. EPA/600/x-85/362.* U.S. Environmental Protection Agency, Environmental Criteria and Assessment Office, Office of Health and Environmental Assessment, Office of Research and Development, Cincinnati, OH.
30. Gaffney, J.S., Marley, N., Martin, R.S., Dixon, R.W., Reyes, L.G., Popp, C.J. 1997. Potential Air Quality Effects of Using Ethanol- Gasoline Fuel Blends: A Field Study in Albuquerque, New Mexico. *Environ. Sci. Technol.*, 31:3053-3061.
31. USEPA. 1993. *Integrated Risk Information System (IRIS) on Acetaldehyde*. U.S. Environmental Protection Agency, Environmental Criteria and Assessment Office, Office of Health and Environmental Assessment, Office of Research and Development, U.S. Environmental Protection Agency, Cincinnati, OH.
32. USEPA. 1987. *Health Assessment Document for Acetaldehyde.* EPA/600/8-86-015A. U.S. Environmental Protection Agency, Environmental Criteria and Assessment Office, Office of Health and Environmental Assessment, Office of Research and Development, Research Triangle Park, NC.
33. CARB. 1993. *Acetaldehyde as a Toxic Air Contaminant: health assessment for the stationary source division.* California Air Resources Board.
34. USEPA. 1993. *Integrated Risk Information System (IRIS) on Toluene*. U.S. Environmental Protection Agency, Environmental Criteria and Assessment Office, Office of Health and Environmental Assessment, Office of Research and Development, U.S. Environmental Protection Agency, Cincinnati, OH.
35. ATSDR. 1992. Toxicological Profile for Toluene. *Agency for Toxic Substances and Disease Registry*, U.S. Public Health Service, U.S. Department of Health and Human Services, Atlanta, GA.
36. CEC. *Evaluating the Cost and Supply of Alternatives to MTBE in California's Reformulated Gasoline*; California Energy Commission: Sacramento, CA, 1998.
37. Rodriguez, R. 1997. MTBE in Groundwater and the Impact on the City of Santa Monica Drinking Water Supply, in *Technical Papers of the 13th Annual Environmental Management and Technology Conference West*, Nov. 4-6, 1997.
38. Werner, I., Hinton, D.E. 1998. Toxicity of MTBE to freshwater organisms, in *Health and environmental assessment of MTBE, vol. 4.* UC Toxics Research and Teaching Program, UC Davis.
39. Couch, A., Young, T. 1998. Failure rate of Underground Storage Tanks, in *Health and environmental assessment of MTBE, vol. 3.* UC Toxics Research and Teaching Program, UC Davis.

ARSENIC IN GROUNDWATER: A WORLDWIDE THREAT TO HUMAN HEALTH

S. MAJID HASSANIZADEH

The arsenic contamination of groundwater in the Bengal Delta Plains is undoubtedly the greatest environmental health disaster of the century. Around one hundred million people in West Bengal and in Bangladesh are drinking waters that are contaminated with arsenic in concentrations far above acceptable levels. Millions of people have already been diagnosed with symptoms of arsenic toxicity. Elevated arsenic concentrations in groundwater have been observed in several parts of the world. Arsenic, particularly in the reduced form arsenite, is extremely toxic (teratogenic) and may cause neurological damage at aqueous concentrations as low as 0.1 mg/L. The sources of arsenic in groundwater are mostly of natural origin. In this presentation, an overview of the occurrence and extent of this problem will be given. Natural sources of contamination are listed. Major areas of the world where arsenic in groundwater is found and the extent of contamination are mentioned. A brief description of the consequences of arsenic poisoning is given. The arsenic contamination in Bangladesh is discussed and presented in more detail.

ARSENIC GEOCHEMISTRY AND REMEDIATION USING NATURAL MATERIALS

DAVID I. NORMAN
Department of Earth and Environmental Science, New Mexico Institute of Mining and Technology, Socorro, New Mexico 87801, USA, tel. 505-835-5404, fax 505-835-6436, email: dnorman@nmt.edu

INTRODUCTION

There is an arsenic health crisis in Bangladesh and India. More than 7,000 thousand are diagnosed with arsenicous and 1,700 cases of arsenic related melanomas are identified. These numbers far under report the problem because less then 1% of the Bangladeshi and Indian villages have been surveyed. There are 20,000,000 to 70,000,000 exposed persons drinking water at levels between 0.05 ppm and 0.5 ppm arsenic. From the exposure to date it is estimated that between 200,000 and 2,000,000 will develop arsenic related cancers. The longer the world waits to reduce arsenic exposure the greater will be the tragedy. A companion paper lists other less-developed countries with arsenic exposure problems that, like Bangladesh and India, are related to development of shallow water supply wells. This paper outlines arsenic geochemistry and toxicity, and a simple point-of-use arsenic remediation device that uses natural material.

This workshop paper is organized differently from a scientific paper. The intention is to provide the reader with summary information and an extensive bibliography. Ten topic areas discussed are: 1) arsenic geochemistry; 2) arsenic mobility in aqueous systems; 3) arsenic in plants and animals; 4) arsenic toxicity; 5) geological environments associated with anomalous arsenic-bearing waters and soils; 6) source and mechanisms for anomalous arsenic-bearing ground waters; 7) field measurement of arsenic and arsenic species; 8) large-scale arsenic treatment methods; 9) proposed solutions to the Bangladesh arsenic problem; and 10) arsenic treatment using natural materials. The bibliography is divided into the same ten topic areas. There are few cited references in the text except information that is found in only one paper.

ARSENIC GEOCHEMISTRY

Arsenic is a group 15 (old CAS group V) element that occurs under phosphorus and above antimony in the periodic table. It has a similar chemistry to both elements, and substitutes for phosphorus in organic molecules. In nature arsenic has oxidation states of

+5, +3, and 0. Common arsenic minerals are arsenopyrite (FeAsS), orpiment (As_2S_3), and realgar (AsS). Native arsenic is known but of rare occurrence.

Arsenic oxides representing the two positive oxidation states are As_2O_5 and As_2O_3. They are highly soluble in water; the solubility of As_2O_5 is 150 gm in 100 gm water at room temperature. The oxides react with water forming arsenic acid H_3AsO_4 (As(V)) and arsenious acid $HAsO_2$ (As(III)), which are arsenic compounds found in groundwater.

Arsenic average crustal abundance is about 1 mg/kg (ppm). Concentrations above the crustal average occur in shale - that averages 12 ppm As, in coal - where concentrations commonly are 5 to 10 ppm, and in iron formations - that have concentations of 10 to 700 ppm. Arsenic rarely occurs in rock as an arsenic mineral, rather it is present substituting for iron in iron-bearing minerals. Of particular note are arsenic concentrations in pyrite (FeS_2) that ranges from 100 to 5,000 ppm.

ARSENIC MOBILITY IN AQUEOUS SYSTEMS

Arsenic in surface and groundwaters occurs principally as arsenic acids. At the normal range of groundwater pH, about 4 to 9, arsenic occurs as an anionic (negatively charged) inorganic As(V) species and as uncharged As (III) arsenious acid (Table 1). Two organic species MMA and DMA (Table1) are found in environments with biological activity, but are rarely detected in groundwater.

Arsenic is a mobile species in surficial environments. This is because arsenic freed from crustal rock by weathering does not commonly precipitate as an insoluble mineral. Limiting arsenic in solution is:

- Adsorption on to colloidal-size oxide and hydroxide minerals that form in soils and occur in recently deposited sediments,
- Precipitation of arsenic sulfide minerals and pyrite,
- Adsorption onto organic compounds and formation of organo-metallic compounds in organic compound-rich formations, and
- Reduction to arsine gas.

The higher arsenic concentrations in soils that average 7.5 ppm, are attributed to sorption processes. The high arsenic concentrations in the sedimentary rock shale and ironstone are explained by arsenic adsorption onto colloid surfaces during sedimentation.

Table 1. Principal Arsenic Aqueous Species.

Inorganic Species			Organic Species		
Compound Name	Formula	pK	Compound Name	Formula	pK
Arsenious Acid As(III)	$HAsO_2$	9.22	Monomethyl-arsinic acid As(V)	CH_3-H_2AsO_3	3.41 8.18
Arsenic Acid As(V)	H_3AsO_4	2.22 6.98 11.4	Dimethylarsinic acid As(V)	$(CH_3)_2$-H_2AsO_3	1.56

Generally, as rock weathers, arsenic and other mobile elements such as sodium are transported in solution by surface and groundwaters to oceans. Common river and groundwater arsenic concentrations are 0.5 to 2 ppb, and Na/As ratios are about 23,000, which is similar to ratios in crustal rocks. Anomalous concentrations of arsenic, which are more common in groundwaters, range from 5 to >10,000 ppb. Table 2 gives some examples together with the percentage of the four common aqueous species.

Table 2. Arsenic Concentrations and Speciation of Selected Wells.

Source	As(T) ppb	As(III) %	As(V) %	MMA %	DMA %	Reference
Bangladesh and West Benegal Shallow Tube Wells	50 to >5,000	>80	<20	nr	nr	Personal communication
Nova Scotia I	8,000	56.3	43.8	nd	nd	Irgolic (1982)
Nova Scotia II	630	49.2	50.8	nd	nd	Irgolic (1982)
Barefoot, Alaska	3,100	77.4	22.6	nd	nd	Irgolic (1982)
Yenshei II, Taiwan	1,100	2.2	98.2	nd	nd	Irgolic (1982)
Yenshei I, Taiwan	850	2.7	98.8	nd	nd	Irgolic (1982)
Antofagasta, Chile (untreated)	750	2.1	98.7	nd	nd	Irgolic (1982)
Antofagasta, Chile (treated)	410	0.7	99.3	nd	nd	Irgolic (1982)
Hinckley, Utah	180	5.6	94.6	nd	nd	Irgolic (1982)
Delta, Utah	20	50	50	nd	nd	Irgolic (1982)
Hanford, CA (Well #19)	90	100	0	nd	nd	Clifford (1986)
San Ysidro, NM (Well #4)	250	100	0	nd	nd	Clifford (1991)
San Ysidro, NM (Well #1)	88	35.2	64.8	nd	nd	Clifford (1991)

nr = none reported nd = none detected

ARSENIC IN PLANTS AND ANIMALS

We are exposed to arsenic in a variety of plants, animals, and anthropogenic sources. Shellfish have arsenic concentrations between 10 and 30 ppm. In shellfish and other marine animals arsenic occurs in the compounds arsenobetaine ($(CH_3)_3AsCH_2COOH$) and arsenocholine ($(CH_3)_3As(CH_2)_2OH$) that are quickly excreted with no known health effects. Plants can contain concentrated inorganic arsenic. We have measured amounts up to about 10 ppm in ash from plants growing in arsenic rich soils. Speciation indicates the occurrence of both As(III) and As(V) inorganic species. Edible seaweed may have up to 200 ppb inorganic arsenic. Arsenic concentations about 17 ppb are measured in German beer. There is little data on the amount and forms of arsenic in the most commonly consumed fruits and vegetables, and how concentrations are related to soil arsenic. Other sources of arsenic are air that has on average 0.003 mg As/m^3 and anthropogenic arsenic. Arsenic was used as a green pigment in paints and wall paper, and was used in embalming. Arsenic was and still is widely used as a herbicide and pesticide. Herbicide use is on railroad beds and to defoliate cotton prior to picking. The common green-colored treated wood used for landscaping and power poles has about 500 ppm arsenic. A single landscaping timber contains enough arsenic to kill about 30 people. As a result of the many avenues of exposure the human body has an arsenic concentration about 50 ppb.

ARSENIC TOXICITY

Arsenic has long been used as a poison for rats, mice, and coyotes. It was a favorite poison for homicide because arsenic is deadly at concentations so low that it could not be detected. Toxic amounts by ingestion are 0.5 mg/kg/day, from for example drinking a liter of water that has 50 ppm arsenic. Animals are less susceptible to arsenic poisoning; their toxic amount by ingestion is 10 mg/day/kg.

Chronic health problems in humans are evident at exposures of 0.02 mg/kg/day. A 10 kg child may experience arsenic health problems by ingesting 0.2 mg As/day. Examples are drinking two liters of 100 μg/kg water/day or ingesting 200 mg of soil containing 1,000 mg/kg As/day. (The U.S. EPA suggested amount of dirt ingestion by children to be used in risk assessment is 200 mg of soil/day.) Health problems in a 70 kg adult require ingestion of at least 1.4 mg As/day. The World Health organization recommends arsenic intake of < 2 μg /kg/day and drinking water that has less than 10 μg/kg (ppb) arsenic. The EU standard for drinking water to go into effect in 2003 is 10 ppb. The USA Academy of Sciences recommends a drinking water limit of 10 ppb, and the proposed USA EPA standard for drinking water is 5 μg /kg/day. Drinking waters with > 100 ppb arsenic is associated with a progression of chronic problems that is manifest as skin mottling; then skin lesions and growths, keratoses of the hands and feet, and circulation problems leading to gangrene; and then skin and organ cancers. At lower levels of exposure, studies suggest elevated rates of skin and organ cancer, elevated rates of cardiovascular disease, and increased rates of birth defects. The risk associated with low exposures to arsenic is not well quantified at present. However, all analyses published on the low dose response relationship for arsenic internal cancers are consistent with response being linear with no threshold.

Animal studies show that inorganic As(III) is 10 times more poisonous than inorganic As(V), and that As(V) is up to 1,000 times more toxic than DMA or MMA. However, none of the arsenic standards is species based. In part this is due to the fact that arsenic speciation is not included in the data bases used for toxicology estimates, and in part due to the questions about the applicability of animal studies to humans because of their differing susceptibilities to arsenic poisoning. Further, Cullen and Reimer[1] show that during human metabolism of arsenic, species proceed from As(V) to As(III) to As(III)MMA to As(V)MMA to DMA, after which DMA is excreted. The compound As(III)MMA is the most poisonous form of arsenic. If and when it is demonstrated that As(V) is less toxic than As(III) there will be a need for field arsenic speciation methods. For water supplies in which As(III) dominates, oxidation of As(III) to As(V) can be accomplished with UV light, a few days exposure to the atmosphere and sunlight, or by adding a small amount of bleach. Hence, a large reduction in risk could be accomplished with simple methods.

GEOLOGICAL ENVIRONMENTS ASSOCIATED WITH ANOMALOUS ARSENIC-BEARING WATERS AND SOILS

In most of the world groundwater and surface water arsenic concentrations are less than 2 ppb. Anomalous ground water arsenic concentrations are associated with several geological environments:

- Young volcanic rocks & sediments derived from them
- Geothermal waters
- Young basin sediments
- Pyrite-bearing shales & sediments derived from them
- Metamorphic gold deposits
- Sulfide mineral deposits

Elevated arsenic concentrations in the Western U.S. are associated with 30 Ma (million years) volcanic rocks (examples on Table 2 are wells in Utah and California). In areas that have experienced volcanic activity in the past few million years, for example most of the Pacific Rim, there is a more severe ground arsenic problem (an example is Chile, Table 2). Geothermal fluids associated with magma intrusions commonly have 10 ppm or more arsenic. These convectively driven fluids rise upwards and mix with surface waters creating a large halo of arsenic-rich ground waters. Examples on Table 2 are the San Ysidro and Taiwan well waters. Surface waters draining geothermal area are also affected.

SOURCE AND MECHANISMS FOR ANOMALOUS ARSENIC-BEARING GROUND WATERS

The pathway of arsenic from volcanic rocks to ground water in not well understood. Arsenic in geothermal waters is hypothesized to be due to volatile transfer from a magma to circulating ground water. Field evidence indicates most volcanic rocks were in contact with hydrothermal waters, hence it seems logical that volcanic rocks sorbed arsenic during hydrothermal activity, and this arsenic is later released to groundwater.

Young basin sediments like in Bangladesh and in eastern China have anomalous arsenic concentrations. There is no comparable arsenic groundwater problem in the Amazon, Nile, or Mississippi estuaries, which makes the process in Bangladesh enigmatic. The Bangladesh groundwater arsenic problem is confined to shallow depths, which suggests that arsenic release is related to interaction with surface waters. There are two hypotheses. Arsenic release is attributed to reduction of sediments by organic material during burial. Iron (III) in hydroxide colloids that are strong arsenic sorbing agents are thought to undergo reduction to Fe(II) thus releasing sorbed arsenic. Others think arsenic is released by oxidation of pyrite. This is believed to be related to the lowering of the water table by tube wells as well as by irrigation and dams upstream that have lowered flow on the major rivers including the Ganges and Brahmaputra. There is data that support both hypotheses including anomalous iron in Bangladeshi groundwaters, and the occurrence of partially oxidized arseniferous pyrite in shallow aquifer sediments.

The arsenic source is much clearer in areas of anomalous groundwater arsenic associated with pyrite-bearing shales, gold-arsenic metamorphic deposits, and sulfide mineral deposits. Examples in Table 2 are groundwaters in Alaska and Nova Scotia. High-arsenic concentrations occur in shallow groundwaters at depths where oxidation of sulfide minerals is taking place. Gold-arsenic mineralization that occurs along structural belts may be kilometers wide. Soils in this environment may have 1,000 ppm or more arsenic. Along Ghana goldbelts Norman et al[2]. have recognized arsenic skin problems even though water concentations are an order of magnitude less than in Bangladesh. The best explanation is that the arsenic dose is from both water and ingestion of soil.

FIELD MEASUREMENT OF ARSENIC AND ARSENIC SPECIES

In order to screen the 4 to 7 million home tube wells in Bangladesh and India, a simple, inexpensive, safe, low tech method of measuring arsenic is needed, capable of being used by minimally trained field agents. The field analytical method now used is an addition of a strong acid to arsenic-bearing water that results in formation of arsine gas. Arsine is measured with a mercury reagent that colors on contact with arsine. Arsine is very toxic and there is valid concern that accidents can occur as well as concern about the hazards associated with handling strong acids and mercury compounds. The Arsenator[3.] is a field portable device that does the same process coupled with a photo sensor that quantifies the amount of arsenic. Its negatives are the reagent expense that are several dollars per analysis, the problems and cost with importing the flammable reagents used by the Arsenator, the working range of 0.5 to 50 ppb that requires dilution or sample splitting to measure amounts above 50 ppm, and maintenance of the apparatus. We borrowed an Arsenator in use in rural Ghana, and discovered it was giving erroneously low readings because windows in the detector were dirty.

The latest development is a simple, rugged method that requires no electrical power. Hach[4.] has announced development of an arsine method that uses a 50 ml bottle and strips of paper that are sensitive to arsenic. The device minimizes exposure to arsine, uses shippable reagents in premeasured pouches, and is sensitive over the range of 10 to 500 ppb arsenic.

Field arsenic speciation may be very important when it is decided that there is a major difference between toxicity of As(II) and As(V) species. Speciation has to be done in the field because there is no accepted means of sample preservation, and As(III) may oxidize to As(V) within days or hours. Based on experience in Mexico, I am of the opinion there is a major difference in the toxicity of the species. At La Primavera 1,000 ppb arsenic stream water is consumed with no apparent health affects. In the Torreon district that is known for mineral deposits, arsenicous is a common problem associated with 250 ppb arsenic wells. Arsenic at La Primavera is all As(V). Presumably Torreon well waters, like other ground waters near mineral deposits, have As(III) > As(V).

Field speciation is accomplished by ion exchange. Clifford[5] and we have developed methods to do this. Our method is in the process of being patented and is undergoing EPA certification, so I can give few details. Both methods take just a few minutes to perform. Clifford's method determines As(III) by difference, ours separates each species and in addition gives total arsenic. We have the ASK2 method for two species (As(III) and As(V)) measurement, the ASK3 that separates As(III), As(V), and organic species, and ASK4 that separates As(III), As(V), DMA and MMA. Some details of the ASK2 procedure are in Miller et al[6].

LARGE-SCALE ARSENIC TREATMENT METHODS

There are a number of methods available for removing arsenic from municipal water supplies. The cost depends on arsenic concentrations and the required lower limit. Methods for arsenic removal are well known and can be divided into the basic process involved: ion exchange, reverse osmosis, and sorption. Variations of the later process are being tested on a large scale in England, which uses a bed of sorption bed Fe-Mn oxides. The bed can be regenerated thereby lowering costs. In Albuquerque, New Mexico a pilot plant is being tested that injects ferric chloride into the water stream, and then removes flocculated iron hydroxide colloids and sorbed arsenic by filtration. Sorption works best on charged species, hence provisions are made to oxidize As(III) in the two pilot plants. Activated alumina is also a good arsenic sorption agent. With all methods there is concern over hazardous waste generation, and what to do with it. Chlorination and sand filtration performed in most municipal water plants can reduce arsenic. In this process iron is oxidized and flocculates adsorbing arsenic. Filtration removes the iron and arsenic. An example is the Antofagasta, Chile water system (Table 2).

PROPOSED SOLUTIONS TO THE BANGLADESH ARSENIC PROBLEM

There are a number of solutions put forth to solve the Bangladesh arsenic problem. Identifying good wells with < 50 ppb arsenic from hazardous wells is being attempted. Low arsenic well pumps are painted green and bad wells are marked with red. The problem with this approach is that there are more than 4 million wells to test and > 70 %

of the wells in So. Bangladesh are bad. But this is only a temporary solution; revisiting green-marked wells a year later shows that many have hazardous arsenic levels. Producing drinking water by solar distillation and rainwater harvesting is possible, but not entirely practical because of problems with speed and lack of rainfall during the dry season. Community water treatment, and drilling deep wells are very expensive solutions. Numerous point-of-use methods have been proposed and are listed on the Harvard arsenic web page[7]. Most are neither practical nor have been field tested. Khan et al.[8] have extensively tested the 3-kalshi (3-pot) method that uses iron fillings, charcoal, fine sand, coarse sand and wood shavings. It however is slow, and clogged up in field trials.

ARSENIC TREATMENT USING NATURAL MATERIALS

The ideal arsenic filter has to have the following qualities:

- Be inexpensive
- Be easy to make and use
- Must work quickly
- Be simple and robust
- Use local materials
- Have large tolerance ranges
- Be culturally acceptable, and
- Produce good tasting, clear water

With this end in mind, we worked on a filter that uses iron concretions common in tropical lateritic soils. Laterite from several areas in Ghana and Brazil has been tested and shown to work well (Fig. 1). A series of experiments shows that a filter can be made with laterite concretions crushed to 3 or 4 mm size with flow rates on the order of a liter/minute. Breakthrough occurs between 100 and 1,000 bed-volumes for water containing 100 ppb arsenic. The smaller number is associated with near metallic concretions (Fig. 1); better sorption occurs with less hardened, more porous concretions. Breakthrough is gradual, and total breakthrough has not been observed.

Our preliminary work indicates an arsenic filter could be fabricated in a bucket that would:

- Provide drinking water for a family for several months
- Have flow rates up to 0.5 l/min
- Would reduce arsenic concentration by 99%
- Could be fabricated by children, and
- Could be scaled upwards to supply arsenic-free water for a village.

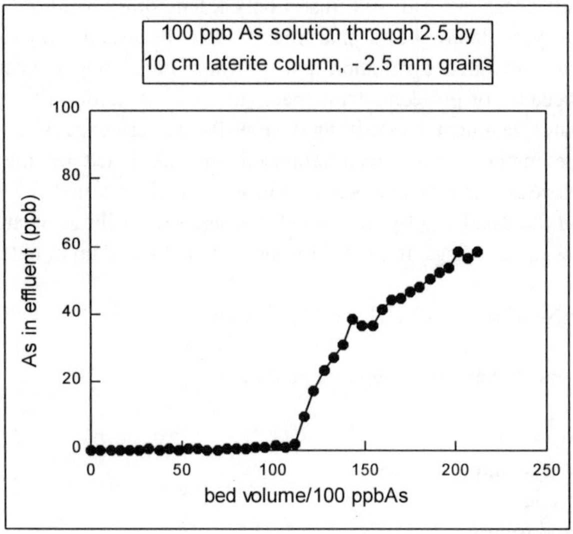

Fig. 1. An arsenic breakthrough curve performed by adding a 100 ppb arsenic solution, pH = 6 to a 2.5 by 10 cm arsenic bed sized with 3 mm window screen. The bed was a very dense, almost metallic Ghana laterite concretion. Effluent is plotted against bed volume because it is easier to scale column experiments to other size devices by use of this unit. The nick in the curve is where the experiment was paused for a lunch break.

Advantages of laterite as a sorption agent are that it is plentiful and costs nothing. It operates at a size that can be made in villages using window screen. Residence time are the order of five minutes, hence high flow rates can be used. It works so well that a meter-size box will produce almost arsenic-free water for 100 man/yr. consumption. We have done preliminary tests with As(III) and it appears to adsorb as well as As(V), which we do not understand.

Field trials were done in Ghana June, 2000 to test: 1) if a bucket filter can be easily made with local materials; 2) that it would work with tropical ground waters that have ten times the silica as temperate climate water; 3) there were no hidden problems with the method; and 4) that the product would be palatable with no off taste and color.

The laterite filter idea was tested in Bopo, a rural village in Ghana of about 1,000 inhabitants that has with 30 to 60 ppb arsenic wells. A bucket filter was made by collecting laterite concretions from farm fields nearby, then having villagers crush the laterite and size it with window screen purchased in the village market. The screen size was 4 mm, which is larger than the 1/8 inch (about 3 mm) screen we had used in our laboratory tests. A hole the size of a Bic pen was cut out of the center bottom of a 20 liter plastic bucket that was purchased in the market. The hole was covered with three layers of screen, and about 8 liters of sized iron concretions was placed in the bucket to a depth about 20 cm. Water flowed through the bucket and laterite at a rate of 0.56 l/min—a bit faster than flow in most electric coffee makers. That flow rate was maintained for about 6.5 hours during which about 220 liters of water passed through the filter. Table 3 documents the well water chemistry and that of the effluent.

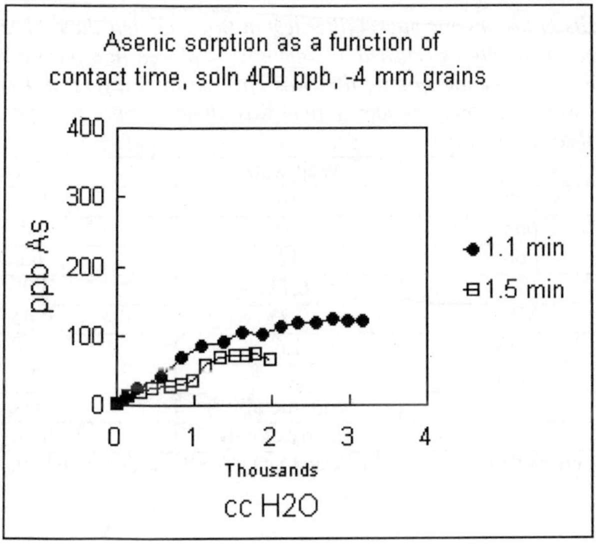

Fig. 2. Arsenic effluent plotted for two column experiments done with differing residence times. The ideal residence time is 15 minutes for 4 mm grains to insure 99% arsenic adsorption. The experiment was done to quantify sorption when a bucket filter is poorly constructed. The bed material is Ghana Bopo laterite sized with a 4 mm widow screen, and the test solution is 400 ppb arsenic in Socorro tap water adjusted to a pH = 6.

The filter worked remarkably well. Unexpectedly the concentration of iron was reduced by a factor of 20, which removed the fetid iron smell. In addition the mild turbidity (cloudiness) in the well water was absent in the effluent. There was no off taste to the effluent, in fact the townspeople thought it much better than the raw water. At the end of the day we were brought some water that was claimed turned food black when used for cooking. It has 14 ppm iron, and the filter decreased the iron level to about 0.1 ppb. There was no indication that the high level of silica in Ghana ground water poisoned iron oxide surfaces. The precipitation of iron probably will increase the life of the filter by providing fresh iron hydroxide sorption surfaces.

We did not run the experiment to breakthrough because that would have taken weeks considering the arsenic concentrations in the well water tested. Our objectives were met. It was easy to make an arsenic filter with local laterite in a village setting. The effluent water was excellent. We ran enough water through the pail (about 55 man-days supply) to verify that the device does not clog. High-iron, high-silica, As(III) dominant water like that reported in Bangladesh was run through the filter, and it worked better than expected.

Table 3. The results of the arsenic bucket filter test in Bopo, Ghana, June 11, 2000. Eight liters of laterite iron concretions crushed to - 4mm were placed in a plastic bucket with about a 5 mm hole in the center bottom that was covered by 3 layers of 4 mm window screen. Flow rate was 0.56 l/min, residence time was about 14 minutes, and 220 l were continuously run through the filter.

Variable		Well water	Effluent
Total As	ppb	30	0.6
As(III)	ppb	20	nd
As(V)	ppb	12	nd
pH		6.11	5.90
Eh	MV	0.27	0.43
Fe	ppm	1.6	0.08
SiO_2	ppm	95	99
smell		fetid, metallic	none
turbidity		slightly cloudy	none

nd = none detected; precision for As +/- 15%, Fe and SiO_2 +/- 5%; As detection limit 0.5 ppb

REFERENCES

1. Cullen, W.R., and K.J. Reimer. 1989. Arsenic speciation in the environment. Chem. Rev.89:713-764.
2. David I. Norman, Ph.D., Gregory Miller, Bret Andrews, Theresa Apodaca, Greta Balderrama, Thresa Benson, Carl Brady, Suzanne Conrad, Peter Conrad, Farah Donahue, Creighton Edington, Deborah Haggerton, Kevin Jarigese, Carla Ludwig, Cate Maley, Gillian Sherwood, Wayne Sherwood, Steve West; Henry Appiah, Jarvis Ayamsegna, and Robert Nartey, 2000, Aresenic in Ghana, West Africa Groundwaters: www.cudenver.edu/as2000.
3. The Arsenator: www.arsenator.com
4. Arsenic Measurement: www.hach.com
5. Clifford, D.A., L. Ceber and S. Chow. 1983. Separation of Arsenic (III) and Arsenic (V) by Ion Exchange. Proceedings 1983 AWWA Water Quality Technology Conference, Norfolk, VA, pp. 223-236, AWWA Denver, CO, December 1983.
6. Miller, G.P., D.I. Norman, and P.L. Frisch. 2000. A comment on arsenic species separation using ion exchange, Water Res. Vol. 34, No. 4, pp. 1397-1400.

7. Harvard Arsenic Site:
http://phys4.harvard.edu/~wilson/arsenic_project_introduction.html
8. Khan A.H., Rasul, S.B., Munir, A.K.M., Habibuddowla, M., Alauddin, M., Newaz, S.S., and Hussam, A., 2000, Appraisal of a simple arsenic removal method for groundwater of Bangladesh, Journal of Environmental Science and Health Part A-Toxic/Hazardous Substances & Environmental Engineering: V. 35 pp. 1021-1041

BIBLIOGRAPHY

ARSENIC GEOCHEMISTRY

Anderson, L.C.D. and K.W. Bruland. 1991. Biogeochemistry of arsenic in natural waters: The importance of methylated species. Environ. Sci. Technol. 25:420-427.

Braman, R.S. and C.C. Foreback. 1973. Methylated forms of arsenic in the environment. Science. 182:1247-1249.

Bright, D.A., M. Dodd, K.J. Reimer. 1995. Arsenic in subArctic lakes influenced by gold mine effluent: the occurrence of organoarsenicals and 'hidden' arsenic. The Science of the Total Environment 180(1996) 165-182.

Cullen, W.R., and K.J. Reimer. 1989. Arsenic speciation in the environment. Chem. Rev. 89:713-764.

Dealy, J.M. and D.S. Sheppard. 1996. Whangaehu River, New Zealand: geochemistry of a river discharging from an active crater lake. Applied Geochemistry. 11:447-460.

Eaton, A., H.C. Wang, and J. Northington. 1998. Analytical Chemistry of Arsenic in Drinking Water. AWWA Research Foundation and American Water Works Association, Denver.

Kimball, A.K., R.E. Broshears, K.E. Bencala, and D.M. McKnight. 1994. Coupling of hydrologic transport and chemical reactions in stream affected by acid mine drainage. Environ. Sci. Technol., v. 28, no. 12, pp. 2065-2073.

Langmuir, D. 1997. Aqueous Environmental Geochemistry. Prentice Hall, New Jersey.

Livesey, N.T. and P.M. Huang. 1981. Adsorption of arsenate by soils and its relation to selected chemical properties and anions. Soil Sci. 131:88-94.

Malotky, D.T. and M.A. Anderson. 1976. The adsorption of the potential determining arsenate anion on oxide surfaces. Colloid and Interface Science. Vol. 4. Milton Kerker (ed.).

Manning, B.A. and S. Goldberg. 1996. Modeling competitive Adsorption of Arsenate with Phosphate and Molybdate on Oxide Minerals. Soil Soc. Am. J. v.60, p. 121-131

Nimick, D.A., 1996. Madison and upper Missouri River arsenic southwestern Montana, July 1993 through July 1996. Montana Department of Natural Resources and Conservation and U.S. Geological Survey (MT150).

Nimick, D.A. 1998. Arsenic hydrogeochemistry in an irrigated river valley: a reevaluation. Groundwater, v. 36, no. 5, pp. 743-753. September-October, 1998.

Norman, D.I. and Bernhart, C., 1982, Assessment of geothermal reservoirs by analysis of gases in thermal waters: New Mexico Energy Institute, EMI-2-68-2305, 129 p.

Norman, D.I., J.N. Moore, and J. Musgrave. 1997. Gaseous species as tracers in geothermal systems. Proceedings from the 22nd Workshop on Geothermal Reservoir Engineering, Stanford, California, January 27-29, 1997.

Oliver, J.T., M.K. Birmingham, A. Bartova, M.P. Li, and T.H. Chan. 1973. Methylated Forms of Arsenic in the Environment. Science, Vol. 182, December, pp. 1247-1251.

Onysko, S.J. and R.L. McNearny. 1997. GIBBTEQ: A MINTEQA2 thermodynamic error detection program. Ground Water, Computer Notes, Vol. 35, No. 5, September-October 1997, p. 912-914.

Oscarson, D.W., P.M. Huang, and W.K. Liaw, 1981. Role of manganese in oxidation of arsenite by freshwater lake sediments. Clays and Clay Minerals. 29(3):219-225.

Oscarson, D.W., P.M. Huang, C. Defosse, and A. Herbillon. 1981. Oxidative power of Mn(IV) and Fe(III) oxides with respect to As (III) in terrestrial and aquatic environments. Nature. 291:50-51.

Oscarson, D.W., P.M. Huang, W.K. Liaw, and U.T. Hammer. 1983. Kinetics of oxidation of arsenite by various manganese dioxides. Soil Sci. Soc. Am. J. 47:644-648.

Oscarson, D.W., P.M. Huang, and W.K. Liaw. 1980. The oxidation of arsenite by aquatic sediments. J. Environ. Qual. 9(4):

Pierce. M.L. and C.B. Moore. 1980. Adsorption of arsenite on amorphous iron hydroxide from dilute aqueous solution. Enivron. Sci. Technol. 14:214-216.

Takamatsu, T., Kawashima, M., Koyama, M., 1985. The role of Mn^{2+}-rich hydrous manganese oxide in the accumulation of arsenic in lake sediments. Water Res. 19, 1029-1032.

Tessier, A., D. Fortin, N. Belzile, R.R. DeVitre, and G.G. Leppard, 1996. Metal Sorption to diagenetic iron and manganese oxyhydroxides and associated organic matter: Narrowing the gap between field and laboratory measurements. Geochim. Cosmochim. Acta, 60(3):387-404.

Tessier, A., P.G.C. Campbell, and M. Bission. 1979. Sequential extraction procedure for the speciation of particulate trace metals. Analytical Chemistry. 51(7):844-851.

ARSENIC MOBILITY IN AQUEOUS SYSTEMS

Aggett, J. and G.A. O'Brien. 1985. Detailed model for the mobility of arsenic in lacustrine sediments based on measurements in Lake Ohakuri. Environ. Sci. Technol. 19:231-238.

Anderson, M.C., J.F. Ferguson, and J. Gavis. 1976. Arsenate adsorption on amorphous aluminum hydroxide. Journal of Colloid and Interface Science. 54(3):391-399.

Brockbank, C.I., G.E. Batley, and G.K-C Low. 1988. Photochemical decomposition of arsenic species in natural water. Environmental Technology Letters. 9:1361-1366.

Deuel, L.E., and A.R. Swoboda. 1972. Arsenic Solubility in a reduced environment. Soil

Sci. Soc. Amer. Proc. 36:276-278.

Elkhatib, E.A., O.L. Bennett, and R.J. Wright. 1984. Kinetics of arsenite sorption in soils. Soil Sci. Soc. Am. J. 48:758-762.

Ford, C.J., J.T. Byrd, J.M. Grebmeier, R.A. Harris, R.C. Moore, S.E. Madix, K.A. Newman, and C.D. Rash. 1996. Final project report on arsenic biogeochemistry in the Clinch River and Watts Bar Reservoir, Volume 1 ORNL/ER-206/V1/H3.

Hemond, H.F., 1995. Movement and Distribution of Arsenic in the Aberjona Watershed. Environmental Health Perspectives. Vol 103, Supp. 1, February.

Hering, J.G., and J. Wilkie. 1996. Arsenic geochemistry in source waters of the Los Angeles Aqueduct. Preliminary report to the Water Resources Center, University of California, Davis.

Hess, R.E. and R.W. Blanchar. 1976. Arsenic stability in contaminated soils. Soil Sci. Soc. Am. J. 40:847-852.

Holm, T.R., M.A. Anderson, R.R. Stanforth, and D. G. Iverson. 1980. The influence of adsorption on the rates of microbial degradation of arsenic species in sediments. Limnol. Oceanogr. 25(1):23-30.

Howard, A.G., M.H. Arbab-Zavar, and S. Apte. 1982. Seasonal variability of biological arsenic methylation in the estuary of the river Beaulieu. Marine Chemistry. 11:493-498.

Irgolic, K.J., 1982. Speciation of arsenic compounds in water supplies. In: USEPA EPA-600/S-1-82-010.

Irgolic, K.J. 1994. Determination of Total Arsenic and Arsenic Compounds in Drinking Water. Arsenic Exposure and Health. Science and Technology Letters. Northwood, pp. 51-61.

Korte, N.E. and Q. Fernando. 1991. A review of As(III) in groundwater. Critical Reviews in Environmental Control. 21:1-39.

Millward, G.E., H.J. Kitts, L. Ebdon, J.I. Allen, and A.W. Morris. 1997. Arsenic in the Humber Plume, U.K. Continental Shelf Research, Vol. 17, No. 4, pp. 435-454.

Millward, G.E., H.J. Kitts, S.D.W. Comber, L. Ebdon, and A.G. Howard. 1996. Methylated arsenic in the southern north sea. Estuarine, Coastal and Shelf Science. 43:1-18.

Onken, B.M. and D.C. Adriano. 1997. Arsenic Availability in Soil with Time under Saturated and Subsaturated Conditions. Soil Science Society of America Journal,Vol. 61, pp. 746-752.

Seyler, P. and J.M. Martin. 1989. Biogeochemical processes affecting arsenic species distribution in a permanently stratified lake. Environ. Sci. Technol. 23(10):1258-1263.

Takamatsu, T., H. Aoki, and T. Yoshida. 1982. Determination of Arsenate, Arsenite, Monomethylarsonate, and Dimethylarsinate in Soil Polluted with Arsenic. Soil Science, Vol. 133, No. 4, pp. 239-246.

ARSENIC IN PLANTS AND ANIMALS

Helgesen, H. and E.H. Larsen. 1998. Bioavailability and speciation of arsenic in carrots grown in contaminated soil. Analyst. 123:791-796.

Maher, W.A. 1981. Determination of inorganic and methylated arsenic species in marine organisms and sediments. Analytica Chimica Acta. 126:157-164.

Nriagu, J.O. and J.M. Azcue. 1989. Food contamination with arsenic in the environment. National Water Research Institute. Burlington, Ontario, Canada. February, 1989.

Rittle, K.A., J.I. Drever, P.J.S. Colberg. 1995. Precipitation of arsenic during bacterial sulfate reduction. Geomicrobiology Journal. 13:1-11.

Small, T.D., L.A. Warren, E.E. Roden, and F.G. Ferris, 1999. Sorption of Strontium by Bacteria, Fe(III) Oxide, and Bacteria-Fe(III) Oxide Composites. Environ. Sci. Tech 33:4465-4470.

ARSENIC TOXICITY

Buchet, J.P. 1994. Inorganic Arsenic Metabolism in Humans. Arsenic Exposure and Health. Science and Technology Letters, Northwood, pp. 181-189.

Chavez V., A.C.P. Hidalgo, E. Tovar, and F.B.M. Garmilla. 1964. Estudios en una comunidad con arsenicismo cronico endemico. Salud Publ. Mex. Mayo-Junio. VI:435-44

Chen, S.L., S.R. Dzeng, M.H. Yang, K.H. Chiu, G.M. Shieh, and C.M. Wal. 1994. Arsenic species in groundwaters of the Blackfoot disease area, Taiwan. Environ. Sci. Technol. 28(5):877-88

Davis, A., M.V. Ruby and P.D. Bergstrom. 1992. Bioavailability of arsenic and lead in soils from the Butte, Montana mining district. Environ. Sci. Technol. 26:461-468.

Del Razo J., L.M., L.H. Hernandez G., G.G. Garcia-Vargas, P. Ostrosky-Wegman, C. Cortinas de Nava, and M.E. Cebrian. 1994. Urinary excretion of arsenic species in a human population chronically exposed to arsenic via drinking water. A pilot study. Arsenic Exposure and Health. Science and Technology Letters. Northwood, pp. 91-101.

Journal of AWWA. 1994. In search of an arsenic MCL. Journal AWWA. September 1994. p. 43.

Maiorino, R. M. and H. V. Aposhian. 1985. Dimercaptan metal-binding agents influence the biotransformation of arsenite in the rabbit. Toxicology and Applied Pharmacology. 77:240-250.

Mushak, P. 1994. Arsenic and Human Health: Some Persisting Scientific Issues. Arsenic Exposure and Health. Science and Technology Letters. Northwood, pp. 305-318.

Ng, J.C., S.M. Kratzmann, L. Qi, H. Crawley, B. Chiswell, and M.R. Moore. 1998. Speciation and absolute bioavailability: risk assessment of arsenic-contaminated sites in a residential suburb in Canberra. Analyst. 123:889-892.

Pontius, F.W., K.G. Brown and C.J. Chen. 1994. Health implications of arsenic in drinking water. Journal AWWA. September 1994. pp. 52-63.

Species Differences in the Metabolism of Arsenic. Arsenic Exposure and Health. Science and Technology Letters. Northwood, pp. 171- 179.

U.S. Environmental Protection Agency. 1996. Research plan for arsenic in drinking water. Board of Scientific Counselors. Review Draft. December 1996. pp. 1-96.

U.S. Environmental Protection Agency. 1992a. Test methods for evaluating solid waste, physical/chemical methods, EPA/SW-846/92.

U.S. Environmental Protection Agency. 1992b. RCRA groundwater monitoring: Draft technical guidance. US Environmental Protection Agency, EPA/530-R-93-001. Nov. 1992.

U.S. Environmental Protection Agency. 1997. Arsenic in drinking water: occurrence of arsenic. Office of Water. http://www.epa.gov/OGWDW/ars/ars5.html

U.S. Environmental Protection Agency. 1998. Arsenic in drinking water: drinking water standards development. Office of Water. http://www.epa.gov/OGWDW/ars/ars2.html

U.S. Environmental Protection Agency. 1998. IRIS substance file: arsenic, inorganic. Integrated Risk Information System. http://www.epa.gov/ngispgm3/iris/subst/0278.htm

U.S. Environmental Protection Agency. 1998. IRIS substance file: arsenic, inorganic. Integrated Risk Information System. http://www.epa.gov/ngispgm3/iris/subst/0278.htm Vahter, M. 1994.

GEOLOGICAL ENVIRONMENTS ASSOCIATED WITH ANOMALOUS ARSENIC BEARING WATERS AND SOILS

Azcue, J.M., J.O. Nriagu. 1993. Arsenic forms in mine polluted sediments of Moira Lake, Ontario. Environ Int. 19(4):405-416.

Chapin, C.E. and N.W. Dunbar. 1995. A regional perspective on arsenic in waters of the Middle Rio Grande Basin, New Mexico. Proceedings of the 39th Annual New Mexico Water Conference. WRRI Report No. 290, p. 257-27

Korte, N.E., 1991. Naturally occurring arsenic in the groundwaters of the midwestern United States. Environ. Geol. Water Sci. 18:137-141.

Maher, W.A. 1984. Mode of Occurrence and Speciation of Arsenic in some Pelagic and Estuarine Sediments. Chemical Geology, 47(1984/1985) 333-345.

Mahood, G.A., A.H. Truesdell, and L.A. Templos M. 1983. A reconnaissance geochemical study of La Primavera geothermal area, Jalisco, Mexico. Journal of Volcanology and Geothermal Research. 16:247-261.

Prol-Ledesma, R.M., S.I. Hernandez-Lombardini and R. Lozano-Santa Cruz, 1996. Chemical variations in the rocks of the La Primavera geothermal field related with hydrothermal alteration. In Press, University of Mexico, Mexico City.

Ramirez-Silva, G.R. 1981. Informe climatologico de la Zona Geotermica La Primavera-San Marcos-Heveres de la Vega, Jalisco. Informe 16-81. Comision Federal de Electridad, Mexico, Suberencia de Estudios Geotermicos, Departmento de Exploracion, May, 1981.

Ramirez-Silva, G.R. 1982. Hidrologia superficial y subterranea en las Zonas Geotermicas La Primavera-San Marcos-Heveres de la Vega, Jalisco. Informe 19-82. Comision Federal de Electricidad, Mexico, Suberencia de Estudios Geotermicos, Departmento de Exploracion, April, 1982

Reid, J. 1994. Arsenic occurrence: USEPA seeks a clearer picture. Journal AWWA. September 1994. pp. 44-51.

Smedley, P.L. 1996. Arsenic in rural groundwater in Ghana. Journal of African Earth Sciences. 22(4):459-470.

Sonderegger, J.L. and T. Ohguchi, 1988. Irrigation related arsenic contamination of a thin, alluvial aquifer, Madison River Valley, Montana, U.S.A.. Environ. Geol. Water Sci. V. 11, No. 2, p. 153-161.

Soussan, T. 1997a. Arsenic study nearly finished. Albuquerque Journal, January 21, 1997, Sec. C, p.1.

Soussan, T. 1997b. Arsenic levels in river over Isleta standard. Albuquerque Journal, December 13, 1997, Sec. C, p.1.

Stauffer, R.E. and J.M. Thompson. 1984. Arsenic and antimony in geothermal waters of Yellowstone National Park, Wyoming, U.S.A. Geochimica et Cosmochimica Acta. 48:2547-2561.

Thompson, J.M. 1979. Arsenic and fluoride in the upper Madison river system: Firehole and Gibbon rivers and their tributaries, Yellowstone National Park, Wyoming, and southeast Montana. Environ. Geol. 3:13-21.

U.S. Geological Survey. 1994. Arsenic contamination in the Whitewood Creek-Belle Fourche River-Cheyenne River System, Western South Dakota, Bibliography of Publications From the Toxic Substances Hydrology Program. U.S. Geological Survey Open-File Report 94-91.

Welch, D. 1999. Arsenic Geochemistry of Stream Sediments Associated with Geothermal Waters at the La Primavera Geothermal Field, Mexico. Masters Thesis, New Mexico Institute of Mining and Technology, Socorro, New Mexico.

SOURCE AND MECHANISMS FOR ANOMALOUS ARSENIC-BEARING GROUND WATERS

Aurillo, A.C., R.P. Mason and H.F. Hemond. 1994. Speciation and fate of arsenic in three lakes of the Aberjona watershed. Environ. Sci. Technol. 28:577-585.

Baker, L.A., T.M. Qureshi, and M.M. Wyman. 1998. Sources and mobility of arsenic in the Salt River watershed, Arizona. Water Resources Research. 34(6):1543-1552.

Bhattacharya, P., A. Sracek, and G. Jacks. 1998. Groundwater arsenic in Bengal delta plains - testing of hypotheses. Dhaka Conference on Arsenic. February, 1998

Bowell, R.J. 1992. Supergene gold mineralogy at Ashanti, Ghana: Implications for the supergene behavior of gold. Mineralogical Magazine. 56:545-560.

Bowell, R.J. 1994. Sorption of arsenic by iron oxides and oxyhydroxides in soils. Applied Geochemistry. 9:279-286.

Bowell, R.J., N.H. Morley, and V.K. Din. 1994. Arsenic speciation in soil porewaters from the Ashanti Mine, Ghana. Applied Geochemistry. 9:15-22.

Christensen, O.D. 1980. Trace element geochemical zoning in the Roosevelt hot springs thermal area, Utah. 3rd International Symposium on Water Rock Intaction. Edmundton, Canada. July, 1980. pp. 121-122.

Criad, A. and C. Fouillac. 1989. The distribution of arsenic(III) and arsenic(V) in geothermal waters: examples from the Massif Central of France, the island of Dominica in the Leeward Islands of the Caribbean, the Valles Caldera of New Mexico, U.S.A., and southwest Bulgaria. Chemical Geology. 76:259-269.

Das, D., G. Samanta, B.K. Mandal, T.R. Chowdhury, C.R. Chanda, P.P. Chowdhury, G.K. Basu and D. Chakraborti. 1996. Arsenic in groundwater in six districts of West Bengal, India. Environmental Geochemistry and Health. 18:5-15.

Robinson, B. 1995. The distribution and fate of arsenic in the Waikato River System, North Island, New Zealand. Chem Speciation Bioaval, v7, No.3, p. 89-97.

Sadiq, M. 1997. Arsenic chemistry in soils: an overview of thermodynamic predictions and field observations. Water, Air, and Soil Pollution, v. 93, pp. 117-136.

Sakata, M. 1987. Relationship between adsorption of arsenic(III) and boron by soil and soil properties. Environ. Sci. Technol. 21:1126-1130.

FIELD MEASUREMENT OF ARSENIC AND ARSENIC SPECIES

Clifford, D.A., L. Ceber and S. Chow. 1983. Separation of Arsenic (III) and Arsenic (V) by Ion Exchange. Proceedings 1983 AWWA Water Quality Technology Conference, Norfolk, VA, pp. 223-236, AWWA Denver, CO, December 1983.

Clifford, D. and C.C. Lin, 1991. Arsenic (III) and arsenic(V) Removal from drinking water in San Ysidro, New Mexico. USEPA Project Summary, EPA/600/S2-91/011, June 1991.

Edwards, M. 1998. Considerations in As analysis and speciation. Journal AWWA. Vol. 90, No. 30.

Ficklin, W.H. 1983. Separation of arsenic(III) and arsenic(V) in ground waters by ion exchange. Talanta. 30(5):371-373.

Ficklin, W.H. 1990. Extraction and Speciation of Arsenic in Lacustrine Sediments. Talanta. Pergamon Press. Vol. 37, No. 8, pp. 831-839.

Grabinski, A.A. 1981. Determination of arsenic(III), arsenic(V), monomethylarsonate, and dimethylarsinate by ion-exchange chromatography with flameless atomic absorption spectrometric detection. Analytical Chemistry. 53:966-968.

Hasegawa, H., Y.S. Sohrin, M. Matsui, M. Hojo, and M. Kawashima. 1994. Speciation of Arsenic in Natural Waters by Solvent Extraction and Hydride Generation Atomic Absorption Spectrometry. Analytical Chemistry, Vol. 66, No. 19, pp. 3247-3252.

Hasegawa, H., M. Masakazu, S. Okamura, M. Hojo, N. Iwasaki, and Y. Sohrin. 1999. Arsenic Speciation Including 'Hidden' Arsenic. Applied Organometallic Chemistry. Vol. 13, p. 113-119

Hem, J.D. 1970. Study and interpretation of the chemical characteristics of natural waters. U. S. Geological Survey. Water-Supply Pap. 1473. 363 p.

Irgolic, K. J. 1994. Determination of Total Arsenic and Arsenic Compounds in Drinking Water. Arsenic Exposure and Health. Science and Technology Letters. Northwood, pp. 51-61.

Soto, E.G., E.A. Rodriquez, P.L. Mahia, S.M. Lorenzo, and D.P. Rodriquez. 1995. Ion-exchange Method for Analysis of Four Arsenic Species and Its Application to Tap Water Analysis. Analytical Letters, Vol. 28, No. 15, pp. 2699-2718.

LARGE-SCALE ARSENIC TREATMENT METHODS

Cadena, F. and T. L. Kirk. 1996. Arsenate precipitation using ferric iron in acidic conditions. New Mexico Water Resources Research Institute Technical Completion Report No. 293, New Mexico State University, Las Cruces, NM, 22 p.

Cheng, R.C., S. Liang, H.C. Wang, and M.D. Beuhler. 1994. Enhanced coagulation for arsenic removal. Journal AWWA. September 1994, pp. 79-90

Edwards, M. 1994. Chemistry of arsenic removal during coagulation and Fe-Mn oxidation. Journal AWWA. September 1994, pp. 64-78.

Forstner, U. and I. Haase. 1998. Geochemical demobilization of metallic pollutants in solid wasted-implications for arsenic in waterworks sludges. Journal of Geochemical Exploration, v. 62, pp. 29-36.

Frost, R.R. and R.A. Griffin. 1977. Effect of pH on adsorption of arsenic and selenium from landfill leachate by clay minerals. Soil Sci. Soc. Am. J. 41:53-57.

Gupta, S.K.and K.Y. Chen. 1978. Arsenic removal by adsorption. Journal of the Water Poll. Control Fed., March 1978, p. 493-506

Hounslow, A.W. 1980. Ground-water geochemistry: arsenic in landfills. Ground Water. 18:331-333.

Los Angeles Department of Water and Power . 1997. Arsenic removal strategies. LADPW. http://www.ladwp.com/bizserv/water/quality/topics/arsenic/arsenic.htm

Los Angeles Department of Water and Power. 1997. Arsenic general information. LADPW. http://www.ladwp.com/bizserv/water/quality/topics/arsenic/arsenic.htm

McNeill, L.S. and M. Edwards. 1995. Soluble arsenic removal at water treatment plants. Journal AWWA. April 1995. pp. 105-113.

Merkle, P.B., W. Knocke, D. Gallagher, J. Junta-Rosso, and T. Solberg. 1996. Characterizing filter media mineral coatings. Journal AWWA. December 1996. pp. 62-73.

Scott, K.N., J.F. Green, H.D. Do and S.J. McLean 1995. Arsenic removal by coagulation. Journal AWWA. April 1995. pp. 114-126.

PROPOSED SOLUTIONS TO THE BANGLADESH ARSENIC PROBLEM AND POINT OF USE DEVICES

Bhattacharya, P.,M. Larrson, A. Leiss, G. Jacks, A. Sracek, and D. Chatterjee. 1998. Genesis of arseniferous groundwater in the alluvial aquifers of Bengal delta plains and strategies for low-cost remediation. Dhaka Conference on Arsenic. February, 1998.

Clifford, D. and C.C. Lin, 1991. Arsenic (III) and arsenic(V) Removal from drinking water in San Ysidro, New Mexico. USEPA Project Summary, EPA/600/S2-91/011, June 1991.

Harvard Arsenic Site: http://phys4.harvard.edu/~wilson/arsenic_project_introduction.html

Khan, A.H., Rasul, S.B., Munir, A.K.M., Habibuddowla, M., Alauddin, M., Newaz, S.S., and Hussam, A., 2000, Appraisal of a simple arsenic removal method for groundwater of Bangladesh, Journal of Environmental Science and Health Part A-Toxic/Hazardous Substances & Environmental Engineering: V. 35 pp. 1021-1041

Robinson, B. 1997. Silica interference in the precipitation of arsenic on iron oxides. Proc. Geothermal Reservoir Eng. Workshop, Stanford University (in press).

EPA/600/S2-85/094, September 1985.

Rogers, K.R. 1990. Point-of-use treatment of drinking water in San Ysidro, NM. USEPA Project Summary. EPA/600/S2-89/050, March 1990.

Rubel, F., Jr.and S.W. Hathaway. 1985. Pilot study for removal of arsenic from drinking water at the Fallon, Nevada, naval air station. USEPA Project Summary.

ARSENIC TREATMENT USING NATURAL MATERIALS

Hingston, F.J., A.M. Posner, and J.P. Quirk. 1974. Anion adsorption by goethite and gibbsite II. Desorption of anions from hydrous oxide surfaces. Journal of Soil Science. 25(1):16-26

Sadiq, M. 1997. Arsenic chemistry in soils: an overview of thermodynamic predictions and field observations. Water, Air, and Soil Pollution, v. 93, pp. 117-136. Sakata, M. 1987. Relationship between adsorption of arsenic(III) and boron by soil and soil properties. Environ. Sci. Technol. 21:1126-1130.

Spackman, L.K., K.D. Hartman, J.D. Harbour, and M.E. Essington. 1990. Adsorption of oxyanions by spent western oil shale. I. Arsenate. Environ. Geol. Water Sci. 15(2):83-91.

3. BIOTECHNOLOGY — TRANSGENIC PLANT VACCINE

SAFETY CONSIDERATIONS WHEN PLANNING GENETICALLY MODIFIED PLANTS THAT PRODUCE VACCINES

FRANCESCO SALA
Department of Biology, University of Milano, Via Celoria 26, 20133 Milano, Italy (e-mail: francesco.sala@unimi.it)

INTRODUCTION

Genetic engineering, combined with conventional breeding, is offering new powerful possibilities to modify plants and, thus, to face specific and novel needs. Up to recently, most, if not all, applications have been in the food industry. Main engineered plants have been maize, soybean, tomato. These, together with engineered cotton, are presently the most widely cultivated transgenic crops in the World. Engineered forest and cultivated trees will be soon ready for cultivation.

Presently cultivated transgenic plants exploit the great potential for genetic manipulation to enhance productivity by conferring resistance to diseases, pests, new herbicides and environmental stresses. Recently, a rice cultivar with a modified seed composition (high provitamin A and iron content) has been produced[1].

New traits are being introduced in ornamental plants. Plant "factories" are being designed for the production of molecules for the chemical industry, of pharmaceuticals or of other beneficial compounds. Genetic modification of endogenous metabolism and gene inactivation are promising important applications. Transgenic plants may also become drug-delivery devices with the most important vaccines being made in edible fruits. Encouraging results along this line have already appeared in the literature[2,3].

WHY MAKE VACCINES IN PLANTS?

There are several reasons why medical doctors are asking plant biotechnologists to try and produce vaccines in plants. The most relevant of these are summarized in Table 1.

Table 1. Advantages offered by the production of a vaccine in transgenic plants.

- It is free from animal (or human) viruses;
- May reduce cost of vaccination to socially acceptable levels;
- Is suitable for local production in developing countries;
- Does not depend on the existence of "cold-lines" necessary for vaccine conservation in developing countries.

As outlined in Table 2, this field of application is strictly dependent on the collaboration between medical doctors and plant molecular biologists.

Table 2. Steps in the development of plants that are genetically modified in order to produce vaccines for medical use.

- Evaluate the medical problem,
- Select the appropriate gene(s) for plant transformation,
- Select the appropriate gene promoter and expression signals,
- Decide the appropriate site of gene expression (nucleus, chloroplast or mitochondrion),
- Verify the efficiency of the biosynthetic pathway (productivity/plant weight),
- Evaluate the final plant product in therapy,
- Evaluate its social acceptability.

When considering the points of this table, I suggest that the final one should be faced first before planning transgenic plants for vaccine production: we should be able to give acceptable answers to the public concern, be it due to rational arguments or to irrational fears. We should also consider that, in this case we are faced both with more general objection concerning, *per se,* the acceptability of transgenic plants, and with ethical issues raised by the production of "food" that has medicinal effects on humans.

Permits for field trials and commercialisation will be granted, especially in the European Community, where this concern is stronger, only if sufficient answer is given to public concern.

Here below, I shall discuss acceptability of transgenic plants for human health and for the environment. Ethical considerations related to the acceptability of plants to make new vaccines and drugs will be introduced and discussed by other presentations in this meeting.

HOW SAFE IS SAFE ENOUGH IN PLANT GENETIC ENGINEERING ?

The long tradition of plant breeding and mutant induction and selection has steadily improved human nutrition and welfare through plant genetic alteration and adaptation to agricultural and industrial needs. This has not been exempt from risks: any new hybrid, by bringing together two full genomic sets, may express unexpected and undesired traits (e.g., production of toxins which were not produced by the parental plants) and any new mutant can carry a number of uncontrolled and potentially risky mutations besides the one(s) selected. But this has traditionally been perceived by the public as entailing minimum risk and high advantage to humanity.

Perception of risks in the case of transgenic plants is different: they are asked to be fully safe for human health and for the environment. In particular the European Community asks scientists to give full assurance that transgenic plants are absolutely free from risks. The answer is no, we cannot give full assurance. Many of the alleged risks have no scientific bases, but others are real. Thus, are transgenic plants acceptable?

All technological developments bring benefits to mankind but are accompanied by risks. Penicillin saves people but sometimes kills due to anaphylactic shock, electricity is extremely dangerous and driving a car is even more. Even sitting in a room is dangerous: the roof may fall down. What makes technology acceptable is the rationalisation of the ratio risks vs. benefits.

New technologies raise both concern and expectations and modern biotechnology is no exception. Käppeli and Auberson[4] stated that: "Better clarity might be achieved in the discussion on transgenic plant safety once it is recognized that potential harm from unexpected plant phenotypes has always existed in traditional plant breeding and that the purpose of selection has been to eliminate any potentially harmful progeny. A biosafety line could therefore be defined from the abundance of experience in plant selection technology, scientific knowledge about the evolutionary significance of plant genomic plasticity and understanding of the role intended for recombinant DNA techniques in plant breeding programmes". On these grounds, the authors proposed that: "The accepted background level of safety in plant modification could be used to define the safety baseline for recombinant DNA modification of plants and to evaluate the tolerability of potential deviations from background levels". A realistic proposal is that we accept transgenic plants if their ratio risks vs. benefits is equal or better than that accepted in traditional agriculture: we should not ask transgenic plants to be fully safe, but rather that they are demonstrated to have an acceptable ratio of risk vs. benefits.

But, frequently, public attitude to the safety of genetically engineered products in general, and food in particular, is not rational in a strictly scientific sense. While the European Community has practically been forced by critics of genetic engineering to stop commercialisation of transgenic maize, soybean and other plants, the USA agricultural industry succeeded in persuading national regulatory agencies that their products are safe to grow.

Evaluating risks of transgenic plants has now become a most difficult task of regulation on both sides of the Atlantic. We are in a critical moment of agriculture, in which the past enthusiasm for chemical herbicides, insecticides and fertilisers has turned into concern for their environmental and health price and in which the hope that these chemicals could solve the problem of nutrition in developing countries has been abandoned. The public fear that this may turn out to be the case also for plant genetic manipulation.

Enhancing the scientific evaluation of risks and benefits of transgenic plants is essential, but is not the whole solution. Just as necessary is the creation of trust. It is that which the European consumers, in particular, appear to lack. The deep-rooted cultural fears of genetic manipulations, together with the past experience of the aggressiveness of some agri-business companies, have contributed to the success of the fight against the "Frankestein food".

As a consequence, the primary duty of scientific researchers, especially of those in public institutions, is that of providing the basic scientific knowledge for the evaluation of present and future risks. But an important task is also that of offering scientific alternatives to irrational fears. An example of the latter, which is discussed below, is the exaggerated fear that antibiotic-resistance genes may be passed to enteric bacteria and even to man upon eating plants carrying these marker genes. In this, as in other cases, the task of the researcher is to show how science can address public concern by offering alternative solutions.

PUBLIC CONCERN AND SCIENTIFIC ANSWERS ON TRANSGENIC PLANTS

Acceptability of transgenic plants is questioned, especially in the European Community, owing to possible adverse effects on human health and on the environment. Of relevance is also the perception that the agri-industry may exert excessive control on their development and exploitation all over the World, including developing countries. Topics of public concern are listed in Table 3.

Table 3. Main topics raising public concern about the use of transgenic plants in agriculture.

Effects on human health	
•	Immediate, medium and long-term effects
Environmental impact	
•	Escape of foreign genes through pollen dispersal
•	Escape of transgenic plants through seed dispersal
•	Modification of the soil microflora and fauna

The public and consumers are composed of non-experts: the average level of technology-related information held by the general public is very low[5]. In general, objections to the transgenic technology depend on the nature of the application rather than on the technological manipulations *per se*.

As a consequence, debate is on the final product, while no public concern has ever been expressed on scientific or methodological options such as the choice of the experimental protocols used for the transformation procedure. Are they characterized by intrinsic risks? Are any of them more acceptable than the others?

SAFETY CONSIDERATIONS ON THE WAY TRANSGENIC PLANTS ARE CONSTRUCTED

Since the first demonstration that foreign genes from any source, cloned in bacterial plasmids, can be transferred to plant cells by *Agrobacterium tumefaciens*, several other approaches have been proposed and utilised. These are summarized in Table 4.

Table 4. Approaches to transfer foreign genes into plants. A presentation of recent advances in plant transformation technology may be found in Reference[6].

1.	Infection with *Agrobacterium tumefaciens*,
2.	Bombardment with accelerated particles,
3.	Gene transfer into protoplasts,
4.	Electroporation of protoplasts, intact cells or embryos,
5.	"Floral dip" approach.

Based on the large experience in hundreds of laboratories all over the World and on considerations that are intrinsic to the gene transfer methodologies, risks are limited to rare potential cases of gene inactivation due to positioning of the foreign gene within or near active cellular genes. Extremely more frequent are cryptic gene inactivations and activations in breeding and in mutant induction. Thus, none of the presently utilized approaches to gene transfer in plants appears to be more acceptable than the others.
Their common feature is that they integrate the gene in the nuclear genome and that, when this happens, the gene is as stable as the other genes in the genome and is inherited as a Mendelian trait.

The different approaches may integrate multiple copies of the gene, although plants with a single copy may be selected by the subsequent molecular analysis. Site of integration may be perfectly determined by molecular analysis but integration is at random genomic positions, as homologous recombination at specific loci is still laborious. If appropriately planned, gene integration may be targeted to the chloroplast genome by homologous recombination. Of course, in the latter case the inheritance will be in most cases, maternal or, in few cases, paternal, depending on whether chloroplasts are inherited through ovules or pollen grains.

In any case, gene expression can be constitutive or inducible, depending on the selected promoter sequence and the gene product may be targeted to different plant sites and organelles, depending on the presence of a "transit" sequence.

Recent refinements of the transformation procedure now allow the use of DNA sequences containing exclusively linear arrangements of promoter-gene-terminator. This avoids the use (and integration) of carrier plasmid DNA, which has been a must until recently.

Another common feature is that all presently available transformation procedures depend on the availability of protocols to differentiate plants from the original selected transgenic cell. The "floral dip" approach, which is based on immersion of floral buds into an *A. tumefaciens* suspension, may dispense from this necessity. However, this has presently been used only with the model plant *Arabidopsis thaliana*[7].

Phenomena of somaclonal variation have been demonstrated in transgenic plants[8]. These are manifested as transposon activation, gene silencing, gene amplification and other types of genomic changes. But these events have been shown to be the same as those that naturally occur in plants as an answer to biotic or abiotic stress. When discussing this phenomenon, Walbot and Cullis[9] proposed that the plant genome, at variance from the animal one, should be considered "plastic": being unable to move, plants adapt to the changing environment by changing their genomic structure.

TRANSGENIC PLANTS AND HUMAN HEALTH

Many of the risks that are attributed to transgenic plants are actually common to all cultivated plants. Health and environmental problems have always accompanied agriculture. But transgenic plants have an extra factor of risk, the foreign gene. May this represent a serious danger to humans? Many fears may not have a scientific base, but scientists have the duty to face them and find appropriate acceptable alternatives.

Here are examples of alleged accusations towards transgenic plants:

Allergenic properties: the foreign gene has been accused of being a potentially allergenic factor. Indeed the gene could code for a protein with allergenic properties. Many proteins are known in nature, and in our food, to cause allergies. In the case of a foreign gene, these properties should be verified by analysing the physical and chemical characteristics of the foreign protein. The effects of the foreign gene on the production of endogenous allergens should also be assessed and ELISA and RAST assays used on the final transgenic plants to confirm assumptions. Furthermore, transgenic plants could be planned where an antisense sequence complementary to an allergenic gene is integrated. This approach is expected to eliminate the incidence of allergenics in our food.

Antibiotic resistance: this is a major issue: the large majority of transgenic plants presently cultivated in the World are endowed with a gene carrying resistance to an antibiotic, usually neomycin and kanamycin. The rational for its use was that this gene provides a selection system for co-transformed plant cells (carrying the gene of interest plus the gene for antibiotic resistance). This is perceived as a possible cause of antibiotic resistance in humans following transfer of the foreign gene from transgenic food to enteric bacteria and, perhaps, to the human genome. The allegation has no scientific bases: our gut is endowed with 10^{14} enteric bacteria belonging to at least 300 different species. Natural mutation frequency for bacterial genes is 10^{-7}. This means that, at any time, 10^{7} bacteria are neomycin-resistant mutants. Even assuming that a resistance gene present in an edible transgenic plant (for instance tomato), and endowed with promoter and terminator regions specific for plants, migrates and integrates into the genome of an enteric bacterium, this would simply be summed up to those already present in the gut. Furthermore, it is well recognised that it is the selective pressure imposed by the use and abuse of antibiotic in therapy (and the use of antibiotics as food additives in livestock nurseries) that determines the selective pressure for resistant microrganisms.

Nevertheless, this is a typical case in which it is strongly advisable to give an answer to public concern by proposing alternative solutions. Novel marker genes are already available and are based on the production of fluorescent products ("green fluorescent protein") or of an enzyme that enables the plant cell to grow on a sugar (mannose) not usually utilised by plants. Genes of interest and marker genes may also be integrated in different chromosomes so that, upon sexual reproduction, individual plants without the marker gene may be selected ("outsegregant approach"). In other cases, such as in the production of herbicide-resistant plants or plants resistant to specific toxins, selection can be directly performed in the presence of the herbicide or of the toxin. In the case of herbicide-resistance genes it is argued that it might be transferred by out-crossing into weeds. A clear cut approach to overcome all concerns is just to remove the selectable marker gene upon its exploitation in the selection step[10]. This has recently been shown to be possible through intrachromosomal recombination[11], and recommended especially for vegetatively propagated species, where the "outsegregant approach" may not be convenient.

35S promoter and tumors: in 1999, Ho et al.[12] raised concern over the effect on human health of the spread, by horizontal gene transfer, of transgenic viral promoters. By examining the safety implication of the presence of recombination spots on the base sequence of the cauliflower mosaic flower promoter (CaMV 35S), which is used in practically all current transgenic crops released commercially, these authors strongly suggested, as a precaution measure, that all transgenic crops containing CaMV 35S or similar promoters should be immediately withdrawn from commercial production, open field trials and sale.

This allegation does not have solid scientific bases: every day we eat, with our vegetables, billions of plant viruses, including CaMV. If horizontal gene flow could occur so easily, then our genome would be filled up with plant genes and promoters. The same is true for animal food and genes. Research on this subject is shedding light on the mechanisms (use of nucleases, other tools ?) by which each species defends its own genome from those used as food or the invading ones.

These and other considerations on the effect of transgenic plants on human health should make us confident that there is no scientific demonstration that safety of transgenic food is different from that of traditional food. The official controls imposed by laws of all countries in the world on transgenic food (but not on traditional food) before commercialisation add additional warranty to this conclusion.

TRANSGENIC PLANTS AND THE ENVIRONMENT

Agriculture has always had a negative impact on environment and biodiversity. Forests have been destroyed and monoculture has been introduced as a means to produce more with less effort. New species have been moved through Continents and this has frequently had adverse effects on local biodiversity, as well as on soil microflora and fauna. The knowledge of this has made us more careful with transgenic plants. But, again, transgenic plants have an extra factor of concern, the foreign gene. May this be of danger to the environment? Can the foreign genes be transferred to sexually compatible plants? Could transgenic seed dispersal endanger biodiversity? Are there strategies or tools to avoid these problems?

Escape of foreign genes through pollen dispersal

Plants in the environment may be sexually compatible with transgenic plants. Thus, it is feared that transgenic pollen may transfer the foreign gene to these plants and create "super-weeds" or otherwise modified plants.

An excellent discussion on this topic has been produced by Daniell[13]. An example is that of the transfer of a "terminator gene" (a gene that induces sterility in the progeny) to a sexually compatible plant. A second example is that of the transfer of an herbicide-resistance gene to weeds. Herbicide-resistant populations of weeds have already reduced the utility of some herbicides in traditional crops and have caused to use different herbicides.

However, as summarized in Table 5, there are restrictions to the success of gene transfer through pollen dispersal. Thus, in every case, pollen dispersal range should be accurately determined: it could reach distances of kilometres (case of maize) or be reduced to a few centimetres (case of rice). Rice and tomato are essentially self-pollinating, while maize is not. Maize has no sexually compatible weeds in Europe, while soybean has.

Table 5. Conditions for the transfer of foreign genes to neighbouring plants through pollen.

1.	Pollen grains must reach a sexually compatible plant,
2.	Cross pollination will not occur if the species is strictly autogamous,
3.	The expression of the foreign gene must give a selective advantage.

The foreign gene should also give an evolutionary advantage to the receiving plant. For instance, in the case of the "terminator gene" the resulting plants would be sterile, and thus unable to produce seed progeny if reproduction is exclusively through seeds (as in cereals), but could be invasive if the plant is capable of intensive vegetative multiplication (as in many weeds).

Strategies should be worked out in all cases in which gene transfer through pollen dispersal cannot be ruled out. Table 6 summarises the most relevant approaches to the problem.

Daniell et al.[14] verified the potentiality of the integration of the foreign gene into the chloroplast by reporting the genetic engineering of herbicide (glyphosate) resistance by stable integration of a petunia gene into the tobacco chloroplast genome. An important advantage of chloroplast transformation is the high gene expression due to the very high copy number (5,000-10,000) of chloroplast genomes in photosynthetic plant cells, while copy number of genes integrated in the nucleus vary from 1 to 50 if multiple integration occur. Furthermore, because the transcription and translation machinery of the chloroplast is prokaryotic in nature, herbicide-resistant genes of bacterial origin can be expressed at extraordinary high levels in chloroplast.

When the Bt-gene was engineered into the tobacco chloroplast genome, protoxin production was produced at 20- to 30-fold higher levels than nuclear transgenic plants[15].

Table 6. Strategies to avoid cross pollination.

1.	Integrate the foreign gene into the chloroplast genome. *Rationale:* Most crop plants are characterized by maternal inheritance of chloroplasts,
2.	Use male-sterile transgenic plants. *Rationale:* may be used when seeds are not the major product (as in poplar, sugarcane, bananas, but not in cereals).
3.	Release allogamous fertile plants in regions where sexually compatible plants are absent.

Escape of foreign genes through seed dispersal

Transgenic crop plants will spread their seed in the environment. However, it is documented that cultivated plants are very poor competitors to wild plants. They have been selected by breeders for traits which have an agricultural value (dwarfism, high yield, public acceptance of the commercial product) but carry many traits (sensitivity to biotic and abiotic stresses) which make them non competitive in the natural environment. Plants in natural conditions have to face a much stronger competition than in the protected agricultural field.

In some cases the use of sterile transgenic plants may radically solve the problem and also provide benefits to the populations. This is the case, for instance, of transgenic poplar. This plant is routinely reproduced by cuttings. The co-transformation with a gene of interest (for instance a *Bt*-gene) and a gene that induces sterility would have beneficial effects on the environment (no cross fertilisation with natural poplar) and on human health (no more allergies due to pollen dispersal).

Thus, the situation must be evaluated case by case, but in most cases, seed dispersal will not turn out to be a problem.

Effects of transgenic plants on natural habitat and biodiversity

Agriculture is not nature! Since it appeared, and at an increased rate in the last century, agriculture meant destruction of forest land, reduction of biodiversity and environmental pollution. In recent decades, the increased awareness of these negative aspects led the public opinion to ask for the development of environmentally friendly approaches to agriculture. It is no surprise that these requests are even more strongly expressed in the case of transgenic plants.

A clear answer should be given to the public concern that transgenic plants may reduce biodiversity. A first and relevant problem is due to the fact that there are two types of biodiversity that are usually confused by the public. The first is the one that exists in natural habitats and that is frequently threatened by a large array of human activities. By increasing productivity per unit of land, biotechnology may be of help in returning agricultural land to forests (at least in developed countries). The second type of biodiversity is referred to diversity of varieties within each cultivated species. In this case it is clear that a transgenic plant is, *per se*, an addition to the number of available varieties, not a limitation: restriction of biodiversity of products on the market is most frequently due to commercial needs rather than to the work of geneticists and bioengineers.

Modification of the soil microorganism (bacteria and fungi) and fauna (larvae) population

There is concern for those transgenic plants that, by excreting the new protein in the soil, may interfere with organisms living in the rhyzosphere. Saxena et al.[16] suggested that this may be the case for maize-*Bt,* whose roots may excrete the *B*t-toxin thus interfering with soil insects. However, in that case experimental results were confined in the laboratory. No field data were produced.

It is important that more conclusive data are produced on this specific topic and that other transgenic plants are tested for their effect on the soil organisms (insects, mycorizzal fungi, bacteria). It is also important that these tests are carefully planned: soil of transgenic crops should not simply be compared to that of non-transgenic crops. It is extremely unlikely that an agricultural soil retains the original natural equilibrium. If we find any change from non-transgenic to transgenic, are we actually moving from one artificial situation to another ? Why should we prefer the one with non-transgenic plants?

If this risk is verified, than it could be faced with the use of inducibile promoters that will allow expression of the gene only when needed.

The agricultural environment has frequently been altered by the use of chemicals (insecticide, fungicides, fertilisers, phytoregulators and others). Many transgenic plants are planned to reduce or eliminate the use of these chemical. Thus, careful analysis should also be performed to verify if the cultivation of these plants gives real advantages to soil microorganisms and fauna.

CONCLUSIONS

The best argument in favour of transgenic plants is the precision by which they are altered by introducing one or a few genes, by comparison to classical plant breeding and mutagenesis. This is what makes scientists confident of the fact that, with transgenic plants, a unique possibility is offered to plan genetic manipulations and predict with sufficient confidence their effect on humans and environment.

As Bengtsson[17] stated, "If gene technology is to be presented as a clean technology, then it must be clean", and "Setting high standards for new transgenic plant varieties is not only a question about human health. It is also a way to protect a vital new technology against short-sighted uses that may later lead to severe setbacks".

Table 7 summarises the main steps and questions that should be analysed to answer rational and irrational fears, before embarking in a project aimed at the production of transgenic plants for commercial use.

Table 7. Experimental details and steps that need careful planning before embarking in the production of transgenic plants for commercial use.

1.	Gene source (animal, fungus, bacterium, plant),
2.	Type of gene construct (gene sequence and expression factors),
3.	Site of gene integration (nucleus or chloroplast),
4.	Tissue and timing of gene expression in the plant,
5.	Level of gene expression,
6.	Quantity of gene product,
7.	Adverse environmental effects (gene flow to other plants, biodiversity),
8.	Social acceptability (risk perception, tangible benefits).

The source of the gene to be transferred is a typical case of irrational fear. Animal, plant or fungal genes use a universal genetic code. It is the global organisation of genes that make an individual develop into an animal or a plant, not the use of animal or plant genes. Minor differences in gene sequence are only due to evolutionary divergence. The question of whether a strawberry transformed with a pig gene can be eaten by a vegetarian has no scientific base. But, if this does not convince the non-experts, then more acceptable applications of plant genetic engineering should be offered, considering that presently transgenic plants carrying foreign genes derived from plants show the best acceptance. This is the case, for instance, of the above mentioned glyphosate-resistant tobacco plants, whose resistance gene was isolated from petunia. A second example is the use of a gene, named B32, which was isolated from maize and is now being transferred into rice to confer resistance to important fungal diseases.

Finally and very important, in the discussion on the acceptability of transgenic plants, it should be made clear that this should not be intended as a unique case to be globally accepted or rejected. Rather, acceptability should be considered separately for each new transgenic plant.

Sufficient warranty to the public should be given by the fact that, for the first time in the history of agriculture, a novel plant (if transgenic) has to undergo a complete set of tests, and go through severe scientific evaluation, including clinical tests, before being legally accepted for cultivation. This is not done, until now, for any new variety produced with traditional genetic tools!

REFERENCES

1. Ye, X., Al-Babili, S., Klöti, A., Zhang, J., Lucca, P., Beyer, P., Potrykus, I. (2000) "Engineering the provitamin A (B-carotene biosynthetic pathway into (carotenoid-free) rice endosperm" Science 287: 303-305.
2. May, G.D., Afza, R., Mason, H.S., Wieko, A., Novak, F.J., Arntzen, C.J. (1995) Generation of transgenic banana (*Musa acuminata*) plants via A*grobacterium* mediated transformation. Bio-Technology 13: 486-492.
3. Yusibov, V., Modelska, A., Steplewski, K., Agadjanyan, M., Weiner, D., Hooper, D.C., Koprowski, H. (1997) Antigens produced in plants by infection with

chimeric plant viruses immunized against rabies virus and HIV-1. Proc. Natl. Acad. Sci. USA 94: 5784-5788.
4. Käppeli, O., Auberson, L. (1998) "How safe is safe enough in plant genetic engineering" Trends in Plant Science 3: 276-281.
5. Urban, D. (1996) "Quantitative measure of public opinions on new technologies. Scientometrics 35: 71-77.
6. Hansen, G., Wright, M.S. (1999) "Recent advances in the transformation of plants". Trends in Plant Sci. 4: 226-231.
7. Clough, S.J., Bent, A.F. (1998) "Floral dip: a simplified method for Agrobacerium-mediated transformation of Arabidopsis thaliana" Plant J. 16: 735-743.
8. Sala, F., Arencibia, A., Castiglione, S., Christou, P., Zheng, Y., Han, Y. (1999). "Molecular and field analysis of somaclonal variation in transgenic plants". In: Altman, A. et al. (eds.). Plant Biotechnology and In Vitro Biology in the 21st Century. Kluwer Academic Publishers. The Netherlands. pg. 259-262.
9. Walbot, V., Cullis, C. (1983) "The plasticity of the plant genome - Is it a requirement for success" Plant Mol. Biol. Rep. 1: 3-11.
10. Puchta, H. (2000) "Removing selectable marker genes: taking the short cut" Trends in Plant Sci. 5: 273-274.
11. Zubko, E., Scut, C., Meyer, P. (2000) "Intrachromosomal recombination between attP regions as a tool to remove selectable marker genes from tobacco transgenes". Nature Biotech. 18: 442-445.
12. Ho, M.W., Ryan, A., Cummins, J. (1999) Cauliflower mosaic viral promoter - A recipe for disaster" Microb. Ecol. in Health and Disease 11: 1-8.
13. Daniell, H. (1999) Environmentally friendly approaches to genetic engineering. In Vitro Cell. Dev. Biol. 35: 361-368.
14. Daniell, H., Datta, R., Varma, S., Gray, S., Lee, S.B. (1998) Containment of herbicide resistance through genetic engineering of the chloroplast genome. Nature Biotechnology 16: 345-350.
15. Kota, M., Daniell, H., Varma, S., Garczynski, S.F., Gould, F., Moar, W.J. (1999) Overexpression of the *Bacillus thuringiensis* (*Bt*) Cry2Aa2 protein in chloroplast confer resistance to plants against susceptible and *Bt*-resistant insects. Proc. Natl. Acad. Sci. USA 96: 1840-1845.
16. Saxena D., Flores S., Stotzky G. (1999) "Insecticidal toxin in root exudates from Bt corn" Nature 402: 480.
17. Bengtsson B.O. (1997) "Pros and cons of foreign genes in crops" Nature 385: 290.

PURIFIED CHOLERA TOXIN B SUBUNIT FROM TRANSGENIC TOBACCO PLANTS POSSESSES AUTHENTIC ANTIGENICITY

XIN-GUO WANG, GUO-HUA ZHANG, RONG-XIANG FANG
Laboratory of Plant Biotechnology, Institute of Microbiology, Chinese Academy of Sciences, Beijing 100080, P.R. China
CHUAN-XUAN LIU, YAN-HONG ZHANG, CHENG-ZU XIAO
Department of Cell Engineering, Institute of Biotechnology, Beijing 100071, P.R. China

ABSTRACT

Cholera toxin B subunit (CTB) mature protein was stably expressed in transgenic tobacco plants under the control of CaMV 35S promoter and TMV Ω fragment. Fusion of the PR1b signal peptide coding sequence to the CTB mature protein gene increased the expression level by 24-fold. The tobacco-synthesized CTB (tCTB) was purified to homogeneity by a single step of immunoaffinity chromatography. The purified tCTB is predominantly in the form of pentamers with molecular weight identical to the native pentameric CTB, indicating the PR1b-CTB fusion protein has been properly processed in tobacco cells. Futhermore, we have shown that the antigenicity of the purified tCTB is indistinguishable from that of the native CTB protein by immunodiffusion and immunoelectrophoresis.

Keywords: transgenic plant; cholera toxin B subunit; purification; antigenicity

INTRODUCTION

Cholera poses a continuous threat to human health, especially to the vast population in developing world[13]. Practical and cost-effective vaccines against cholera, especially oral vaccines, are urgently needed. The nontoxic cholera toxin B subunit (CTB) has been shown to be an important component of the vaccine in a field trial when mixed with a killed-whole cell vaccine strain[4,5]. Furthermore, it can function as an effective carrier to facilitate induction of mucosal immune response and immunological tolerance to polypeptides to which CTB is coupled either chemically or through gene fusion technology[3,6,8,18,19]. Production of CTB in plants offers several advantages over the conventional fermentation systems, including a lower cost in large-scale production and providing a more stable environment for storage of the heat-labile CTB. In addition, CTB produced in edible plants may serve as an oral vaccine that is easy to administer[20].

CTB has been expressed in transgenic potato leaf and tuber tissues at a level of 0.3% of total soluble plant protein[2]. CTB protein accumulated in potato tubers formed predominantly a pentameric structure and retained its native antigenicity and the binding capacity for G_{M1}-ganglioside, the mammalian cell membrane receptor of cholera toxin (CT). Oral administration of transgenic potato tissues to mice induced both mucosal and serum CTB-specific antibodies and reduced diarrhea caused by CT[1]. In this study, we report the expression of CTB in transgenic tobacco plants and the purification of CTB protein (tCTB) from transgenic leaf tissues by a single step of immuno-affinity chromatography. We have shown that the purified tCTB retained the pentameric structure and possessed the authentic antigenicity.

MATERIALS AND METHODS

Construction of Plant Expression Vectors

The plant binary vector pBin438, a derivative of pBI121(Clontech), contains a duplicated CaMV 35S promoter and the tobacco mosaic virus (TMV) Ω sequence to drive the expression of inserted genes[16]. It was used to create the CTB expression vectors pBI-CTB and pBI-SPCTB. The CTB mature protein coding sequence (309 bp) was amplified and modified by PCR from the plasmid pUC19-CTB which harbors a 2.4 kb *Xba*I-*Eco*RI fragment of the CT operon encompassing the entire CTB coding sequence[17]. Two PCR primer sets, i.e. Set 1: 5′ primer 1 (5′-AGGATCCACCATGACACCTCAAAATATTAC-3′) and 3′ primer (5′-AGTCGACTTAATTTGCCATAC-3′), and Set 2: 5′ primer 2 (5′-AAGTACTCCTCAAAATATTAC-3′) and 3′ primer (the same as in Set 1), were used in amplification. The PCR products were cloned into pGEM-T vector (Promega), resulting in pGEB1 and pGEB2 respectively. The CTB sequence in pGEB1 was cut out with *Bam*HI and *Sal*I whose recognition sequences are included at the 5′ ends of 5′ primer 1 and 3′ primer, respectively, and inserted into pBin438 to form pBI-CTB. The CTB sequence in pGEB2, which contains a *Sca*I site at the 5′ end, was first fused in-frame to the 3′ end of the tobacco pathogenesis-related 1b (PR1b) signal peptide (SP) coding sequence (90 bp) in the plasmid pBIPR1b through the filled-in *Mlu*I site (ACGCG). The PR1bSP-CTB fusion sequence was then moved to pBin438 as a *Bam*HI-*Sal*I fragment to produce pBI-SPCTB. The CTB sequence in pBI-CTB and PR1bSP-CTB fusion sequence in pBI-SPCTB was confirmed by DNA sequencing.

Tobacco Transformation

Binary vectors pBI-CTB and pBI-SPCTB prepared from *E. coli* XL1-blue cultures were separately transferred into *Agrobacterium tumefaciens* strain LBA4404 by electroporation. Plasmids from LBA4404 transformants were prepared and verified by restriction digestions. Tobacco (*Nicotiana tabacum* cv K326) leaf discs were transformed by co-cultivation method[11] and transgenic plants were selected on medium containing 300 mg/L kanamycin. Transformed plants were confirmed by PCR assay and southern blot analysis.

Determination of CTB Protein Level in Transgenic Tobacco Plants

CTB expression level in individual tobacco plants was determined by a quantitative ganglioside-dependent ELISA assay. Tobacco leaves were collected from aseptically grown plants or greenhouse plants. Leaf samples (50-100 mg) were ground in 500 μL PBST buffer (10 m*M* PBS pH 7.4, 1 m*M* PMSF, 1% 2-mercaptoethanol, 0.1% TritonX-100). Insoluble plant debris was removed by centrifugation at 13,000 rpm at 4°C for 10 min, and the supernatant was used for analysis. Total protein concentration of the leaf extracts was determined using Coomassie dye-binding assay (Bio-Rad), using bovine serum albumin (BSA) as a standard. For CTB ELISA, the microtiter plate was coated with 2 μg/well of monosialoganglioside-G_{M1} (Sigma G 7641) in 100 μL of 0.05 *M* carbonate buffer (pH 9.6) and blocked with 1.5% BSA. Then serial diluted leaf extracts (100 μL/well) and a series of dilutions of bacterial CTB (Sigma C 9903) solution were added and incubated at 37°C for 1 h. After the plate was washed three times with PBST, 100 μL/well of rabbit anti-CT serum (1:5,000, Sigma C 3062) was added and incubated at 37°C for 1 h, following by incubation with goat anti-rabbit IgG conjugated to horseradish peroxidase (1:10,000, Sigma A 6154) (100 μL/well) at 37°C for 1 h. After washing, the color was developed with 3,3′,5,5′-tetramethyl benzidine dihydrochloride (TMB) and the absorbance was measured in a Model 550 microplate reader (Bio-Rad), operated according to the manufacturer's instructions.

Purification of tCTB Protein

Transgenic tobacco leaf samples of greenhouse-grown plants were homogenized in ice-cold extraction buffer (10 m*M* PBS pH 6.0, 1m*M* PMSF, 0.1% Triton X-100, 1% 2-mercaptoethanol) in a glass homogenizer. Insoluble plant tissue was removed by centrifugation for 15 min at 10,000*g* at 4°C. CTB protein was purified from crude plant proteins by affinity chromatography. Rabbit anti-CT IgG was purified from rabbit anti-CT serum by a batch method of DEAE-cellulose 52 (Whatman) chromatography[21]. Rabbit anti-CT IgG (10 mg) was coupled to 1 g of CNBr-Sepharose 4B as described by the manufacturer (Pharmacia) and the treated Sepharose particles were packed into a chromatography column. The clarified tobacco leaf extract containing CTB was filtered through 0.8 μm membrane and loaded onto the column. After washing with PBS, the CTB protein was eluted with 0.1 *M* glycine-HCl buffer (pH 2.8) and neutralized to pH 7.4 with 1*M* Na_2CO_3 followed by dialysis against 10 m*M* PBS.

SDS-PAGE and Immunoblot

Purified tCTB was analyzed by 12% SDS-PAGE either loaded directly on the gel or boiled for 3 min prior to electrophoresis. Gels were stained with Coomassie blue or blotted using a semidry blot apparatus onto PVDF membrane (Millipore) in transfer buffer (25 m*M* Tris-HCl pH 8.3, 192 m*M* glycine, 1% SDS, 20% methanol). The blot was blocked for 1 h in TSET buffer (20 m*M* Tris-HCl pH 7.5, 150 m*M* NaCl, 1 m*M* EDTA, 0.1% Tween-20) containing 3% BSA and subsequently incubated for 1 h in a 1:5,000 dilution of rabbit anti-CT serum in TSET buffer plus 1% BSA. The blot was washed 3 times for 10 min each in PBST, and finally incubated in a 1:5,000 dilution of goat anti-rabbit IgG conjugated to alkaline phosphatase (Promega) in TEST buffer containing 1% BSA for 1 h. Color development was performed using BCIP and NBT (Promega).

Immunodiffusion and Immunoelectrophoresis

Immunodiffusion and immunoelectrophoresis were carried out following the method described previously[21]. For double immunodiffusion, 1% of agarose in PBS (pH 7.4) was melted and poured onto pre-cooled slides on a leveled surface. Holes with diameter of 3 mm were punched and 10 μLof rabbit anti-CT serum (1:10, Sigma C 3062) or 10 μL of CTB protein (bacterial CTB or tCTB, each in 0.05 μg/μL) were separately added into the holes. The slide was then placed in a humid chamber and incubated overnight at 37°C. Gel slides for immunoelectrophoresis were prepared as for immunodiffusion. Two holes 1.5 cm apart were made on the gel with hypodermic needles. One hole was filled with 15 μL of bacterial CTB (0.05 μg/μL) and the other with 15 μL of tCTB (0.05 μg/μL). After electrophoresis for 1.5 h with 10 mA current in barbitone buffer, a 3 mm x 5 cm trough lying in between the two holes was made and filled with rabbit anti-CT serum (1:10). The slide was incubated overnight in a humid chamber at 37°C.

RESULTS AND DISCUSSION

CTB Plant Expression Vectors

The structures of the T-DNA regions of CTB plant expression vectors pBI-CTB and pBI-SPCTB are depicted in Figure 1. In these two constructs, the CaMV 35S promoter with a duplicated enhancer[12] is used to drive the transcription of the CTB and the SPCTB genes and the tobacco mosaic virus RNA Ω fragment serves as a translational enhancer for the transcripts[9]. pBI-CTB contains the mature CTB coding sequence with addition of a sequence ACCATG 5′ to the first codon ACA. Nucleotides ATG would serve as the translation start codon and ACC provide part of the nucleotide context for favorable translational initiation[14]. In pBI-SPCTB, the mature CTB coding sequence is fused to the 3′ end of the sequence encoding the tobacco PR1b signal peptide through the half site of *Sca*I, ACT, a silent mutant of the native first codon ACA. Use of the tobacco PR1b signal peptide rather than the bacterial CTB leader peptide, is based on the fact that the PR1b signal peptide functions efficiently in secretion of a heterologous protein in plant[15] and that the fusion protein is likely processed upon secretion[7]. It was reported that the bacterial CTB leader peptide was not removed from the CTB protein when expressed in potato plants[2] and retention of the CTB leader sequence might interfere the oral immunogenecity of the plant-derived CTB protein[1].

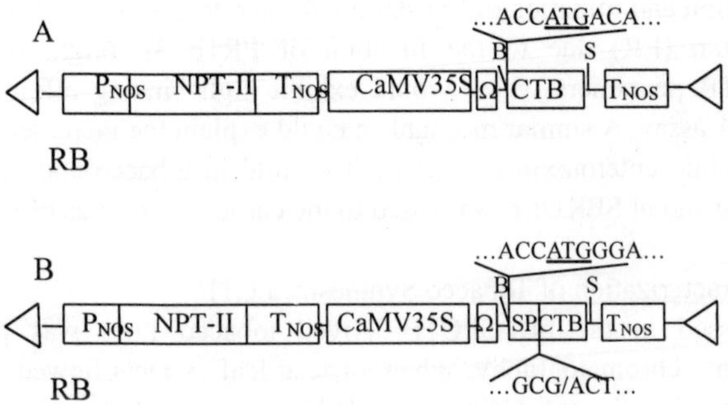

Fig. 1. Structure of the T-DNA regions of binary vectors pBI-CTB (A) and pBI-SPCTB (B). LB: left border sequence; RB: right border sequence; Pnos: nopaline synthase promoter; Tnos: nopaline synthase terminator; NPT-II: neomycin phosphotransferase gene; CaMV 35S: cauliflower mosaic virus 35S promoter with doubled enhancer sequences; Ω: the 5′ untranslated leader sequence of tobacco mosaic virus RNA; CTB: CTB mature protein coding sequence; SP: PR1b signal peptide coding sequence. Sequences around the translation initiation codon ATG (underlined) and the sequence at the junction of SP and CTB are shown above and below the diagrams respectively. Restriction sites of BamHI (B) and SalI (S) used for insertion of the genes are also shown.

<u>CTB Expression Level in Transgenic Plants</u>

Thirty-seven and 42 kanamycin-resistant tobacco plants were obtained after transformation with pBI-CTB and pBI-SPCTB, respectively. Integration of the T-DNA regions into the plant nuclear chromosomal DNA in all these plants was verified by PCR assays and further confirmed by southern blot hybridization on some of the transformants (data not shown). The presence of CTB protein in 24 pBI-CTB-transformed plants and 29 pBI-SPCTB-transformed plants was analyzed by ganglioside-dependent ELISA. The results showed that more than 80% of the assayed plants of each group synthesized CTB protein, but the CTB levels in different plants varied significantly, possibly due to the chromosomal position effect of the T-DNA insertion. For each of the constructs, 4 plants with the highest CTB levels were selected and the average amounts of CTB protein were calculated to represent the CTB expression levels in pBI-CTB- and pBI-SPCTB-transformed plants. While CTB protein synthesized in the pBI-CTB-plants accounted for only 0.004% of total soluble leaf protein, the pBI-SPCTB-plants produced CTB protein at a level up to 0.095% of total soluble leaf protein, about 24-fold higher than the pBI-CTB-plants did. Since the same CaMV 35S promoter and TMV Ω fragment are used to control the gene expression in both constructs, it seems unlikely that the elevated expression of CTB observed in pBI-SPCTB-plants is attributed to up-regulation of the gene expression at levels of transcription and initiation of translation. Rather, targeting of CTB to the plant endoplasmic reticulum (ER) due to the function of PR1b SP might facilitate the formation of the CTB pentamers which would exhibit high binding affinity for G_{M1}-ganglioside in ELISA assay. A similar mechanism could explain the increased expression of the *E. coli* heat-labile enterotoxin (LT-B) by 3-4 –fold in tobacco and potato plants when an ER-retention signal SEKDEL was fused to the carboxy-terminus of LT-B[10].

Purification and Characterization of Tobacco Synthesized CTB

CTB protein expressed in the pBI-SPCTB-derived tobacco lines was purified by immunoaffinity column chromatography. When tobacco leaf extracts flowed through the column, CTB was bound to the anti-CT IgG coupled to the resin. The retained CTB was eluted with the glycine-HCl buffer followed by neutralization and dialysis. The amount of the recovered CTB (tCTB) was determined and about 275 ug of tCTB was obtained from 100g of tobacco leaves. The purity and the biochemical property of tCTB were examined by SDS-PAGE along with bacterial pentameric CTB (Sigma C 9903). As revealed by Coomassie blue staining (Fig. 2A), tCTB co-migrated with the native CTB as a single band with a molecular weight of 45.2 kDa under non-denatured condition, and heat treatment of tCTB and bacterial CTB reduced the size of the proteins to 11.6 kDa, expected for a monomeric CTB. The results indicate that a single affinity column chromatography efficiently removed the tobacco plant proteins and tCTB predominantly formed a pentameric structure. Identity of molecular weights between tCTB and bacterial CTB suggests that the PR1b SP-CTB fusion protein was properly processed in tobacco cell. Western blot analysis probed with the rabbit anti-CT serum confirmed the biochemical nature and immuno-reactivity of tCTB (Fig. 2B).

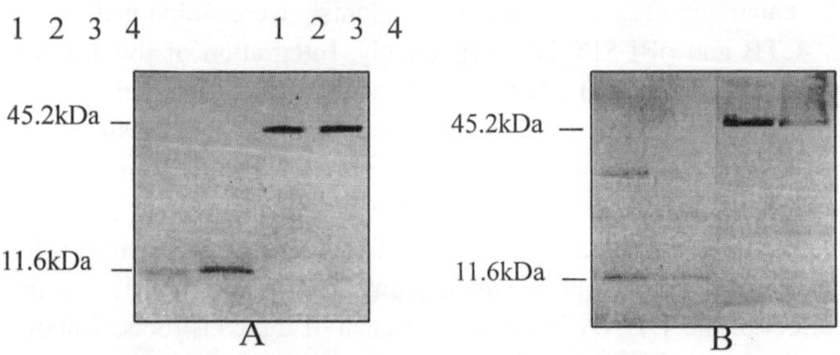

Fig. 2. Characterization of purified tobacco-derived CTB. tCTB (lanes 2 and 4) and bacterial CTB (lanes 1 and 3) were loaded either directly (lanes 3 and 4) or after boiling for 3 min (lanes 1 and 2) on a 12% SDS-polyacrylamide gel. After electrophoresis, the gel was subjected to Coomassie blue staining (A) or western blot (B).

Antigenicity of the Purified tCTB Protein

Tobacco-synthesized CTB possesses the biochemical and immunological properties indistinguishable from the native CTB as revealed by the G_{M1}-ganglioside binding assay and immuno-blot analysis shown above. We have also shown that the purified tCTB was capable of inducing high titer of serum anti-CTB antibody in mice after intramuscular immunization. The mouse anti-CTB antiserum neutralized the cytopathic effect of CT on CHO cells and significantly reduced the fluid accumulation in mouse ileal loop caused by CT (data not shown). We have further tested the antigenicity of the purified tCTB by immunodiffusion and immunoelectrophoresis experiments. The double immunodiffusion results depicted in Figure 3A showed that the precipitation line produced by the purified tCTB versus rabbit anti-CT antiserum fused completely to that by the native CTB versus the same antiserum. It indicates the presence of identical antigenic determinants in tobacco-derived CTB as in the native CTB. The same conclusion can be drawn from the results of immunoelectrophoresis. The precipitation arcs formed by the purified tCTB or the native CTB with the rabbit anti-CT antiserum migrated to the same distance in agarose gels (Fig. 3B).

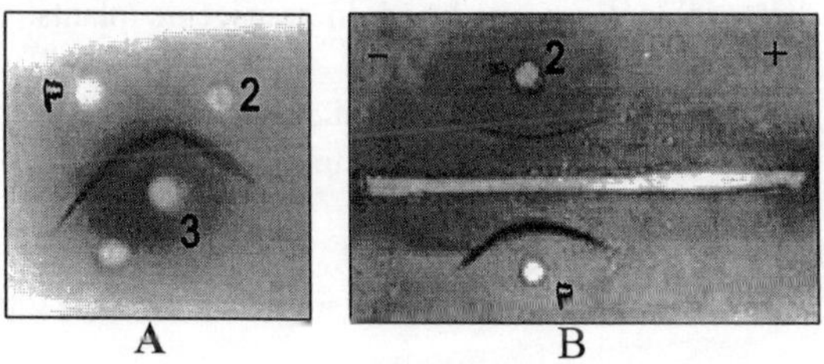

Fig. 3. Antigenicity of purified tobacco CTB. (A) Double immunodiffusion: rabbit anti-CT serum (3) versus bacterial CTB (1) and tCTB (2). (B) Immunoelectrophoresis: rabbit anti-CT serum versus tCTB (1) and bacterial CTB (2).

REFERENCES

1. Arakawa, T., Chong, D.K.X., Langridge, W.H.R. 1998a. Efficacy of a food plant-based oral cholera toxin B subunit. Nat. Biotech. 16:292-297.
2. Arakawa, T., Chong, D.K.X., Merritt, J.L., Langridge, W.H.R. 1997. Expression of cholera toxin B subunit oligomers in transgenic potato plants. Transgenic Res. 6:403-413.
3. Arakawa, T., Yu, J., Chong, D.K.X., Hough, J., Engen, P.C., Langridge, W.H.R. 1998b. A plant-based cholera toxin B subunit-insulin fusion protein protects against the development of autoimmune diabetes. Nat. Biotech. 16:934-938.
4. Clemens, J.D., Sack, D.A., Rao, M.R., Chakraborty, J., Khan, M.R., Kay, B., Ahmed, F., Banik, A.K., van Loon, F.P., Yunus, M. 1992. Evidence that inactivated oral cholera vaccines both prevent and mitigate *Vibrio cholerae* O1 infections in a cholera-endemic area. J. Infect. Dis. 166:1029-1034.
5. Clemens, J.D., van Loon, F., Sack, D.A., Chakraborty, J., Rao, M.R., Ahmed F., Harris, J.R., Khan, M.R., Yunus, M., Huda, S. 1991. Field trial of oral cholera

vaccines in Bangladesh: serum vibriocidal and antitoxic antibodies as markers of the risk of cholera. J. Infect. Dis. 163:1235-1242.

6. Czerkinsky, C., Russell, M.W., Lycke, N., Lindblad, M., and Holmgren, J. 1989. Oral administration of a streptococcal antigen coupled to cholera toxin B subunit evokes strong antibody responses in salivary glands and extramucosal tissues. Infect. Immun. 57:1072-1077.
7. Denecke, J., Botterman, J., Deblaere, R. 1990. Protein secretion in plant cells can occur via a default pathway. The Plant Cell 2:51-59.
8. Dertzbaugh, M.T., Elson, C.O. 1993. Comparative effectiveness of the cholera toxin B subunit and alkaline phosphatase as carriers for oral vaccines. Infect. Immun. 61:48-55.
9. Gallie, D.R., Sleat, D.E., Watts, J.W., Turner P.C., Wilson T.M.A. 1987. The 5 - leader sequence of tobacco mosaic virus RNA enhances the expression of foreign gene transcripts *in vitro* and *in vivo*. Nucl Acids Res 15:3257-3273.
10. Haq, T.A., Mason, H.S., Clements, J.D., Arntzen, C.J. 1995. Oral immunization with a recombinant bacterial antigen produced in transgenic plants. Science 268:714-719.
11. Horsch, R.B., Fry, J.E., Hoffmann, N.L., Wallroth, M., Eichholz, D., Rogers, S.G., Fraley, R.T. 1985. A simple and general method for transferring genes into plants. Science 227:1229-1231.
12. Kay, R., Chan, A., Daly, M., McPherson, J. 1987. Duplication of CaMV 35S promoter sequences creates a strong enhancer for plant genes. Science 236:1299-1302.
13. Kaper, J.B., Morris, J.G., Levine, M. 1995. Cholera. Clinical Microbiol. Rev. 8:48-86.
14. Kozak, M. 1986. Partial mutations define a sequence flanking the AUG initiator codon that modulates translation by eukaryotic ribosomes. Cell 44:283-292.
15. Lund, P., Dunsmuir, P. 1992. A plant signal sequence enhances the secretion of bacterial ChiA in transgenic tobacco. Plant Mol. Biol. 18:47-53.
16. Li, T.Y., Tian, Y.C., Qin, X.F., Mang, K.Q., Li, W.G., He, Y.G., Shen, L. 1994. Transgenic tobacco plants with efficient insect resistance. Science in China 37:1479-1487.
17. Shi, C.H., Cao, C., Zhig, J.S., Li, J.Z., Ma, Q.J. 1995. Gene fusion of cholera toxin B subunit and HBV PreS2 epitope and the antigenicity of fusion protein. Vaccine 13:933-937.
18. Sun, J.B., Holmgren, J., Czerkinsky, C. 1994. Cholera toxin B subunit: an efficient transmucosal carrier-delivery system for induction of peripheral immunological tolerance. Proc. Natl. Acad. Sci. USA 91:10795 - 10799.
19. Sun J.B., Rask, C., Olsson, T., Holmgren, J., Czerkinsky, C. 1996. Treatment of experimental autoimmune encephalomyelitis by feeding myelin basic protein conjugated to cholera toxin B subunit. Proc. Natl. Acad. Sci. USA 93:7196-7201.
20. Walmsley, A.M., Arntzen, C.J. 2000. Plants for delivery of edible vaccines. Curr. Opin. Biotech. 11:126-129.

21. Xiong, L.S., Ma, Q.J., Zhang, Y.H. 1990. The purification of cholera toxin B subunit from carrying recombinant plasmid containing *E. coli* strain. Chinese Biochem. J. 6:27-31.

DEVELOPMENT OF PLANT VACCINES: THE POINT OF VIEW OF THE MUCOSAL IMMUNOLOGIST

JEAN-PIERRE KRAEHENBUHL
Swiss Institute for Experimental Cancer Research, Institute of Biochemistry, University of Lausanne, CH 1066 Epalinges, Switzerland. Phone: (41 21) 692 58 56. Fax: (41 21) 652.69 33. Email: jean-pierre.kraehenbuhl@isrec.unil.ch

INTRODUCTION

Plant genetic engineering is a rapid expanding field and represents a promising avenue for the production of recombinant vaccines. Recombinant proteins and plant viral vectors have already been produced in plants and tested in animal and human clinical trials (for review see[1,2]). Plants can be used to produce plant-based vaccines either in the form of subunit vaccines or recombinant pathogenic plant viruses for active immunization or as antibodies for passive protection. Edible vaccines, however, share with food antigens a number of properties that if not taken into consideration may trigger unwanted reactions. The plant-based vaccines as all vaccines must be antigenic and immunogenic and trigger long lasting effector/memory cells that mediate protection against the pathogen. Compliance remains a problem, especially in developing countries, if protection required several boost administrations with plant-based subunit vaccines. The dosage is also an issue. Indeed depending on the nature and the dose of orally administered antigens, systemic and local immune unresponsiveness can be induced rather than protective immune responses. It should be emphasized that the vast majority of foreign antigens in the intestine are derived from food and the commensal microbial flora, and these generally do not trigger defensive immune responses in spite of the fact that such antigens regularly enter the mucosa. This is because mucosal antigen-presenting cells, lymphocytes and even the epithelium itself play important but poorly understood roles in modulating immune responses to incoming antigens. Indeed, a major role of the mucosal immune system is to down-regulate or suppress immune responses to food antigens and commensal bacteria. The exact sites or mechanisms of this "oral tolerance" are still controversial and have been reviewed elsewhere[3,4]. Finally, the route of administration of the vaccine determines where immune effector/memory cells are targeted and where they mediate protection. It is the aim of this presentation to briefly review some of these aspects which are important for the design of efficient orally administered plant-based vaccines.

SAMPLING OF ANTIGENS, PATHOGENS AND VACCINES AT MUCOSAL SURFACES

The sequence of events involved in processing and presentation of foreign antigens by professional antigen-presenting cells, and the responses and interactions of local lymphocytes that lead to production of effector and memory cells, are likely to be similar in the mucosal and systemic branches of the immune system. However, induction of mucosal immune responses is complicated by the fact that antigens and microorganisms on mucosal surfaces are separated from cells of the mucosal immune system by epithelial barriers. To mount protective mucosal immune responses, samples of the external environment on mucosal surfaces must be delivered to the immune system without compromising the integrity and protective functions of the epithelium[5]. Antigen sampling strategies at diverse mucosal sites are adapted to the cellular organization of the local epithelial barrier (Fig. 1).

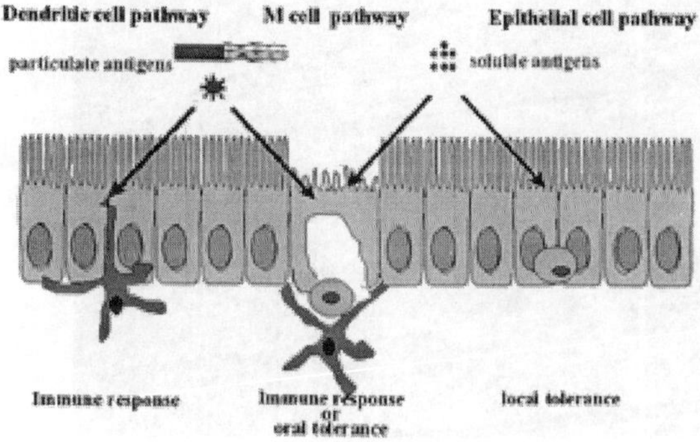

Fig. 1. Particulate antigens are usually taken up by dendritic cells (red cells) and M cells but not by epithelial cells (blue), while soluble antigens are taken up and processed by epithelial cells. The outcome of the immune response depends on which cells sees initially the antigen.

In stratified epithelia (skin, vagina, oral cavity) but also in simple epithelia (airways, gut) motile dendritic (or Langerhans) cells move into the epithelial layer, where they may obtain samples to carry back to local mucosal lymphoid tissues or distant lymph nodes (Fig. 2). In simple epithelia where intercellular spaces are sealed by tight junctions, specialized epithelial cells, the M cells, present in the epithelium overlying lymphoid tissue (appendix, tonsil crypts, Peyer's patches, colonic follicles) transport samples of lumenal material directly to mucosa-associated lymphoid tissue (MALT). Antigens and pathogens that cross epithelial barriers may be released at the basolateral side of the epithelium and taken up and carried by dendritic cells into local organized MALT and/or to draining lymph nodes or spleen[6]. The apical membranes of M cells are designed to facilitate adherence and uptake of antigens and microorganisms, and these cells take up macromolecules, microorganisms and particles by multiple mechanisms[7].

FATE OF ANTIGENS IN ORGANIZED MALT

M cells provide a pathway across the epithelial barrier through vesicular transport activity, but there is little known about the fates of specific antigens and pathogens that enter this pathway. Immediately under the FAE in the so-called "dome" region that caps the underlying lymphoid follicle is an extensive network of dendritic cells and possibly macrophages, intermingled with $CD4^+$ T cells and B cells that appear to be derived from the underlying follicle[6].

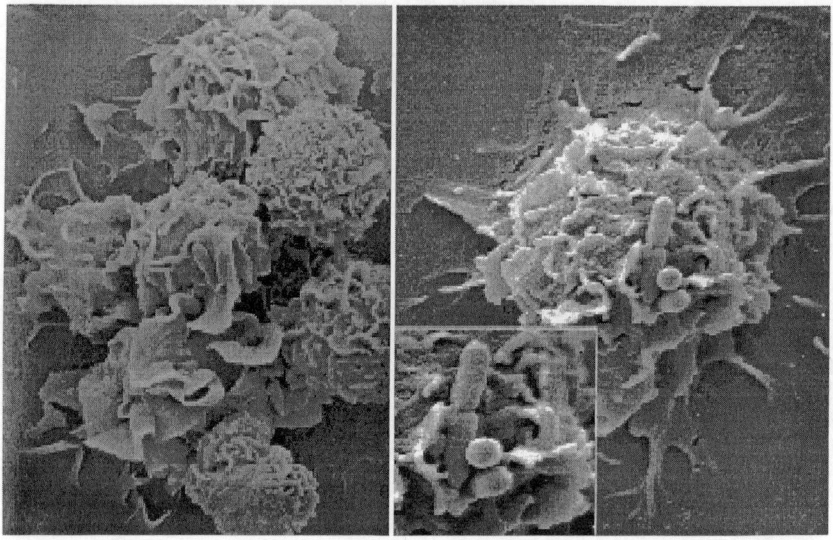

Fig. 2. Uptake of *Salmonella typhimurium* dendritic cells
left : wild type
right : attenuated vaccine strain.
Such dendritic cells form a network in the dome region of MALT structures
Scanning electron microscopy: Courtesy Florence Niedergang.

The dome region has all the earmarks of an active immune inductive site, where endocytosis and killing of incoming pathogens as well as processing and presentation of antigens occurs. A recent confocal light microscopic study detected live, attenuated *Salmonella typhimurium* in dendritic cells of the dome region after oral administration[8] (Fig. 2).

However, there is little information about the processing of nonliving macromolecules, particles, killed microbes and mucosal vaccines in this tissue, and the migration patterns of antigen-containing DCs out of the dome region is in need of further investigation.

The local signals that govern migration of cells into the subepithelial dome region or M cell pocket are unknown, but recent studies suggest that chemokines play a role. *In situ* hybridization showed that the human CC chemokine MIP 3α is produced by intestinal FAE epithelial cells but not villus cells of both humans and mice[9]. This chemokine thus is the first protein shown to be expressed specifically by FAE epithelial cells. The fact that MIP 3α has selective chemotactic activity for naive B and T lymphocytes and dendritic cells that express CCR6 receptors, and that $CCR6^+$ cells are present immediately under the FAE, suggests that MIP 3α is important for maintenance of mucosal antigen sampling functions[10].

INDUCTION OF IMMUNE RESPONSES IN MUCOSAL TISSUES

Following stimulation by antigens and T helper cells, naive B cells in organized mucosal lymphoid tissues (MALT) of the gut, the airways or the oropharyngeal cavity move to the germinal center. There they clonally proliferate, undergo affinity maturation, first by somatic hypermutation which generates variability in B cell receptors and second by selection of those with highest affinity for the antigen. Selection of cells bearing these mutated receptors by antigen occurs on the surface of the follicular dendritic cell, a process which rescues cells expressing high affinity Ig receptors from apoptosis (for review see[11]. In MALT germinal centers, B lymphocytes undergo isotype switch and differentiate further into B cells that express IgA receptors[12]. MALT $CD4^+$ T cells have been shown to promote IgA isotype switch of IgM-bearing B cells[13]. Mucosal adjuvants, including cholera toxin and *E. coli* heat labile toxin are known to facilitate switch[14]. Subsequently B lymphocytes differentiate into effector or memory cells following contact with T helper lymphocytes and CD40-CD40 ligand interactions (Liu et al., 1991).

In MALT stimulated B and T cells acquire a mucosal homing program (Fig 3). The effector and memory lymphocytes lose their adhesion to stromal cells, leave organized-MALT structures and enter the blood stream via the lymph. Depending on the mucosal site at which priming takes place, different homing receptors are expressed by B lymphocytes. Virtually all IgA and even IgG-antibody secreting cells detected after peroral and rectal immunization expressed α4β7 integrin receptors, while only a minor fraction of these cells expressed the peripheral L-selectin receptor. In contrast, circulating B cells induced by intranasal immunization co-expressed L-selectin and α4β7 receptors[15].

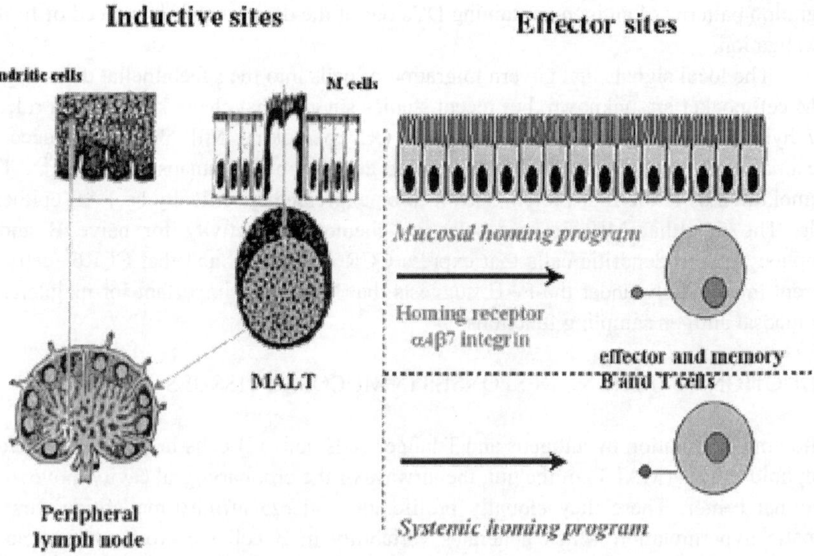

Fig. 3. The type of immune response depends on where antigens are processed and presented to the immune system. If antigens are processed in an inductive lymphoid tissue close to the mucosal epithelium, the immune effector and memory cells acquire a homing program that sends them back to mucosal sites. If antigens reach a lymphoid organ (peripheral lymph node) distant from the mucosal epithelium, the homing program that is acquired allows the effector/memory cells to recirculate in peripheral lymph nodes and eventually in skin but not in mucosal tissues.

Effector and memory B cells are able to home to distant mucosal tissues or return to MALT structures (Fig. 3). The lymphocytes expressing mucosal α4β7 homing receptors interact with post-capillary venule endothelial cells bearing mucosal addressins on their lumenal surfaces[16]. After migration into the lamina propria, effector B lymphocytes differentiate into antibody-secreting plasma cells. This process is regulated by cytokines from T lymphocytes as well as epithelial cells. In the intestinal mucosa, the number of plasma cells producing IgA exceeds those producing all other immunoglobulin isotypes[17]. In the mucosal environment, all plasma cells irrespective of their immunoglobulin isotype express J chain, the small polypeptide required for IgA polymerization.

The function of mucosal CTLs in protection against infectious agents has been recently reviewed[6]. Mucosal immunization is also required to trigger mucosal CTLs[18,19].

REGULATION OF IMMUNE RESPONSES IN MUCOSAL TISSUES

That ingestion of antigens elicits immune responses different from those associated with systemic immunization was recognized at the beginning of the century. Its immunological nature was established much later (Fig. 4). Antigen uptake in mucosal tissues may result in the development of immunity, tolerance, or both depending on the physical-chemical nature of the antigen and where antigen presentation takes place. Deletion[20], anergy of antigen-specific T cells[21], and/or expansion of cells producing immunomodulating cytokines (IL-4, IL-10 and TGF-β)[22] have been linked to decreased T cell responsiveness. Since both serum and cells can transfer tolerance from tolerized animals, it is possible that humoral antibodies, circulating undegraded antigens, tolerogenic protein fragments and cytokines may act synergistically to confer T cell unresponsiveness.

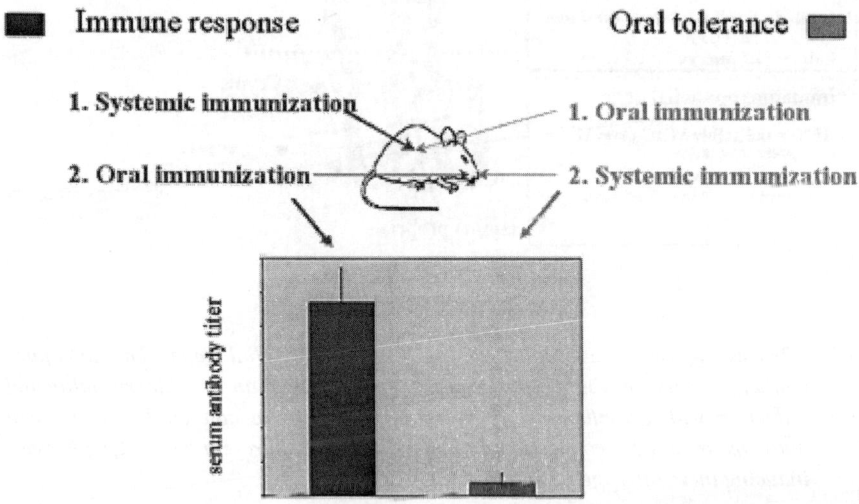

Fig. 4. Oral tolerance. Systemic immunization followed by oral immunization using ovalbumin as an antigen induces a strong systemic antibody and T cell response. In contrast oral immunization followed by systemic immunization induces a state of unresponsiveness, the so-called oral tolerance.

Little is known about the molecular mechanisms whereby antigens administered mucosally can induce local and/or systemic tolerance. On mucosal surfaces, antigens encounter multiple factors including proteases, acids, salts, and detergents that can alter their native conformation and expose new epitopes. The observation that mucosally-induced systemic tolerance depends on an intact epithelial barrier[23] suggests a central role for the epithelium. Antigens sampled from the lumen by intestinal enterocytes are usually soluble molecules that can diffuse through the glycocalyx[24]. Non-classical MHC class I

(CD1d) molecules expressed by enterocytes in the intestine may present these antigens to subsets of CD8^{+} regulatory IELs known to induce local unresponsiveness[25]. Epithelial enterocytes are also known to produce cytokines such as IL-10 and TGF β which are particularly efficient at suppressing the inductive phase of CD4^{+} T cell-mediated responses (Fig. 5).

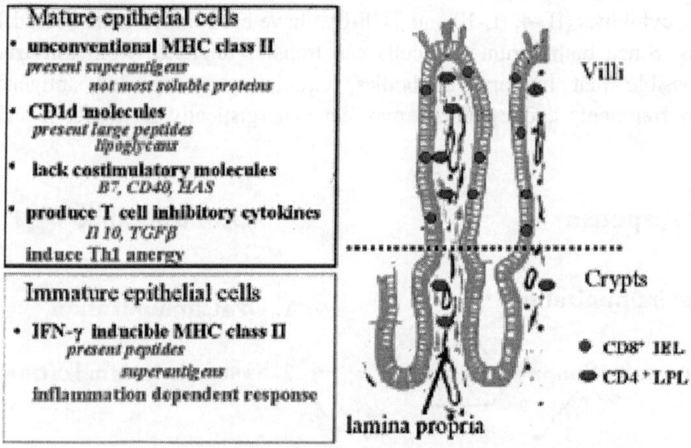

Fig. 5. Processing and presentation of antigens by epithelial cells. The immature epithelial cells in the crypts express the presentation machinery following induction with proinflammatory cytokines. The mature epithelial cells present antigens in a non conventional manner to CD8 intra-epithelial lymphocytes inducing there tolerogenic activity.

In addition to epithelial cells and T cells, B cells[26], T cells[3] and dendritic cells[27] have been proposed as important players in induction of oral tolerance. Dendritic cells in mucosal tissues such as the Peyer's patches and mesenteric lymph, the intestinal lamina propria and the airways stimulate rather than suppress immune responses. Interestingly, LPS, which is known to cause the rapid exit of dendritic cells from mucosal tissues, has also been shown to enhance tolerance induction[28]. Recently, a distinct dendritic cell subset has been shown to endocytose apoptotic intestinal cells and transport them to T cell areas in the draining mesenteric lymph nodes[27]. This suggests a role for dendritic cells in inducing and maintaining peripheral self-tolerance. It was recently observed that expression of ligands of the notch pathway in dendritic cells can induce naive peripheral CD4^{+} T cells to become regulatory cells that inhibit primary and secondary immune responses[29]. This is the first demonstration of a molecular mechanism that may underlie the induction of tolerance.

IMMUNOLOGIC MEMORY IN MUCOSAL TISSUES

While systemic infections usually induce long-lasting protective immunity and prolonged serum antibody titers, mucosal antibody responses are usually relatively short-lived[30]. This may be due to differences in the maturation and selection of B cells in mucosal tissues compared to other peripheral compartments. Long-lived memory B cells and plasma cells have recently been isolated from spleen and bone marrow[31,32]. The nature of the signals and micro-environmental factors that promote survival of specific B cells have yet not been identified, but there is evidence that follicular dendritic cells (FDC) play a crucial role. FDCs in germinal centers of MALT organs express MAdCAM-1 that could recruit $\alpha4\beta7$-expressing B lymphocytes[33]. Antigen-specific $\alpha4\beta7^{high}$ B lymphocytes, a memory phenotype, have been detected in Peyer's patches and lamina propria of mice[34] but how long these cells persist in the gut remains unclear. Additional factors such as chemokines and/or survival signals might be required for the retention and maintenance of specific B cells and antibody-secreting plasma cells in MALT compartments. Taken together, the ability of a vaccine to induce mucosal B cell memory responses seems to be dependent on systemic B cell and T helper cell priming.

REGIONAL NATURE OF MUCOSAL IMMUNE RESPONSES

In both mice and humans, the secretory immune response to foreign antigens and microorganisms may be detected at the mucosal site where the antigen was initially taken up, and also in distant mucosal and glandular secretions[35]. This phenomenon reflects the dissemination via the bloodstream of effector and memory cells from the site of antigen exposure into widespread mucosal and glandular connective tissues, where they differentiate into plasma cells that produce dimeric IgA. This has been termed the "common mucosal immune system" and has led to the idea that immunization at one mucosal site could induce protective secretory immunity in mucosal tissues throughout the body[36]. Indeed, although oral immunization results in antigen uptake only at inductive sites of the oral cavity and upper intestine, it can elicit antibodies not only in salivary and intestinal secretions but also in mammary gland and vaginal secretions. However, there is increasing evidence that local exposure to antigen can result in much higher levels of specific secretory IgA (sIgA) in the region of exposure than at distant sites. Even within the GI tract, administration of antigen into proximal small intestine, distal small intestine, colon or rectum evokes highest levels of specific secretory IgA in the segment of antigen exposure[37]. Such observations have led to testing of rectal and vaginal immunization strategies for vaccines against sexually transmitted diseases. The rectum appears to be a particularly effective inductive site, consistent with the fact that M cells and lymphoid aggregates are numerous in the rectal mucosa. Rectal immunization of mice, rhesus macaques and humans generated high levels of specific antibodies in local rectal secretions[38,39]. Conversely, in rhesus macaques and humans the vaginal immunization route effectively induced local immune responses in the female genital tract[40,41]. In mice, vaginal immunization evoked local and systemic immune responses against live pathogens but not against nonliving antigens[37]. There is great current interest in the nasal immunization route. Nasal immunization has been shown to produce impressive systemic immune responses, as well as local secretory responses in the upper respiratory tract and in the female genital tract as well. Nasal immunization has been used experimentally to confer protection against vaginal mucosal challenge by Herpes simplex virus-1[42].

CONCLUSIONS AND PERSPECTIVES

Development of plant-based vaccines represents a promising approach for the prevention of infectious diseases. The rate-limiting step in the rational design of a vaccine, including plant-derived vaccines, is to identify the immunological correlate of protection, in order to trigger the appropriate arm of the immune system. For the most prevalent infectious diseases, including tuberculosis, AIDS, *Helicobacter pylori* –associated ulcer disease and gastric cancers, human papilloma-induced cervical cancers, there is a desperate need to identify such correlates. Plant-derived antigenic proteins have already been shown to delay or prevent the onset of disease in animal models and have proven to be safe and functional in human clinical trials. Future areas of research should further characterize the induction of mucosal immunity or induction of oral tolerance. Appropriate crop species will have to be identified and developed for the production of subunit vaccines while their use for the delivery of animal and human vaccines remains more problematic for the reasons discussed above, including compliance, dosage and oral tolerance among others.

ACKNOWLEDGMENTS

I am grateful to the current and former members of my laboratory. I also wish to thank my collaborators, including Dr Hans Acha Orbea from the Ludwig Institute, Lausanne Branch, Dr Denise Nardelli-Haefliger, André Blum, Giuseppe Pantaleo from the Centre Hospitalier et Universitaire Vaudois in Lausanne, Dr. Armelle Phalipon and Philippe Sansonetti at the Pasteur Institut in Paris, Dr. Pierre Michetti from the Beth Israel Hospital in Boston, Dr. Marian Neutra and her colleagues at the Children's Hospital, Harvard Medical School in Boston, who have contributed to the work summarized in this review. The author is supported by the Swiss National Science Foundation Grant 31-56936-99 and the Swiss League against Cancer Grant SKL 635-2—1998.

REFERENCES

1. Ma, J.K. & Vine, N.D. Plant expression systems for the production of vaccines. *Curr. Top. Microbiol. Immunol.* **236**, 275-92 (1999).

2. Walmsley, A.M. & Arntzen, C.J. Plants for delivery of edible vaccines. *Curr. Opin. Biotech.* **11**, 126-9 (2000).
3. Mowat, A.M. & Weiner, H.L. Oral tolerance: physiological basis and clinical applications. in *Mucosal Immunology* (eds. Ogra, R. *et al.*) 587-618 (Academic Press, New York, 1999).
4. Mayer, L. Oral tolerance: New approaches, new problems. *Clin. Immunol.* **94**, 1-8 (2000).
5. Neutra, M.R., Pringault, E. & Kraehenbuhl, J.P. Antigen sampling across epithelial barriers and induction of mucosal immune responses. *Annu. Rev. Immunol.* **14**, 275-300 (1996).
6. Kelsall, B. & Strober, W. Gut-associated lymphois tissue: antigen handling and T cell responses. in *Mucosal Immunology* (eds. Ogra, R. *et al.*) pp 293-318 (Academic Press, New York, 1999).
7. Kraehenbuhl, J.P. & Neutra, M.R. Epithelial M cells: structure and function. *Annu. Rev. Cell Develop. Biol.* **16**, 301-332 (2000).
8. Hopkins, S., Niedergang, F., Corthésy-Theulaz, I.E. & Kraehenbuhl, J.P. A recombinant *Salmonella typhimurium* vaccine strain is taken up and survives within murine Peyer's patch dendritic cells. *Cell. Microbiol.* **2**, 56-68 (2000).
9. Tanaka, Y. *et al.* Selective expression of liver and activation-regulated chemokine (LARC) in intestinal epithelium in mice and humans. *Eur. J. Immunol.* **29**, 633-642 (1999).
10. Cook, D.N. *et al.* CCR6 mediates dendritic cell localization, lymphocyte homeostasis, and immune responses in mucosal tissue. *Immunity* **12**, 495-503 (2000).
11. MacLennan, I.C., Liu, Y.J. & Johnson, G.D. Maturation and dispersal of B-cell clones during T cell- dependent antibody responses. *Immunol.Rev.* **126**, 143-161 (1992).
12. Cebra, J.J., Logan, A.C. & Weinstein, P.D. The preference for switching to expression of the IgA isotype of antibody exhibited by B lymphocytes in Peyer's patches is likely due to intrinsic properties of their microenvironment. *Immunol.Res.* **10**, 393-395 (1991).
13. Kawanishi, H., Saltzman, L.E. & Strober, W. Mechanisms regulating IgA class-specific immunoglobulin production in murine gut-associated lymphoid tissues. I. T cells derived from Peyer's patches that switch. *J.Exp.Med.* **157**, 433-450 (1983).
14. Lycke, N. & Strober, W. Cholera toxin promotes B cell isotype differentiation. *J.Immunol.* **142**, 3781-3787 (1989).
15. Quiding-Jarbrink, M. *et al.* Differential expression of tissue-specific adhesion molecules on human circulating antibody-forming cells after systemic, enteric, and nasal immunizations. A molecular basis for the compartmentalization of effector B cell responses. *J. Clin. Invest.* **99**, 1281-6 (1997).
16. Butcher, E.C. & Picker, L.J. Lymphocyte homing and homeostasis. *Science* **272**, 60-66 (1996).

17. Brandtzaeg, P. *et al.* The B-cell system of human mucosae and exocrine glands. *Immunol. Rev.* **171**, 45-87 (1999).
18. Klavinskis, L.S. *et al.* Mucosal or targeted lymph node immunization of macaques with a particulate SIVp27 protein elicits virus-specific CTL in the genito-rectal mucosa and draining lymph nodes. *J. Immunol.* **157**, 2521-7 (1996).
19. Belyakov, I.M. *et al.* Induction of a mucosal cytotoxic T-lymphocyte response by intrarectal immunization with a replication-deficient recombinant vaccinia virus expressing human immunodeficiency virus 89.6 envelope protein. *J. Virol.* **72**, 8264-8272 (1998).
20. Chen, Y.H. *et al.* Peripheral deletion of antigen-reactive t cells in oral tolerance. *Nature* **376**, 177-180 (1995).
21. Whitacre, C., Gienapp, C., Orosz, I.E. & Bitar, D. Oral tolerance in experimental autoimmune encephalomyelitis. III. Evidence for clonal anergy. *J. Immunol.* **147**, 2155-2163 (1991).
22. Chen, Y. *et al.* Regulatory T cell clones induced by oral tolerance: suppression of autoimmune encephalomyelitis. *Science* **265**, 1237-1240 (1994).
23. Bruce, M.G., Strobel, S. & Hanson, D.G. Transferable tolerance for cell-mediated immunity after feeding is prevented by radiation damage and restored by immune reconstitution. *Clin. Exp. Immunol.* **70**, 611-618 (1987).
24. Kaiserlian, D. Antigen sampling and presentation in mucosal tissues: Epithelial cells. *Current Topics in Microbiology and Immunology* **236**, 55-78 (1999).
25. Blumberg, R.S. *et al.* Antigen presentation by intestinal epithelial cells. *Immunol Lett.* **69**, 7-11 (1999).
26. Czerkinsky, C., Sun, J.B. & Holmgren, J. Oral tolerance and anti-pathological vaccines. *Curr. Top. Microbiol. Immunol.* **236**, (79-92 (1999).
27. Huang, F.P. *et al.* A discrete subpopulation of dendritic cells transports apoptotic intestinal epithelial cells to T cell areas of mesenteric lymph nodes. *J. Exp. Med.* **191**, 435-443 (2000).
28. Khoury, S.J., Lider, O., Al-Sabbagh, A. & Weiner , H.L. Suppression of experimental autoimmune encephalomyelitis by oral administration of myelin basic protein. III. Synergistic effect of lipopolysaccharide. *Cell. Immunol.* **131**, 302-310 (1990).
29. Hoyne, G.F. *et al.* Serrate 1-induced Notch signalling regulates the decision between immunity and tolerance made by peripheral CD4(+) T cells. *Internat. Immunol.* **12**, 177-185 (2000).
30. Belyakov, I.M., Moss, B., Strober, W. & Berzofsky, J.A. Mucosal vaccination overcomes the barrier to recombinant vaccinia immunization caused by preexisting poxvirus immunityProc. Natl. AScad. Sci USA **96**, 4512-4517 (1999).
31. Manz, R.A., Thiel, A. & Radbruch, A. Lifetime of plasma cells in the bone marrow. *Nature* **388**, 133-4 (1997).
32. McHeyzer-Williams, L.J., Cool, M. & McHeyzer-Williams, M.G. Antigen-specific B cell memory: Expression and replenishment of a novel B220(-) memory B cell compartment. *J. Exp. Med.* **191**, 1149-1165 (2000).

33. Szabo, M.C., Butcher, E.C. & McEvoy, L.M. Specialization of mucosal follicular dentritic cells revealed by mucosal adressin-cell adhesion molecule-1 display. *J. Immunol.* **158**, 5584-5588 (1997).
34. Williams, M.B. *et al.* The memory B cell subset responsible for the secretory IgA response and protective humoral immunity to rotavirus expresses the intestinal homing receptor, alpha(4)beta(7). *J. Immunol.* **161**, 4227-4235 (1998).
35. McDermott, M.R. & Bienenstock, J. Evidence for a common mucosal immunologic system I. Migration of B immunoblasts into intestinal, respiratory, and genital tissues. *J.Immunol.* **122**, 1892-1898 (1979).
36. McGhee, J.R. *et al.* The mucosal immune system: from fundamental concepts to vaccine development. *Vaccine* **10**, 75-88 (1992).
37. Haneberg, B. *et al.* Induction of specific Immunoglobulin A in the small intestine, colon-rectum, and vagina measured by a new method for collection of secretions from local mucosal surfaces. *Infect.Immun.* **62**, 15-23 (1994).
38. Lehner, T. *et al.* T- and B-cell functions and epitope expression in nonhuman primates immunized with simian immunodeficiency virus antigen by the rectal route. *Proc.Natl.Acad.Sci.USA.* **90**, 8638-8642 (1993).
39. Kozlowski, P.A., Cu-Uvin, S., Neutra, M.R. & Flanigan, T.P. Comparison of the oral, rectal, and vaginal immunization routes for induction of antibodies in rectal and genital tract secretions of women. *Infect. Immun.* **65**, 1387-94 (1997).
40. Ogra, P.L. & Ogra, S.S. Local antibody response to poliovaccine in the human female genital tract. *J. Immunol.* **110**, 1307-1311. (1973).
41. Lehner, T. *et al.* Induction of mucosal and systemic immunity to a recombinant simian immunodeficiency viral protein. *Science* **258**, 1365-1369 (1992).
42. Parr, M.B. & Parr, E.L. Mucosal immunity in the female and male reproductive tracts. in *Handbook of Mucosal Immunology* (eds. Ogra, P.L. *et al.*) 677-690 (Academic Press, New York, 1994).

PLANT-DERIVED ORAL VACCINES; FROM CONCEPT TO CLINICAL TRIALS

CHARLES J. ARNTZEN, PH.D.
Florence Ely Nelson Presidential Chair in Plant Biology, Arizona State University and *President Emeritus*, Boyce Thompson Institute for Plant Research, Inc.
Plant Biology Department, Arizona State University, Tempe, AZ 85287-1601
Office telephone: (1) 480-727-7322. e-mail: charles.arntzen@asu.edu

AGRICULTURAL BIOTECHNOLOGY PRODUCTS; TODAY AND TOMORROW

Genetically modified crops, which are now commercially available, have largely been created to provide farmers with production advantages, such as reduction in use of costly pesticides, easier weed control, or protection from virus diseases. In coming years, however, many more diverse products will be possible. For example, public funding and philanthropic sources have helped develop plants that are enriched in essential micronutrients to alleviate Vitamin A and iron deficiencies (two major problems in the developing world). Other studies are defining the constituents of plant foods which may have beneficial value as "nutraceuticals;" for example, to provide anti-cancer benefits (or other positive health values). It is likely that traditional and crop breeding will provide new varieties of our food crops which will have direct benefit to human well-being.

Another area in which plant biotechnology is likely to have a global impact in the next decade is in the area of new vaccine technology. It is estimated by the World Health Organization that more than 5 million children in developing countries die each year from common diseases; the most dominant are diarrhea and respiratory infections. Although preventative medicine has proceeded rapidly in the last decade as biotechnology has been applied to create new vaccines, the new products are comparatively expensive for lesser developed countries. For this reason, a novel strategy has been developed for vaccine production that uses transgenic plants to both manufacture and deliver oral vaccines.

EDIBLE VACCINES

Plant-derived edible vaccines are based on genes from human pathogens introduced into transgenic plants, resulting in the plant producing proteins that mimic subunits of the pathogenic organism. The resulting plant material, when provided to mice as food, acted as an oral vaccine. Approval from the US Food and Drug Agency (FDA) was obtained to conduct three human clinical trials (two for prototype diarrhea prevention, and one against Hepatitis B infection).

The concept of "edible vaccines" is designed to stimulate mucosal immunity, which is important to block disease causing agents that enter at mucosal surfaces such as the enteric, respiratory, and genito-urinary tracts. Mucosal immunity is best achieved by delivery of vaccines at mucosal surfaces. The digestive process may sometimes limit the amount of orally delivered vaccine that reaches the gut immune system. To deal with this eventuality, we have studied vaccine adjuvants that may enhance the immunogenicity of orally delivered antigens. Bacterial enterotoxins such as the E. coli heat-labile toxin (LT) are strong mucosal adjuvants, but they cause diarrhea and thus are unsuitable for use in humans. Mutated forms of LT (mLT) show greatly lowered toxicity but retain adjuvant activity. We have, therefore, produced mLT in plants to increase the potency of co-expressed, orally delivered vaccines. LT is a multi-subunit protein complex assembled from one A subunit and five B subunits. The nontoxic LT-B pentamer targets LT to mucosal cells via its specific binding to the cell surfaces.

The research on plant-based vaccines has progressed to the point that three human clinical trials have been conducted in the United States. These were conducted after the U.S. Food and Drug Agency evaluated the protocols and gave their approval of the proposed protocols. Vaccines to prevent diarrhea were chosen for the first two studies since this is the cause of approximately 2.5 million cases of infant mortality on an annual basis, with most deaths occurring in the developing world. Both of these human studies have now been completed in "Phase I trials" that have verified the safety and efficacy of the approach, and results have been published. A third trial has evaluated potatoes which were genetically modified to produce the Hepatitis B surface antigen (HBsAg). HBsAg is currently used in a commercially available injectable vaccine, but has been evaluated for its effectiveness as an oral vaccine in the current trials.

Fig. 1. Human clinical trials of uncooked potatoes. Volunteers were given small plastic bags containing peeled, washed potatoes cut into small cubes. Each volunteer gave blood samples on a weekly basis to allow monitoring of the appearance of antibody secreting cells, and the presence of antibodies (IgG and IgA).

To accomplish oral immunization of infants using transgenic food, it will be necessary to select an appropriate crop plant which can be grown in most developing countries, and which is eaten uncooked (to avoid destruction of the vaccine proteins by heat). Efforts are underway to develop both tomatoes and bananas for this purpose. Current research is identifying ways to prepare a dry formulation of vaccine-containing tomato extract using common food processing technology, and to cause the appropriate proteins to accumulate in the banana fruit so that infants could be fed an "edible vaccine" in a banana baby food puree. In both cases, the desired outcome is agriculture and food-based technologies, which are readily available in all developing countries.

A primary goal of developing plant-based vaccines has been to reduce the cost of a dose of vaccine. Based upon results already in hand, it is possible to make cost estimates for the production in tomatoes. For antigens tested to date, we have achieved levels of expression of 0.1% protein or higher (based upon measurements of total soluble protein in the fruit). If the minimal level of 0.1% were to be achieved under agricultural production conditions (which is very likely), we can use production cost values from existing agriculture (see Table 1). In the United States, tomatoes are either grown for fresh markets or for harvesting to be processed into soups or sauces. Because the latter is less labor intensive, the cost of production is less. For actual vaccine production, the costs may be somewhere between the fresh and processed categories (due to possible needs for hand harvesting to obtain uniform antigen levels at specific ripening stages). However, it should be emphasized that the cost estimates of US$0.0025-0.035 per dose are much higher than would be likely in developing countries where labor costs are significantly less. Clearly, it should be possible to create plant containing material at less than US$0.01!

Table 1. Cost of production of tomatoes in the United States

	Fresh Market	Processed Market
U.S. Acreage	134,000 acres	320,000 acres
Average Yield	11.8 tons/acre	30.7 tons/acre
Average Value	$709 per ton (70¢ per kg)*	$54 per ton (5¢ per kg)**

Derived cost of vaccine production in tomatoes: *3.5 cents per dose
**1 cent per 4 doses

(calculation based upon 0.1% protein expression level; excludes costs of processing, quality assurance, packaging, marketing and distribution)

The use of transgenic plants to produce and deliver oral vaccines also has applicability for novel strategies for disease prevention in animals, thereby improving the safety of our food supply, and stability of animal production.

ADDITIONAL READINGS

Haq T.A., Mason H.S, Clements, J.D., Arntzen, C.J. 1995 Oral immunization with a recombinant antigen produced in transgenic plants. bacterial protein in transgenic plants. Science 268:714-716.

Mason, H.S., Ball, J.M., Shi, J.-J., Jiang, X., Estes, M.K., and Arntzen, C.J. 1996. Expression of Norwalk virus capsid protein in transgenic tobacco and potato and its oral immunogenicity in mice. Proc. Natl. Acad. Sci. USA 93:5335-534.

Arntzen, C.J. 1997. High-tech herbal medicine: Plant-based vaccines. Nat. Biotechnology 15:221-222.

Wong, S.Y., K.S. Ho, H.S. Mason, C.J. Arntzen. 1998. Edible Vaccines. *Science & Medicine*, 5: 36-45.

Arntzen, C.J. 1998. Pharmaceutical foodstuffs-Oral immunization with transgenic plants. Nature MedicineVaccine Supplement, 4:502-503.

Tacket, C.O., Mason, H.S., Losonsky, G., Clements, J.D., Levine, M.M., C.J. Arntzen. 1998. Immunogenicity in humans of a recombinant bacterial antigen delivered in a transgenic potato. Nature Medicine, 4:607-609.

Palmer, K.E., Arntzen, C.J., G. Lomonossoff. 1999. Antigen Delivery Systems - Transgenic Plants and Recombinant Plant Viruses. In: *Mucosal Immunology* (2nd edition) (P.L. Ogra, J. Mestecky, M.E. Lamm, W. Strober, J. R. McGhee, J. Bienenstock, eds.) Ch. 49:793-807. Academic Press, San Deigo, CA.

Walmsley, A.M., C.J. Arntzen. 2000. Plants for delivery of edible vaccines. Current Opinion in Biotechnology 2000. 11:126-129.

Tacket, C.O., Mason, H.S., Losonsky, G., Estes, M.K., Levine, M.M., C.J. Arntzen. 2000. Human immune responses to a Novel Norwalk virus vaccine delivered in transgenic potatoes. The Journal of Infectious Diseases. 182:302-305.

Dr. Charles J. Arntzen is the Florence Ely Nelson Presidential Chair in Plant Biology at Arizona State University, and the *President Emeritus* and a Research Project Leader of the Boyce Thompson Institute for Plant Research, Inc.—a not for-profit corporation affiliated with Cornell University. Dr. Arntzen's career spans industry, academia, and government service. He has held positions with the USDA, with The DuPont Company as Research Director, and as Deputy Chancellor for Agriculture at Texas A&M University. He is a member of the U.S. National Academy of Sciences, and a foreign member of the National Academy of Sciences of India. He has served on the editorial board of *SCIENCE*, as chairman of the U.S. National Institute of Health's National Biotechnology Advisory Board and on scientific advisory boards of biotechnology companies.

4. ENERGY

STATUS OF MAGNETIC FUSION RESEARCH

J. ONGENA
Laboratoire de Physique des Plasmas-Laboratorium voor Plasmafysica Association "EURATOM-Belgian State" Ecole Royale Militaire-B-1000 Brussels -Koninklijke Militaire School. Partner in the Trilateral Euregio Cluster
F. WAELBROECK
Institut für Plasmaphysik, Forshungszentrum Jülich, D-52425 Jülich, Germany

INTRODUCTION

Nuclear fusion is one of the few options to sustain the long term energy needs of our modern society in an environmentally friendly way[1]. It would provide humanity with an unlimited energy source offering many advantages: compact, no production of reactive or greenhouse gases and a controllable amount of nuclear waste with a half-life of only some decades[2]. It is however one of the most difficult projects ever undertaken by mankind.

Two approaches are being pursued at present, one based on inertial confinement of the hot fuel and the other on magnetic confinement. As the European Union focuses on magnetic fusion research, being totally civilian research, we will focus in this paper on the latter.

The knowledge acquired from the different tokamaks around the world enables us now to design and construct a prototype fusion reactor, capable of delivering 500-1000MW output power from fusion reactions in pulses from 500 to 1000s. For the first time in history, mankind is now capable of extracting energy in a controlled way from fusion reactions at temperatures 6–10 times hotter than the centre of the sun!

CONTROLLED THERMONUCLEAR FUSION

Fusion reactions

Fusion reactions provide the energy of our sun and the stars in the universe. One of the basic reactions in all stars, fusion of protons in the so-called p-p reaction chain, has such a low probability to occur, that it is not suited for use on earth. That it is nevertheless a useful process for stars is because sufficient reactions occur in their enormous volume to keep them hot. E.g. in our sun, each second a staggering 600 million tons of hydrogen is fused to form 596 million tons of helium; 4 million tons of mass thus disappear per second, being completely converted in energy. On earth, fusion reactions with a higher reaction probability have to be used, in order to obtain a power density useful for economical purposes. Possible fusion reactions for this purpose are those involving the hydrogen isotopes deuterium (^2H or D), tritium (^3H or T) and the stable helium isotope ^3He as given in the Table below :

$$D + T \rightarrow {}^4He\ (3.5\ MeV) + n\ (14.1 MeV)$$
$$D + D \rightarrow {}^3He\ (0.82\ MeV) + n\ (2.45\ MeV) \quad [50\%]$$
$$T\ (1.01\ MeV) + H\ (3.02\ MeV) \quad [50\%]$$
$$D + {}^3He \rightarrow {}^4He\ (3.6\ MeV) + H\ (14.7\ MeV)$$

The 'easiest' reaction is the so-called D-T fusion reaction between deuterium and tritium with a reaction rate at the temperatures of interest which is about 10^{26} times larger than that of the p-p reaction. The particles resulting from the D-T reaction are a helium nucleus (^4He) and a neutron (n) (Fig.1). The difficulty in realising any fusion reaction is the mutual repulsion of the nuclei, due to their positive electric charge. The fusion process would be impossible if it were not for the strong attractive nuclear forces, which dominate at very short internuclear distances. If the nuclei are highly energetic they will be able to overcome the (long-range) Coulomb repulsion and reach the realm of the strong nuclear force, where they can fuse.

The diagram in Figure 1 shows immediately the large potential of fusion:

1. Enormous energy gain: the reacting particles have an energy of several keV, the reaction products energies in the MeV range, i.e. 1000 times larger. Per unit of mass we gain 5 times more energy than in ^{235}U fission reactions.
2. Inexhaustibility: the reacting particles are easily found on earth, or can be easily produced from other elements which are abundant. Especially for the D-T reaction under consideration: deuterium can be obtained cheaply from ordinary water (nearly 1/7000th of all water on earth is heavy water), and tritium can be produced from neutron irradiation of lithium, according to the reactions :

   $$^7Li + n \rightarrow {}^4He + T + n - 2.47\ MeV$$
   $$^6Li + n \rightarrow {}^4He\ (2.05\ MeV) + T\ (2.73\ MeV)$$

   The reserves of Li on earth are large enough for several thousands of years of energy production by the D-T reaction at the current rate of <u>total</u> energy consumption.
3. Energy independence: due to the abundance of the fuel, there is no dependence on a fuel supplier, enhancing geopolitical stability.
4. Clean reaction: the reaction product is mainly ^4He which is a stable, non-radioactive and inert chemical element, being a noble gas. Thus no acid-forming reactions and no destruction of the ozone layer is possible with the

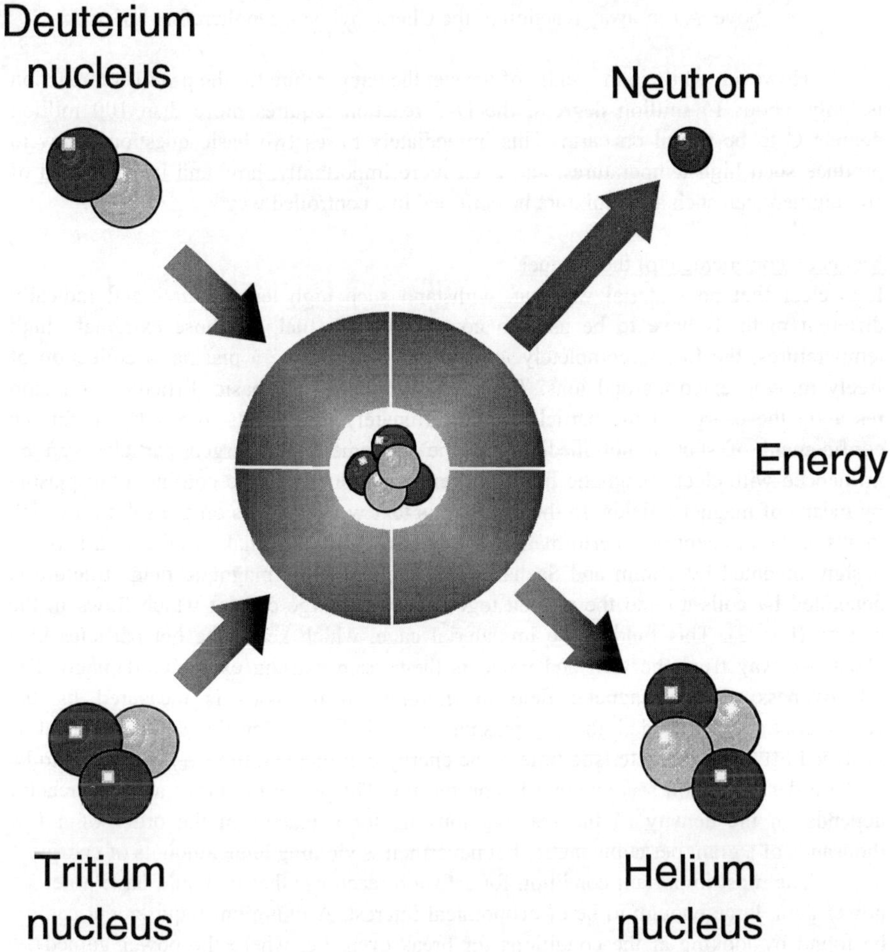

Fig. 1. Schematic view of the D-T fusion reaction

'ash' of the fusion reaction. In addition, ^4He is a monatomic gas, and therefore cannot play any role in enhancing the Greenhouse effect.

5. Inherent safety: there is no neutron multiplication in the fusion reactions listed above. A run-away reaction of the Chernobyl type can therefore not occur.

However, while in the centre of the sun the temperature for the p-p fusion reaction is 'only' about 15 million degrees, the D-T reaction requires more than 100 million degrees C to be useful on earth. This immediately raises two basic questions: how to produce such high temperatures, and even more importantly, how and in what kind of containment can such a hot mixture be confined in a controlled way ?

Magnetic confinement of the hot fuel

It is clear that no material wall can withstand such high temperatures and radically different methods have to be used to confine the hot fuel. At those extremely high temperatures, the fuel is completely ionised and has become a plasma, a collection of freely moving electrons and ions. The property causing the basic difficulty of fusion research, the charge of the particles, thus fortunately also points to a solution for the confinement of such a hot medium, as the movement of charged particles can be influenced with electromagnetic fields. Different solutions exist to confine a hot plasma by means of magnetic fields. In the rest of this text we will focus on the solution which yields at the moment best performance. This device is the tokamak, originally a Russian design, invented by Tamm and Sacharov. In a tokamak the magnetic field structure is generated by coils around the plasma together with a large current which flows in the plasma (Fig. 2). This field is the immaterial cage, which keeps the hot particles in a ghost-like way from the wall and prevents them from causing excessive damage. The effectiveness of the magnetic field to minimise heat losses is measured by the characteristic time needed for the plasma to cool down after the source of heat is switched off. This characteristic time is the energy confinement time τ_E, which has to be in the order of several seconds in a fusion reactor. The power output of a fusion reactor depends on the density of the reacting ions n_i, for a reactor in the order of a few thousands of a gram per cubic metre, but nevertheless yielding huge amounts of energy.

The most important condition for a fusion reactor is that it should have a net power gain, large enough to be of economical interest. A minimum requirement can be found by looking at the conditions for break-even, i.e. where the power gained from fusion reactions equals the power needed to heat up the plasma. This is given by the so-called Lawson criterion, which in its original form reads:

$$n_i\tau_E \geq 2\ 10^{20}\ \mathrm{m^{-3}\ s} \quad \text{at } T_i = 10\text{-}20\ \mathrm{keV}$$

where n_i is the central ion density and T_i is the central ion temperature of the plasma. An alternative formulation of this criterion which is often used (but not totally equivalent) is given by:

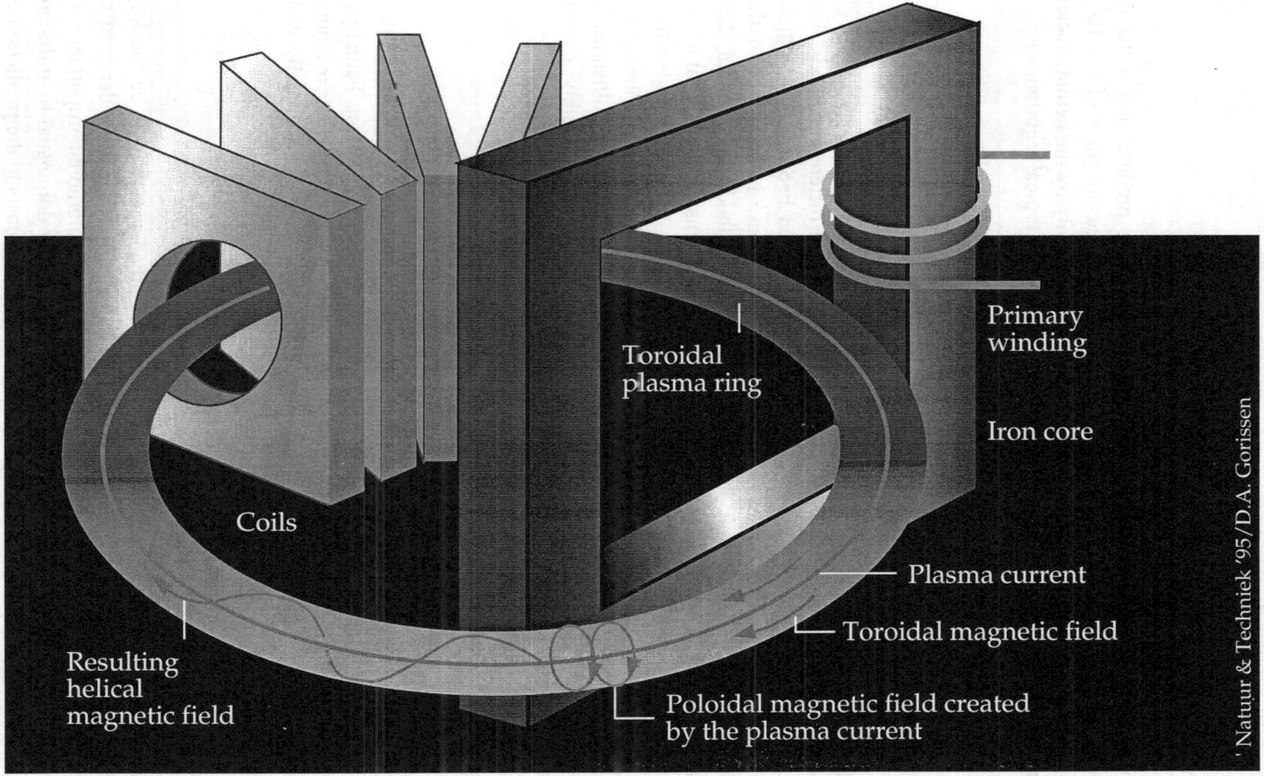

Fig. 2. In the tokamak configuration, the plasma is the single-loop secondary winding of a huge transformer. A current pulse in the primary winding induces a current in the plasma ring. The (poloidal) magnetic field generated by this current combines with the toroidal magnetic field of the toroidal field coils, with a resulting helical magnetic field, necessary for a stable confinement of the plasma.

$n_i\ T_i\ \tau_E \geq 2\ 10^{21}\ m^{-3}\ s\ keV$
or
$p\ \tau_E \geq 10$ bar s

The product $n_i\ T_i\ \tau_E$ is usually referred to as the 'fusion triple product'. For a fusion reactor based on the D-T reaction, the triple product must exceed the value of 8×10^{21} m^{-3} s keV to reach ignition, i.e. a self-sustained plasma, heated only by the helium nuclei released in the fusion reactions. This translates into the following typical requirements:

Central ion temperature:	$T_i = 10 - 20$ keV
Central ion density:	$n_i = 2 - 3 \times 10^{20}\ m^{-3}$
Energy confinement time (global):	$\tau_E = 2 - 4$ s

Heating of the fuel

The required high temperatures are reached by a combination of different methods. The plasma current provides a first method: the hot plasma is heated because of its ohmic resistance (Joule effect). This however is limited by the decrease in resistance of the plasma with increasing temperature, and additional heating methods are required. The different methods can be divided into two groups: injecting beams of fast neutral atoms or electromagnetic waves into the plasma. In the first case, fast atoms penetrate unimpeded the magnetic fields which confine the plasma, are ionised upon entering the hot fuel and subsequently transfer their energy to the rest of the plasma by collisions. In the second method, the energy of electromagnetic waves with a suitably chosen frequency (one of the different resonance frequencies of the plasma) is absorbed by certain classes of particles, which then become very hot, and subsequently transfer their high energy in collisions to the rest of the particles in the plasma.

Plasma configuration

Different plasma configurations are possible and currently under investigation. One option is a material limiter, where a solid structure determines the plasma boundary. Another option is a poloidal magnetic divertor (see Fig. 3), in which the outer magnetic surfaces are opened (by means of so-called divertor coils) and intersect eventually with target plates away from the main plasma.

REQUIREMENTS FOR A FUSION REACTOR

For the design of a next generation machine, a sufficiently large τ_E will be the key requirement determining the machine size. A detailed knowledge of the dependence of τ_E on the plasma parameters and machine size is therefore of paramount importance in fusion research. Unfortunately, the value of τ_E is to a large extent dominated by turbulent processes, which makes it rather difficult to give a firm physical basis to the prediction of this value. The problem is tackled experimentally by an approach much like wind tunnel experiments in the aviation industry. Experimental data for τ_E are collected from many

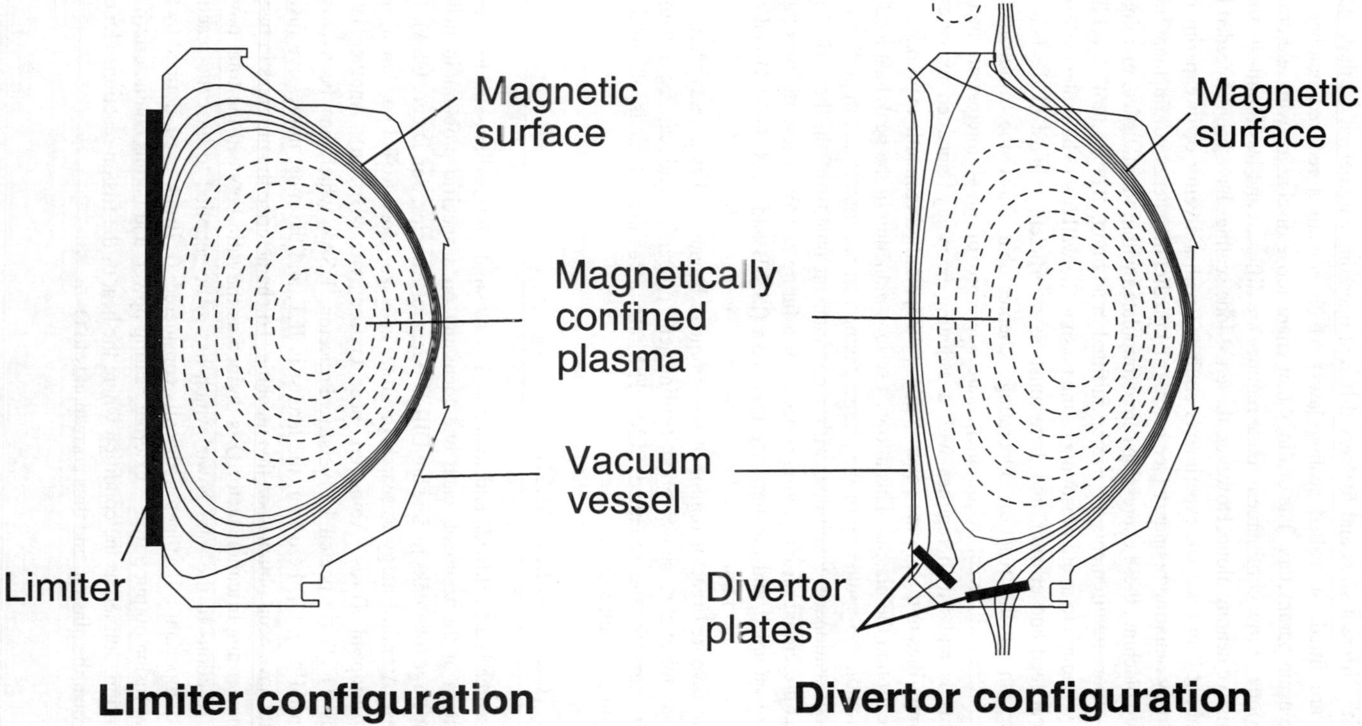

Fig. 3. Illustration of a limiter and divertor plasma

machines with different sizes and for very different operating parameters. To these data expressions are fitted (so called scaling laws) which contain a reduced number of significant plasma parameters. The scalings that characterise discharges with additional heating (L-Mode or low confinement mode regime) are different and less optimistic than those for resistive heating alone. However, these L-Mode scalings have been exceeded by a factor of about 2 in tokamak experiments equipped with a divertor or in experiments with a uniformly radiating boundary (see below). To predict the energy confinement time for a next step machine, these expressions are extrapolated within the statistical margins.

Many more requirements have to be fulfilled in addition. The impurity level has to be kept low enough to avoid poisoning of the plasma, the wall has to be protected from excessive heat load and erosion, and there must be an efficient exhaust of the helium particles–the ash of the reaction. Operating schemes that allow the simultaneous realisation of these different requirements are therefore of great interest to fusion research. This is realised in regimes with a radiating boundary, obtained by careful seeding of impurities in the plasma edge. In this regime, a radiating mantle is produced in an edge zone around the plasma. This allows a serious reduction in the peak heat load to and in the erosion and sputtering of the plasma facing components. It can in addition be obtained under stationary conditions, with a confinement quality of the best H-Mode plasmas, thus presenting an integrating concept for a future reactor. These regimes have been obtained on small and medium size tokamaks (ISX-B and TEXTOR-94) and the extrapolation of these regimes to larger tokamaks is currently a subject of intense research. The favoured reactor regime at this moment, because it has already shown to work in large tokamaks, shows good confinement properties and can be obtained stationary, is currently the so-called ELMy H-Mode, a regime which is obtained in tokamaks equipped with a divertor.

STATUS OF TOKAMAK RESEARCH

Tokamak research is a worldwide endeavour, and many small and medium-size tokamaks are in operation at the moment, each one focussing on a specific problem in fusion research. The large tokamaks to date are DIII-D (General Atomics, San Diego, USA), JT-60U (Japanese Atomic Energy Research Institute, in Naka, close to Tokyo, Japan) and JET (Joint European Torus, Abingdon, near Oxford, Great-Britain). Another large tokamak was TFTR (Princeton University, Princeton, USA), which has been closed recently (April 1997). Of these four, the largest is JET, and the most impressive fusion plasma results have been obtained on this machine. To characterise the progress in fusion research, a power amplification factor Q is defined as the ratio between the total power from fusion reactions to the total power which has to be supplied to heat the reaction mixture. Two important milestones are usually considered: (i) Break-even, defined by the condition that fusion output power = external heating power, and is thus characterized by Q=1; (ii) ignition, defined as the condition where the heat of the fusion reactions alone is sufficient to heat the plasma, and thus corresponds to $Q=\infty$.

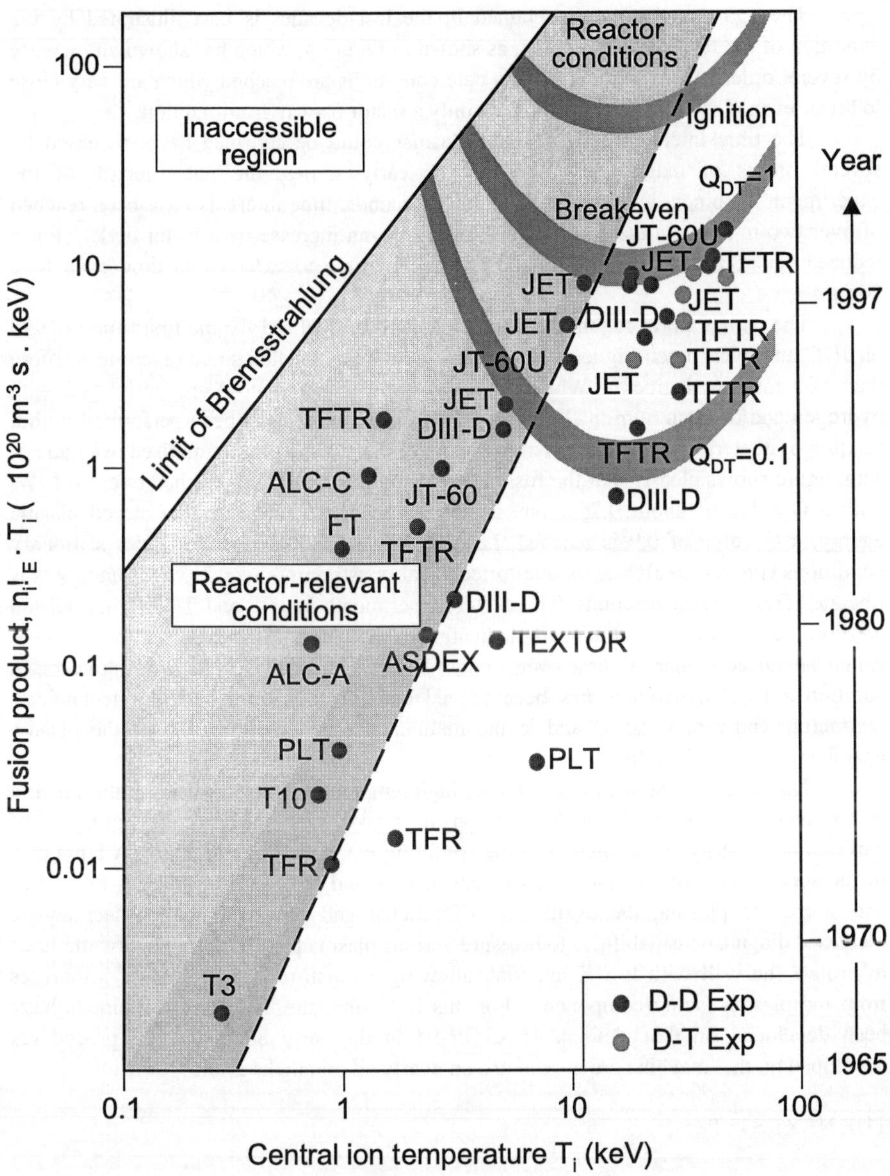

Fig. 4. Overview of the evolution of the fusion product (version central ion temperature) during the last 35 years for a number of tokamak devices worldwide.

The enormous progress obtained in the last decades is best illustrated by the evolution of the fusion triple product, as shown in Figure 4, which has shown an increase by several orders of magnitude. To this date conditions are reached which are very close to break-even (Q=0.6-0.7), and which are only a factor 6 away from ignition.

The time intervals during which plasmas could be confined have increased by several orders of magnitude. While in the early sixties, the pulse length of the experiments on pinch devices was some microseconds, time intervals have been reached of over two minutes in the device Tore-Supra[3], i.e. an increase by a factor of 10^7. For a reduced set of plasma parameters it has been possible to extend this time even to 2 hours[4].

The temperatures required for fusion have been realised for the first time in 1990 on JET, and even much higher temperatures have been reached since (even up to more than 500 million degrees[5]). While most experiments up to now are performed with hydrogen and/or deuterium, in JET and TFTR experiments have been performed with a mixture of deuterium and tritium. An overview of these results is summarised in Figure 5. This figure shows clearly that the fusion power output in JET has reached over 16 MW, with a Q value of about 0.7; if one takes into account changes in the stored plasma energy, a Q value of 0.9 is reached, i.e. very close to break-even)[6]. Under stationary conditions (limited in JET to 5s due to technical constraints), over 21MJ of energy was liberated from fusion reactions. These D-T experiments in JET and TFTR[7] in addition confirm the possibility of alpha particle heating of the plasma. It is expected that with so-called 'advanced scenarios' these results could be even superseded in the coming years. In addition a large experience has been gained on JET in tritium handling technology (extracting and reprocessing) and in the maintenance and modification of the plasma chamber by remote handling.

The progress obtained in physics of high temperature plasmas is not only due to a better understanding of the underlying physical processes linked to the transport of particles and energy perpendicular to the confining magnetic field structure. A large part of the success is owing to the development of new and advanced techniques to control and shape the plasma, the availability of reliable and powerful heating techniques, increased diagnostic capabilities to measure various plasma parameters, and new methods to protect the wall with low Z material, allowing a much reduced influx of impurities from the plasma facing components. For this last point, the necessary techniques have been developed on the tokamak TEXTOR-94 in the early 80s[8], and the procedures developed on that machine are now in use on nearly all tokamaks in the world[9].

THE NEXT STEP

Scaling laws obtained from experimental data of many tokamaks are now so much developed that an extrapolation to the working parameters of a reactor like device becomes possible. On this basis a next step tokamak device has been designed, the International Thermonuclear Experimental Reactor, or ITER, originally a co-operation between the European Union, Japan, the Russian Federation and the United States.

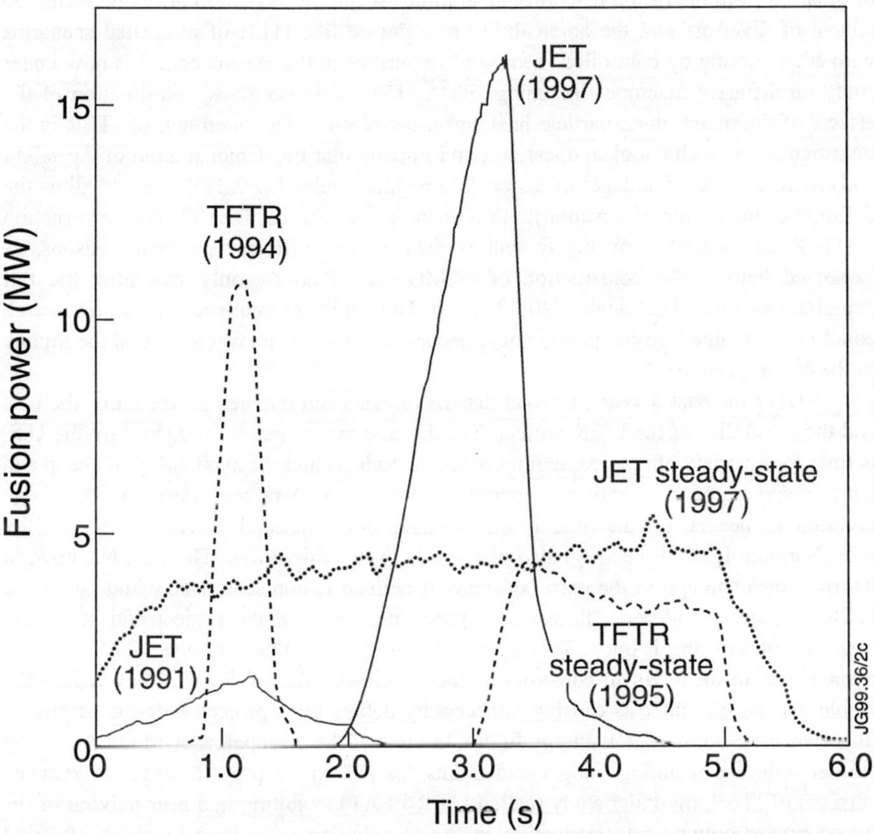

Fig. 5. Time traces of the fusion power released in different high performance deuterium-tritium experiments in TFTR and JET.

Originally, ITER was designed for a fusion power output of 1500 MW and burn duration of 1000 seconds[10]. The power output of ITER thus comes close to that of conventional oil or coal fired power stations. ITER would allow studying and optimising heat and particle exhaust under reactor conditions, more in particular also the exhaust of helium originating from the fusion reactions. Operational scenarios have been developed with the advent of divertors and the applicability to a device like ITER of integrated scenarios with edge cooling by controlled seeding of impurities in the plasma edge[11] is now under study on different machines including JET[12]. ITER will also allow investigation of the effects of dominant alpha particle heating on the plasma. The main aim of ITER is the investigation of technological questions, and in particular the demonstration of the safety and environmental advantages of fusion. The results obtained with ITER should allow the defining of the design of a demonstration fusion power station, DEMO. The construction of ITER could start immediately, but is delayed for different (political) reasons, as explained below. The construction of DEMO can of course only start after the full exploitation of the ITER device (20-30 years). Thus, if the current pace in fusion research could be maintained, fusion power could become available somewhere around the middle of the coming century.

Over the past 3 years, a broad discussion has been pursued on the aims, the cost and the feasibility of the ITER project. The discussion was mainly triggered by the U.S. partner for a variety of reasons, mainly reducing budgets and the availability of cheap and large reserves of conventional energy resources in America. Possibly the large investments needed for the (mainly on hydrogen bomb research oriented) laser facility NIF (National Ignition Facility) also play a role in this discussion. The U.S. has now, to its own detriment and to the regret of many American fusion scientists, withdrawn from ITER. Experts world-wide, however, express their faith in the project and stress the necessity of carrying it out[13]. This is reflected in the fact that the other ITER partners remain firm in their will to construct a reactor relevant fusion device. They regard the whole discussion rather as causing unnecessary delays for a project which is of crucial importance for our common energy future. In view of the disappearance of the American partner with the resulting budget constraints for the whole project, currently reduced versions of ITER are under study (called ITER-FEAT), resulting in a near halving of the fusion power output and a reduction in the Q value from ignition to about 10. First estimates indicate that it will be possible to construct such a device (which will address a reduced set of technical questions) with the currently available budgets. This would enable us to build nevertheless a next step machine, an absolute necessity to keep the current momentum in the important and complex field of fusion research.

The will to go on with fusion research is clearly expressed in several instances over the past months: (i) The reorganization of the European Fusion Programme under EFDA (see Section 6) in view of the preparation for a next step device; (ii) An upgrade for the JET tokamak, the JET-Enhanced Performance or JET-EP (where in essence the additional heating power up will be roughly doubled to 50MW) has been agreed and is in full preparation. This upgraded facility should be ready mid 2003 and is foreseen to run until December 2006; (iii) ITER-Canada[14] has proposed Clarington (Ontario, close to

Toronto) as a possible site for ITER-FEAT (December 1999) and the Commissariat à l'Énergie Atomique has expressed its readiness to offer Cadarache (France) as an ITER-FEAT site (June 2000); (iv) The European Council has approved in a very recent meeting (Nov. 2000) a mandate for the EU Commission to negotiate the creation of an international framework in view of the preparation of a legal entity for ITER, its construction and its exploitation. All these facts are reason for optimism for the future of fusion research, and everybody hopes that the last few years, which caused so many unnecessary difficulties, can be quickly forgotten! In this context it is interesting to note that in Nov. 2000, also the European Physical Society has officially underlined "The importance of European Fusion Energy Research" in a position paper[15].

REFORM OF THE EUROPEAN FUSION PROGRAMME TO PREPARE FOR THE NEXT STEP

Since about a year ago, the EU Fusion Programme has restructured several of its fusion activities in order to prepare in an integrated manner for ITER-FEAT. The operation of JET, the EU participation in ITER-FEAT, and the European activities in fusion technology are now grouped under EFDA (European Fusion Development Agreement). Since the beginning of 2000, the tokamak JET is no longer operated as a single entity, but by a collaborative effort of the Fusion Associations throughout Europe. In practice this means that experiments now are proposed by teams of researchers from all over Europe, spending part of their time at JET to execute their proposal, and after that return home to continue further analysis. So far three campaigns have been successfully completed with further experimental campaigns planned in the coming months and years. Several new results have been obtained already under this new organisation[16], among others, the realisation of plasmas with simultaneously high density and high confinement leading to parameters which are very close to those required for the reference scenario for ITER-FEAT! It goes without saying that these results, which constitute important inputs to the ITER design team, show the success of the new JET organisation and are a clear expression of the strong will and determination of the EU Fusion Community to pursue the fusion endeavour for the benefit of future generations!

REFERENCES

1. J. Raeder *et al.*, "Safety and Environmental Assessment of Fusion Power (SEAFP)", European Commission, Report EURFUBRU XII-217/95 (June 1995).
2. S. Barabaschi (red.) "Fusion Programme Evaluation 1996", European Commission, Report EUR 17521, ISBN 92-827-9325-7 (December 1996).
3. B. Saoutic *et al.*, Fusion Energy 1996, 1, 141 (IAEA, Wien 1997).
4. S. Itoh *et al.*, Fusion Energy 1996, 3, 351 (IAEA, Wien 1997).
5. S. Ishida *et al.*, Fusion Energy 1996, 1, 315 (IAEA, Wien 1997).

6. JET team, 17th IAEA Fusion Energy Conference, Yokohama, Japan, 19-24 Oct. 1998 (paper IAEA-F1-CN-69/EXP1/08).
7. K.M. McGuire *et al.*, Fusion Energy 1996, 1, 19 (IAEA, Wien 1997).
8. F. Waelbroeck, J. Winter *et al.*, J. Vac. Sci. Technol. A2, 1521 (1984).
9. J. Winter, J. Nucl Mat, 176-177, 14-31 (1990).
10. R. Aymar, "The ITER Project", Fusion Energy 1, 3 (IAEA, Wien 1997).
11. A. Messiaen, J. Ongena et al., Phys. Rev. Lett., 77, 2487-2490 (1996).
12. J. Ongena *et al.*, Plasma Physics and Controlled Fusion, 41 (3), A379-A399 (1999).
13. Concluding session, chaired by M. Rosenbluth, of the 1998 International Congress on Plasma Physics (ICPP) combined with the 25th EPS Conference on Controlled Fusion and Plasma Physics (Prague, July 3th 1998).
14. See the website http://www.itercanada.com.
15. Sir Arnold Wolfendale, "European Physical Society: Position Paper. The importance of European fusion energy research", 6 Nov. 2000, EPS, Geneva.
16. J. Pamela *et al.*, Post-deadline presentation at the 18th IAEA Fusion Energy Conference (4-10 October 2000, Sorrento, Italy), IAEA-CN-77, IAEA Vienna.

NEW TRENDS IN RUSSIA'S ENERGY STRATEGY

ANDREI YU.GAGARINSKI
Russian Research Centre "Kurchatov Institute", Moscow, Russia

In 1999, the expected event in the Russian economy occurred: for the first time after a 14-year decrease the energy demand grew by 2.3%. The growth of energy consumption was 90% covered by supplementary electricity produced by NPPs - installed capacity of which has remained unchanged for 7 years.

In the next three or four years another important change in Russia's economy is expected: the country, for the first time in decades, will face an energy deficit and will change from being "energy-redundant" to "energy-deficient". In parallel, according to the data provided by the Ministry of Energy and the Russian Academy of Sciences, even if the level of energy consumption was maintained (and it is expected to increase by 5% annually), the continuing electricity deficit would become a brake for Russian economic development.

Naturally, these events—long predicted by the specialists—have pushed forward the urgent revision of the country's energy strategy, which was last considered and approved at governmental level in 1995. Now the new project of the "Energy Strategy of Russia" is being developed. It will be submitted for governmental consideration at the end of this year. This paper presents an overview of the present state, main preconditions and expected changes in the trends of the country's energy development.

ENERGY SITUATION

Today's pre-crisis situation in the Russian fuel & energy complex (FEC), masked until recently by the energy consumption decrease, results from the previous long-term fast development phase, when the resources and funds were invested, for many years, in the development of capacities without sufficient infrastructure (including resource supply). The dynamics of the primary fuel & energy resources in Russia are given in Figure 1.

In today's economic, environmental and technological conditions—both in Russia and in the whole world—Russian economy is unable to reproduce not only the elements of fuel & energy complex (FEC) at their achieved level of development, but in many cases even to support their operation abilities.

Fig. 1 Fuel production in Russia

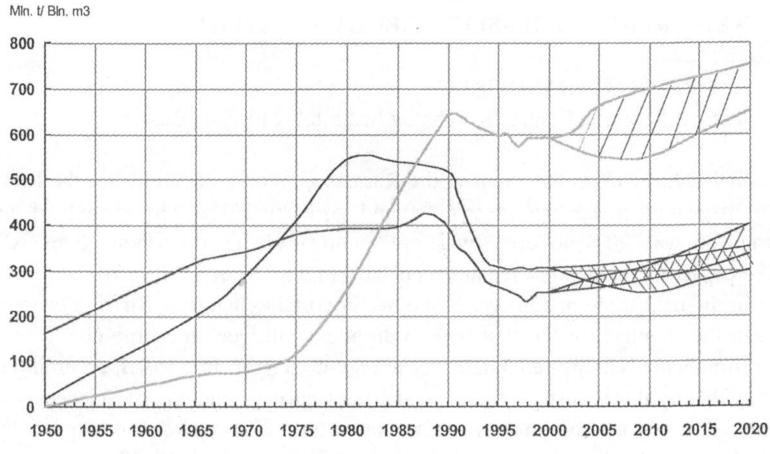

Figure 1. Fuel production in Russia

Having 2.8% of the world's population and 12.8% of the world's territory, Russia also possesses 11-13% of prospected resources and about 5% of proven recoverable reserves of oil (7 billion tons), 42% of the resources and 34% of the reserves of natural gas (about 50 trillion m^3), and about 20% of proved recoverable reserves of coal (about 160 billion tons). Total extraction for the whole history of the resources' use is: oil - about 20% of prospected recoverable resources, and gas - 5%. Extraction supply with proven fuel reserves is estimated at several decades for oil and gas, and much more for coal and natural uranium.

A major potential resource of hydrocarbons for a long-term perspective is represented by Russia's shelf, which is 6 million square km^2, or 20% of the world's ocean shelf. Only 1-2% of oil and gas deposits in the Russian shelf have been investigated; however, large deposits (for example, the Stockman deposit in the Barents Sea) and promising structures have already been found. Over 80% of hydrocarbon resources of the Russian shelf are concentrated in the Arctic seas.

However, the complicated situation today is due to the fact that economic efficiency depends only to a very small extent upon the proved raw reserves, and depends also very little on the amount of prospected resources—which are the real wealth of the country. Efficiency of the present-day economy is to a considerable extent determined by the economy of processing and transportation of fuel, and efficiency of its use for the services in producing "final consumption" products. And here, Russia is far behind the developed countries. Russia's natural resources represent a somewhat hampering factor for its development and prosperity, making it possible to apply the easiest decisions in difficult situations at the expense of the resources' extraction and consumption increase.

With this background, the country has been unable to realize the announced decrease in the energy capacity of the country's economy, which was practically stabilized, exceeding by 20% the already high level of the 80's. (Energy capacity of gross domestic production for the decade since 1990 has increased from 1.27 to 1.44 tce/thousand USD).

No new cheap fossil fuel resources should be expected in Russia in the future. The available power industry structure, based on such resources, will change for objective reasons because there are no finances to develop expensive deposits.

Presently, oil extraction has stabilized at the level of about 300 million tons per year. The rate of the cost-effective reserves exhaustion of the country's exploited deposits has reached 53% (and in the main oil region, West Siberia, 43%). Main oil and gas provinces have reached the last stages of the deposits' development with decreasing output. The time, when giant deposits, providing growth of reserves and decrease of prospecting and extraction expenses, were found, has passed. The part of hardly recoverable reserves has reached about 60% and continues to grow. Growth of proved reserves in the last years doesn't cover current oil extraction.

Basic gas deposits in West Siberia, which in 1999 had provided 72% of gas extraction in Russia, have reached the stage of decreased output and are more than half-exhausted: Medvezhie deposit - for 78%, Urengoi - for 67% and Yamburg - for 46%. By 2020, according to assessments, gas extraction at these deposits will not exceed 80 billion m3, or just 14% of today's extraction output in Russia. As a result of larger amounts of gas extracted compared to an increase of proven reserves, the amount of the latter decreases. In order to maintain today's extraction output only for the period up to 2020, as a minimum, a three-fold increase of investments in the development of new Stockman and Yamal gas deposits would be necessary.

Today's power industry has much more inertia than forty years ago, because it is principally different in terms of its components' capacity level. Energy consumption restructuring would require huge investments. That is why the prompt restructuring of energy production and consumption is impossible. A dramatic downfall of the power industry's reliability and efficiency could be avoided only through the refusal to preserve nothing, which is not vitally important. That means the reduction of oil and gas extraction and consumption levels to such an extent, that the funds released as a result of the refusal of inefficient and excess equipment, could be directed to the more effective operation of the residual facilities.

The inertia of power technologies, resulting from the length of the power installations' lifetime (30-40 years or more), and the long time needed for development of new, more difficult to access deposits (15-20 years) makes it impossible to introduce any cardinal structural changes in the next 15-20 years.

Besides, Russia's export of energy carriers, reaching in the last years of up to 35% of their production (including over 57% of oil and oil products and 34% of natural gas), as estimated, will only grow. This is quite explicable by taking into account a seven-fold price difference between gas for export and for domestic use.

The existing situation is aggravated by the investment and structural crisis in Russia's power industry. The amount of annual investments in the fuel and energy complex in the last years has decreased more than threefold. This has created a real threat for the country's energy security because of the unsatisfactory state of FEC facilities. By 2010, in the European part of Russia, exhaustion of calculated physical resources will reach 50 GW of electricity generating capacities.

In these conditions, at the end of last year, the monopoly Russian gas producer, GAZPROM Concern, officially and rigorously announced the objective of needing to substitute considerable amounts of gas in electricity generation (the expected gas deficit would be over 60 billion m3 already in 2002, which is close to 50% of the amount burned today in the power industry) with alternative energy resources. It should be noted that natural gas provides more than 73% of fuel burned by European Russia's fossil-fueled plants, which exceeds the limit of an admissible energy security level.

SHORT-TERM PERSPECTIVES

Russia's economy is not ready to use its own resources at world market prices; it is even less ready to rely upon imported resources. The existing Western economy would not be suitable for Russia, especially in the period of transition from an FEC structure born in conditions of centralized pricing mechanism to the structure efficient in conditions of market pricing mechanisms. It may be expected that a gradual restructuring of prices for energy carriers would make it possible to create the economic conditions for a future change of the fuel consumption balance towards a decrease of the gas part and, thus, towards energy supply reliability enhancement.

For the next few years, the most probable development of the FEC situation seems to be represented by the scenario providing for decrease of gas extraction to ~ 550-500 billion m^3/year; oil to ~ 250-200 million t/year (with the need to partially substitute gas in electricity generation). Then by 2020 their extraction is expected to increase: gas to 600-650 billion m^3/year, oil - to 300-350 million t/year.

Here it should be noted that the prepared draft of the "Energy Strategy of Russia" presents a more optimistic scenario of the country's energy sector development (Fig.1), especially related to gas (750 billion m^3) - based on availability of "favourable conditions" (primarily, world prices and taxes). It is also worth saying that even in the case of the "favourable" scenario, the role of economically justified technologies of renewable energy resources (except hydroenergy) is limited, by 2020, to 8-20 million of tce (or 0.5-1.0% of primary energy resources).

Naturally, in order to avoid Russia's heading towards an energy crisis, it would be necessary to realize compensatory measures, which would be possible in a short-term perspective and would need large investments.

It is known that Russia possesses a great potential for organizing and technical energy saving. Its realization, according to expert estimations, would make it possible to reduce the current fuel consumption in the country (900 million tce) by 40-50%, with 40% of this economic potential belonging to the fuel & energy complex itself. However, in the short term (till 2005), according to forecasts, an economy of 30-50 million tce is possible - including 20-40 billion kWh of electricity, which corresponds to savings of 6-12 billion m^3 natural gas. Such an energy saving level would demand minimal investments of about 500-800 million USD.

As a basic and relatively promptly realizable measure, the country's "energy headquarters" are now considering the increase of the capacity factor of coal condensation and co-generation plants and reverse (where possible) transfer of gas plants (initially built as coal plants) back to coal.

Estimations show that in case of investments in electricity generating plants of about 1.5 billion USD and, of the same order in coal extraction development, in 3-4 years up to 14-17 billion m^3/year could be substituted. However, here serious technical, economic and environmental problems should be considered, because such a solution contradicts the world trend of reducing the use of coal, as the most hazardous fuel in terms of greenhouse gases' emission. On the other hand, coal rate consumption in Russia's power industry is much less, compared to other countries.

One of the most economically efficient means of local heat & electricity supply of territories, industrial objects and houses is the development of "small" power industries based on steam & gas turbine installations. The advantages of this energy supply method are: the maximum possible efficiency (up to 80% in joint electricity & heat production mode) of the energy carriers' use in the steam & gas turbine cycle; relatively cheap domestic equipment; high environmental parameters and, thus, the possibility to place energy sources in the immediate vicinity of the consumers; module capacity increase and high production readiness of the equipment. All the above considerably reduce the periods of the plants' commissioning, amounts of capital investments and the energy grid operation expenses.

Concerning "Gazprom", for over 20 years it has actively used gas-turbine facilities (GTF) in gas pumping plants. It currently uses about 3000 GTF of over 20 types with unit power of 2.5-25 MW (air open-cycle turbines with ~ 25% efficiency).

The period of a "turn-key" construction of gas-turbine plants is about 1.5-2 years. Estimated level of domestic production cost - 400-500 USD/kW (for Western analogues - 1000-1500 USD/kW).

A longer term, but quite technically feasible perspective, is represented by two proposals put forward by E.P.Velikhov. The first one is related to wind energy use for gas piping via main gas pipelines.

Today, piping of 600 billion m^3 of natural gas per year needs over 50 billion m^3 of high-potential commercial gas burned by gas-piping compression plants, which have the total installed capacity (taking reserves into account) of over 40 GW. Practically all the regions having gas pipelines in their territory also possess the wind energy potential sufficient for local industry energy supply. Serial production of powerful wind energy facilities exists. This could substitute up to 2/3 of the capacities of operating gas-turbine piping plants and, consequently, of gas spending for the particular needs of the gas transportation system[1].

Another known proposal is connected with elimination of gas use for the particular needs of the gas transportation system by abandoning the scheme of distant gas transportation via pipelines, and transition to the scheme of universal electricity production directly in the areas of large-scale gas extraction, with further energy transmission via energy grids. Such a diversification of gas production seems realistic for remote gas extraction centres and in specifically complicated conditions, for example, for the gas deposits in the Arctic Ocean.

Availability of a high-potential fossil energy carrier, natural gas, makes it possible to create a highly efficient compact, combined gas-turbine electricity generation plant with 60% efficiency and 16 GW unit power (which corresponds to 25 billion m^3/year productivity of a gas deposit), and with specific energy and composition parameters on the level of 4 t and 60 m^3 per 1 MW of installed capacity. Such a scheme of energy conversion would give additional gas equivalent savings of over 2.5 billion m^3/year for each gas-turbine plant, taking into account its high efficiency. Such a plant weighing 60 thousand tons could be placed on a sea platform with construction parameters corresponding to operating prototypes.

A brief overview of possibilities related to fossil fuel economy in a short-term perspective should be concluded with large-scale proposals coming from the nuclear power sector.

GREAT EXPECTATIONS OF NUCLEAR POWER

The latest commissioning of a nuclear unit in Russia took place in 1993. In 1998 the Russian government adopted the Program of Nuclear Power Development for 1998-2005 and for the period up to 2010, which provided for a moderate growth of nuclear power capacities (up to 27 - 29 GWe by 2010). However, this Program was so poorly financed, that even the completion of three nuclear units—which were at the stage of high constructional readiness—was practically stopped.

[1] It should be noted that gas-pumping facilities' transfer to electricity feeding would make it possible to save natural gas thanks to NPP electricity. The option of using small nuclear power facilities was considered in this connection.

The situation has considerably changed this year, when nuclear specialists announced that the above problems of the Russian power sector could be solved through nuclear power development, the main reserves of which are as follows:

- Increase of installed capacity factor. By the middle of 2000, Russian NPPs increased this factor by 6% compared to 1999, and it reached 73.4% (with 75-85% according to design, see Fig. 2). In 2000-2001 there are plans to increase energy production on NPPs up to 140 billion kWh due to the realization of the design capacity factor level.
- Extension of operation life. 30 years of NPP operation prescribed by Russian designs reflects the earlier conservative approach to its calculated substantiation, and not a real deterioration. Presently work is underway in order to substantiate the units' service life extension for up to 40-50 years.
- Construction of new nuclear units.

According to the Russian Minatom, nuclear power has considerable resources for growth:

- Reserves of uranium and industrial infrastructure are sufficient for a four-fold increase of existing NPP capacities;
- Existing constructions for NPP units of ~12 GW total capacity, which would require specific capital investments of ~700 USD/kW for completion;
- Available NPP design using domestic equipment, which would require only about 1000 USD/kW for its realization[2];
- Nuclear machine-building reserves make it possible to manufacture up to 4 sets of VVER-1000 unit equipment annually;
- Under Russian projects, 5 NPP VVER-1000 units of the third generation are planned for construction or already underway (2 in China, 2 in India and 1 in Iran).

---

[2] According to the assessments made by independent experts, so small specific capital investments are possible only on the sites, in which considerable funds have been already invested (Russia has 12 such sites, and another two dozens were studied and construction base was prepared there). This correlates with the world level of expenses - for example, 1700 USD/kWe for the new Finnish unit, also supposed for construction on a prepared site.

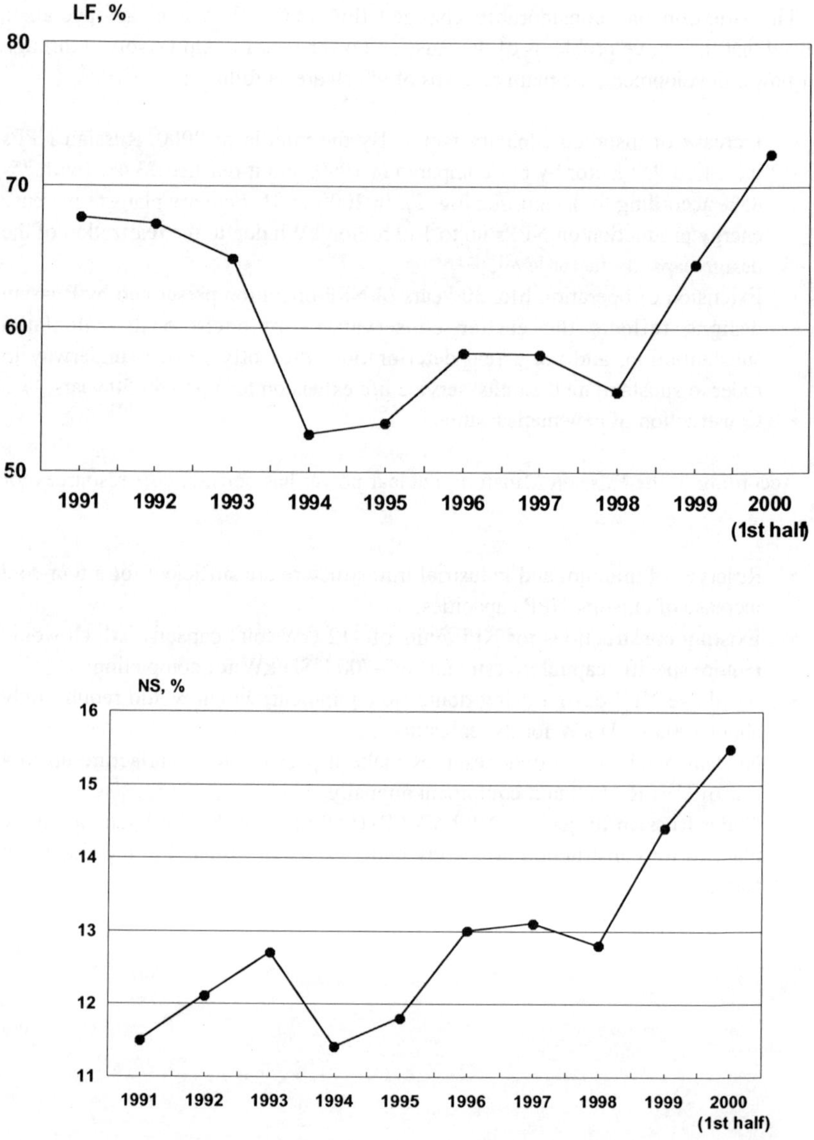

Fig. 2 Nuclear Share of Electricity Production (NS) and Russian NNP's Load Factor (LF) Trends

Relative to these reserves, two scenarios of nuclear power development up to 2020, presented in Table 1, were developed:

Table 1. Scenarios of nuclear power growth.

Parameters	Minimum scenario	Maximum scenario
Capacity factor	Up to 75-82 %	Up to 80-85 %
Extension of design service life of operating nuclear units, years	Up to 40, with additional output of over 950 billion kWh of electricity	Up to 40-50, with additional output of over 2700 billion kWh of electricity
Decommissioning of nuclear units by 2020	6.8 GW	6.8 GW
Growth of NPP capacities, GW:		
2005	Up to 24.2, with energy production of ~ 160 billion kWh	Up to 25.2, with energy production of ~ 172 billion kWh
2010	Up to 31.2, with energy production of ~ 205 billion kWh	Up to 32.0, with energy production of ~ 224 billion kWh
2020	Up to 35.8, with energy production of ~ 235 billion kWh	Up to 50.0, with energy production of ~ 372 billion kWh
Including:		
By 2010:		
NPP installed capacity increase	By 10 GW	By 10.8 GW
Completion of 5 GW of nuclear units	5 units	5 units
New construction of 5-6 GW of nuclear units	5 units	6 units
By 2020:		
Replacing of 6.8 GW of nuclear units	7 units	7 units
NPP installed capacity growth	5 units	20 units

As the Table shows, in the case of a maximum (or minimum) scenario of nuclear power development, the additional amount of substituted natural gas (compared to today's 36 billion m^3/year) would reach 16 (14) billion m^3/year already by 2005, and 32 (26) billion m^3/year - by 2010.

The total requirement of investments for this program is estimated at 2.5 billion USD for the period up to 2005, 12 billion—from 2006 to 2010, and 20 billion—from 2011 to 2020. Minatom proposes the following sources for financing the nuclear energy development:

- increase of electricity tariff;
- partial use of gas export income ("gas dollars");
- legally established tax rebates;
- direct state budget involvement in the nuclear investment program—as part of the financing of NPP safety enhancement measures ;
- providing services to foreign NPPs in the field of spent fuel management.

At the end of May this program was discussed at the session of the new Russian government and received its approval in principle. Planned commissioning of the first unit of the Rostov NPP should be a key event this year.

It should be noted that the strategy of Russia's nuclear power development in the first half of the XXI century, submitted to the government, provides not only for creation of NPPs with a high unit capacity, but also for construction of small nuclear power objects, including stationary and floating power and desalination plants (the floating NPP design was developed and is presently being licensed in Russia), though these developments are postponed till 2030.

At the same time, small nuclear power plants, in the opinion of many specialists, could make a large-scale breakthrough in the power systems of many world regions. Figure 3, taken from the recent report of the World Energy Congress and showing the world energy flows, demonstrates shaded large world areas (which include half of Russia), where the economic development requires supply from small and medium-power energy sources. In conditions of prospected growth of fossil fuel prices and the deficit of investments for building large gas, oil and electricity transportation grids, the basis for their economic development could be created using renewables or nuclear sources of small and medium capacity.

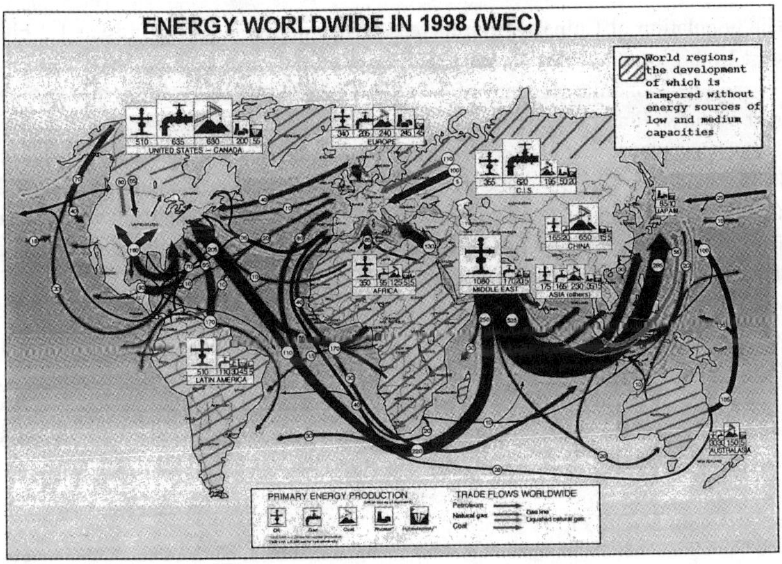

Fig 3. Energy worldwide (WEC) and possible regions for small and medium power systems.

ENERGY PROBLEMS AND PROSPECTS OF CHINA

HUO YUPING
College of Physics and Engineering, Zheng Zhou University, Zheng Zhou, China

ENERGY CONSUMPTION OF CHINA IN THE NEXT CENTURY

The population of China will be increased, inevitably, from the present 1.2 billion to at least 1.6 billion in the next century.

The economic standard of living in China will, inexorably, grow very fast in the next century. Finally, China will be a developed country in the world.

Comparing with other developed countries (U.S.: 11.5 TCE; West Europe: 5.6 TCE; Japan: 5.1 TCE), the energy consumption per person in China in the middle of the next century should be more than 3 TCE. Recently, other Chinese Institutions have given nearly the same values by different ways. This figure has already taken into account the significantly improved energy efficiency.

Therefore, the total energy consumption of China in the middle of the next century should be at least 5 billions tons of oil equivalent (TOE). This is the basic requirement of the modernization of China.

The 1996 energy consumption of China was 1.316 billion TCE, 74.8% of which was coal, only 17.1% was oil, and the electricity generation was 1079 billions KWh. It was already the second largest energy system in the world.

REQUIRED ENERGY SYSTEM

One of the basic difficulties of the sustainable development of China is the required energy system (near 5 billions TCE), and if this was provided mainly by fossil fuel, could not be well supported by the energy resources and could not meet the requirements for a clean environment:

- The oil production rate of 1996 was 157M tons, it was only 16.7% of the total energy production, and the surplus of proved reserves was only 2.25B tons. Everyone agrees that the oil production rate in the future could not be over 200M tons. Including imports, the oil consumption could be less than 10% of our total energy consumption, and could not even meet the future requirements of the transportation and chemical industries. Shortage of liquid fuel in the future is one of the main energy problems of China.

Natural gas provided only 1.9% of the total energy production in 1996, and the proven reserves were 824 billion cubic meters. Though many experts have predicted high growth of the natural gas production in future, it seems that it could not play any significant role (for example, more than 10%) in total energy production.

In 1996, 1.36B tons of coal were produced, which provided 74.6% of the total energy production. The surplus of the proven reserves was 488.7B tons, which seems to be enough for hundreds of years of consumption, but 82.6% of the reserves are distributed in the North-Western part of China, far from the main energy consuming region. Coal will still provide nearly 50% of the total energy production around 2050.

- The possible exploitable amount of hydropower is 3.8 GW, but nearly 68% of this is in the Southwestern region, mostly mountain areas with very low density of population. In 1996, 187TW hydro-electricity were produced, 17.6% of the total. Therefore, in the middle of the next century, the maximum possible produced hydro-electrical power could only be near 15% of the total electrical power (assuming 2000GW, corresponding to near 1 kW for each person), and much less than 10% of the total energy consumption.

There were 7000 kW Solar Power Stations worked in 1996, most of them equipped with crystalline silicon solar cells. Due to the high price of such cell and rechargeable batteries, this type of solar power station can only be used for special purposes. The wind power (56.5 MW in 1996) and geothermo-power (28.6 MW) all are and will be too small, compared to the total energy consumption, and could only be used in special areas. The Biomass energy in 1996 was near 250 million TCE and already produced very serious air pollution and other environment problems. To improve living conditions, most of the biomass fuel will be replaced by other fuels or energy sources.

China now has only two nuclear stations with 2.1GW power, which is 0.9% of the total electrical power. Several nuclear power stations are under construction, and a plan of 20GW has been carried out. But due to worrying over the safety and limited Uranium resources, it is rather difficult to predict further growth.

- China consumed 1.37 billion TCE (76% were coal) in 1996, which is only nearly one fourth of the demanded 5 billion TCE in the future. But the environmental pollution has already been very serious. The annual emission of sulphur dioxide was 23.5MT, and the particulate matter fed into the atmosphere annually was 19MT. 71.7% of the cities in southern China have acid rain, and the acid rain probabilities for several big cities have already been over 90%. Beijing and other six Chinese big cities have been listed in the ten most polluted cities in the world. Near 3 BT of CO_2 were emitted annually. All agree that China should vastly increase the energy production in the next century, but it should not pollute the environment any further.

SOLVING MAIN SCIENTIFIC PROBLEMS

Since the end of 1997, a national scientific program for solving some main scientific problems related to the sustainable development of Chinese economy in the next century, has been founded (near 300 million USD equivalent for the first four years). One part of it has been aimed at the sustainable development of energy production. As a long term scientific program, we have concentrated on three topics:

1. How to efficiently and cleanly use the billions of tons of coal;
2. How to overcome the future shortage of liquid fuel;
3. How to find ways to develop non-fossil energy to a hundred million TCE scale.

Since we need annually to use several billion tons of coal for nearly all the next century, the most important and urgent problem is to find the ways to use it efficiently and cleanly. Several ways have already been used in other countries (such as the U.S., Japan, and so on), but could not be used in China widely, by just transferring the technologies. Up to now, we have approved the following projects:

1. To find a cheap and semi-dry way to clean the smoke gas in the stack, with less, so called, "white pollution;"
2. A new way to improve the stability, and to prevent disasters of the regional or national electrical network;
3. Some principles for energy-saving methods, which could be widely used in China;
4. Basic scientific research for ways to gasify billions of tons coal. Gasification of coal, instead of burning, could perhaps be the main first step to use coal in the future.

To overcome the possible shortage of liquid fuel in the future, China will need at least 400-500 MT oil for each year to the middle of the next century, but less than half could be self-produced. We have approved the following projects:

1. To study the new process forming oil resources within the mainland;
2. To study the process for liquefying natural gas or coal gas;
3. Hydrogen energy, including producing, storing and transferring hydrogen in large-scale, and the fuel cell technologies.

To find ways to develop billion kW non-fossil energy. By many discussions with different people, we have decided to select two approaches: photo-voltaic cell and nuclear fission energy. Two projects have been approved:

1. Cheap and long life solar cell: there are only two kinds of solar cells, the prices of which have the possibility to be reduced to nearly 1 USD/watt. Those are Gratzel cells, so called, made by TiO_2 nano-crystalline firm attached by charge transfer dye, and amorphous silicon firm cell.
2. Study the new type of fission reactor, which should be breeder type; with improved safety and which should be able to treat the long life radioactive wastes by neutrons.

All our programs have emphasized the scientific aspect of the problems, and to solve the energy problem of China from a long-term point of view. We hope our efforts should be valuable also for developing countries. Next year we will begin another five-year plan, and the program could be further enhanced.

5. POLLUTION — BLACK SEA

PROBLEMS OF CONTROL AND RATIONAL USES OF THE BLACK SEA RESOURCES

MIKHAILOV V.I., GAVRILOVA T.A., LISOVSKY R.J.
Ministry For Ecology and Natural Resources of Ukraine, Ukrainian Scientific Centre of the Ecology of Sea (UkrSCES) 89, Frantsuzsky Boul., Odessa, 65009, Ukraine

The Azov-Black Sea basin is a unique warm basin of Ukraine. Its recreational importance is really unique.

Now the Black Sea is an object of economic activity of six independent states (Fig. 1.) The fact is that the states lying on the coast of the Black Sea basin are not rich enough and cannot invest in the development of technologies and waste water treatement plants, and the ecosystem of the sea is in a crisis condition.

UkrSCES, being the main organization of the Ministry for Ecology and Natural Resources of Ukraine on Natural Sea Usage and Regional Activity Centre on Pollution Monitoring and Assessment, constantly carry out complex long term monitoring investigations of the Black and Azov seas. The modern ecological condition of waters of the Odessa Bay, urban beaches, and also water areas of the main ports are of vital interest for ecologists.

For the rescue of the system of the Black Sea, the Convention on the Protection of the Black Sea against Pollution, was signed in 1992 in Bucharest, Romania, which the Ukraine ratified in 1994. In the development of the rules of the Convention in Odessa, a meeting of the Ministers of Ecology of 6 countries took place and the Odessa Declaration in 1993 was signed in Odessa. In order to carry out the Odessa Declaration by the World Ecological Fund, an international program to investigate the ecological problems of the Black Sea was organized.

In order to carry out the program in six countries, different Activity Centres were set up:

- Bulgaria: the Activity Centre on the Environmental and Safety Aspects of Shipping (Varna);
- Georgia: the Activity Centre on the Conservation of Biological Diversity (Batumi);
- Romania: the Activity Centre on Fisheries and other Marine Living Resources (Constanta);

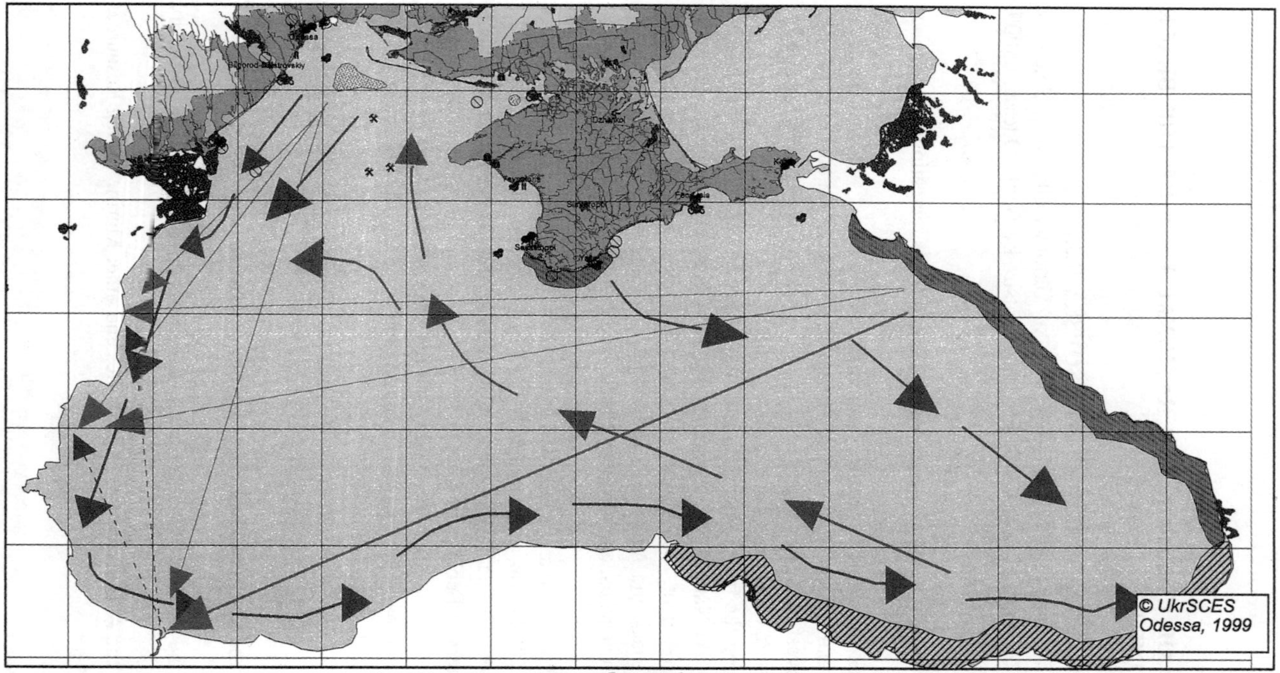

Legend:

ε - sand minings; ⊥ - oil terminals; [oil terminal under construction symbol] - oil terminals under construction; _ - dumping sites; □ - mussel banks; ξ - fish spawning places; 4 - oil storages; ; - ports; [dotted box] phillophora fields; [shaded box] beaches; [stippled box] - general marine ecosystem; ➔ - fish winter migrations; ➔ - fish spring migrations; ➔ - main tanker routes to Bosphorus; ···➔ - tanker routes from Bosphorus; [hatched box] - місця зимування чорноморської хамси [cross-hatched box] - місця нересту чорноморської хамси

Fig. 1. Classification of Black Sea Marine Waters by Usage.

- Russia: the Activity Centre on the Development of Common Methodologies for Integrated Coastal Zone Management (Krasnodar);
- Turkey: the Activity Centre on Control of Pollution from Land Based Sources (Istanbul);
- Ukraine: the Activity Centre on Pollution Monitoring and Assessment (Odessa).

The work of the Activity Centres consists in coordinating the appropriate works on ecological problems of the Black Sea (Fig. 2).

As a result of three-years joint work of all Black Sea countries the basic priorities and prime tasks on the Rehabilitation of the Black Sea ecosystem were determined.

In each country "Hot spots" are determined "which cause up to 85 % of all the Black Sea pollution." For Ukraine, these Hot spots in the Black Sea are allocated as:

- 3 points in the region of Odessa and Ilichevsk: non-perfect waste water treatment plants;
- 5 points in the region of Crimea: absence of modern waste water treatment plants Balaklava, Evpatoria, Yalta, Gurzuf, Sevastopol;
- 1 point in the region of Kerch: ecologically dangerous enterprise Kamuchburunsk;
- 1 point in the region of Krasnoperekopsk: ecologically dangerous Krasnoperekopsky brome plant.

The reconstruction of the above designated plants gives appreciable result in the improvement of the Black Sea ecosystem (Fig. 3).

In 1998, on the basis of investigation in the Black Sea within the International program, a strategic plan of action was prepared and signed by the ministers of ecology of 6 countries. On this basis, each of the countries would prepare national plans of action for the improvement of ecological conditions.

Within the framework of the Ukrainian strategic plan of action "The Concept of the Protection and Rehabilitation of the Environment of the Azov and Black Seas" was prepared. And already by the end of 1999, a Ukraine State program for protection and restoration of the Azov and Black Seas was prepared and coordinated with the Cabinet of Ministers.

Thus a scientific-legal base for implementation of measures is now created which gives the impetus for carrying out the basic measures for the improvement of the Black Sea system.

The analysis of the existing legal base and investigations carried out within the framework of the International programs, show that the priorities for the revival of the Black Sea ecosystem essentially have changed. The data of UkrSCES confirm this. For more precise analysis of the ecological condition of the Black Sea, it is necessary conditionally to divide water areas at some levels, in which various mechanisms of

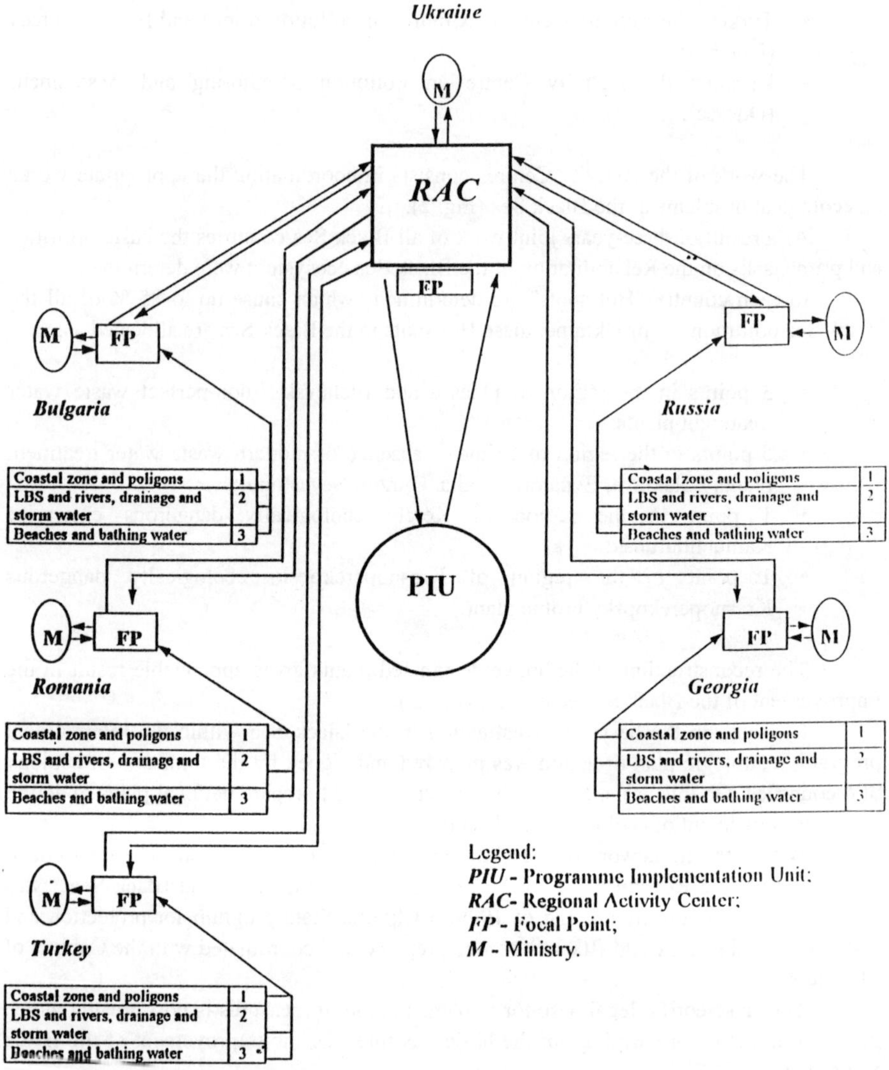

Fig. 2. Flow Diagram of Proposed Data Management System.

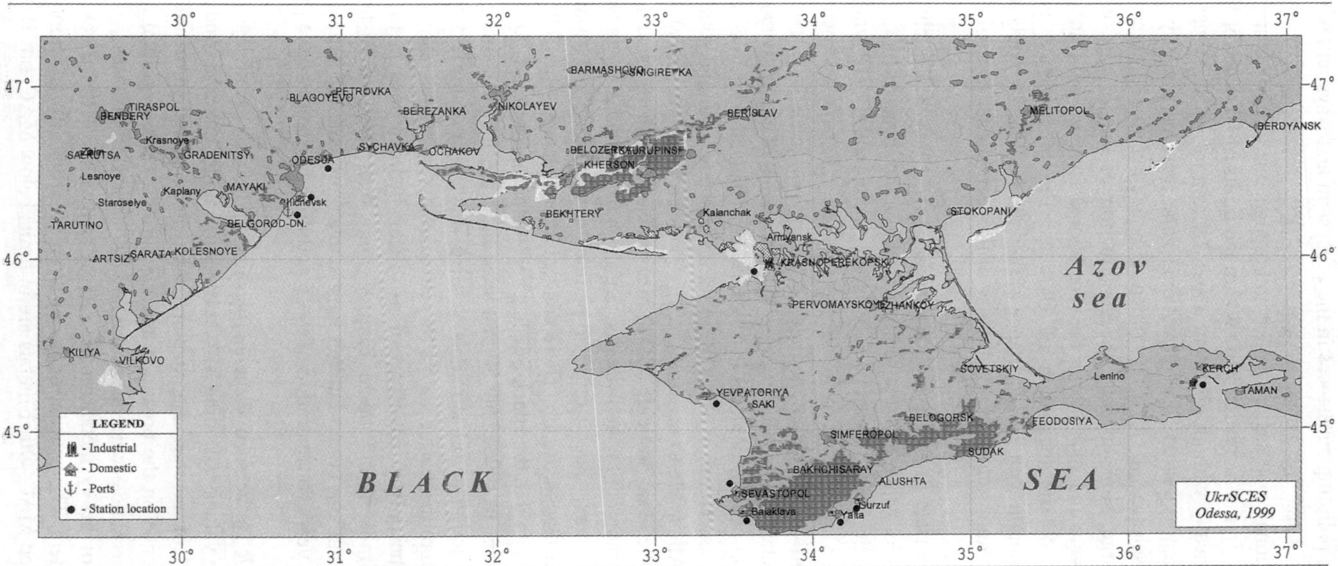

Fig. 3. Hot Spots in the Ukrainian Part of the Black Sea.

receipt of the basic polluting substances in the ecosystem and ways removing them from the ecosystem (Fig. 4 and Fig. 5).

The recreational zone has the largest anthropogenous influence for many reasons. Into the Black Sea (in the recreation zone within the limits of Ukraine for the last years) is discharged practically without treatment about 7.4 mln m^3 of waste water, about 195 mln m^3 not sufficiently treated. The recreational zone receives annually about 31 mln tons of suspended substances, etc. It is pertinent to notice that these figures do not reflect volumes of discharge, since recently the construction of sanatoriums, camping sites, places of public usage and such like in the recreational zone are built unsystematically, with infringement of the legislation of Ukraine. The phenomenon is even more aggravated in connection with the acceptance of the law about land privatization, whereas till now there is no norm-legal base on the use of the recreational zone of the Black and Azov seas.

The modern state of the recreational zone of the Black Sea is characterized by significant pollution of waters, bottom sediments and sand of beaches.

Chlore organic (DDT, HCCH), polychlorine bephynils (PCB) synthetic surface-active agents (SSAA), petroleum hydrocarbons (PH) polyaromatic hydrocarbons (PAH), most dangerous parts of petroleum, and also benz-a-pyren, phenols, dissolved organics, and some heavy metals are practically constant components of coastal waters and bottom sediments. The level of their contents in 1999 has not changed a lot in comparison with 1998. The average concentration of (COP) in waters of the recreational zone makes about 7 ng/l (Maximum Allowable Concentration (MAC) in sea water - absence). In percentage terms, amounts of the basic pesticide (DDT) in some cases reach 40 %, and that gives an indication of the duration of the half-life disintegration of pesticides in sea water.

In the last years, the quantity of petroleum was stabilized in waters of the recreational zone of Odessa region. There concentrations change from 0.00 mg/l (MAC 0.05 mg/l). But the fact that the Black Sea has become a transport corridor for petroleum transported by tankers, in 1998 about 50 mln tons, and plans for 2000 make 84 mln tons. Besides, the construction of oil terminals in all 6 Black Sea countries could result in significant pollution of the marine waters by petroleum hydrocarbons many times over. Concentration of the polyaromatic part of petroleum hydrocarbons in the recreational zone changes in limits from 5 up to 29 μg/l (more than MAC). The largest part of polyaromatic hydrocarbons comes on their most stable representative benz-a-pyren (up to 60 %). For the last years concentration of these polluting substances have practically not decreased.

The synthetic surface-active agents (washing-up liquids) are always present in the recreational zone in values exceeding MAC (100 μg/l), up to 250 μg/l. And recently with the huge quantity of foreign manufactured, physic-chemical properties that have appeared, the influence on the organism and the period of disintegration is not known. The circumstances assume occurrence of unknown allergic skin diseases.

The traces of heavy metals in the recreational zone of the Black Sea are met practically everywhere. The concentrations of arsenic, chrom, litium, strontium, mercury in some cases exceed MAC. Other metals are in limits below MAC, but 10 times exceed

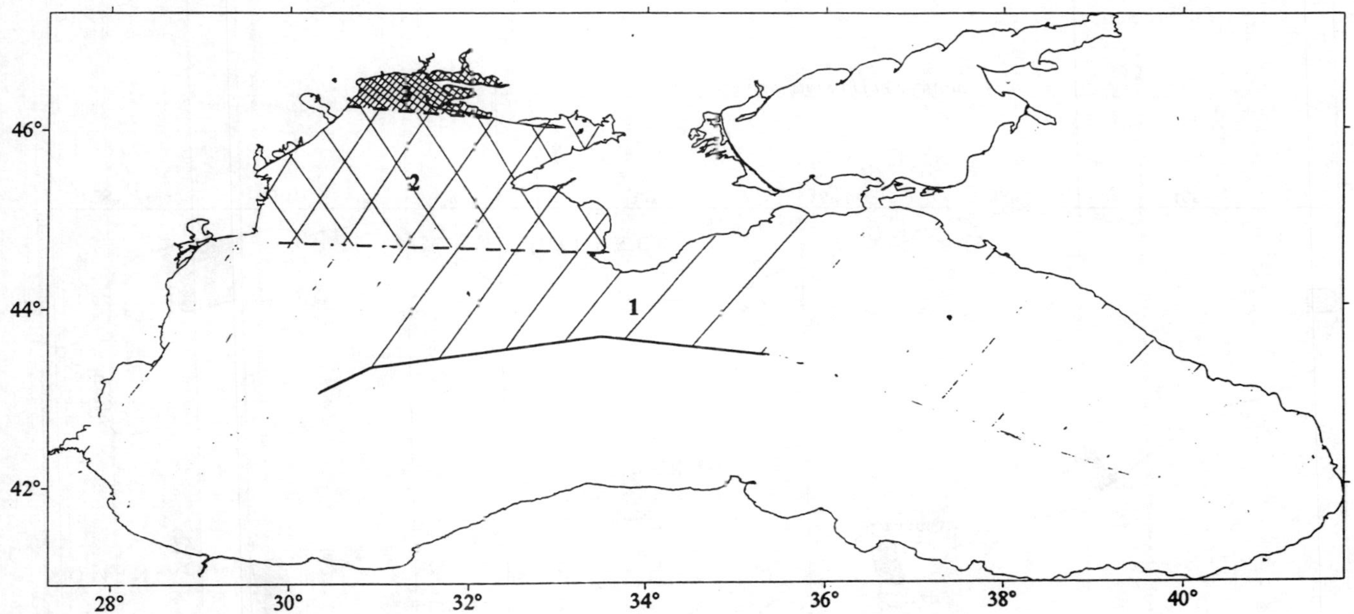

Fig. 4. Levels of Monitoring.
1-Background
2-Regional
3-Lical

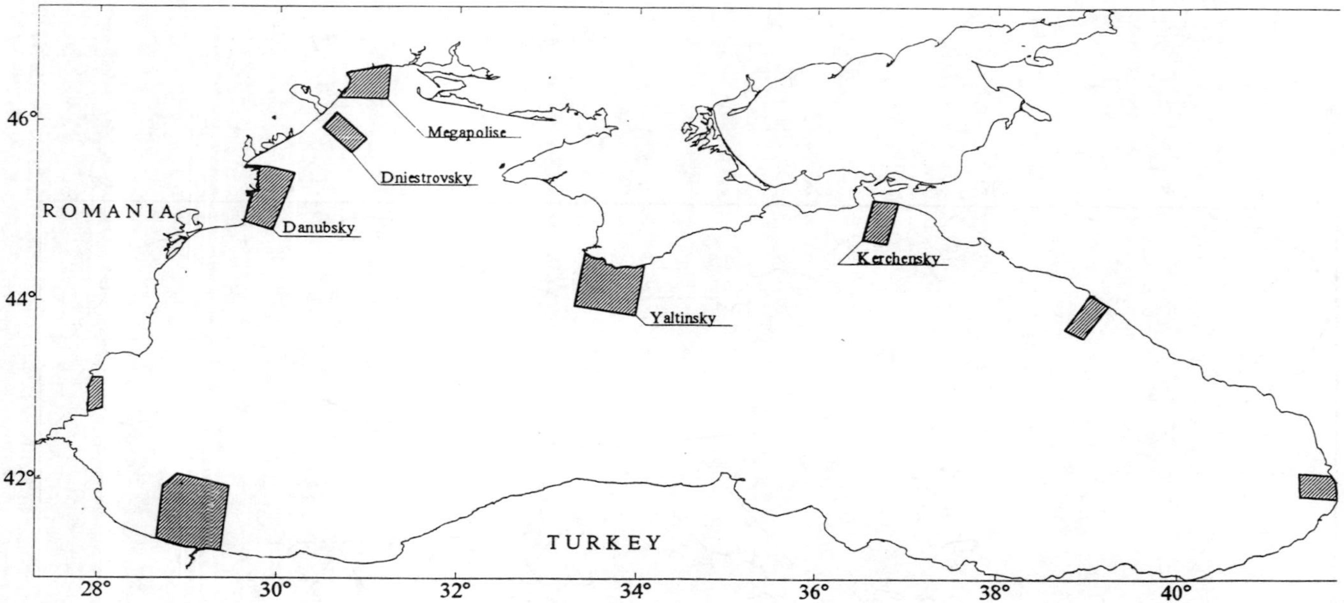

Fig. 5. The Black Sea Monitoring Stations.

the natural contents in the sea environment. In bottom sediments their significant concentration takes place.

Polychlorinebephynils (waste materials of paint industry etc.) in the recreational zone are met everywhere in significant concentrations (more than 25 ng/l), the MAC makes 0 ng/l. This indicates the chronic pollution of the recreational zone by this dangerous substance.

In the water of the recreational zone, dissolved organic substance in huge quantity is present. It proves to be true by high values of oxidizability, exceeding 5 mg O2/l.

Also significant concentrations of phosphorus, nitrogen in the recreational zone, results, at the end to the reduction of oxygen dissolved in water, up to values at which the extensive zones of the dead phenomena and occurrence of hydrogen sulphide are observed.

Thus the recreational zone of the Northwestern part of the Black Sea within the limits of the Odessa region, is in the crisis condition, despite of, that many enterprises being potential polluters, work not in complete capacity (Fig. 6 and Fig. 7).

The Odessa Bay and shelf zone of the sea are also considerably polluted. Practically average concentrations of the basic polluting substances essentially do not differ from the pollution in the shelf zone and Odessa Buy.

In the recreational zone phenols are met everywhere, and the concentrations exceed the MAC more than 20 times.

The shelf zone is also practically polluted by petroleum in concentrations that in some cases exceed the MAC. The significant concentrations are in bottom sediments. Average concentrations of polyaromatic hydrocarbons decrease a little bit.

Heavy metals are met in waters of the shelf zone of the Black Sea in very small quantities.

The significant concentrations of organic substances and biogenic elements PN are met in all areas of the shelf zone. Their significant concentrations cause the phenomenon of eutrophication in the North-Western part of the Black Sea. In 1999, the territory covered by eutrophication in the North-Western part of the Black Sea occupied about 40 %, and brought irreparable harm to biological stocks of the Black Sea, as well as recreational stocks (Fig. 8).

In all regions of the Odessa Buy there is a layer of silt at the bottom, which in some cases exceeds 3 cm. This phenomenon has been observed over the last 10 years and practically destroys all life living at the bottom in our region.

The analysis convincingly specifies the degradation of the Black Sea ecosystem. Despite the reduction of industrial wastes, as the quantity of household discharges and organics are constantly increased, it causes irreparable damage to the ecosystem.

In this connection the question of protection and rational use of resources of the Black Sea should be constantly in the field of attention of all states of the coastal zone, and in the Odessa region, of the services that are responsible for this.

Unfortunately, in the sphere of natural usage in the Black Sea, in the past there was no separate file for the ecology-economic requirements; standards; specifications; norm legal base regulating the economic activity in sea water areas and international

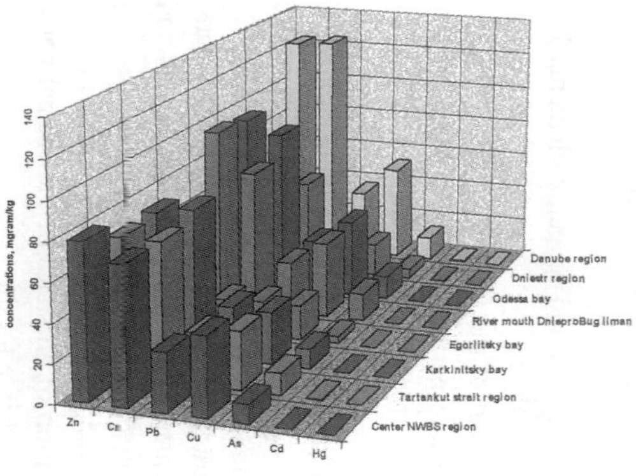

Distribution of toxic metals in bottom sediments

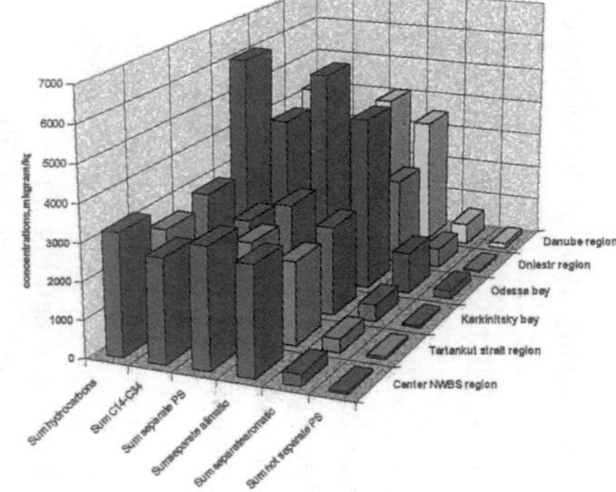

Distribution of oil hydrocarbons in bottom sediments

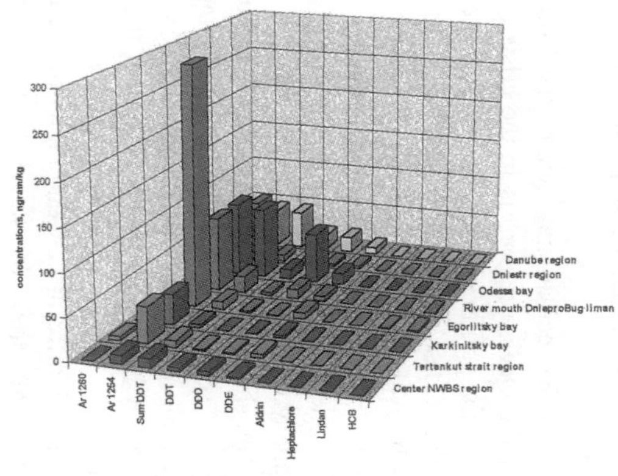

Fig. 6. Distribution of Chlorinated Hydrocarbons in Bottom Sediments North-West Part of the Black Sea (1999).

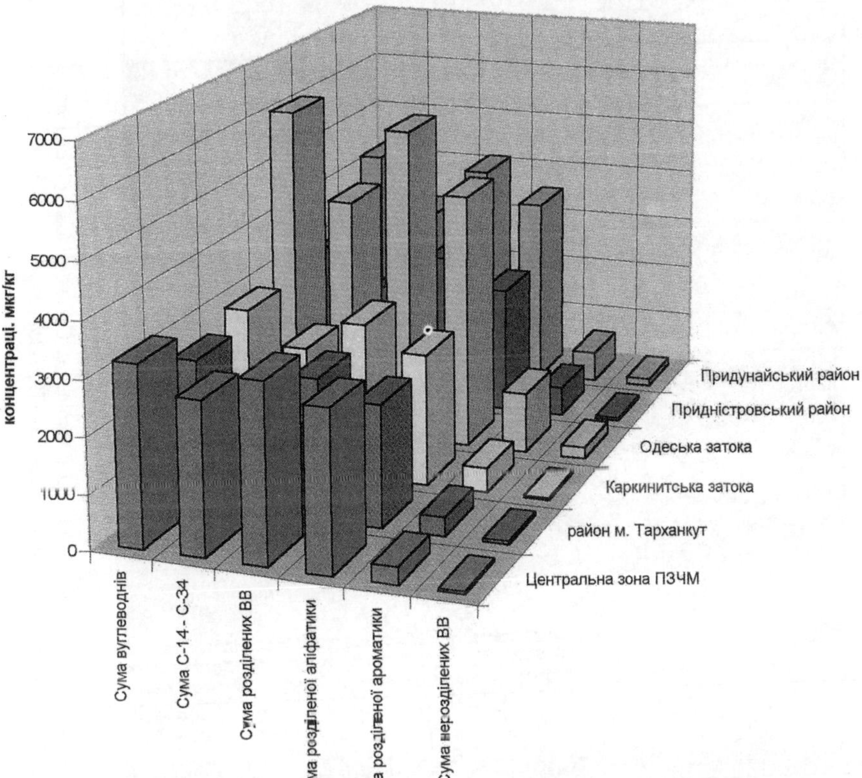

Fig. 7. Distribution of Petroleum Hydrocarbons in Sediment of NWBS (1999p).

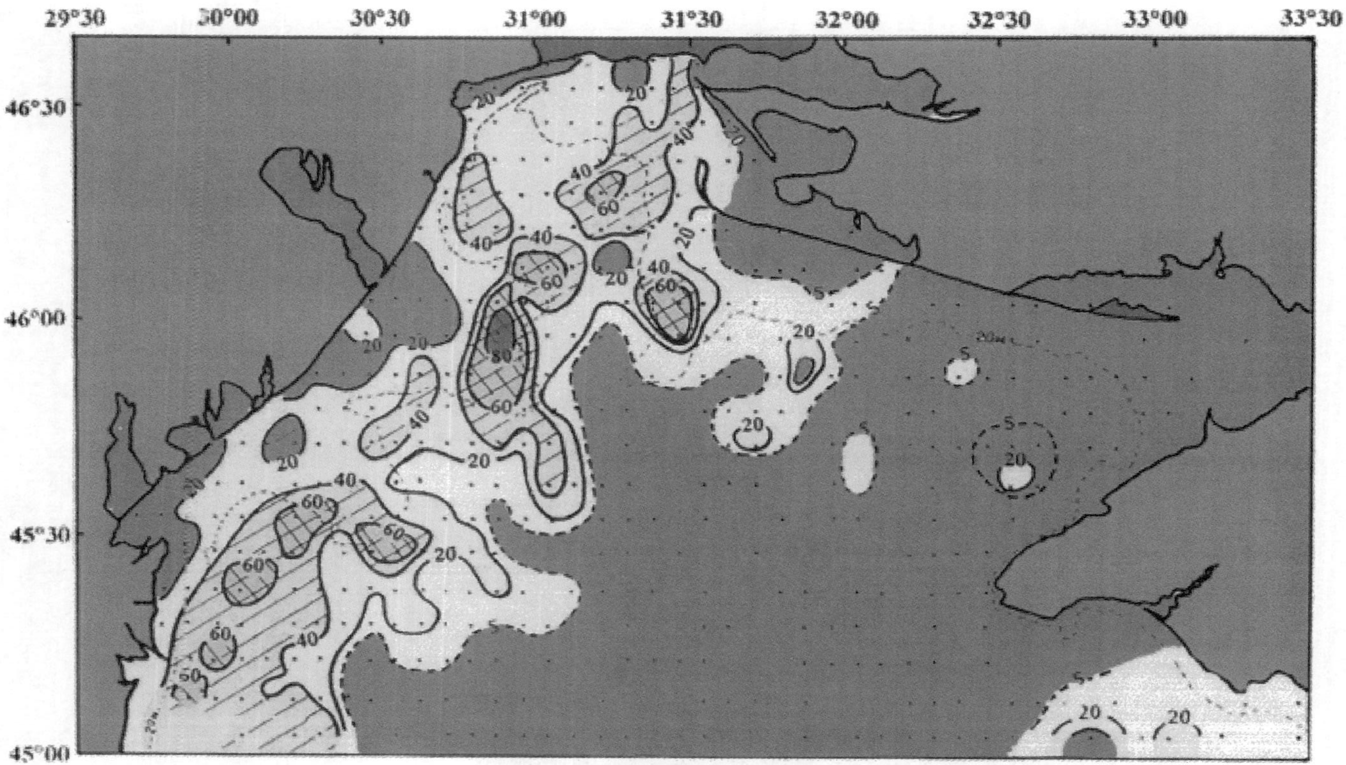

Fig. 8. Frequency of Revealing Hypoxia Phenomenon Cases in Marine Water of the North-Western Part of the Black Sea (1971-1988).

rivers, ensuring rational use of natural sea and river environments in view of the requirements for the protection of the natural environment. An example of this are the infringements from the Romanian side by dumps into the river Danube, as there are no rules of law in Romania for pollution of this kind.

In the Ukraine, the first stage of the legal reform in the sphere of natural usage is completed, that is, confirmed by the Law of Ukraine on Protection of the Natural Environment, Water Code, Law on the State Ecological Expertise of Ukraine. By these documents, the main strategic aim of Ukraine in the protection of the natural environment is:

- Maintenance of ecological safety for the present and future generations;
- Updating and protecting the biodiversity of balance (at local, regional and global levels);
- Rational and complex use of all of the natural environment potential of Ukraine;
- Consecutive decision of problems of development of the economy of Ukraine on the way of achieving complete biodiversity.

In this connection, many ecological tasks have been put before the Ukraine government:

- Improvement of the ecological state of the Dnieper basin and the quality of drinking water;
- Termination of pollution of the Black and Azov seas and improvement of their ecological condition.

At the present stage of socio economic development, conditions and preconditions for a concrete definition of an ecological policy of the states, expansion of the economic methods and ecology-economic specifications in the regulation sea natural usage are already formed.

It predetermines the necessity for the formation of a qualitatively new ecology-economic base and the legal normative base of sea natural usage and decision of problems of prevention of an ecology-economic crisis in the Black and Azov Sea basin. The basic methods of development and the formation of the normative-methodical base of sea natural usage in Ukraine are:

- Classification of natural resources and quality of the sea environment with orientation to international standards. Creation of environment-resource quality of recreational zones of Ukraine and, further, the whole Black Sea;
- Standardization of the system of parameters of monitoring investigation at the international level;
- Perfection of methodical and normative base of ecological regulation of natural usage in the coastal territories and sea water areas;

- Development of scientific-proved criterias for definition of economic and ecological priorities in the sphere of economic activity (various patterns of ownership) on water areas and adjacent territories;
- Development of theoretical and methodical base of formation of the system of payments for deterioration of the quality of natural environment;
- Perfection of theoretical and methodical base formation of payment for pollution of the sea water areas in the view of their legal status, kinds of sources of pollution, specificity of resource of ecological potential of water areas;
- Perfection of the system of payments for uses of sea resources and sea water areas;
- Development of the base of the financial-credit relations in the field of protection and restoration of the Black and Azov seas;
- Development and acceptance of laws, legislative instructions in the sphere of sea natural usage and protection of sea environment.

Thus, one of the first priorities should be:

- Recreational facilities of the Black-Azov seas - food cycles of agriculture, social-ecological sphere of service.

The second group of priorities should be formed by a group of Marine facility cycles (foreign trade, oceanic fishery etc.).

The third group of priorities should contain scientific technology (electronics, instrument making, which exclude discharge of pollution).

The urgent measures concern:

- Inventory and ecological certification of all enterprises and ecological dangerous objects, territories, water areas, regions, cities, adjacent states of the Black-Azov sea basin;
- Development and signing of an International convention on economic zones in the Black and Azov seas;
- Preparation of the Law of the coastal and recreational zone of the sea;
- Ecological norms of the economic activity taking into account the assimilation capacity of separate regions, first of all the zones of recreation, reserved territories and the whole sea by chemical toxic parameters;
- Introduction of a special mode of natural usage within the limits of the 3 kilometer recreational zone of the sea and coast with obligatory state control;
- Bringing the quality of the Black Sea water up to the requirements of international standards.

THE SUBOXIC ZONE OF THE BLACK SEA

Ö. BASTURK
Institute of Sciences, Mersin University, Icel, Turkey
S. TUGRUL AND I. SALIHOGLU
METU Institute of Marine Sciences, P.O. Box 28, Erdemli, 33731 Icel, Turkey

ABSTRACT

The suboxic zone in the Black Sea can be defined as the layer where oxygen and hydrogen sulfide concentrations reach extremely low values, i.e. D.O.<10 ..M and H2S < 10 nM, and do not show any density gradient. The suboxic zone in the Black Sea is a uniquely well defined site for studying oxidation-reduction reactions that are important for suboxic regions throughout the ocean basins and sediments. Using primarily data we present four hypotheses for further study. These hypotheses are; (i) the upward flux of sulfide is oxidized by Mn (III, IV) and Fe(III) species, (ii) Mn species act as a catalyst in which the downward flux of nitrate is reduced by Mn(II) and the upward flux of NH_4^+ is oxidized by Mn (III, IV) species, (iii) sulfide is oxidized anaerobically in association with phototrophic reduction of CO_2 to organic carbon (this hypothesis is an alternative to hypothesis (i)), and (iv) detailed alkalinity profiles can constrain the stoichiometry of suboxic reactions.

INTRODUCTION

The layer of coexistence (C-layer) in which DO and H_2S, suggested to coexist (Sorokin, 1983; Vinogradov and Nalbandov, 1990; Faschcuk and Ayzatullin, 1986) was shown to be analytical artefacts in sampling and analysis procedures (Murray et al. 1989; Tugrul et al. 1992; Bastürk et al, 1994), and has been proven to be unrealistic when one considers the rapid dynamics of the reaction between them (Murray et al. 1995; Millero et al 1991; Gökmen 1996). At least 10 times faster rates of oxidation of H_2S in deep Black Sea waters compared to surface waters, with added H_2S, and the similarities between the rates of oxidation of filtered and unfiltered samples under oxic (150-200 μM DO) (Millero et al. 1991) and suboxic oxygen levels (DO<50 μM)(Gökmen,1997) implied that DO is not a direct oxidant for H_2S, but probably through the coupling with $Mn\text{-}O_2$ and $Mn\text{-}NO_3$ cycles between them. Tebo (1991) measured nearly two orders of magnitude higher Mn(II) residence time in the SOZ of the central parts of the basin (30-90 days) compared to that at the near-shore waters (0.6-1.0 day).

Worth mentioning is the observation of the intense light transmission minimum layer, called the Fine Particle Layer (FPL), not close to the lower boundary of the SOZ, but at deeper density surfaces down to 16.6 surface. At the Sta. L29M46 (Fig. 1), the onsets of NH_4 and Mn(II) start at the same density surface (15.85) and have similar profiles. On the other hand, dissolved oxygen concentration was measured down to 16.4 density surface, below which sulfidic layer has started in contrast to its frequently observed basin-wide positions (DO<10 uM at σ_θ =15.4-15.6; H_2S> 1 uM at σ_θ =16.2 surface). About 20 μM DO concentration was measured at 16.2 density surface which is commonly accepted as the layer of H_2S onset. The well established SOZ layer is very narrow at this station compared to previous boundaries (Murray et al. 1995)($\Delta\sigma_\theta$ = 50±.10). Even though the upper regions of the sulfidic layer were eroded and thus oxygenated down to the 16.4 density surface, oxic and anoxic layers still do not overlap with each other. In other words, C-layer does not exist even under intense mixing conditions.

At the Sta. L34M46, which is about 5 nm north of the Sta. L29M46, the FPL starts to be sliced intermittently by horizontal intrusions of sulfidic water masses from the surrounding in the offshore direction (Fig. 2). There is no single and broad FPL as was observed at coastal station, but rather a series of minima and maxima. Additionally, lower boundary of the FPL further deepens from 16.4 surface down to 16.7 density surface. A pair of peaks at 16.35 and 16.40 surface are followed by a particle-free sulfidic layer, and then another FPL peak at around 16.5-16.6 surfaces. When the vertical distributions of DO and H_2S are examined at this station (Fig. 2), it will be recognised that both properties follow each other in a sequential order, but never overlapping each other. When there is a DO peak, H_2S concentration drops to below the detection limits or vise versa. In contrast to the previous DO profile where water column down 16.4 surface is well oxygenated, DO concentration drops to about 10 uM at 16.3 surface where H_2S concentration is at detectable level. On the contrary, 50 uM DO concentration was detected at around 16.4 surface, followed again by a sulfidic zone. At the interfaces of these oxic/anoxic zones no overlapping of the layers are detectable.

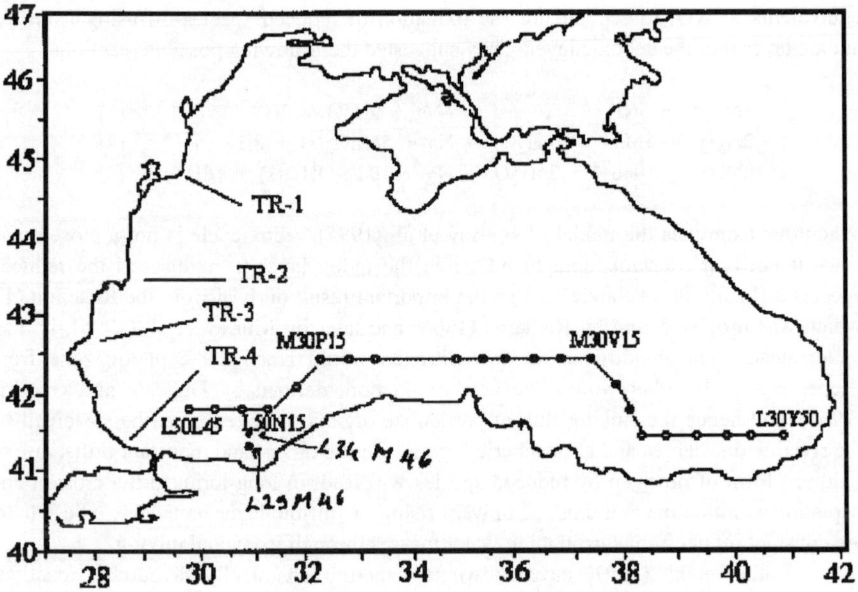

Fig. 1. Locations of the sampling stations in the Black Sea.

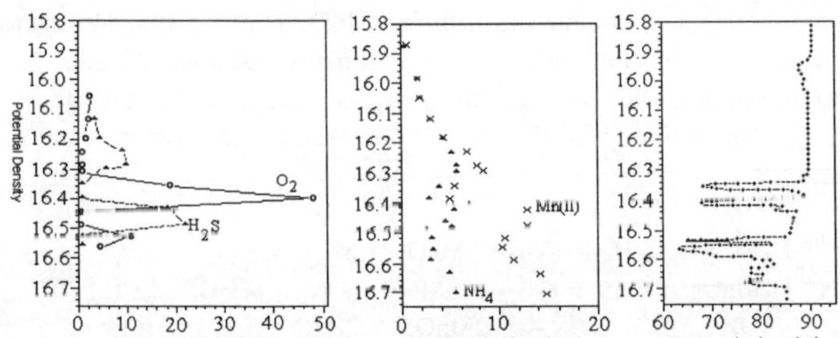

Fig. 2. Anomalous distributions of biochemical parameters within deep fine particle layer observed at St. L34M46 in the Sakarya Canyon region (R/V Bilim July,1997 cruise) (redrawn from Bastürk et al. 1997).

Since NO_3, NH_4, Mn(II) and Fe(II) all decrease to low concentrations at about the same density levels (σ_θ =15.95, 15.95, 15.85 and 16.00, respectively) and equivalence of their electron gradients (downward for NO_3; and upward for the sum of the Mn (II)+Fe(II)+NH_4) have led Murray et al. (1995) to hypothesise that there are sufficient equivalents of NO_3 to account for the oxidation of reduced species diffusing from the anoxic layer into the suboxic layer. They suggested the following possible reactions;

$$3NO_3^- + 5NH_4^+ = 4N_2 + 9H_2O + 2H^+ \quad (1)$$
$$2NO_3^- + 5Mn^{2+} + 4H_2O = N_2 + 5MnO_2(s) + 8H^+ \quad (2)$$
$$2NO_3^- + 10Fe^{2+} + 24H_2O = N_2 + 10\,Fe(OH)_3(s) + 18H^+ \quad (3)$$

When one examines the model of Murray et al. (1995), redox cycle is not a closed one, since it needs a constant input of NO_3 into the redox layer to oxidise all the reduced species diffusing into suboxic zone. One important result derived from the Reaction (1), which was proposed first by Richards (1965) and later by Murray et al. (1995), is that, under steady-state conditions, no new nitrogen would reach the euphotic zone from deeper layer. In other words, the new production, defined by Dugdale and Goering (1967), and hence the sinking flux of particulate organic nitrogen, must be sustained by the riverine discharges and atmospheric inputs. On the other hand, constant utilisation of oxidised form of nitrogen by reduced species will lead, in long-term, to the erosion and depletion of nitracline and thus the upward rising of sulfidic layer boundary, unless there are constant inputs from surrounding water masses through isopycnal mixing.

Luther et al. (1997) have shown two thermodynamically favourable reactions between manganese and nitrogen species: the first one is the reduction of NO_3^- to N_2 by dissolved manganese which is converted to particulate Mn-oxides (Rxn 4), and the second one is the oxidation of Mn(II) in the presence of O_2 to $MnO_2(s)$ (Rxn 6) which oxidizes NH_4^+ and organic nitrogen to N_2 (Rxn 5). Mass-balance calculations show that the oxidation of NH_4 and organic-nitrogen by MnO_2 may be the dominant process producing N_2 in Mn-rich continental margin sediments. Above pH 6.8, the reaction between Fe(III) species and NH_4^+ to form N_2 is shown to be thermodynamically unfavourable (Luther et al. 1997). Rozanov (1996) stated that oxygen is the oxidant of ammonium nitrogen.

$$5Mn^{2+} + 2NO_3^- + 4H_2O \text{ -------> } 5MnO_2 + N_2 + 8H^+ \quad (4)$$
$$2NH_3 + 2MnO_2^- + 6H^+ \text{ -------> } 3Mn^{2+} + N_2 + 6H_2O \quad (5)$$
$$2Mn^{2+} + 4OH^- + O_2 \text{ ------> } 2MnO_2 + 2H_2O \quad (6)$$

A NO_2 maximum (σ_t = 15.85) almost always coincides with the zone of denitrification. It also coincides with the zone of Particulate Mn maximum (σ_t = 15.85)(Murray et al. 95). Coincidence of Mn-Oxide maximum surface with that of NO_2, and positioning of particulate manganese layer between the layer of NO_3 minimum, suboxic DO zone, and particulate Fe oxides suggest that particulate manganese oxide couples the redox processes in the upper layer of the suboxic zone with those in the lower layer by carrying the oxidation potential of nitrate and oxygen to the lower section of SOZ where ammonia is oxidized to N_2 and/or to NO2 plus reduces NO3 to NO2 and/or N2 while itself is oxidized. On the other hand, it is shown that both H2S and Fe(II) reduce MnO2 rapidly (Burdige and Nealson 1986; Lowely and Phillips, 1988)

$$MnO_2 + 2Fe^{2+} + 2H_2O === Mn^{2+} + 2FeOOH + 2H^+ \quad (7)$$

Thus, the oxidized iron readily oxidizes H2S to elemental sulfur.

$$2FeOOH + H_2S ==== 2Fe^{2+} + S^o + 4OH^- \quad (8)$$

Field studies and surveys suggested that NH_4 is oxidized basically by the particulate $Mn0_2$, probably, to N_2 (Bastürk et al. 1997). Observed decrease in the Mn(II) concentration within the oxygenated water layers (see Fig. 2) is considered to be due to the oxidation of Mn(II) back to $Mn0_2$ (s), which in turn oxidizes NH_4+ to N_2, and indirectly the H_2S through redox coupling with iron oxidation. The possible reaction schemes were suggested as below; Bastürk et al (1997)

$$4Mn(II) + 2\,O_2 + 4\,H_2O = 4MnO_2\,(s) + 8\,H^+ \quad (9)$$
$$3Mn0_2(s) + NH_4^+ + 4H^+ = 3\,Mn^{2+} + N0_2^- + 4H_20 \quad (10) \text{ and/or}$$
$$3Mn0_2\,(s) + 2NH_4^+ + 4H^+ = 3Mn^{2+} + N_2 + 6H_20 \quad (11)$$
$$MnO_2 + 2Fe^{2+} + 2H_2O = Mn^{2+} + 2FeOOH + 2H^+. \quad (12)$$
$$2FeOOH + H_2S = 2Fe^{2+} + S^o + 4OH^- \quad (13)$$

As reviewed in the scientific background, the distributions of O2, NO3, NO2, NH4, H2S, Mn(II), Fe(II) and particulate forms of these metals play crucial, but not well defined roles in the suboxic zone redox processes. Therefore, it is critically and vitally important to test some of these hypothesis by performing in situ and lab experiments. More importantly, variations in the kinetic rates of these biochemical redox processes, and their coupling with each other need a detailed field study and experiments.

Field studies and simulated lab experiments which will serve as supplementary data for the above mentioned redox processes should also be done. These studies are summarised below:

- Vertical speciation of redox sensitive metals (mainly Iron and Manganese) in terms of their oxides and sulfide complexes, and x-ray and crystallographic analysis of metal particles;
- Ionic forms of iodine (I^{-1}, I_2 and IO_3^{-1}) within the water column (from surface down to upper layers of anoxic zone) and their role in redox chemistry of above mentioned metal ions.
- Sulfur speciation within the sub-oxic layer of the Black Sea and their relative concentrations (such as elemental sulfur, H_2S, HS^{-1}, $S_2O_3^=$, poly sulfides and thiols).

REFERENCES

Bastürk ,Ö., C. Saydam, I. Salihoglu, L.V. Eremeev, S. Konovalov, A. Stoyanov, A. Dimitrov, A. Cociasu, L. Dorogan and M. Altabet. (1994): Vertical variations in the principle chemical properties of the Black Sea in the autumn of 1991. *Marine Chemistry, 45: 149-165.*

Bastürk, Ö., I.I. Volkov, S. Gökmen, H. Güngör, A.S. Romanov, and E.V. Yakushev (1997): International Expedition on Board R/V Bilim in July 1991 in the Black Sea. *Okenologia, 6: 1997 (in press, in Russian)*

Burdige, D.J. and Nealson, K.H. (1986): Chemical and microbiological studies of sulfide-mediated manganese reduction. *Geomicrobiol. J. 4:361-387.*

Fashchuk, D.Ya. and T.A. Ayzatullin (1986): A possible transformation of the anaerobic zone of the Black Sea. *Oceanology, 26(2):171-173.*

Gökmen, S. (1996). A comparative study for the determination of hydrogen sulfide in the suboxic zone of the Black Sea. Ms. Thesis, Inst. of Marine Science, Erdemli-Icel, Turkey, 156 pp.

Gökmen, S. and Ö. Bastürk (1997): Some remarks on the H_2S removal rates within the suboxic zone of dynamically different regions of the Black Sea. NATO TU-Black Sea Project: Symposium on Scientific Results, Extended Abstracts, p:64. 15-19 June, 1997, Crimea-Ukraine.

Lovely, D.R. and Phillips, E.J.P. (1988): Manganese inhibition of microbial reduction in anaerobic sediments. *Geomicrobiol. J. 6:145-155.*

Luther, G.W.III., T.M. Church and D. Powell (1991): Sulfur speciation and sulfide oxidation in the water column of the Black Sea. *Deep-Sea Res. 38(2A):1121-1137.*

Luther, G.W., B. Sundby, B.L. Lewis, P.J. Brendel and N. Silverberg (1997): Interactions of manganese with the nitrogen cycle: Alternative pathways to dinitrogen. *Geochim. Cosmochim. Acta. 61(19):4043-4052.*

Millero, F.J., S. Hubinger, M. Fernandez and S. Garnett (1987): Oxidation of H_2S in seawater as a function of temperature, pH and ionic strength. *Jr. Environ. Sci. and Tech. 21:439-443.*

Millero, F.J. (1991): The oxidation of H_2S with O_2 in the Black Sea. IN: ***Black Sea Oceanography,*** E. zdar and J.W. Murray (eds), NATO ASI Series C: Volume 351. pp:205-227.

Murray, J.M., H.W. Jannasch, S. Honjo, R.F. Anderson, W.S. Reeburgh, Z. Top, G.E. Friderich, L.A. Codispoti and E. zdar (1989): Unexpected changes in the oxic/anoxic interface in the Black Sea. *Nature, 338: 411-413.*

Murray, J.M., L.A. Codispoti and G.E. Freiderich (1995): Oxidation-reduction Environments: The suboxic zone in the Black Sea. IN: ***Aquatic Chemistry,*** C.P.Huang, C.R.O'Melia and J.J.Morgan (eds), ACS Advances in Chemistry Series No: 244, pp:157-176.

Romanov, A., Ö. Bastürk, S. Konovalov and S. Gökmen. (1997): A comparative study of Spectrophotometric and Iodometric Back Titration methods for hydrogen sulfide determination in anoxic Black Sea waters. NATO TU-Black Sea Project: Symposium on Scientific Results, Extended Abstracts, p:67. 15-19 June, 1997, Crimea-Ukraine.

Rozanov, A.G.,(1996): Redox stratification in Black Sea waters. *Oceanology, 35(4):500-505.*

Tebo, B.M. (1991): Manganese(II) oxidation in the suboxic zone of the Black Sea. *Deep-Sea Res., 38(2A):s883-s905.*

Tugrul, S., Ö. Bastürk, C. Saydam and A. Y lmaz (1992): Changes in the hydrochemistry of the Black Sea inferred from water density profiles, *Nature, 359: 137-139.*

Vinogradov, M.Ye., and Yu.R. Nalbandov (1990): Effects of changes in water density on the profiles of physicochemical and biological characteristics in the pelagic ecosystem of the Black Sea. *Oceanology, 30:567-573.*

BUILDING ENVIRONMENTAL COALITIONS AND THE BLACK SEA ENVIRONMENTAL INITIATIVE

MARIAN KAY THOMPSON
United States Department of Energy

Good morning ladies and gentlemen, I want to thank Professor Zichichi and the World Federation of Scientists for inviting me to speak to you today. I will talk about building coalitions to support environmental stewardship and the Department of Energy's Black Sea Environmental Initiative.

Most people around the world share a concern about the environment. We all want to live on a planet that is free of man-made environmental hazards. We all want to maintain the quality of our air, forests and seas. There is growing concern about the capacity of the environment to provide the goods and services we cannot live without and our role in degrading the ecological systems that we cannot live without. In 1991, the U.S. President's Council on Environmental Quality issued a document called the 'Global 2000 Report.' It offered recommendations on a wide range of issues, including population, food and agriculture, renewable energy resources and increased energy efficiency, biological diversity, coastal and marine resources, water, global pollutants, development assistance, and institutional changes within the U.S. Government. That list is not unlike the WFS's list of global problems.

At the same time, there are competing demands for the resources of society, and legitimate concerns about the rising costs of laws, policies, and regulations designed to minimize environmental damage. We frequently hear that there is competition between the goals of economic development and environmental protection. Having said that, I go back to my previous comment and reiterate that we all share both concern about and commitment to our environment. Politicians, non-governmental environmental organizations, and a growing number of citizens are seeking solutions that provide possibilities for 'sustainable development.'

Those who believe that environmental progress is as essential as economic progress recognize that making changes in policies requires an approach that does not polarize the political landscape. The challenge for the United States and for all nations is to protect and restore the natural environment while providing for the economic needs of our citizens. This requires that society find better ways of reducing the environmental impact of the day-to-day decisions of billions of people—consumers, industrial workers and managers, fishermen, and farmers—this requires that we learn to build consensus.

And this brings me to the issue that is the primary focus of my presentation—building consensus for environmental action and the U.S. Department of Energy's Black Sea Environmental Initiative. Governments cannot monitor and police the billions of daily decisions of its citizens. Effective environmental policies require 'buy-in.'

Environmental policies that are successful over time require the participation, collaboration, and cooperation of:

- Policy makers and administrators in governmental agencies, including federal, state and municipal government organizations;
- Non-governmental groups, and community organizers;
- Manufacturing, commercial industrial, agricultural, transportation and residential sectors;
- Financial institutions;
- The media;
- Citizens likely to be affected by the policies adopted;
- And, even school children, because, they are the decision makers of tomorrow.

Bringing these groups together to make collective decisions is not enough. These groups must have good information on which to base their discussion and to help build consensus. Science and scientists have a significant role to play in this process and I will return to this point later.

This brings us to the Department of Energy's Black Sea Environmental Initiative. The Black Sea is a very unique ecosystem. It has not fared well at the hand of man over the last decades. And new challenges to the sea begin as the large reserves of hydrocarbon deposits from central Asia are transported to world markets across, underneath and around the sea.

As we looked at the work being done within the Black Sea region to encourage environmental stewardship, and important work is being done, and at the other groups that are involved, including the world bank and the international maritime organization, we tried to identify specific areas where we could add value, but not duplicate. We decided to work to:

- Mobilize additional resources to assist the region to prepare for the flow of oil and gas that will expand dramatically over the next several decades;
- And to try and provide information resources to inform all of the different constituency groups I listed about what is going in and around the sea.

With the help of other U.S. government agencies and private sector companies, we began a series of information exchange forums (seminars and workshops) to address issues associated with oil spill response contingency planning.

At the same time we established a web site called the Black Sea environmental information center. When we first began the web site it was focused on oil spill response issues, policies, and technical information, but we very quickly learned that there were other important constituencies who wanted to be part of this effort and we broadened the web site to include information on:

- Oil spill response contingency plans, plus
- Other environmental laws and policies which governments in the region might want to share with the rest of the world; and,
- A database of information on pollution testing from scientific research facilities in Black Sea countries;

We also included a communications component in the web site. We have already conducted several meetings on the web site's chat room, and have plans to conduct the first on-line training session within the next several months on water modeling to predict oil flows. The web site has a place for scientists to post proposals for joint research. Private companies with technologies appropriate to Black Sea environmental problems may also post information on their capabilities and how they may be contacted.

Within the next few months, the web site will be expanded to include information on all the existing petroleum pipelines and proposed additions to the petroleum transportation network surrounding the black sea.

The web site also provides links to other sites, including sites within the region, to increase communication and the flow of information.

The web site has been successful beyond the most ambitious goals we had when we started. We are close to having 100,000 visitors to this web site. That is a remarkable demonstration of interest in the Black Sea and the environmental challenges faced by the countries who share its shores.

I want to return briefly to a point I made earlier—the importance of science in environmental decision making and the development of solutions to environmental problems. Without sound science to explain changes in our environment, we cannot hope to develop policies and programs to address those changes effectively. Without good science, we are making policies in the dark. Scientific research is also critical to developing new technologies to address problems. Even thought there is general recognition of this fact, research laboratories around the world are being threatened by a lack of adequate resources.

The Department of Energy is very pleased to have joined with the World Federation of Scientists to work on putting together the web site data base reporting the pollution testing results from Black Sea marine science research facilities. We are close to completing entering 30 years of pollution testing data from the center of sea ecology in Ukraine. Thanks to the U.S. Department of Defense we have been able to provide computers for this project. We hope to begin entering data from Romanian marine research facilities in January. We are also working with Dr. Ragaini and Professor Martellucci to prepare a proposal that we hope will attract funding to allow the Black Sea research institutes to conduct a three year benchmark study to bring their historical pollution testing data up-to-date. We have recently expanded the web site to provide an opportunity for scientists to post research papers so that they can reach a broader audience. We hope to post our first research papers in January.

This is an exciting project for the Department of Energy. Within the U.S. government it has attracted support from agencies that are known for supporting this kind of project like the United States Agency for Economic Development, but it has also attracted support from the U.S. Navy and the Department of Defense's Partnership for Peace Program, the Department of State, the Department of Interior, and the U.S. Coast Guard, which is the responsible agency for oil spill response in U.S. territorial waters. We have also had strong support from the U.S. oil industry.

As this project proceeds, we want our joint efforts to encourage a stronger coalition in support of environmental stewardship in the Black Sea. We hope that more governments will use the web site to communicate their policies. We hope that the scientific community will use the chat room to share information, and conduct on-line meetings, and strengthen their own network for cooperation. We hope that all of the Black Sea research institutes will contribute to the data base to demonstrate their impressive capabilities to a worldwide audience. We invite NGOS to use the database to develop information that will make their organizations more effective. And soon, we hope to implement our first outreach to school children, when Oak Ridge National Laboratory organizes a joint environmental research project between a school in Oak Ridge, Tennessee and a school in Constanza, Romania in real time over the internet.

We presented the Black Sea environmental information center web site to the workshop on Black Sea pollution when it met earlier this week. The web master, Melissa Lapsa, is here with me and I believe she can be convinced to provide demonstrations to others who are interested in seeing it.

I will conclude by saying that it is a great pleasure for me to be involved with the remarkable people in the World Federation of Scientists, and especially the scientists who are part of the WFS's Subcommittee on Pollution. I want to especially thank Dr. Valeri Mikhailov whose dedication to his science, his institute and the Black Sea has inspired me to work to be part of the solution, not part of the problem.

Thank you very much.

6. AIDS — MOTHER–INFANT HIV TRANSMISSION

THE TRAGEDY OF THE MOTHER TO INFANT TRANSMISSION OF HIV IS PREVENTABLE

GUY DE THÉ
Institut Pasteur, 75015 Paris, France

Among the 15 Planetary Emergencies presented by Professor Antonio Zichichi, the HIV/AIDS epidemic is the most tragic in its human and socio-economical aspects.

Last year, our WFS workshop focused on the urgency to develop an AIDS vaccine adapted to the poorest populations, at highest risk to HIV infection in Africa and Asia. The Erice statement prepared at the end of last year's workshop was formally presented to the French press on December 1 (AIDS Day) at the foot of the Eiffel Tower in Paris, with the symbolic reunion of Robert C. Gallo, Luc Montagnier and William Makgoba (head of the Medical Research Council of South Africa). In fact, the Erice statement has become one part of the statement for action of the International AIDS Vaccine Initiative (IAVI) in New York. Different clinical trials in phases two and three are being implemented in Africa and Asia, aimed at testing the safety and efficacy of different vaccine preparations. As you all know, due to the high genetic mutation rates of HIV, the ideal AIDS vaccine preparation is not yet at hand. But progress is being made to try and develop vaccine preparations aimed at mounting an immune response directed to the less HIV variable regions.

In parallel to such efforts**, a new hope has emerged**. A virucidal drug, named Nevirapine, given in one dose only to pregnant mothers just before delivery, does decrease dramatically the viral load in the circulating white blood cells of the mother, thus nearly annihilating the risk of HIV transmission to the newborn. To assess the importance of such a development, one has to realise that 600,000 newborns are being infected every year by HIV, mostly from their mothers, the large majority being in Subsaharan Africa. Since the beginning of the AIDS epidemic, one can estimate that 4,5 millions infants have been infected by HIV and that 3 million have died, the survivors becoming orphans.

We felt that this new possibility of prevention merited to be discussed and evaluated in Erice by the World Federation of Scientists, within both the PMPs on AIDS and Infectious Diseases and that on Mother and Child Health.

Nathalie Charpak and myself therefore invited, on behalf of the WFS, participants of 7 developing countries (Togo, Uganda, Colombia, Brazil, Senegal, India, Indonesia) together with scientists from France, Sweden and United States.

The Workshop focused on two major issues:

- the feasibility of promoting Nevirapine interventions, in populations where the epidemic is most severe;
- discussion toward a consensus on the controversal issue of breastfeeding after treatment by Nevirapine.

Agreement was easily reached regarding the urgency of promoting, by all possible means, Nevirapine interventions in deliveries occuring in hospitals. This is feasible since the unit cost is only $4 per dose, and the pharmaceutical companies are proposing to provide Nevirapine free of charge to developing countries. It remains however that any antiretroviral interventions do necessitate an organization with proper counselling and follow-up of the mothers and of their newborns. The feasibility of proper antiretroviral treatment of the mother must be assessed and adapted to local conditions.

Why is breastfeeding controversal in such a case? There exists a sizable risk of transmitting the HIV by the breast milk, thus annihilating possibly the benefits of the prevention of viral infection at the time of delivery. But what to do when drinkable clean water is not available, as is the case in very large areas of rural Africa and South-East Asia ? While bottle feeding is the ideal solution in industrialised countries with proper drinkable water distribution, we must be very careful to assess the respective advantages and risks of bottle feeding versus breastfeeding in each area and cultural environment.

The group realised that we lack proper data evaluating the level of risk of transmission of HIV according to the duration of breastfeeding. It is known, for another retrovirus named HTLV (presented here two years ago), that the risk of transmission of the virus is very low during the first two to three months of lactation, rising sharply thereafter. The group therefore urged to conduct epidemiological studies aimed at assessing whether three months of breastfeeding, following Nevirapine, and followed by other antiretroviral therapy, could prevent to a large extent or eliminate the risk of transmission of HIV to the infant.

I shall present the results of this year's Erice workshop at the international meeting of the Institute of Human Virology in Baltimore in September (Organizer: Pr. Gallo). The Website of the International Network for Research on Mother and Child Health, set up by the French and Swedish Academies of Sciences, will be most instrumental in promoting discussions and implementations of Nevirapine interventions around the developing world. In parallel it will stimulate collaborative interventions between African scientists and physicians.

Next year, we plan to have an Erice Workshop focusing on the progress made since last year's workshop concerning the availability of AIDS vaccine preparation adapted to the populations mostly affected in the developing world. A particular focus will be directed toward the proposal of having therapeutic vaccines which, together with Nevirapine intervention and antiretroviral treatment, could represent a most promising global approach to combat the dramatic situation of the mother to child transmission of HIV in the developing world.

HIV AND INFANT FEEDING: SITUATION IN BRAZIL

MARINA FERREIRA REA
Instituto de Saude, Sao Paulo, Brazil email: marifrea@usp.br
phone/fax 55-11-31067328

It is known that the mother-to-child transmission rate during pregnancy, delivery and breastfeeding is around 16-25%, with an average of 19%, according to the only Brazilian published study done in 5 cities of Sao Paulo State (Tess, B. et al, AIDS, 1995). If a mother breastfeeds and is HIV positive, the average additional risk of transmission is 15%. It is also known that there are factors that influence this rate, such as recent infection or severity, STDs, duration of breastfeeding and introduction of complementary food, cracked nipples, C-section, etc. Disruption of epithelial integrity of the gut membrane due to food or fluid other than breast milk might explain the recent data shown that Exclusive Breastfeeding (nothing else but breast milk) is better than breast milk plus any other fluid (Coutsoudis, A. Lancet, 1999).

On the other hand, even today when vertical transmission can be importantly decreased in settings where breastfeeding is not a cultural behaviour and can be safely replaced for the infants of HIV positive mothers, there remains a huge population in Africa, Asia and Latin America where the practice of *no breastfeeding* might jeopardize children's life. In countries of these continents, the simple offer of an alternative food to replace breastfeeding besides increasing the risk of infant mortality by other infectious diseases (Victora, C. et al, Lancet, 1986), cannot be a guarantee that the baby is not getting breast milk. Studies have shown that it is very difficult for a mother to comply with the single use of infant formula (Nduati, R.BMJ,1999), and she can be stigmatized by society if she does not offer her breast.

In Brazil, the ratio male/female of AIDS cases is now 2/1, and therefore, women being more present now than before in the epidemic, particularly during reproductive age, allow for more children to become infected through vertical transmission from mother to child. One important constraint in our reality is the spread of the epidemic in low-income groups, with less education and bad access to antenatal health care. For these women, even the provision of free HIV tests and complete drug treatment, as is the policy in Brazil, does not necessarily achieve better rates of mother-to-child transmission. The provision of condoms has started since the 80's, and recently was improved with the availability of the female condom; to spread condom use, however, is a challenge in a country where female sterilization reached 40 % in some areas and the husbands do not believe their wives need any other protection against STD/AIDS.

Although the majority of the population in higher prevalence areas give birth in hospitals, many of them arrive in the maternity without knowing their HIV status.

In our culture where breastfeeding is the norm, since 1981 a comprehensive breastfeeding programme is ongoing with important results (Rea, MR & Berquo, E Bulletin of WHO, 1990). The comparison of Exclusive breastfeeding for 0-4 months in 1986 with 1996 showed a 10-fold increase (from 4 to around 40%, DHS[1]).

We recommend strongly that the mother feed the baby exclusively on the breast for 6 months and continue breastfeeding up to at least 2 years together with complementary foods.

In the case of HIV+ women, free distribution of commercial infant formula must follow Code[2] recommendations (since 1988 a Brazil sanitary law) to avoid the spill-over effect to the whole population; consequently no donations of infant formula companies are allowed. A possibility of city health authorities to include the formula (or even whole powdered milk) among the list of products distributed for free as part of the PAB (Basic Care Programme) exists, with strict orientation about preparation and administration; however, periodical shortage of this milk normally occurs in the public service. Policy makers are aware that the price of infant formula is high (it might represent one third of minimal salary per month), and it is difficult to keep a mother feeding that specific milk only to that baby instead of diluting it among the other siblings.

We have started recently reviewing our policy regarding the feeding of HIV positive mothers and we are in the process of defining the best way to proceed.

In Brazil, since the beginning of the AIDS epidemic, the policy has been to recommend the mother to have her breast milk dried up in the post-partum period, and prescribe infant formula to the infant. However, how many HIV+ mothers can afford to continue feeding their infants when the free formula is not provided any more? Studies of artificial feeding show that the cost of feeding a baby during the first 6 months are U.S.$176 with infant formula (UNICEF, 1998). One alternative that we are trying to improve is the use of pasteurised human milk donated as extra milk by mothers tested in well qualified human milk banks. This country has a huge human milk banking net (more than 120 banks, 37 only in Sao Paulo), well organised with sanitary monitoring, where all milk samples are pasteurised (62.5 degrees for 30 minutes), therefore killing the HIV virus (besides others). The target is to expand coverage, training more health professionals to deal with the milk, improve the collection and transportation.

We already have 2 successful experiences of 24 hours/day human milk collection at home done by the fire workers in Brasilia and Rio de Janeiro. In these cities, after having been trained, the fire workers to attend the phone call of possible donors and go to their houses to collect the milk and transport it to the nearest banks.

In the process of reviewing our policy we are considering WHO\UNICEF \UNAIDS' (1998) recent recommendation about working on different possibilities for feeding of the infants of HIV positive mothers. The health care provider has to get tools

---

[1] Demographic and Health Survey

[2] International Code of Marketing of Breast Milk Substitutes, a WHO\UNICEF document of 1981.

to make the appropriate decision for the set up he is responsible for. If his decision is infant formula, he must provide it for free for at least 6 months with no interruption. If his decision is to recommend breastfeeding, all orientation and counselling must be provided to guarantee that *exclusive breastfeeding* will be pursued for 6 months, taking care to minimise cracked nipples or any other breast problem. From 6 months on, family food can be recommended.

We are increasing the number of banks and discussing possibilities of making the human milk banking a feasible possibility to provide milk for infants of HIV+ mothers, at least during their hospital stay, combining with alternatives above, and the best way to assess this policy. Other countries with similar urban characteristics, such as Venezuela, have asked for help to get them started with this successful experience.

As a conclusion, we believe that the HIV/AIDS epidemic and the way to feed infants of women that live with the virus is a decision that should be country-cultural based, certainly taking into account the economical and technical resources, but also the community support and the role of women in society. It is urgent to assess the impact and the process of implementation of the different possible approaches and allow the scientific community as well as policy makers to gather and discuss the constraints and the best way to overcome them.

MOTHER TO CHILD TRANSMISSION OF HIV AND PLANS FOR PREVENTIVE INTERVENTIONS: THE CASE OF INDONESIA

HADI PRATOMO
Faculty of Public Health, the University of Indonesia, Depok Campus, West Java, Indonesia

BACKGROUND

The Republic of Indonesia lies on the crossroads between Australia and the continent of Asia, and between the Indonesian and Pacific oceans. It consists of 13,000 large and small islands, makes up an extensive archipelago from west to east as long as the west to east coast of the USA. When the Hindus came to Indonesia about two thousand years ago, they found an indigenous population with a distinctive culture of their own. Around the turn of the 15th century Islam penetrated the country and has now become the dominant religion on the islands. By the end of the 16th century the arrival of the Dutch traders marked the beginning of three centuries of Dutch colonial expansion. After the Japanese occupation between 1942-1945, Indonesia proclaimed its independence on August 17, 1945.

Currently Indonesia is the 4th biggest country in the world with a total population of 240,392 million (1998). Almost half of the population are of reproductive age, about one-tenth are in the under five-year age group, and a very small proportion are in the elderly age group. The majority (65%) of the population live in the rural areas, work in the agricultural sectors and the growth rate of the population is 1,8%[1]. In 1997 it has a GNP of $U.S. 980[2].

With the ongoing economic crisis since the midyear 1997 it became one of the more economically disadvantaged countries. As Indonesia approaches the new millenium, the country faces enormous economic and political change. The current political and economic upheaval leads to social change for a more people-centered, decentralized approach of governance and social development. But the most vulnerable children and women may be left out as politicians focus on the nature of new political and economic structures rather than developmental needs.

CURRENT SITUATION OF HIV/AIDS

HIV infection is becoming a serious viral sexually transmitted infection (STI) for women in Indonesia, where heterosexual spread of AIDS places them equally at risk of contracting the disease. In 1987, the first individual case of AIDS was reported which places Indonesia in the category of countries described as Pattern III by WHO. This refers to areas where HIV was introduced in the 1980s[3].

As of 31 May 2000, 23 out of 27 provinces in Indonesia reported a total of 934 HIV-positive individuals (of which 40% were women), plus another 323 with AIDS (of which 18% were women). A total of 1257 cases of HIV/AIDS were reported which consisted of 74% HIV positive cases and the rest was AIDS cases. The majority of them (73%) were Indonesian. The HIV/AIDS women cases were about half of that of the men. The majority of the cases (83%) were in the reproductive age (15-49 years). Based on the risk factor, it was reported that the majority were heterosexual, then homo/bisexual, substance abusers and lastly perinatal transmission. The corresponding figures were the following: 67%, 10%, 4% and 0,6% respectively[4]. Two provinces with an HIV prevalence rate higher than 1% were Riau (Batam 1% and Tanjung Pinang 3.7%) and Irian Jaya 1.36%. The 1998 national estimate of AIDS prevalence is 0,11%, 1,1% cases per one million population.

In general, Moran concluded that Indonesia is a low-HIV prevalence country. However, in the absence of effective interventions, the spread of HIV is likely to accelerate in the near future, especially among commercial sexual workers (CSW) and their clients [5]. A top referral hospital with an STI clinic in Jakarta has reported prevalence data related to another STI problem that affects women of reproductive age: genital herpes. Among groups of 528 and 489 men tested in 1994 and 1995, the hospital data found 3,8% and 3,1 %, respectively, had genital herpes. In women, who numbered 424 and 462, the prevalence was 2,6% and 2,8%[6]. Another STI, hepatitis B, has been found to be prevalent. The Indonesian Household Survey (IHHS) 1995 identified hepatitis B in 11,5% of pregnant women surveyed; and 20.3% were also positive for hepatitis B antigen[7].

Unprotected sexual activity among married women is very high, as reflected by low level of condom use, less than 1% over the past decade (0,8% in 1991, 0,9% in 1994 and 0,7% in 1997). In addition, the discontinuation rates for first year of use for condom users were the highest, namely 38%[8].

The consequences of STIs for the health and social well-being of women and their children are frequently devastating. Nonetheless, as many as half of all women with STIs may be asymptomatic. There are indications that the prevalence of clamydia infection, gonorrhea, trichomoniasis and genital herpes among women in general (housewives) is much greater than commonly assumed. For example, one study of primary health care facilities for family planning in North Jakarta found that almost 40% of 486 female family planning clients who were screened by laboratory testing were positive for one or more infections and 14,4% had one or more Sexually Transmitted Diseases (STDs)[9].

Existing information from Puskesmas records showed only syphilis and gonorrhea in an absolute number of cases. This figure does not yield accurate estimates of the prevalence of syphilis in pregnant women. Caution should be exercised in interpreting sentinel surveillance data that showed a decline in the prevalence of syphilis in pregnant women from 1,15% in 1995/96 to 0,55% in 1997/98[10]. Currently there is no good nationwide surveillance of HIV among the pregnant women and MCH services in the country.

In 1992, a study on Knowledge, Attitude, Practice and Belief (KABP) toward HIV/AIDS among perinatal health care providers in the country was conducted. Although their knowledge and attitude toward HIV/AIDS were sufficient, their practice of universal precautions was below standard[11].

To anticipate the first delivery of the HIV positive pregnant woman, the Deptartment of Obstetrics and Gynecology in cooperation with the Department of Child Health and Working Group on AIDS has developed guidelines for attending delivery for HIV positive pregnant women (Dept Ob-Gyn, 1996)[12]. The first case of an HIV infected Indonesian baby girl born from an HIV positive mother was reported at Cipto Mangunkusumo (RSCM) Hospital, Jakarta. The full term baby was born on July 20, 1996 with a body weight of 3,380 gram, body length 49 cm, and Apgar score of 9/10. The mother was known as an HIV infected woman 4 years ago without any symptoms . The mother received AZT 5X100 mg/day orally until the baby was born. After delivery the mother continued to have AZT 300 mg every 3 hours orally. The baby was given AZT 5 mg orally every 6 hours for 6 weeks. The infant was in good health. She was fed with infant formula and the mother took bromocriptine orally to suppress the lactation. At 3 days old, the blood was examined for DNA HIV by means of PCR and HIV culture. Both results indicated positive. The baby was planned to be followed up at the public health center and periodically requested to visit the outpatient child health clinic of the hospital[13].

The Pelita Ilmu Foundation (PIF), a non-government organization from Jakarta was informed by the district hospital in which the above pregnant woman (with HIV positive) lived. The community where the woman lived were against her presence as they were scared of contracting the HIV/AIDS. Both during pregnancy and after delivery the presence of the HIV positive woman and her baby was rejected by the surrounding community. The volunteers of the PIF established its field office in the village and performed extensive community awareness on HIV/AIDS. Demonstrations by the volunteers in front of the public in which they kissed, hugged and cared for the cute HIV positive newborn convinced the people of the importance of socially accepting them in the community[14].

Up to June 2000, the PIF reported caring for 225 people living with AIDS. About 27,5% of them were drug users and about the same proportion were from the low socioeconomic level and 13% were married. Currently 8 of them have delivered their babies and 3 of them are currently pregnant. The AZT treatment is generally considered expensive[15].

There was an observation that substance abuse is becoming more common and there are about 4-10 new HIV/AIDS cases every week (from observation of private practice). It was also reported that many perinatal health care providers in the hospitals are still not ready for assisting delivery of HIV positive women[16]. In addition, the Merauke district hospital in Papua reported that they took care of 153 HIV/AIDS cases[17].

To achieve the Healthy Indonesia 2010, one of the recent policies of the government is to strengthen inter-sectoral cooperation and promote healthy behavior (life style), self-reliance of the community and partnership with private sectors. One of the main programs of the Ministry of Health (MOH) is controlling communicable diseases. However, at the same time there are many communicable diseases to control namely Dengue haemmorhagic fever, TBC, malaria and HIV/AIDS. Due to limited resources of the government, the prevention of HIV perinatal transmission is not one of the priorities[18].

Since 1994 the Indonesian Society for Perinatology (Perinasia) put HIV perinatal transmission issues on the agenda of the National Congress (every 3 years). In 1995, this society received a grant from the World Bank to produce video programs to support training for Universal Precautions particularly for Preparation of attending delivery of HIV positive pregnant woman. In addition, (with the support of PATH/USAID) it also developed a Lactation Management training module. Recently the organization was assigned by MOH/WHO to develop IEC & counseling and health services to adolescents in the public health center and received permission from WHO to translate two publications relevant to HIV and Infant Feeding.

The Yayasan Citra Usadha initiated the prevention of HIV/AIDS among the adolescents in Bali using both in and out group approaches[19]. In the recent recommendations of the 2nd National Meeting on HIV/AIDS, the importance of strengthening the empowerment of the community, particularly women and adolescents, as part of the prevention and management of HIV/AIDS in the future[20], was stressed.

CONCLUSIONS

In a big country with a relatively low level of HIV prevalence it seems that this condition is not conducive for encouraging the readiness of perinatal health care providers. Continuous effort for creating awareness, skills and compliance with universal precautions among the perinatal health providers is a must activity. The unavailability of good and reliable surveillance among pregnant women prevents identifying the exact magnitude of the problem. The high cost of the therapy is still a problem since it is not a priority of the government. The government has limited funds and at the same time all donors are not interested in providing funds for curing the cases.

The existing health services are not fully ready for the prevention and management of HIV perinatal transmission. At the same time the HIV positive pregnant women and their infants are likely not be socially acceptable to their community. Several NGOs are trying to initiate programs in combating the problems.

Perinasia, a NGO which is concerned with perinatal health, has taken several initiatives in programs relevant to issues of reproductive health but has not included HIV/AIDS and HIV perinatal transmission. An alternative vaccine for HIV which is effective, inexpensive, acceptable and widely available may benefit the women at risk in the country.

FUTURE PLANS FOR PREVENTIVE INTERVENTIONS

The following plans are proposed for future programs:

1. With the support from WHO, Prevention of Mother-to-Child Transmission of HIV-1 (PMCT) will be discussed during the upcoming 7th National Congress of the Indonesian Perinatal Society. A follow-up to initiate a pilot test of PMCT in selected high endemic areas such as in Merauke, Batam and Jakarta should be made.
2. Socialization of guidelines: HIV and Infant Feeding (A guide for health care managers and supervisors and guidelines for decision makers) should be made through the same society. Revise the existing Lactation Management module with the incorporation of more recent guidelines for breastfeeding of HIV positive women.
3. Revise the existing video program on preparation of attending delivery of HIV positive pregnant women so it becomes an effective media in supporting the training of Universal Precautions. Conduct continuous training on Universal Precautions for perinatal health care providers to make them constantly aware, and adhere to compliance of the implementation in daily routine work.
4. Expand programs for the prevention of perinatal problems through reaching the target audience of adolescents to better prepare them to become responsible parents. This could be made through integration of HIV/AIDS, substance abuse awareness, counseling and prevention of the existing Adolescents Reproductive Health Module recently being implemented in the Public Health Center in an urban area of Jakarta.
5. Support the innovative community-based programs by other NGOs to remove the barriers for social acceptance of persons with HIV/AIDS and encourage the buddies program for them.

REFERENCES

1. Unicef, 1996. World Situation of Children, 1996. Translated into Indonesian by R.F. Maulany, PT Intergraphika, Jakarta.
2. Pusat Data Kesehatan, Departemen Kesehatan RI (Center for Health Data, Ministry of Health, Republic of Indonesia), Profil Kesehatan Indonesia 1999 (Indonesian Health Profile, 1999). Appendix, Jakarta, 1999.
3. Brookmeyer, R and Gail, M.H. AIDS Epidemiology: A Quantitative Approach, Oxford University Press, New York, 1994.
4. Ministry of Health (MOH), Director General Communicable Diseases Control and Environment Health (DG, CDC EH), Monthly Report as of May, 2000. Analyzed and quoted by Support, a publication by Pelita Ilmu Foundation.
5. Depkes (MOH). Ditjend PPM and PLP. (DG, CDC EH) Epidemiologi Penyakit Menular Seksual , HIV/AIDS dan Perkembangan Data Infeksi HIV/AIDS di

Indonesia. (Epidemiology of Sexually Transmitted Diseases including HIV/AIDS and Recent Data of HIV/AIDS in Indonesia). Paper presented at the Seminar Commemorating World AIDS Day, Jakarta, December 8, 1998.

6. Moran, J. S. The Epidemiology of HIV and other STDs in Indonesia. An overview article prepared for USAID's External Review Team, Jakarta, February, 1999.
7. Fakultas Kedokteran Universitas Indonesia/ Rumah Sakit Umum Pusat Nasional (FKUI/RSUPN), Dr Cipto Mangunkusumo (RSCM). "RSCM Records Summary" in Insidens Penyakit Menular Seksual (Incidence of Sexually Transmitted Diseases), 1994-1995.
8. Departemen Kesehatan, Badan Penelitian & Pengembangan Kesehatan (Ministry of Health, Center for Research & Development). "Survei Kesehatan Rumah Tangga" (Indonesian Household Survey) 1995, Jakarta, 1997.
9. Central Bureau of Statistics, State Ministry of Population/ Family Planning Coordinating Board and Institute for Health Research & Development. Indonesia Demographic and Health Survey, 1997. Summary Report, Jakarta, September, 1998.
10. Iskandar, Meiwita B.; Paten, J; Qomariyah, S.N.; Vickers, C.; and Indrawati, S. "Difficulties of relying on public health approaches for detecting cervical infection among family planning clients: The case of primary care in Indonesia", Working paper, The Population Council, Jakarta, September, 1998.
11. Departemen Kesehatan, Badan Penelitian & Pengembangan Kesehatan (Ministry of Health, Center for Research & Development). "Survei Kesehatan Rumah Tangga" (Indonesian Household Survey) 1995, Jakarta, 1997.
12. Pratomo, Hadi; Chair, Imral; Notoatmodjo, Sukidjo et al "Studi tentang Pengetahuan, Sikap terhadap HIV/AIDS dan Praktek Pencegahan Risiko Tertularnya di Kalangan Petugas Pelayanan Perinatal di 5 RS Pendidikan dan Rujukan di Indonesia" (Study on the Knowledge, Attitude, Beliefs and Practice concerning Prevention of HIV/AIDS among Perinatal Health Care Providers in 5 Teaching Hospitals in Indonesia), Indonesian Epidemiology Network, 1994 (1), pp 34-43.
13. Dept of Obstetrics & Gynaecology, FKUI-RSCM. "Pertolongan persalinan pada Ibu HIV Positif" (Attending Delivery of HIV Positive Pregnant Women at Cito Manngunkusumo Hospital, Jakarta"), in Lokakarya Pencegahan dan Penanggulangan Kehamilan/Persalinan dengan HIV positif dalam Sistem Pelayanan Kesehatan di Indonesia (Workshop in Prevention and Management of Pregnancy/Delivery of Women with HIV Positive in the Health Care System in Indonesia), Bali September 20, 1997 the Indonesian Society for Perinatology (Perinasia), Bali Chapter, Denpasar.
14. Matondang, Corry S.; Wisnuwardhani, Siti D.; Suradi, Rulina et al. "A Case of HIV Infected Child Born to HIV Positive Mother", Paediatr Indones, 1996; 36:216-220.

15. Djauzi, Samsuridjal and Habsyi, Husein "Pengalaman memberi dukungan pada Ibu dan Anak Seropositif di Pedesaan" (Experience in Supporting Mother and Child with Seropositive in the Village) in Lokakarya Pencegahan dan Penanggulangan Kehamilan/Persalinan dengan HIV positif dalam Sistem Pelayanan Kesehatan di Indonesia (Workshop in Prevention and Management of Pregnancy/Delivery of Women with HIV Positive in the Health Care System in Indonesia), Bali September 20, 1997 the Indonesian Society for Perinatology (Perinasia), Bali Chapter, Denpasar.
16. Sanggar Kerja Pelita Ilmu (Pelita Ilmu Foundation Workshop), Report up to June 2000. No date.
17. Djurban, Zubairi. Personal communication, June 14, 2000.
18. Anonymous. "Penatalaksanaan Penderita HIV/AIDS di Rumah Sakit Umum Daerah Merauke" (Management of HIV/AIDS Cases at Merauke District Hospital), presented at the 2nd National Meeting of HIV/AIDS, July 17-20, 2000, Jakarta.
19. Departemen Kesehatan RI (Dep Kes RI – Ministry of Public Health). Rencana Pembangunan Kesehatan Menuju Indonesia Sehat 2010 (Health Development Plan toward Healthy Indonesia 2010), Jakarta, October, 1999.
20. Merati, Tuti Parwati "Upaya Pencegahan HIV/AIDS pada Remaja di Bali" (Prevention Program of HIV/AIDS among Adolescents in Bali), in Lokakarya Pencegahan dan Penanggulangan Kehamilan/Persalinan dengan HIV positif dalam Sistem Pelayanan Kesehatan di Indonesia (Workshop in Prevention and Management of Pregnancy/Delivery of Women with HIV Positive in the Health Care System in Indonesia), Bali September 20, 1997 the Indonesian Society for Perinatology (Perinasia), Bali Chapter, Denpasar.
21. Nasser, M. Rekomendasi Pertemuan Nasional II HIV/AIDS (Recommendations of the 2nd National Meeting on HIV/AIDS), Jakarta, July 17-20, 2000.

TOWARD PHARMACOLOGICAL DEFEAT OF THE THIRD WORLD HIV-1 PANDEMIC

LOWELL WOOD
Long Range Foundation, Palo Alto, California; Hoover Institution, Stanford University; Lawrence Livermore National Laboratory, Livermore, California

Breakout of HIV-1 infection into major population groups in sub-Saharan Africa several years ago and, more recently, in southern and eastern Asia is estimated by the U.N. to presently involve at least 30 million cases, a number which may double or triple in the current decade, depending on currently uncertain actions by impacted governments. Since >90% of these cases will receive at most palliative treatment, the resulting scale of human suffering and death over the coming 1-2 decades, in the surprise-free scenario for the HIV-1 pandemic, will exceed that of World War II or the Black Death of the 14th Century. It is therefore of very considerable interest to consider radical approaches to curative-level treatment of Third World cases, therapies with few if any correlates in the First World.

In the present work, HIV-1 parasitism of an adult human is modeled computationally in considerable detail, in order to provide a generally applicable platform for quantitatively evaluating novel therapeutic avenues. The infected human in this model is partitioned in coupled compartments of various cellular populations and virions, and each of the several stages in the retroviral replication cycle in various cell types is followed in time in each compartment, under the effects of ever-more impaired immunological pressure, as well as those of imposed time-dependent pharmacological stresses including cytokines. Escape mutations of the virus are modeled in reasonable fidelity, and kinetics of the entire resulting viral quasispecies and of both parasitized and responding host cellular populations followed.

The resulting model reproduces all published aspects of HIV-1 parasitism of an adult human to well within observational variability, including both transient and sustained suppression of viral populations under varieties of pharmacological stress, failure of monocomponent and intermittent polycomponent antiviral therapy, successful and unsuccessful prophylaxis immediately post-infection, slowly- and non-progressing cases, immune system recovery and collapse, post-therapeutic viremic rebound, and quasispecies hyperproliferation and T-cell population-crash during ARC/AIDS.

This model is exercised to explore the parameters of successful HIV-1 infection-clearing therapies in Third World medical and economic contexts. Intensive, multi-axis pharmacological suppression of net "gain" in the viral replication "loop" over a few dozen viremia e-folding times, accompanied by cytokinetic hyperstimulation of the proviral-carrying PBMC population, is the key feature of these (possibly single-dose) therapeutic schemes of few dozen dollar per patient likely cost. Typical modeling results are presented of such approaches to rescue of the Third World's HIV-1-infected population.

7. TRANSMISSIBLE SPONGIFORM ENCEPHALOPATHY

IATROGENIC CREUTZFELDT-JAKOB DISEASE IN THE YEAR 2000

PAUL BROWN, M.D.
Laboratory of Central Nervous System Studies, National Institute of Neurological Disorders and Stroke, National Institutes of Health, Bethesda MD

As the cause of approximately 250 deaths during the past 15 years, iatrogenic Creutzfeldt-Jakob disease (CJD) hardly qualifies as a planetary emergency, yet it has not escaped the critical attention of both the general public and their governments, and it is largely due to the cooperative efforts of governmental regulatory agencies and the medical research community that the historical iatrogenic causes of CJD have all but disappeared, and new causes have been avoided.

As shown in Table 1, the great majority of iatrogenic cases of CJD have resulted from unsuspected contamination of either human growth hormone (extracted from cadaveric pituitary glands) or dura mater grafts (also prepared from human cadavers). The geographic distribution of these and other iatrogenic cases is illustrated in Table 2, and roughly corresponds to the level of wealth and sophistication brought to medical practice in different nations of the world. Those with sufficient resources to underwrite the expense of using growth hormone and dural grafts in the treatment of hormone deficiency or neurosurgical repairs have reaped these unwanted fruits of high-tech medicine.

But this same sophistication has also brought its solutions. The problem of contaminated growth hormone was solved in 1985 by the replacement of cadaveric by recombinant hormone, and thus eliminated the risk of not recognizing CJD as a cause of death in the harvested cadaver. Without this advance, we would still be in something of a quandry about continuing to use cadaveric hormone, although the inclusion of a processing step with 6M urea would have probably been adequate to reduce any infectivity to negligible levels.

The problem of contaminated dura mater grafts has been solved by a combination of four measures: careful neuropathological and immunological screening of cadaver brains before selection as donors, individual processing of each donor graft, processing the graft with sodium hydroxide (thereby reducing any potential infectivity), and utilization, where possible, of cadaveric graft substitutes such as fascia lata or synthetic materials.

It is perhaps surprising that more cases of CJD have not resulted from corneal grafts, which are known to be infectious in patients with CJD, or from contamination of neurosurgical instruments used on patients with CJD. Evidently, corneal donor screening and routine instrument decontamination procedures have been adequate to prevent further iatrogenic disease from these causes. As for general surgery and other types of grafts or organ transplants, infectivity in non-neural tissues is so irregularly present, and when present, levels of infectivity are so low, that the occurrence of iatrogenic disease transmission in these situations must be close enough to zero as to be undetectable.

There remains the problem of variant CJD (vCJD), the probable result of the consumption in Britain and France of beef contaminated by brain or spinal cord tissue during the slaughter of cows with bovine spongiform encephalopathy (BSE). This variant form of CJD shows enough clinical, pathological, and biological peculiarities to invite speculation about whether tissues in patients incubating the disease might be more infectious than tissues in patients with sporadic CJD. In particular, vCJD can be distinguished from sporadic CJD by the presence of PrP (the pathologic amyloid isoform of 'prion protein' that is a surrogate marker for infectivity) in lymphoreticular organs such as spleen, tonsils, and appendix, all of which are interactive with circulating blood.

If the outbreak of vCJD does not greatly exceed its current incidence - 10 to 15 cases per year since 1994 - the problem will not be serious; however, if the current incidence represents the leading edge of an epidemic of vCJD in thousands of people who are only now incubating the disease, the issue becomes urgent, and the temporary blood and tissue donor exclusion measures already taken by some countries with respect to donors who have resided in Great Britain will need to be continued and perhaps even expanded to include countries in which only one or two cases of vCJD have surfaced (France and Ireland), or in which BSE has occurred (many European countries). However, we have at present no pre-emptive solution for the possibility of iatrogenic spread of disease from pre-clinical vCJD cases via cross-contamination of surgical instruments, or invasive medical procedures such as endoscopy and catheterization.

Much hinges on the hope that vCJD will not turn out to be a major epidemic, and on the successful conclusion of a search for both a diagnostic blood screening test to permit the identification of pre-clinical infections, and an effective therapy to interrupt the disease process before neurological symptoms appear. Both topics are presently the subject of intense research in many laboratories of the world.

Reprint requests: Dr. Paul Brown
Building 36, Room 4A-05
National Institutes of Health
36 Convent Drive, MSC 4122
Bethesda, MD 20892-4122
Tel: (301) 496-5292 Fax: (301) 496-8275
e-mail: brownp@ninds.nih.gov

Table 1. Summary of iatrogenic cases Creutzfeldt-Jakob disease from all causes (4 July 2000)

Mode of Infection	Number of Patients	Agent entry into brain	Median incubation period (range)[1]	Clinical signs on presentation[2]
Corneal transplant[3]	3	Optic nerve	16,18,320 mos	Dem/Cereb
Stereotactic EEG	2	Intra-cerebral	16, 20 most	Dem/Cereb
Neurosurgery	5	Intra-cerebral	17 most (12-28)	Vis/Dem/Cereb
Dura mater graft	114	Cerebral surface[4]	6 yrs (1.5-18)	Cereb (Vis/Dem)
Growth hormone	139	Hematogenous (?)	12 yrs (5-30)	Cerebellar
Gonadotrophin	4	Hematogenous (?)	13 yrs (12-16)	Cerebellar

[1]Calculated from the mid-point of treatmetn to the onset of disease.

[2]Dem = dementia; Cereb = cerebellar signs; Vis = visual signs

[3]One definite, one probable, and one possible case.

[4]In two cases, dura was used to embolize vessels of non-CNS tissues, rather than as intracranial grafts.

Table 2. International distribution of iatrogenic cases of Creutzfeldt-Jakob disease (4 July 2000)

		Surgical procedures			Hormone therapy	
	Dura mater grafts	Surgical Instruments	Sterotactic EEG needles	Corneal transplants	Growth hormone	Gonadatropin
Argentina	1					
Australia	4				1	4
Austria	1					
Brazil					1	
Canada	4					
Croatia	1					
France	8	1			74	
Germany	4			1		
Holland	2				1	
Italy	4					
Japan	67			1		
New Zealand	1				5	
Spain	6					
Switzerland	1		2			
Thailand	1					
United Kingdom	6	4			35	
United States	3			1	22	
Worldwide totals	114	5	2	3	139	4

INFECTION CONTROL GUIDELINES FOR TSEs IN HOSPITALS AND HOME CARE SETTINGS

MAURA N. RICKETTS
World Health Organization, Geneva, Switzerland

BACKGROUND

Transmissible spongiform encephalopathies (TSEs), also known as prion diseases, are fatal degenerative brain diseases that occur in humans and certain animal species. They are characterized by microscopic vacuoles and the deposition of amyloid (prion) protein in the grey matter of the brain. All forms of TSE are experimentally transmissible.

TSE agents exhibit an unusual resistance to conventional chemical and physical decontamination methods. They are not adequately inactivated by most common disinfectants, or by most tissue fixatives, and some infectivity may persist under standard hospital or healthcare facility autoclaving conditions (e.g. 121°C for 15 minutes). They are also extremely resistant to high doses of ionizing and ultra-violet irradiation and some residual activity has been shown to survive for long periods in the environment. The unconventional nature of these agents, together with the appearance in the United Kingdom, Republic of Ireland and France of a new variant of CJD (vCJD) since the mid- 1990s, has stimulated interest in an updated guidance on safe practices for patient care and infection control. The World Health Organization guideline on the prevention of iatrogenic and nosocomial exposure to TSE agents was prepared following the WHO Consultation on Caring for Patients and Hospital Infection Control in Relation to Human Transmissible Spongiform Encephalopathies, held in Geneva from 24 to 26 March 1999. The meeting was chaired by Dr Paul Brown. Dr Martin Zeidler and Dr Maurizio Pocchiari kindly agreed to be Rapporteurs.

The full guideline is available at:
http://wwwstage.who.int/emc-documents/tse/whocdscsraph2003c.html

The document is open for discussion until September 30th, after which it will be published in hard copy for distribution.

HAZARD IDENTIFICATION AND RISK REDUCTION

TSEs are not known to spread by contact from person to person, but transmission can occur during invasive medical interventions. Exposure to infectious material through the use of human cadaveric-derived pituitary hormones, dural and cornea homografts, and contaminated neurosurgical instruments have caused human TSEs.

When considering measures to prevent the transmission of TSE from patients to other individuals (patients, healthcare workers, or other care providers), it is important to base the measures upon the known and limited ways in which TSEs are transmitted between humans. Risk is dependent upon three considerations:

- the probability that an individual has or will develop TSE;
- the level of infectivity in tissues or fluids of these individuals;
- the nature or route of the exposure to these tissues.

From these considerations it is possible to make decisions about whether any special precautions are needed.

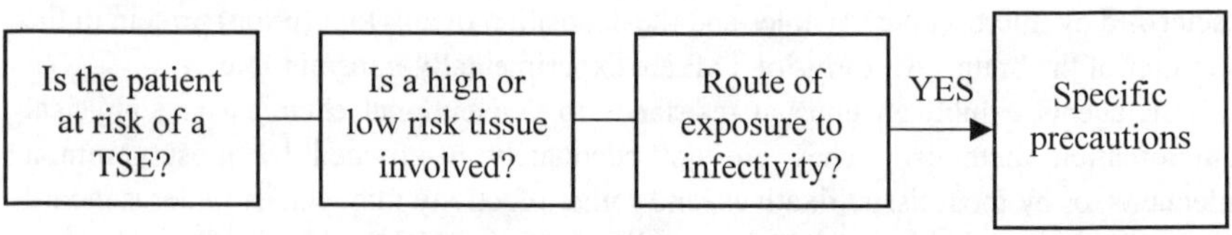

EVALUATING RISK IN PATIENT POPULATIONS

Where a clinician diagnoses or suspects the diagnosis of CJD or another TSE, the patient must be considered to be a risk for transmitting infection. Where a person has a history of exposure to dura mater, cornea or human pituitary hormones, they are considered to be 'at risk' for transmitting infection, however the risk is dependant upon the risk level of the tissues, as discussed next. Where persons have a family history of TSEs, or carry a genetic marker of inheritable TSEs, without any clinical signs of TSEs, it proved impossible to be infinitive – alternative approaches can be taken, also discussed in the following section. Finally, vCJD was not the subject of the consultation; the published guideline discusses this issue in an appendix.

EVALUATING RISK OF TISSUES

Infectivity levels in tissue were classified in one of three categories – high, low or no detectable infectivity – as per Table 1, below.

Table 1 Distribution of infectivity in the human body[1].

Infectivity Category	Tissues, Secretions, and Excretions	
High Infectivity	Brain Spinal cord Eye	
Low Infectivity	CSF Kidney Liver Lung Lymph nodes/spleen Placenta	
No Detectable Infectivity	Adipose tissue Adrenal gland Gingival tissue Heart muscle Intestine Peripheral nerve Prostate Skeletal muscle Testis Thyroid gland Blood[2]	Faeces Milk Nasal mucous Saliva Semen Serous exudate Sweat Tears Urine

EVALUATING RISK OF PROCEDURES

Not all clinical or medical procedures carry a risk of transmission of TSE. Routine patient activities do not require any sort of special precautions for TSEs. However, neurosurgical and ocular surgical procedures are the highest risk procedures and it is essential that appropriate precautions be taken for any person known, suspected or at risk of a TSE if they undergo such procedures. If a person is at risk for familial TSEs, as was noted earlier, there was no consensus on whether precautions should be taken for neurosurgical or ocular surgical procedures. Regarding other surgical procedures, special precautions are recommended for persons known or suspected of TSEs, however, it was acknowledged that the less rigorous methods could be used and that special precautions were unnecessary for persons at risk. Regarding dental procedures, no consensus was reached, however some potential interventions can be taken and are described in the full document.

No special precautions are needed for routine laboratory procedures, with the exception of where CSF is being handled. Similarly, when high and low infectivity tissues are being examined (i.e. pathology), special handling is required. It is best if high risk tissues are handled only in specialized facilities with experience, training and specialized equipment.

Autopsy requires special precautions, however can be conducted without undue risk. Mortuary handling of bodies may require alterations if the brain pan is open or if an autopsy was conducted, otherwise the body is handled as per routine procedures.

SPECIFIC PRECAUTIONS FOR TSEs

The safest and most unambiguous method for ensuring that there is no risk of residual infectivity on contaminated materials is to discard and destroy them by incineration. While this strategy should be universally applied to those devices and materials that are designed to be disposable, it was also recognized that this may not be feasible for many devices and materials that were not designed for single use. For these situations, the autoclave/chemical methods recommended below appear to remove most and possibly all infectivity under the widest range of conditions. Incineration remains the most suitable method for disposing of all waste and of contaminated tissues.

Those surgical instruments that are going to be re-used may be mechanically cleaned in advance of subjecting them to decontamination. Mechanical cleaning will reduce the bio-load and protect the instrument from damage caused by adherent tissues. If instruments are cleaned before decontamination, the cleaning materials must be treated as infectious waste, and the cleaning station must be decontaminated by one of the methods listed below. The instruments are then treated by one of the decontamination methods recommended below before reintroduction into the general instrument sterilization processes.

Incineration

1. Use for all disposable instruments, materials, and wastes.
2. Preferred method for all instruments exposed to high infectivity tissues.

Autoclave/chemical methods for heat-resistant instruments

1. Immerse in sodium hydroxide (NaOH)[3] and heat in a gravity displacement autoclave at 121°C for 30 min; clean; rinse in water and subject to routine sterilization.
2. Immerse in NaOH or sodium hypochlorite[4] for 1 hr; transfer instruments to water; heat in a gravity displacement autoclave at 121°C for 1 hr; clean and subject to routine sterilization.
3. Immerse in NaOH or sodium hypochlorite for 1 hr.; remove and rinse in water, then transfer to open pan and heat in a gravity displacement (121°C) or porous load (134°C) autoclave for 1 hr.; clean and subject to routine sterilization.
4. Immerse in NaOH and boil for 10 min at atmospheric pressure; clean, rinse in water and subject to routine sterilization.
5. Immerse in sodium hypochlorite (preferred) or NaOH (alternative) at ambient temperature for 1 hr; clean; rinse in water and subject to routine sterilization. Autoclave at 134°C for 18 minutes.
6. Autoclave at 134°C for 18 minutes.[5]

Chemical methods for surfaces and heat sensitive instruments

1. Flood with 2N NaOH or undiluted sodium hypochlorite; let stand for 1 hr.; mop up and rinse with water.
2. Where surfaces cannot tolerate NaOH or hypochlorite, thorough cleaning will remove most infectivity by dilution and some additional benefit may be derived from the use of one or another of the partially effective methods listed in the guideline.

Autoclave/chemical methods for dry goods

1. Small dry goods that can withstand either NaOH or sodium hypochlorite should first be immersed in one or the other solution (as described above) and then heated in a porous load autoclave at 121°C for 1 hr.
2. Bulky dry goods or dry goods of any size that cannot withstand exposure to NaOH or sodium hypochlorite should be heated in a porous load autoclave at 134°C for 1 hr.

SUMMARY

The principle recommendations of the guideline are contained in the following table.

Table 2. Decontamination levels for different risk categories.

Patient category	Tissue category	Decontamination options
Confirmed or suspect cases of TSE	High infectivity	Specific Decontamination Methods for TSEs
	Low infectivity	Specific Decontamination Methods for TSEs (but note that CSF, and peripheral organs and tissues are regarded as less infectious than the CNS)
Persons with known prior exposure to human pituitary derived hormones, cornea or dura mater grafts	High infectivity	Specific Decontamination Methods for TSEs
	Low Infectivity	Routine cleaning and disinfection procedures
Members of families with heritable forms of TSE	High Infectivity	No consensus was reached. The majority felt that TSE decontamination method should be used, but a minority felt this was unwarranted.
	Low Infectivity	Routine cleaning and disinfection procedures
All of the above categories	No detectable Infectivity	Routine cleaning and disinfection procedures
Confirmed or suspect cases of vCJD	All tissue categories	Specific Decontamination Methods for TSEs

REFERENCES

1. Assignment of different organs and tissues to categories of *high* and *low infectivity* is chiefly based upon the frequency with which infectivity has been detectable, rather than upon quantitative assays of the level of infectivity, for which data are incomplete. Experimental data include primates inoculated with tissues from human cases of CJD, but have been supplemented in some categories by data obtained from naturally occurring animal TSEs. Actual infectivity titres in the various human tissues other than the brain are extremely limited, but data from experimentally-infected animals generally corroborate the grouping shown in the table.
2. Experimental results investigating the infectivity of blood have been conflicting, however even when infectivity has been detectable, it is present in very low amounts and there are no known transfusion transmissions of CJD.
3. Unless otherwise noted, the recommended concentration is 1N NaOH.
4. Unless otherwise noted, the recommended concentration is 20,000 ppm available chlorine.
5. In worse-case scenarios (brain tissue bake-dried on to surfaces) infectivity will be largely but not completely removed.

8. LIMITS OF DEVELOPMENT — MEGACITIES

MEGACITIES: WATER AS A LIMIT TO DEVELOPMENT

WILLIAM J. COSGROVE
President, Ecoconsult Inc., Montreal, Canada

INTRODUCTION

This paper presents the lessons learned in the consultative process that led to the World Water Vision[1]. It is shared with participants in spirit of the Erice International Centers–that its author may learn from participants here. The Vision exercise was open-ended, and hence its outputs did not focus on the needs of megacities. Nevertheless, the knowledge that it generated is relevant to the development of megacities. This presentation highlights those areas of greatest relevance to this seminar. The Vision was debated in dozens of sessions of interested stakeholders in The Hague in March 2000. This presentation concludes with summaries of discussions of a group that discussed the issues confronting megacities and of a group that discussed "Scientists on Water and Knowledge".

TODAY'S WATER CRISIS—AND TOMORROW'S

The evolution of man as a species is relatively new to the planet in geological terms. While the first single-cell organisms appeared some 3.5 billion years ago, the human species evolved only about 100,000 years ago. Our impact on the surrounding environment probably was not significantly different from that of other species until about 10,000 years ago when we developed tools, learned that we could cultivate our own food instead of just gathering it, and began migrating long distances. Since then, we have demonstrated the evolving distinctiveness of our species. Increasingly we find ways to transform the natural resources of the planet to meet not only our basic life-sustaining needs of food and water, but to improve the quality of our human existence. We continuously seek to improve our physical comfort and to satisfy our intellectual, cultural and social needs. Ultimately we seek security from this way of life.

Until a century ago, with a few local exceptions, our behavior continued to have little impact on the environment. This situation changed drastically in the past century. During that period the world's population more than tripled, placing unprecedented demands on natural resources to provide sustenance and shelter. At the same time, we developed new processes to produce goods and services that are perceived to improve the quality of life. These placed new demands on our limited natural resources, both non-renewable and renewable. The result has been exponentially increasing demand on the services provided by the land, air and water of the planet. Under current trends, these demands will continue to increase to satisfy the life-sustaining needs of the still growing global population and to improve the quality of life not only for them, but also for the large majority of mankind who can only dream of such an existence.

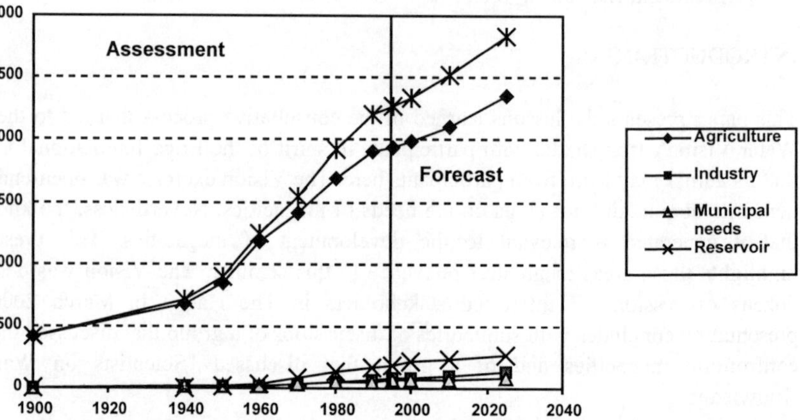

Fig. 1. Water Consumption – after Shiklomanov 2000.

There is a water crisis today – even though this crisis is not about a lack of water to satisfy our needs, but about managing water so badly that billions of people – and the environment – suffer badly.

During the 20th century the world population tripled, but water use for human purposes multiplied six-fold! The most obvious uses of water for people are drinking, cooking, bathing, cleaning, and—for some—watering family food plots. This domestic water, though crucial, is only a small part of the total–an estimated 350 cubic kilometers in 1995 (Shiklomanov[2]). Worldwide, industry uses about twice as much water as households, mostly for cooling in the production of electricity. Far more water is needed to produce food and fiber (cereals, fruits, meat, and cotton)–2500 cubic kilometers in 1995. (Fig. 1). We are not sure how much water must remain in our ecosystems to maintain them, but indications are that we are approaching – and have surpassed in many places the limits of how much we can divert.

THE WORLD WATER VISION EXERCISE

Participants at the First World Water Forum – held in Marrakech, Morocco, in 1997 and sponsored by the World Water Council – recognized the coming crisis. Some of the contributing factors they identified included:

a) Water scarcity (and the opposite – floods)
b) Lack of accessibility
c) Water quality deterioration
d) Fragmentation of water management
e) Decline of financial resources
f) Lack of awareness by decision-makers

To begin to address the crisis the Council called for a World Water Vision. Its purpose would be to increase awareness of the water crisis and develop a widely shared view of how to bring about sustainable use and development of water resources (Cosgrove[3] and Rijsberman).

The World Water Vision exercise launched in August 1998 had as its objectives to:

- "develop knowledge on what is happening in the world of water regionally and globally, and on trends and developments outside the world of water which may affect future water use;
- based on this knowledge, produce a consensus on a "Vision" for the year 2025 that is shared by water sector specialists and decision makers in the government, the private sector and civil society;
- raise awareness of water issues among the general population and decision-makers in order to foster the political will and leadership necessary to achieve the Vision; and
- utilize the knowledge and support generated to contribute to the Framework for Action developed by the Global Water Partnership".

The World Water Vision exercise drew on the accumulated experience of the water sector, particularly through sector visions and consultations for Water for People (Vision 21), Water for Food and Rural Development, Water and Nature, and Water in Rivers. Professionals and stakeholders from different sectors have developed integrated regional visions through national and regional consultations. These covered Arab countries, Australia, Canada, Central America and the Caribbean, Central Asia, Central and Eastern Europe, China, the Mediterranean Basin, the Nile Basin, North America, the Rhine Basin, Russia, South America, South Asia, Southeast Asia, Southern Africa, and West Africa. In addition, there were a series of special projects on Inter-basin Water Transfers; River Basin Management; A Social Charter for Water; Water, Education, and Training (WET); Water and Tourism, Water and Sovereignty, and Mainstreaming Gender Issues.

The participatory process that led to the World Water Vision made it special. From August 1998 up to the opening of the Second World Water Forum, some 15,000 women and men at the local, district, national, regional, and global levels shared their aspirations and developed strategies for the sustainable use and development of water resources. The Internet made these consultations possible in a short timeframe. As the Vision evolved, more networks of civil society groups, non-governmental organizations (NGOs), women, and environmental groups joined the consultations that influenced the World Water Vision. The diverse backgrounds of participants—authorities and ordinary people, water experts and environmentalists, government officials and private sector participants, academics and NGOs—offered a wide range of views.

BUSINESS-AS-USUAL WILL LEAD TO SEVERE WATER STRESS

As part of the World Water Vision exercise a Scenario Development Panel of 14 distinguished water experts, modelers and futurists, co-chaired by Commission chairman Ismail Serageldin and Frank Rijsberman, Deputy Director of the Vision Unit, developed three global level water scenarios (Gallopin and Rijsberman, 2000). They described scenarios for Business as Usual (BAU); Technology, Economics and the Private Sector (TEC); and Values and Lifestyles (VAL). Simulation models[4] were subsequently used to explore these scenarios. The basic data set for renewable water resources availability and use (domestic, industrial and agriculture) at the national level was provided by the State Hydrological Institute of Russia (Shiklomanov, 2000).

The BAU scenario shows that because of population growth, the global average annual per capita availability of renewable water resources is projected to fall from 6,600 cubic meters today to 4,800 cubic meters in 2025. Given the uneven distribution of these resources, some 3 billion women and men will live in countries—wholly or partly arid or semiarid—will have less than 1,700 cubic meters per capita, the quantity below which one starts to suffer from water stress. Also by 2025 about 4 billion people, or more than half the world's population, are estimated to live in countries where more than 40% of renewable resources are withdrawn for human uses—another indicator of high water stress under most conditions (Alcamo et al., 2000).

Under business as usual, with present policies continued, in developed and transition-economy countries economic growth to 2025 tends to increase water use (Gallopin and Rijsberman[5]). But more efficient water use and the saturation of water demands in industry and households can offset this increase. In addition, the amount of irrigated land stabilizes, and water for irrigation is used more efficiently. As a result, total water withdrawals can – and should – decline. Extrapolating current trends on water quality does not present a rosy picture, however.

Higher incomes and providing increased access in developing countries lead to greater household water use per capita multiplied by the greater number of people. Meanwhile, economic growth expands electricity demand and industrial output, leading to a large increase in water demand for industry. Even though water may be used more efficiently in households and industry, pressures to increase water use overwhelm these efficiency improvements. Providing food to the growing population and ending hunger will remain the largest challenge in terms of quantities of water demand.

The result is a projected large increase in water withdrawals in the agricultural, domestic and industrial sectors of the developing world, in response to rising population and industrialization, and higher consumption from higher incomes. Adding together the trends in developed and developing countries under business as usual increases global water withdrawals from 3,800 cubic kilometers in 1995 to 4,300–5,200 cubic kilometers in 2025. The difference largely depends on how much irrigated agriculture does or does not expand.

This increase in water withdrawals, implies that water stress is projected to increase significantly in more than 60% of the world, including large areas of Africa, Asia, and Latin America (Alcamo[6]).

The total withdrawals of water in *Europe* are growing slowly or not at all as households, industry and agriculture become more water-efficient. The per capita use of water in households goes up slightly with the economic growth of the Business-as-Usual scenario between 1995 and 2025, while the amount of water used by industry per megawatt-hour goes down because of greater recycling and other efficiency improvements. The amount of irrigated area stabilizes and new technologies improve the efficiency of irrigation systems so that there is also a decline in the amount of water used per hectare during this period. Although water withdrawals go down, the pressure on water resources continues to be high in some areas because of the density of population and industrial activity. Hence, some river basins remain in the high stress category where there is sharp competition between industrial, domestic and some agricultural water users for available water resources.

Under the Business-as-Usual scenario, domestic water withdrawals in *Sub-Saharan Africa* increase from about 10 km^3 per year in 1995 to 42 km^3 per year in 2025. This is because higher income leads to higher per capita water use, even though technology tends to improve the efficiency of water use. For example, in 2025, domestic water use in West Africa is 34 m^3/cap-year which is a factor of 2.1 over its 1995 value, but still far below the Western European level in 1995 (105 m^3/cap-year). In this part of Africa, industrial output and related water use also increase from about 3 to 16 km^3/year between 1995 and 2025.

Because of abundant rainfall, it is likely that there will be enough water to cover the increase in domestic and industrial water use. Instead, the question is whether water distribution systems can be expanded fast enough to fulfill the needs of growing population and industry. To cover the growth in water withdrawals noted above, the capacity of municipal water withdrawals must be expanded by about 5.5 % per year, and industrial withdrawals must be expanded 7.1 % per year.

In *South and East Asia*, the extent of irrigated area under the Business-as-Usual scenario grows only slightly between 1995 and 2025, while irrigation efficiency improves. The net effect is a decrease in water used for irrigation from 1359 to 1266 km^3 per year. At the same time, strong economic growth between 1995 and 2025 leads to more material possessions and greater water use in each household, which increases water withdrawals for domestic use from 114 to 471, km^3 per year. This economic growth also requires larger quantities of water for Asian industry, and so water withdrawals for industry increase from 153 to 263 km^3 per year. The sum of these trends is an overall increase in water withdrawals between 1995 and 2025. Hence the pressure on water resources becomes even greater than that already experienced in 1995, when about 6.5 million km^2 of river basin area were under high water stress. This increases to 7.9 million km^2 in 2025. The number of people living in these areas also grows tremendously from 1.1 to 2.4 billion during this period.

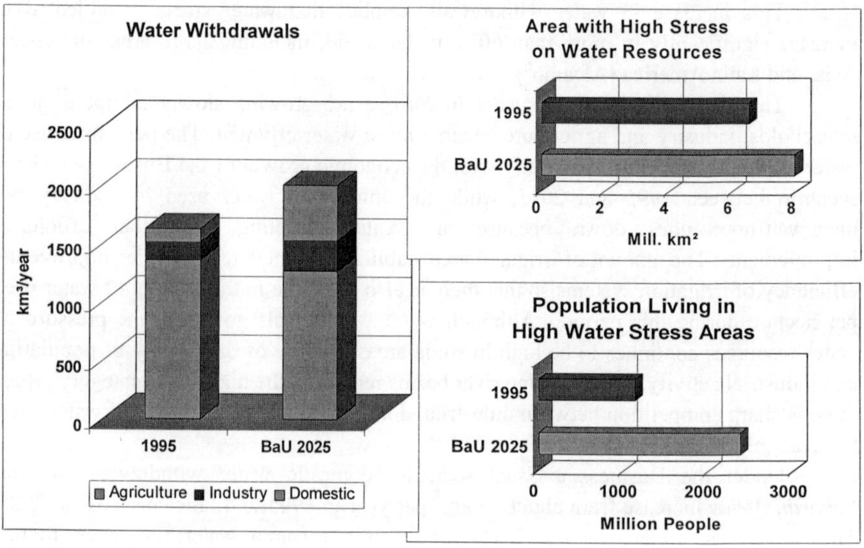

Box 1. South and East Asia: Water Resources Uses and Trends
Source: WaterGAP 2 calculations (Alcamo, et al.)

MOVING FROM CRISIS TO VISION: TURNING POINTS

Whether the water crisis will deepen and intensify—or whether key trends can be bent or turned towards sustainable management of water resources—depends on many interacting trends in a complex system. Real solutions require an integrated approach to water resource management.

Crucial issues that may provide levers for very different futures include:

- Limiting the expansion of irrigated agriculture.
- Increasing water productivity.
- Developing biotechnology for agriculture.

- Increasing storage.
- Reforming water resource management institutions.
- Increasing co-operation in international basins.
- Valuing ecosystem functions.
- Supporting innovation.

Participants in the Vision process examined all of these approaches. They came to the conclusion that we can manage them to achieve the Vision they have for a water secure world.

A VISION FOR WATER AND LIFE IN 2025

How then, will the water world look in 2025? Almost every woman and man, girl and boy in the world's cities, towns, and villages will know the importance of hygiene and enjoy safe and adequate water and sanitation. People at the local level will work closely with governments and non-governmental organizations, managing water and sanitation systems that meet everybody's basic needs without degrading the environment. People will contribute to these services according to the level of service they want and are willing to pay for. With people everywhere living in clean and healthy environments, communities and governments will benefit from stronger economic development and better health.

<u>New management—transparent and accountable</u>
Water services will be planned for sustainability, and good management, transparency, and accountability will be standard. Inexpensive options for providing water-efficient equipment will be widely available. Rainwater harvesting will be broadly applied. Municipal water supply will be supplemented by extensive use of reclaimed urban wastewater for non-potable uses (and even for potable uses in seriously water-short urban areas). On small islands and in some dry coastal areas, desalination will augment the water supply. Many cities and towns will use low- or no-water sanitation systems, for which communities and local authorities will manage the collection and composting services.

<u>Secure and equitable access to and control of resources</u>—and fair distribution of the costs and associated benefits and opportunities derived from conservation and development—will be the foundation of food and water security. Overcoming sector-oriented approaches and developing and implementing integrated catchment management strategies will be supported by wider social and institutional changes. Many government institutions will have recognised the value of the groundwork of grass-roots community-based initiatives at the turn of the century —and built on this extensively. All new central government policies and legislation will be subject to ex-ante assessment of their impacts on the different types of stakeholders and beneficiaries. Private and public institutions will be more accountable and oriented towards the local delivery of services and

conservation of ecosystems than they are today. They will fully incorporate the value of these services in their cost-benefit analysis and management.

More power for communities

At local levels, the empowerment of women, traditional ethnic groups, and poor and marginalized women and men will start making local communities and weak nations stronger, more peaceful and more capable of responding to social and environmental needs. The institutional structures, including river basin commissions and catchment committees, will actively support the equitable distribution of goods and services derived from freshwater ecosystems. Both spouses will be members with voting rights in water user associations in farming communities. Clear sets of property and access rights and entitlements will ensure that individuals, companies and organisations holding those rights meet their associated responsibilities.

Higher crop yields

Extensive field research on water management policies and institutions in developing countries early in the 21st century will focus on bringing average yields closer to what was being achieved by the best farmers. Closing the yield gap will make the rural livelihoods of poor women and men much more sustainable. Countries that have a basic policy of food self-sufficiency and the capability to implement that policy will increase their yields and production. They will do so by increasing the productivity of water through technical and institutional innovation, up to economic and technical limits. India and China will be among them.

Drawing on technological innovations, as well as traditional knowledge, large improvements will be made in agriculture. Genetically modified crops may be initially introduced on a small scale, given the lack of public and political support. The biggest advances in food production in the century's first decade would be on plant improvements through tissue culture and marker-aided selection, crop diversity (especially relying on locally adapted indigenous varieties), and appropriate cropping techniques and soil and water conservation. By 2025, the industry will have demonstrated its responsibility and gained credibility, and the use of genetically modified crops will become common and greatly increase the reliability of crops in drought-prone regions.

More efficient use

There is likely to be a 10% increase in water withdrawals and consumption to meet agricultural, domestic, and industrial requirements. Nevertheless, food production will increase 40%. This will be possible—in part—because people recognize that water is not only the blue water in rivers and aquifers, but also the green water—in soil. Recognition of its crucial role in the hydrological cycle will help make rain-fed agriculture more productive while conserving aquatic and terrestrial ecosystems.

Only a small percentage of water delivered to the domestic and industrial uses will be consumed by evaporation—most will be returned after proper treatment to the ecosystems from which it is drawn. Domestic and industrial water reuse will be common, and new methods of ecosanitation not dependent on water as a carrier will be applied in many areas to reduce pollution and make full use of human waste as agricultural fertilizer. Natural and artificial wetlands will be commonly used to improve polluted waters and treat domestic effluents. Countries that face water scarcities early in the century will invest in desalination plants—or reduce the amount of water used in agriculture, transferring it to the other uses, and importing more food.

Table 1. Renewable water use in the World Water Vision.

Water use	**1995 (cubic kilometers)**	**World Water Vision 2025 (cubic kilometers)**	**Percentage increase 1995-2025**
Agriculture			
Withdrawal	2,500	2,650	6
Consumption	1,750	1,900	9
Industrial:			
Withdrawal	750	800	7
Consumption	75	100	33
Municipal:			
Withdrawal	350	500	43
Consumption	50	100	100
Reservoirs (evaporation)	200	220	10
Total:			
Withdrawal	3,800	4,200	10
Consumption	2,100	2,300	10
Groundwater Overconsumption	200	0	

Source: Cosgrove and Rijsberman (2000)

<u>Less pollution – more recharge</u>
Concerns about polluting groundwater through leaching nitrates and other chemicals will be addressed. Restrictions will be placed on fertilizers, pesticides, and other chemicals in recharge areas after research on maximizing the rate of recharge and controlling pollution. Ideally, the recharge areas will not be used for any other purpose. But in densely population areas, land will simply be too valuable to be set aside for this single use.

<u>Healthier catchments</u>
Management of our water in 2025 will be based on recognizing the environmental goods and services that healthy catchments provide. Catchments require constant maintenance, which will be provided largely by local communities through erosion control, water quality protection, and biodiversity conservation, among others. Strategic or unique

natural ecosystems will be highly valued. And conservation programs, including protected areas, will reflect the needs and involvement of the local communities that depend on them.

<u>More innovation</u>

Innovation in most areas of water resource management—supported by the best of science and traditional knowledge—will accelerate significantly. It will also support development and management of freshwater and related ecosystems. Scientific analysis and modern technologies will provide an analytical perspective to problem solving. Traditional knowledge, the wealth of many generations of water resource management, will also be a natural part of decision-making and management. The dialogue between scientists and the holders of traditional knowledge will form a cornerstone for many innovative resource management practices.

<u>Smarter investments</u>

Investments in cleaner technologies and reduced water and wastewater use will continue to help many industries lower their production costs while reducing their effluent taxes. Development investments will be based on economic valuations and linked to compliance with the international environmental assessment and management standards.

<u>Better governance</u>

Governance systems in 2025 will facilitate transboundary collaborative agreements that conserve freshwater and related ecosystems and maintain local livelihoods. Management and decision making will generally take place at the level where they are most effective and efficient, helping to set up more open dialogue, information exchange, and cooperation. Despite huge efforts, transboundary conflicts will still be the most difficult water resource issues to resolve in 2025.

SOME IMPLICATIONS FOR URBAN AGGLOMERATIONS

The equivalent of the total growth of 1.5 billion people in the world population over the next 25 years will be added to the urban population. Half of this number will be born in urban areas, the other half will move there from rural areas. In the human settlement context, the urbanization process provides perhaps the best-integrated picture of the interplay of forces driving unsustainable consumption. Urban areas have become the engines of growth for developing countries, currently contributing to more than half of gross domestic product. Driven by an ever-increasing population and increased economic activity, cities are consuming resources and generating waste at a much higher pace than the national average.

It is also in the cities that the income disparities and poverty stand out in sharp contrast between wealthy neighborhoods and squatter settlements, exacerbating unsustainable consumption patterns and attendant environmental degradation. The highest levels of resource use and waste generation tend to occur in the wealthiest cities and among wealthier groups within cities.

Cities in Watersheds

Spreading urban areas occupy increasingly larger percentages of the basins or catchments within which they are situated. As they grow, they create impervious areas that block groundwater recharge and modify the hydraulic behavior of basins, accentuating both floods[7] and water scarcity. Water pollution from human activity makes water unsuitable for many downstream uses. They become increasingly dependent on upstream areas to provide them with freshwater for human use and to maintain the ecosystems within and surrounding them. They therefore have reason to be concerned in turn about upstream land use changes. Falkenmark[8] has coined the term *hydrosolidarity* to refer to this upstream-downstream interdependence and the cooperative integrated management required. Failure to care for the health of watersheds and recharge areas could result in the collapse of the surrounding ecosystem, and certainly will cause increased financial, economic and social costs.

In a discussion of issues facing the city of Metropolitan São Paulo and the surrounding areas, Braga[9] has suggested that the following water management principles must prevail if one wishes to address the issue of hydrosolidarity. These include:

- Water quantity and quality cannot be treated separately.
- Dominant alternatives must be identified so that trade-offs can be made in an optimal way. This means that a user-friendly decision support system must be developed to support the decision-making process at the basin level.
- Proper water use implies charging for abstraction and charging for returning used water to natural water bodies. A corollary is that everyone has to have authorization from the established government to perform such abstraction and return.
- An adequate legal apparatus is required to govern the process. Even this will not produce results unless there is a willingness to negotiate. Generally such willingness will not be present unless each interested group sees clearly the possibility of trading off benefits and losses in the negotiation.

Privatization and commercialization

The above issues were not the ones that attracted attention during the session on Water and Megacities at the Second World Water Forum in The Hague. The over 200 persons who participated in that session vigorously examined whether a privatization option was the most appropriate solution at city level. While there were good examples of private sector successes, there were also many failures. Further, the normative statement that the public sector is inefficient was challenged with examples such as Singapore. It was also highlighted that the private sector option was not restricted to large multi-nationals, but is also available to the small-scale community based entrepreneurs.

Serious concerns were raised that this basic need and national resource could be considered a commercial commodity. Further concerns were raised around the corruption incidents that were related to some privatization deals. At the same time it was recognized that substantial investment was required. Government has to create an enabling environment for the participation by both private and public partners at all levels.

The conclusion reached was that all the options and combinations should be considered in a rational manner. The warning was sounded that the use of privatization options in the absence of a strong functional regulatory environment would not be prudent for any country, but perhaps especially for developing countries. The goal should always be to ensure that water services in the urban environment are not only accessible spatially, but economically and socially as well.

Framework for Action

The Global Water Partnership followed up on the Vision exercise by producing a Framework for Action. In this[10] they propose four things to do to meet the challenge of urbanization:

1. Governments, UN organizations and donors to strengthen and expand existing initiatives such as the African Water Utilities Partnership, Sustainable Cities Program and Cities Alliance.
2. Municipal authorities to integrate water planning with urban spatial and economic planning
3. Governments to set policy incentives for, and designers to develop, innovative technical solutions such as solar power, desalination and bio-remediation.
4. Municipal authorities to prepare plans for wastewater and solid waste disposal and treatment close to the source of pollution with maximum financially feasible involvement of the stakeholders.

All of these issues will be discussed and debated at the conference "Frontiers in Urban Water Management: Deadlock or Hope?" convened by UNESCO and the Académie de l'Eau to be held in Marseilles, France 18-20 June 2001.

SCIENTISTS AND WATER

About 110 persons attended a special session at the Second World Water Forum entitled "Scientists on Water and Knowledge". A wide ranging discussion led by water scientists discussed the major and growing gaps in our knowledge, reached some conclusions and made recommendations for action.

Conclusions

- Science, including physical, economic, political and social science, when wisely applied, can contribute significantly to saving lives and ensuring sustainable development. Scientific knowledge and the ability to use it is therefore not a luxury, but a necessity and it is cost-effective to promote science and to apply its findings.
- The acquisition of the necessary knowledge requires the scientific study and the presentation of the results in a form that can be used for specified purposes, together with the employment and training of the staff needed to apply the knowledge correctly and wisely.
- There are gaps in our knowledge in many critical areas. The society in which we live is changing, as is the climate and many other factors, and it will never be possible to claim that we know all that is necessary. Science therefore has a continuing responsibility to society to interpret the facts and provide rational advice on which wise decisions can be based.

Actions

- Increased integration within the sciences and improved co-operation between them and those who frame and implement policy.
- Increased efforts to collect, store and analyse data so as to provide the scientific and decision making communities with the critical information that are needed to address water problems.
- Development of indigenous scientific capabilities in developing countries so that they might know better and thus have control of their own water resources and aquatic environment.
- Encouragement of the scientific community to be more actively involved in public debate and of policy makers to heed the advice of scientists.
- Support for new initiatives at international level, such as Hydraulics for Environment, Life and Policy (HELP), so as to bring countries and disciplines together in a common search for knowledge.
- Recognition that combinations of factors can differ in time and place and that solution to problems must often be tailored to specific situations, thus requiring local scientific studies.
- Establishment of regional data bases on groundwater and, if possible, a global centre of information and expertise.

These conclusions and actions may serve to nourish the discussions of our meeting in Erice.

THE WAY FORWARD

As planned, the Vision Unit of the World Water Council closed down its operations on June 30. Its website and files were transferred to the offices of the World Water Council in Marseilles. We can proudly say that thousands of people working together have indeed achieved the objectives of the Vision exercise, for:

Forty reports were produced from the various regional and sector consultations. A common data base on water supply and demand was agreed by modelers who had until then been using different sources. Forecasts were produced of water and food availability for the Year 2025 based on different scenarios. All of this information was shared and discussed by 5500 participants at the Second World Water Forum.

Following four drafts commented on by hundreds of participants – both individuals and organizations, the Vision Unit produced a report that reflected the views of all consulted. It included a CDROM with the full reports of all of the consultations (Cosgrove. and Rijsberman for the World Water Council, 2000). The World Water Commission produced its independent report (A Water Secure World: Vision for Water, Life and the Environment). A further volume (Rijsberman ed., 2000) describing the scenario process and modeling and providing other background information will be released later in 2000.

Over 600 journalists attended the Second World Water Forum. Thanks to the efforts of media consultants, the Vision message was carried to all corners of the globe before and during the Forum by all major television networks and all major newspapers.

The reports of the many groups participating in the vision exercise, including that of the World Water Commission were discussed at a Ministerial Conference in The Hague by 600 delegates from 140 countries, including 120 ministers. They influenced the Ministerial Declaration issued at the conclusion of the session and increased the commitment of a number of countries and donors. In addition, the results of the Vision exercise are being incorporated, through the Global Water Partnership, into the ongoing development of frameworks for action at the national level.

Much effort was made to include all elements of society in the visioning process. It is a measure of the importance of the exercise that at the conclusion of the Second World Water Forum a number of NGOs issued a declaration that expressed their disappointment at not having been involved and indicated their desire to be included in follow up activities.

Providing six times more water now than a hundred years ago has had significant impacts on people and the environment. *The cup is half-empty.*

An unacceptably large portion of the world population–one in five–does not have access to safe and affordable drinking water, and half the world's people do not have access to sanitation. Each year 3–4 million people die of waterborne diseases, including more than 2 million young children who die of diarrhoea (WHO, 1999). More than 800 million people, 15% of the world's population and mostly women and children, get less than 2,000 calories a day. Chronically undernourished, they live in permanent or intermittent hunger.

Much economic progress has come at the cost of severe impacts on natural ecosystems in most developed and transition economies. Half the world's wetlands were lost in the 20th century, causing a major loss of biodiversity. Many rivers and streams running through urban centers are dead or dying. Major rivers from the Yellow River in China to the Colorado in North America are drying up and barely reach the sea.

Most governments heavily subsidize water services - irrigation water, domestic and industrial water supply, wastewater treatment. This is done for all the right reasons (providing water, food, jobs) but with perverse consequences. Users do not value water provided free or almost free - and so waste it. Water conservation technologies do not spread. There are insufficient incentives for innovation.

Unregulated access, affordable small pumps, and subsidized electricity and diesel oil have led to over-pumping groundwater for irrigation and to groundwater tables falling meters per year in key aquifers. As much as 10% (or some 200 cubic kilometers) of global annual water consumption may come from depleting groundwater resources.

In most countries water continues to be managed by a highly fragmented set of institutions sector-by-sector, ineffective for allocating water across purposes. Processes do not provide for effective participation of other stakeholders in decision-making and management. These deficiencies pose major obstacles to integrated water resource management.

Clearly continuing to do business as usual will lead to many more national and regional crises with global implications.

But the cup can also be seen as half-full. A major investment drive, the International Drinking Water Supply and Sanitation Decade (1981–90) and its follow-up—led by national governments and supported through international organizations—ended with safe and affordable drinking water for 80% of the exploding world population and sanitation facilities for 50%.

Major investments in wastewater treatment over the past 30 years have halted the decline—even improved—the quality of surface water in many developed countries. Food production in developing countries has kept pace with population growth, with both more than doubling in the past 40 years.

In perhaps the biggest achievement of the century, rising living standards, better education and other social and economic improvements have finally slowed population growth.

Gleick[11] has pointed out that as traditional approaches to water supply are becoming less appropriate or more expensive, unconventional methods are receiving more attention. He cites the concept and practice of transporting fresh water in large ocean-going plastic bags; large and small-scale desalination technology; water reclamation and reuse; and fog collection. He notes that more and more cities are discovering that wastewater can be a resource, not a liability, for purposes ranging from irrigation to drinking. Matching water demands with available waters of different quality (as practiced in Tunisia) can reduce water supply constraints, increase system reliability, and solve costly wastewater disposal problems.

During the Vision exercise thousands of people around the world have looked the possible crisis in the face and have seen that another future is possible. They have proposed the many actions described in this paper to make their water visions come true.

CONCLUSION

To conclude: there is a water crisis, but it is a crisis of management. We have badly threatened our water resources with bad institutions, bad governance, bad incentives, and bad allocations of resources. Participants in the Vision exercise recognised the crisis, but they envisioned a better world, and developed a strategy for making it happen. As put by the Secretary General of the United Nations: " …none of this will happen without public awareness and mobilisation campaigns, to bring home to people the extent and the causes of the current and impending water crisis" (Annan[12]). It begins with launching a movement to move from vision to action—by making water everybody's business, including that of the scientists gathered here.

ACKNOWLEDGMENTS

The World Water Vision is a programme of the World Water Council. It was carried out in co-operation with a large number of partners who were responsible for some 40 sector visions, regional visions and special studies. Over 200 people commented on early drafts of the World Water Vision reports. More than 15 thousand people contributed to the overall World Water Vision development process. More than 5.500 participated in the Second World Water Forum. The author is deeply indebted to all those people, too many to list here, who provided valuable inputs comments and insights. The remaining errors and omissions are, of course, the responsibility of the author. The Netherlands Ministry of Foreign Affairs was the principal source of funding of the World Water Vision project.

REFERENCES

1. Cosgrove, W.J., and Frank. R. Rijsberman for the World Water Council. 2000. "World Water vision: Making Water Everybody's Business." London. Earthscan.
2. Shiklomanov, I.A.. 2000. "World Water Resources and Water Use: Present Assessment and Outlook for 2025." In: Rijsberman, F.R., ed. World Water Scenarios: Analyses. Forthcoming, Earthscan Publications Ltd., London.
3. Cosgrove, W.J. and Frank R. Rijsberman. 1998. "Creating a vision for water, life and the environment". Water Policy. 1(1998):115-122.
4. The WaterGAP model of the center for Environmental Systems Research of the University of Kassel, Germany (Alcamo et al., 2000); IMPACT of the International Food Policy Research Institute in Washington (Rosegrant and Ringler, 2000); and PODIUM of the International Water Management Institute in Colombo, Sri Lanka (IWMI, 2000)

5. Gallopin, G.C. and F.R. Rijsberman. 2000. "Three Global Water Scenarios". International Journal of Water (in press).
6. Alcamo, J., T. Henrichs, T. Roesch. 2000. World Water in 2025: "Global Modeling and scenario Analysis for the World Commission on Water for the 21st Century". In: Rijsberman, F.R., ed. World Water Scenarios: Analyses. Forthcoming, Earthscan Publications Ltd., London.
7. Flooding in urban areas is not receiving nearly the attention it deserves. In India alone during the monsoons of 1999 millions of people from the cities of Delhi, Mumbai, Calcutta, Chennai, Ahmedabad, and Bangalore were forced to move from their homes and hundred died. A key reason is the increasing occupation of the low-lying lands within the cities, especially by the poor. In mountainous areas, these same poor tend to occupy steep and unstable slopes surrounding the cities. High intensity rainfall brings disastrous consequences in both cases.
8. See, for example, Falkenmark, Marlene. 2000."Competing Freshwater and Ecological Services in the river Basin Perspective: An expanded conceptual framework". Water International, Vol. 12 No.2, pp. 172-177.
9. Braga Jr., Benedito P.F. 2000. "The Management of Urban Water conflicts in the Metropolitan Region of São Paulo. Water International, Vol. 12 No.2, pp. 208-213.
10. Global Water Partnership. 2000. "Towards Water Security: A Framework for Action". GWP. Stockholm.
11. Gleick, Peter H. 2000. "The World's Water 2000-2001". Island Press. Washington D.C.
12. Annan, Kofi A. 2000. We the Peoples: the role of the United Nations in the 21st Century.Millenium Report of the United Nations Secretary General. United Nations, New York.

DELHI: A THIRSTY CITY BY THE RIVER

K.C. SIVARAMAKRISHNAN,
Professor, Centre for Policy Research, Delhi

Delhi today is a Megacity of more than 12 million people. Half of it is comprised in the "Union Territory", an area of 1500 sq.km for which the Central Government assumes much of the administrative responsibility. Close to 9.5 million people live in this territory. An estimate in preparation for the census of 2001 has reported the present figure as closer to 14 million. An area of another 1,700 sq.km provides a ring around this territory with half a dozen towns of rapid growth, which together have a population of more than 3 million. Beyond this ring lies an area of another 27,000 sq.km identified by the planners as the National Capital Region with a population of about 5 million. Much of this region is still countryside but there are another six important urban centres providing a range of functions from manufacturing to agricultural marketing. The population of each of these centres ranges from two to four hundred thousand. With the surrounding agricultural land, much of it irrigated and productive, the NCR, consisting of the union territory, the metropolitan area and the surroundings, is one of the largest urbanised regions in the world comprising more than 30,000 sq.km and is home to some 18 million people.

Within this region, the main city of Delhi is a fast growing, fast changing entity spreading its fingers beyond the metropolitan area and reaching towards some of the other towns in the capital region in a radical fashion. Its growth rate has averaged above 30% during the past three decades. Density has increased tenfold and it is now around 13,000 sq.km in the union territory and is growing in the metropolitan ring. The Megacity is therefore a contemporaneous event, growing in size and shape.

DELHI HISTORY

Delhi has a long and hoary past. Since memory, there has always been a Delhi: not one, but seventeen in seven sites of proximity, which have lasted for a few years or a few centuries. Gerald Breese, Urban Planner and historian and leader of a group assembled in 1957 to prepare a plan for Delhi, captures the long sweep of Delhi history as follows:

"Destiny and Delhi have for well over a thousand years been part and parcel of one another. Evidence from written and unwritten history confirms the connection. No matter at what point the record is examined it becomes clear that Delhi's unique situation vis-a-vis the Indian sub-continent has inevitably embroiled it in conquest for control.

Why does Delhi have such a long and storied past? Geography and the ambitions of men provide an explanation. Delhi occupies a site that has been the repeated focus of invasion. It lies south of the Himalayan Mountains and so has been generally protected from that direction. But it is also only somewhat over two hundred miles from passes through the westerly Hindukush, Sulaiman and Kirthar mountain ranges, whose main passes provide access to the Delhi which also lies at the beginning of the Upper Ganges Valley. It appears to have been relatively easy for invaders to have crossed the Punjab Plains of the Indus Valley to seize the city that provided the key to domination of lands beyond.

Over and over through time invasion and conquest have accompanied the pendulum's swing. The city's inhabitants have been exposed to an endless succession of death and destruction, pestilence and privation, fire and famine, and even earthquakes and exoduses. But always there has been a Delhi rising, Phoenix-like, from the onslaughts and the remains left in their wake. If there ever were a city whose past is inextricably intertwined with its present, it is Delhi."

A CITY BY THE RIVER

The river Yamuna, second longest in the country has been a major reason for one city after another to rise on its banks. At least in the past, the river held the promise of plentiful supply. The occasional changes in its course have also forced the shifting of the sites from time to time. Delhi's other attraction has been that it is at the entrance to the vast and fertile Gangetic plain and the trading and the textile realms of Bengal to its east. Yet all the rulers were not entirely happy with Delhi. Akbar the great Mughal preferred Agra, two hundred kilometers downstream of the same river. So did Jehangir, his son. Shahjehan the grandson regarded as a 'city-planner among kings and a king among planners' built the vast Red Fort in Delhi, the Great Jama Masjid and "the magnificent city of Shahjehanabad with wide streets and parks within a wall whose circumference was four miles". But the great builder himself did not derive much happiness from his city. The yearning for his beloved queen Mumtaz Mahal could find expression only in Taj Mahal, raised not in Delhi but in Agra. However the city Shahjehanabad bearing his name prospered and declined alternatively. There were earthquakes in 1782, 1803 & 1829. The British arrived in 1803 when Shahjehanabad's population was estimated to be around 150,000. Between all these vicissitudes of life the city traded, artisans flourished, writers and poets found a receptive court and the survivors of the Mughal dynasty seemed to hold a sway at least over the hearts of their subjects, if not the land and its revenue, which had already passed into the control of the East India Company.

DELHI AND THE BRITISH

Then came 1857 and the Sepoy Mutiny as the English called it or the 'First war of Independence' as the Indians claim. Gun power prevailed. The last of the Mughals, Bahadur Shah was banished to Burma after his two young sons were executed by Canon fire. The British crown took over the reins of the government from the Company but the capital still remained in Calcutta, a trading settlement that the English had set up in 1697, which was now a centre of industry and an entrepot through which the wealth of India was siphoned off to the west. On December 12, 1911 George V of England was crowned the Emperor of India at a great Darbar where erstwhile rulers, nobles and merchant princes of India assembled to acknowledge the Empire and pay homage to it. Yet another dynasty was to be established in Delhi: the British succumbed to the power of the saying "he who rules Delhi, rules India". The Emperor duly announced that the capital of the country would be transferred from Calcutta to Delhi.

"The Delhi Town Planning Committee was recruited in January 1912 with Edwin Lutyens as the chief planner. Between April and May the planning team surveyed on foot, car and elephant back, an expanse of 25 sq.miles, south of Shahjehanabad for setting up the new capital and a cantonment. The plan that emerged qualified to be the shortest ever published for a capital city. "It is a perfect example of a western transplant that bears little or no resemblance to the cultural environment in which it is placed: a more un-Indian plan could be scarcely imagined. Its creators considered it a garden city, the first for India. It was sanitised and separated symbolically from Shahjehanabad by an open space."

The Central point of the plan was the Viceroy's house and two blocks of government offices. The theme of Colonial rule was evident. "Liberty will not descend to the people" admonished the writing over the archway of one of the entrances. The people must raise themselves to liberty". East and in front of the Vice Regal's house a vista was provided. On either side, hexagons, triangles, circles and squares were arranged to connect the dwellings. Indian princes, shorn of power but seeking imperial patronage, were "invited" to build their Delhi residences, rich and imposing, but modest in relation to the Vice-Regal's house. Bungalows were to be constructed for major and minor officials of the government strictly in order, descending in rank but increasing in distance. The lowest in rank would have the farthest to travel to work. But golf clubs and a race course were provided for the Colonials. Road connections to old Delhi and the region were ignored or inadequately provided. No tramway, by then a well-known feature of most European cities, was provided as they "would not be likely to give a satisfactory return". The land was retained in government ownership. Lutyens' Delhi vastly altered Delhi's characteristics.

World War I greatly slowed down construction. In 1923, there was a major shift in policy, precipitated by a request to locate the University not in New Delhi but north of the old city. Some thinking began on better links with the old city. In the meantime ribbon development was creeping between the old and the new. The new capital was officially opened in 1932, planned for 3,200 acres and a maximum of 65,000 persons. The old city already had some 300,000. Together the two crossed half million by 1941.

INDEPENDENCE AND AFTER

Independence in 1947 also brought the partition of the country and waves of refugees. At least half a million came to Delhi. By 1961 Delhi's population was 2.6 million and ten years later, 4 million. The 1981 and 91 census showed further increases to 6.2 and 9.4 million. When the census is taken in 2001, Delhi's population is likely to exceed 14 million. In these 3 decades the population had grown by 255%. As Delhi grew, so did its environs.

Refugees and the requirements of the capital of an independent country were not the only causes for this growth. There have been major shifts in the functions of the city as well.

From a seat of government it has become a centre of industry. The number of industrial units rose from 26,000 in 1971 to 126,000 in 1996 and the employment from 215,000 to about 1.4 million.

From a trading post on the crossroads and a supply base for its immediate hinterland, Delhi has become a major hub of wholesale trade in the country. Nearly half of the food grains, fuel and oils, construction materials and as much as 80% of the fruit and vegetables flowing into Delhi are not for Delhi's consumption but for export to points beyond the national capital region.

Employment in government and quasi-government offices has risen from 125,000 in 1961 to over 600,000 in 1991.

The total workforce in Delhi now exceeds 3 million, of which 30% is in Trade and Transport, another 25% in government and quasi-government jobs, 25% in manufacturing and 20% in other categories.

MIGRATION

Delhi today is a magnet for migration. Its per capita income of Rs.19,800 ($450) is more than double the national average of Rs.9300 ($211). The northern Indian states of Bihar, Uttar Pradesh, Madhya Pradesh and Rajasthan are the most populous states of India with high demographic growth. Set in this high growth area, Delhi has been a natural destination for migrants. Between 1981 and 1991, half the increase in Delhi population was due to migration estimated around 1.6 million. Half of them came from UP, another 11% from Bihar, nearly 2,000 km away while 18% came from the neighbouring states of Haryana and Rajasthan. Estimates of migration are not easy to establish in India. Origins and destinations of migrants are not easily available or published unlike other aspects of census data. Inter-state migration in particular has been difficult to assess because of fears that open access to and publicity of such data may aggravate ethnic and linguistic differences within the country. But the inflow into Delhi is measured by various proxies like the issue of ration cards, voter identity slips and so on. According to one estimate the annual migration into the Delhi union territory is about 160,000.

Unlike so many other places in the country, Delhi can claim no natural advantage for industry. It has no raw materials: nor is it a major market for industry. But as the nation's capital, Delhi was well provided with infrastructure services. At least until recently its water supply, roads, transport and communication have been better compared to other states. Using its influence as the seat of the national government it has been able to draw most of the electricity it needs from the regional grid although neighbouring states suffer power cuts of 20 to 40%. The disparate tax regime in the region has also helped Delhi. Taxes on sales, motor vehicles, and fuel are 2 to 10% less than in the neighbouring states. There is no tax levied on food grains, which has needlessly made Delhi into a wholesale trade centre. Services are heavily subsidised. All these factors have contributed to making the Delhi magnet even more attractive. Today it is under severe strain, unable to service what it has attracted.

THE IMPACT OF DELHI'S GROWTH

Shelter for the growing population has been under severe strain though Delhi has perhaps done more than any other state in housing. For those employed by the government there are large housing estates comprising some 250,000 houses, all stratified and unattractive, but made available at heavily subsidised rents and services. The Delhi Development Authority, set up to monitor and enforce the master plan, became a major real estate developer itself and has built about 280,000 dwelling units. Significant as these numbers are, the private sector housing is nearly twice this number. It is estimated that there are altogether some 1.4 million formal housing units where about 6 million people reside. That still leaves over 3 million people to fend for themselves. Delhi, like other Indian cities, has its own variety of informal or substandard housing. There are the old decrepit buildings in the walled city, officially classified as slums. There are resettlement colonies, crammed and tiny dwelling houses built on small plots provided by the government. And then there are squatters, mostly on public land, living in shacks and other make-do structures. It is estimated there are at least 1100 such shanty settlements comprising some 600,000 dwellings.

Slums are not exclusive to Delhi but their growth and proliferation are more recent here than elsewhere. Most of the squatter settlements have been on government owned lands designated for specific projects but these lands were poorly identified and hardly guarded. Squatting on such lands was easy. The Authorities felt that providing services to much squatter settlements would legitimise the squatting and also pre-empt the land from the proposed projects. Alternate accommodation or developed sites was therefore adopted as the preferred approach for slum removal. Between 1960 and 85, about 250,000 families were provided. However due to political pressure and administrative weakness, the alternative house sites or accommodation became virtual give aways. This itself is cited as one of the reasons for the rapid growth of squatter settlements on public lands in Delhi.

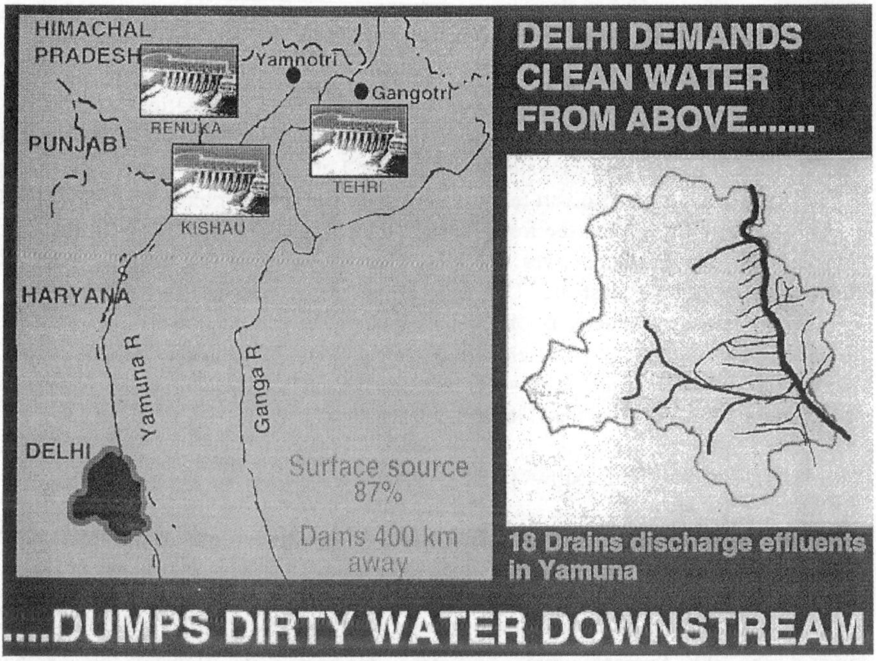

WATER AS A DEFINING LIMIT

Water has always been a question mark over Delhi. Excepting Shahjehanabad other Delhi's have had to contend with a shifting river and a tenuous supply, though during the monsoon the river rise was ominous and flooding frequent. The current demand is of 4,200 MLD as against the supply of 3,000. Surface water is the principal source. Nearly half the raw water supply is taken from the upper Ganga canal and other 40% from Bjakra Dam storage and Yamuna. Only a little more than 10% of the total supply is taken from groundwater. Consumption has increased four times in the last two decades, most of it for domestic use. Industry's use is only 14%. Unfortunately as much as 26% of the treated water is "unaccounted", a euphemism for leakage, wastage and other losses in distribution.

FRESHWATER SUPPLIES

Delhi's water supply has to reckon with problems of both quantity and quality. Much of the flow in Yamuna, regarded as one of the great rivers of the country, fed partly from the snows of the Himalayas, but mainly the streams below, is abstracted for irrigation, upstream of Delhi. The problem is severe during the so-called lean season or dry months of the year. Nearly 85% of the annual run off in the Yamuna basin is of 'monsoon' flows during the five months of June to October. The balance of 15% is the flow during the remaining seven months. It is also during this period that water for irrigation has to compete with other uses such as domestic, industrial, pollution control ecological needs, etc. Abstraction of water for irrigation has a long history. Work on the Right (West Bank) Canal began at Tajewala (see map) as early as 1356 AD by Feroze Shah II, a pre-moghul ruler. It was remodelled and extended by 1892. The Left Bank Canal was completed by 1852. Together these canals divert the lean season flow in the river almost entirely. Delhi has had to depend mainly on the fresh flows into the river below the Tajewala Barrage. Yamuna being an interstate river, the allocation of water to the riparian states of Himachal, Haryana, Rajasthan, UP and Delhi is determined through interstate agreements. The physical delivery of water itself is regulated through a complex network of canals, feeders and escape channels. Calculations of demand, actual supply, evaporation and other water losses enroute are matters of frequent controversy and reconciliation through political and judicial interventions.

POLLUTION AND CONTAMINATION

The problem of quantity is further compounded by pollution and contamination of various channels of delivery as also the main river itself. From time to time water supply treatment plants in Delhi have had to shut down as the raw water became heavily polluted and the treatment processes and equipment could not cope. There were fourteen such shut downs during 1994. In recent years the frequency has come down. However the quantum and range of pollutants discharged into the Yamuna River and the canal system upstream of Delhi continue to be a matter of concern. The West Bank or Western Yamuna canal for example receives nearly 80 million litres of industrial wastes containing heavy metal residues like cadmium, nickel etc. Fertiliser and pesticide residues from the intensively cultivated agricultural area are another problem. Pollution and contamination thus undermines the availability of fresh water and adds considerably to the cost of treatment. Unmindful of this high cost it has to bear, Delhi itself discharges about 2,000 MLD of untreated wastewater into the river downstream. This is nearly two thirds of the water it receives from the Yamuna system and other sources upstream.

WATER SUPPLY AND DISTRIBUTION

Delhi City has 4 major water supply treatment plants, which together have a capacity of about 2,400 million litres per day. Work in progress for two locations and proposed in another three will add another 1,300 million litres. In addition the system also makes use of radial or Ranney wells along the bed of the river Yamuna and tube wells in different locations which together yield another 300 million litres. This of course excludes the water, which is extracted by private parties from tube wells for which no reliable estimates are available. The technology is conventional based on flocculation and rapid gravity filters with chlorination. The total OQM cost, including interest and depreciation is about Rs. 1,200 million ('92 figures) against which the revenue from tariffs was less than a third at about Rs.360 million.

DEMAND SUPPLY GAP

Period	Population (million)	Demand (MLD)	Supply (MLD)	Gap (MLD)
1951-56	2.17	670	270	400
1961-66	3.29	1020	580	440
1990-91	9.42	2900	2070	830
1998-99	12.85	4000	2600	1400
1999-00	13.41	4200	2900	1300

IS THE DEMAND REALSITIC ?

ARE THE SUPPLY DATA CORRECT ?

IS PER CAPITA SUPPLY FIGURE ACCEPTABLE?

The bulk of the water supply distribution is through direct connections, which number about 100,000. The distribution system is highly iniquitous. The average per capita supply for the city as a whole is 225 litres per capita per day, but this is no more than an arithmetical figure based on the assumed quantity of water production divided by the estimated population. The distribution can be broadly divided into 3 categories. Out of the total estimated population of 13 million, 6.2 million who are in planned areas receive nearly 1,400 million litres, i.e. about 55%. The innumerable unauthorised colonies, slum, slum clusters, etc. where nearly 5 million people live, receive about 240 million litres which is just 10% of the supply. The rest is supplied to the so-called resettlement colonies where slum dwellers and persons displaced by public acquisition of land have been resettled. Even this broad three-fold categorisation obscures several realities. In many slum clusters, per capita availability may be less than 20 to 30 litres. Conversely in many affluent areas, treated water availability exceeds 400 litres per capita per day in the small British built area of Delhi and the Cantonment, and is used for a variety of household purposes including gardening.

Wastage is a significant item in water supply systems in South Asia. In Delhi it is estimated that out of 2,700 million litres produced, nearly 600 million litres, i.e. about

22%, is lost in leakage, transmission losses and overflow. The city administration has now taken up an extensive repair and rehabilitation programme for plugging these leaks and repairing the distribution networks.

While waste, unaccounted-for water, and the serious inequities in distribution continue to be major unresolved problems, Delhi's thirst and demand for additional water are ceaseless. Delhi is seeking to more than double its water availability by negotiation for shares in the releases from dams under construction/proposed, such as the Tehri Dam on the Ganga and the Kishau and Renuka on Yamuna's tributaries. All three are capital intensive, financially constrained and environmentally controversial mega projects, which may take another couple of decades to be functional, even if commenced now. All three dams are also more than 300 km away from Delhi and the conveyance of water will need significant changes and augmentation of the existing supply routes.

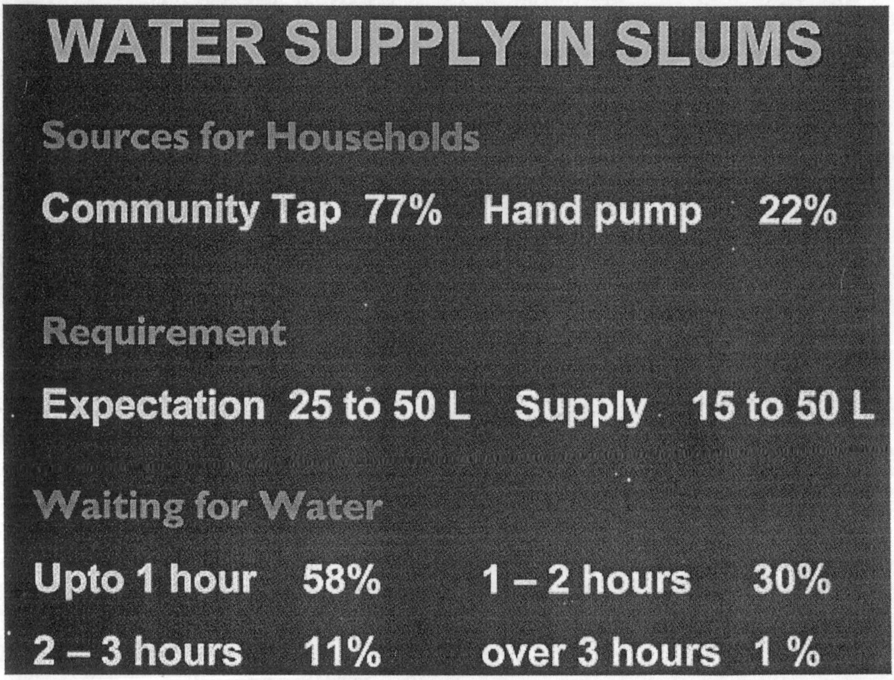

The Delhi Water Supply and Sewerage Disposal Undertaking (DWS&DU) used to be responsible for bulk water supply as well as the sewerage system in the entire municipal corporation area. The Delhi Municipal Corporation itself was set up in 1958. Its jurisdiction of about 1,400 sq.km covers most of the Union Territory of Delhi. The New Delhi Municipal Committees area of 43 sq.km accommodates the President's residence, central government offices and residences and several public institutions like the Museum, art galleries, parks etc. The Cantonment Board looks after the military area of another 43 sq.km. These bodies receive their bulk water from DWSDU. This undertaking itself was a subsidiary of the Municipal Corporation. However, the undertaking has recently been made into an autonomous water supply and sewerage board, called the Delhi Jal Board.

Elsewhere in the national capital region, water supply and sewerage is handled by state level para-state organisations, though in some cases individual municipalities handle the distribution. In terms of quantity of water supplied and distributed this is not a high figure.

WASTEWATER TREATMENT AND DISPOSAL

As many as 17 drains discharge about 2,700 litres of wastewater every day into Yamuna which is as much as its present water supply from the public system. This is not just domestic sewage but a combination of sewage and industrial wastewater. As regards treatment, Delhi itself was one of the earliest in the country to set up a comprehensive sewerage system. A major sewage treatment plant at Okhla with a capacity of 650 million litres, one of the largest in Asia, was set up as early as 1936. While the technology was the conventional activated sledge process, the Okhla plant was one of the first in the country to recover methane gas from sewage and use it to fire dual fuel engines which generated a part of the electricity required for operating the plant. Five other treatment plants were set up in the following years with a total capacity of 1,300 million litres.

Discharge of wastewater has been a major public concern, particularly in the context of the polluting of the Yamuna River which is considered one of the important holy rivers in the country. In 1994-95, following public interest litigation, the Supreme Court of the country directed that sewage treatment capacity of the Delhi City should be increased to about 2,200 million litres. Since then the Delhi Jal Board has been engaged in constructing 16 additional sewage treatment plants in various locations. Out of these, 9 have been completed already and the rest are targeted to be completed next year. All the STPs continue to use conventional technologies such as trickling filter or activated sledge method. Oxidisation ponds have been used in one location with a capacity of 27 million litres, but due to the high cost of land, this is not a favoured option in Delhi.

The irony of the situation is that while the sewage treatment capacity is being stepped up considerably, the sewerage network itself, particularly the trunk sewers, are in a dilapidated condition. It is estimated out of 140 km length of trunk sewers, for a length of 53, the siltation is as much as 70%. In another 7 km the trunk sewers have virtually collapsed. The rehabilitation of the sewer network has therefore become a major challenge.

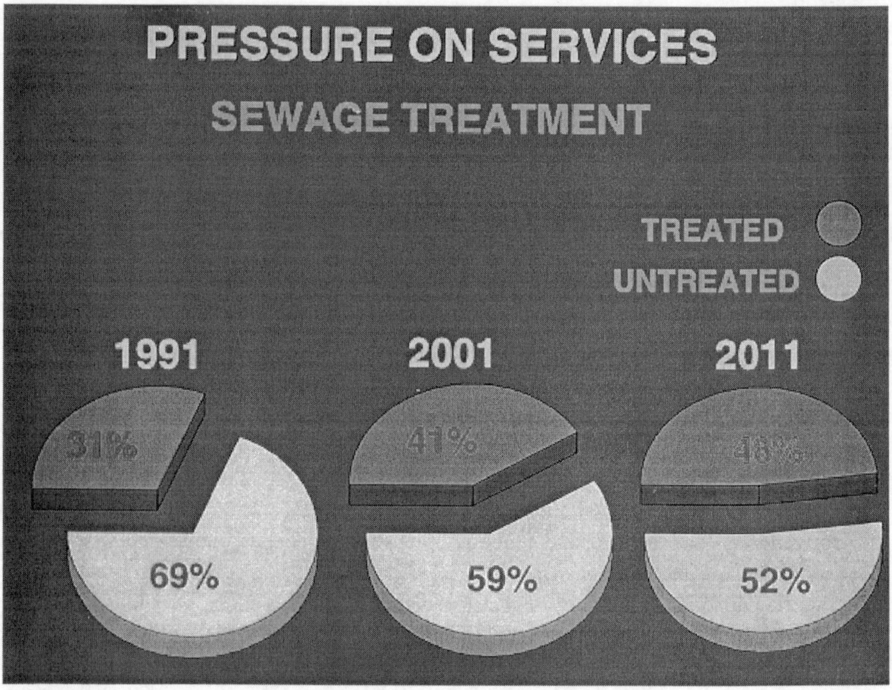

So far as industrial wastewater is concerned, within the city of Delhi most of the units are medium or small scale without pre-treatment or treatment facilities of their own, and which discharge their effluents into open drain and municipal sewers. Only a small percentage of these industrial effluents are treated as part of the municipal sewerage system. The rest find their way into the river through the open drains. Outside of Delhi within the national capital region, large scale industries have their own treatment plants but the majority, which are in the medium and small scale, discharge untreated or partly treated waste into open drains and other water courses.

The Central Pollution Control Board under the Ministry of Environment and Forests has prescribed elaborate standards for treatment of different kinds of effluents and has prescribed end of the pipe norms for effluent discharge. The enforcement of these standards however, is left to the Pollution Control Boards of the different states. It is common knowledge that enforcement has been slack. As far as the municipal authorities are concerned they have not been assigned any role in formulating or enforcing the standards or monitoring the treatment facilities.

RIVER ECOLOGY

In recent years, pollution of water and air have become the subject of frequent judicial intervention. As a result of public interest litigations filed before the Supreme Court, the Court has given directions from time to time about domestic as well as industrial wastewater discharge and their treatment. The Court has also directed that a minimum flow should be maintained in the river Yamuna for ecological purposes. The Ministry of Environment at the national level has been enjoined to formulate specific proposals in this regard. For the past two years, the existing interstate agreements for sharing of the Yamuna water, the augmentation of wastewater treatment, their capacities, compliance with norms for discharge of effluents, monitoring of the performance of utilities and regulatory authorities etc., have all become important tasks. However, the overall shortage of water, competing demands of riparian states as well as multiple uses within each area have stood in the way of consensus and a common programme to go ahead. Furthermore, as far as the public are concerned the gross inequities in water supply distribution and the fact that polluters have managed to escape penalties under law have not helped in building the public support needed for effective pricing and demand management. Determining where public interest lies and safeguarding that interest has been a very difficult task.

STRAINING THE ENVELOPE OF SUSTAINABILITY

Delhi's woes are not limited only to water and wastewater. The city has been adding one motorised vehicle, every six minutes, every day, for the past few years. It has 3 million now which is more than what Bombay, Madras and Calcutta together have. About 67% of the vehicles in Delhi are low cost, but are highly polluting scooters and three wheelers. In Delhi the vehicle population ratio is nearly 1:3. Where private vehicles proliferate, public transport suffers. Total of transit trips in Delhi today is 12 million: 62% of that is performed by buses which have to compete for road space with private vehicles whose number is ten times more though their passengers are ten times less.

Delhi is more fortunate in its road space which is 16% of the total land use. Yet its 25,000 kms of roads are not enough to cope with the rising number of vehicles. Average travel speed is already down from 18 km/h in 1994 to 15 now. Two things seem certain. At the present rate of increase, the total number of vehicles is expected to reach 4.5 million in 5 years and 6 million in 10 years. Vehicles will then crawl at 5 km/h: the pollution load from vehicle emission, which is already 3,000 tons per day, will go up to 7,000.

Delhi needed a mass transit system years ago. After several expert group reports and feasibility studies, work finally started last year. The Mass Rapid Transit System is a combination of elevated, surface and underground railway lines totalling 55 kms. Being built at a cost of U.S. $15 billion, the system will have a capacity of 40,000 passengers per hour. Unfortunately it is a stand-alone system unconnected to the region's existing and proposed rail lines.

Running out of water, transport and clean air to breathe are therefore real challenges that Delhi has to face. Added to this is the growing incidence of crime and violence. Patricide and fratricide were frequent among Delhi's princes coveting the throne. Invaders like Timur and Nadir Shah have laid the city bare with death and devastation. Violence among communities has been a chronic feature before and after independence. The Father of the Nation, Mahatma Gandhi, hailed as an apostle of peace, was assassinated in this city. Indira Gandhi fell to the bullets of her own guards some forty years later. Revenge and retribution, violence and vendetta have lain dormant in Delhi spouting ever so frequently and put down ever so briefly. Prosperity, rising incomes and the long arms of law under the central government were expected to ensure law and order. But crime in Delhi has grown. The incidence of murder and armed robberies rise by 10% to 11% annually: burglaries and theft by 7%. The number of criminal offences registered in 1995 was 75,000, about 30,000 more than in 1998. Flouting law and using political patronage to escape its consequences is now a part of Delhi culture.

IS DELHI THEN DOOMED?

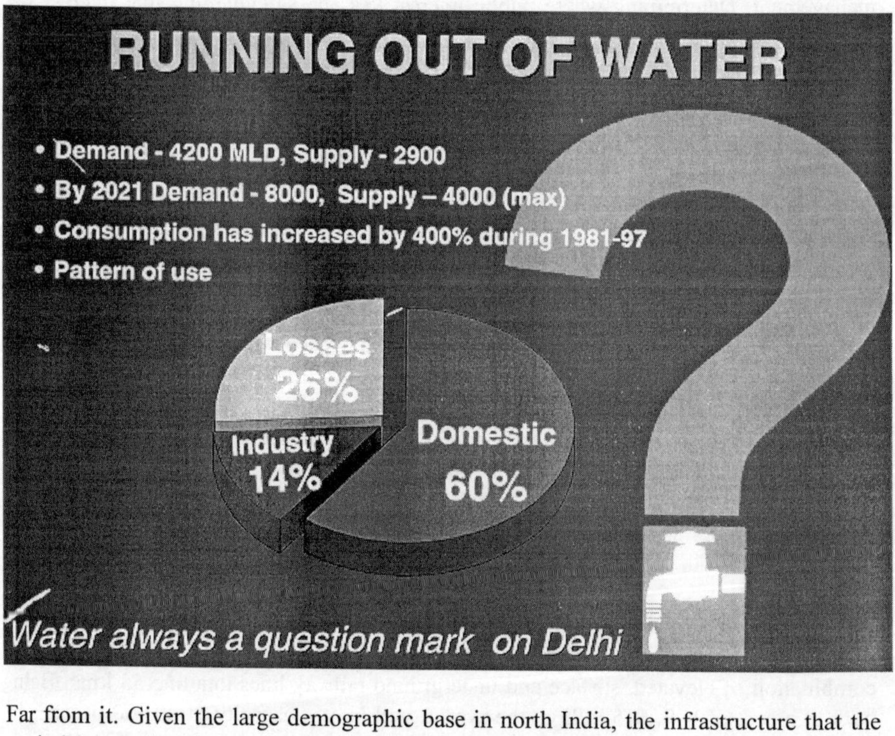

Far from it. Given the large demographic base in north India, the infrastructure that the capital has gathered over the decades, its easy access to the country and the world, and the aggressive entrepreneurship of its people, there will always be a Delhi. It has the money and the means to set itself right. Indeed the search for prescription has been perennial. After Lutyen's plan of 1912, a review in 1938 by the Delhi Development Committees and the recommendation of the Birla Committee in 1951, an Interim General Plan was prepared in 1956. This was a prelude to the preparation of a comprehensive master plan to become effective in three to four years. The Ford Foundation offered the government a team of consultants to help its Town and Country Planning organisation. They and the Indian experts worked hard to prepare the draft Master Plan for Delhi which was published in 1960.

THE 1961 MASTER PLAN

The plan recognised that Delhi could not remain only as the seat of administration and that its function would expand to include industry and commerce. It was however anti large industry and hoped these would be located in the Capital Region rather than the city. Migration was expected and even encouraged by earmarking zones where annually, 70,000 immigrants could come and put up cheap houses in a layout of some standards. Obnoxious Industries would be shifted to other locations. Railway yards would be shifted to the other side of the river and terminals to the periphery. Shopping and marketing would be decentralised through 15 district centres. Employment within Delhi would be kept within limits by shifting offices to the metropolitan area around the city.

THE SET UP FOR NON-GOVERNANCE

Unfortunately, Delhi's administrative set-up has been a "grand design" for non-governance. A 'union territory' from 1957 to 1993, the administration was handled directly by a Chief Commissioner and later a Lt. Governor appointed by the Central Government. Law and order, revenue and land management functions were his major responsibilities. He was also the chairman of the Delhi Development Authority set up under a special act in 1957 to implement and enforce the master plan. A metropolitan council, as a significantly reduced version of a provincial assembly functioned from 1966 to 1993 consisting of elected councillors. The Council was assigned some functions like health and education. The Delhi Municipal Corporation itself had been established in 1884 as a partly elected, partly nominated civic body with a limited range of municipal functions. City planning and land development was not one of them. Additionally, a New Delhi Municipal Committee performed some municipal functions for the limited area of the government houses, offices and residences in the area, broadly corresponding to what Lutyens had planned. The Cantonment was yet another separate entity.

In 1993 Delhi became a separate state with a legislative assembly of 70 elected members and a Council of Ministers headed by a Chief Minister answerable to the Assembly. The form is similar to what prevails in other parts of the country, but law and order and land are still subjects handled by the Central Government directly through the Lt. Governor. The Delhi Municipal Corporation area has been enlarged to cover most of Delhi's urban limits. It now has 134 elected councillors. The New Delhi Municipal

Committee and the Cantonment continue as separate entities. Delhi also has its share of members in the national Parliament. The 7 MPs, 70 Assembly Member and 134 councillors occupy the same political turf. So do the central, state and the municipal governments. Who does what is a perpetual mystery to the public.

PLANNING FOR THE REGION

Against this background the concept of the National Capital Region stands out as bold and imaginative. The need for regional planning was recognised in the stages of the '61 Master Plan. A region of some 30,000 sq.km was identified which, according to the 1961 census, had a population of 10.7 million.

Projections indicated that by the end of the century there could be 32 million people in the region, half of whom would be living in urban areas. Planners considered possible patterns of urban growth in the region such as satellite towns, additional metropolitan centres and multi town development. A radial corridors plan putting together the strong points of the other three was agreed upon. The Ministry of Works, Housing and Urban Development took the lead. Its Town and Country Planning organisation published a draft NCR Plan in 1971 with the lofty objective of "maximising growth by a planned development of a National Capital Region and also to mould and refashion the region both physically and economically for a further realisation of wider and deep social values".

It took fourteen more years for the National Capital Region Planning Board Act to be passed by the Parliament with the concurrence of the neighbouring states of Haryana, Uttar Pradesh and Rajasthan. The Central Government Urban Affairs Minister is the Board's Chairman. The Chief Ministers of the four states including Delhi, the Central Government Ministers of Railways, Transport, Power and Telecom, the Lt. Governor of Delhi, and the ministers of urban development from the participating states are its members.

The NCR Board obviously does not lack in rank or status. But what does it seek to do and what has it achieved?

Population: The '61 Master Plan projected that Delhi's population would be 5.5 million by 1981 but it reached 6.2. A revised plan published in 1990 projected the figure to 12.8 million by the end of the century. It has already crossed that point. The NCR Planners had earlier proposed it should be kept at 11 and future increases deflected to the region. That plan sought to develop 6 towns in the metropolitan area and 8 towns beyond, on a priority basis, and provide better road connections between these priority towns, regional rail links which would by-pass Delhi, a rapid transit system with and between the towns in the metropolitan area, improved power supply and better telecom.

Industry and employment: NCR Planners had also argued that Delhi does not need more industry. Pollution load is already severe and public interest litigation has prompted the courts to order the closure and shifting of 40,000 noxious industries. Yet Delhi would like to keep them by changing its land use plans and setting up new industrial zones.

Disparate tax regime: NCR proposes a Common Economic Zone which would end the disparities in taxes which attract activities into Delhi and aggravate the pressures. Recent efforts to rationalise taxes on sales in the whole country may also help deal with this problem.

Future growth: Within the Common Economic Zone, Delhi would have restricted growth, the metropolitan area a normal growth and the rest of the region an accelerated growth. To help in the process the NCR Board has assembled a modest fund to enable the participating states to develop the priority towns.

THE PROSPECTS

The vision of a Common Economic Zone however is not fully shared. The four participating states continue to maintain a competitive rather than a complimentary approach. Delhi in particular is loath to any suggestions to reform its tax regime and increase the rates to bring about parity in the region. The other states are attracted by the prospects of growth in the region but would like the central government to bear the costs of developing the infrastructure. Some small steps have been taken to involve the private sector and take up joint projects for infrastructure. But these efforts are limited to parts of the metropolitan area and the scale hardly matches the need.

While the multiple governments and their institutions in the region share the fear that Delhi's growth may not be sustainable, they continue to wrangle about the means to spread that growth more rationally. Vexed by their inaction and prompted by various public interest cases, the High Court and the Supreme Court have intervened repeatedly. As mentioned before, such intervention has been prominent in dealing with water and wastewater.

In Oct '96 the Supreme Court ordered the closure of 39,000 industrial units operating illegally in the residential areas. In subsequent decisions the court has banned the supply of leaded fuel in Delhi and ordered that commercial vehicles more than 15 years old should not ply in Delhi. In related judgements, the Courts ordered the executive to take up a programme for increasing public buses and restrain licensing of private cars not complying with Euro I and Euro II emission norms. The Supreme Court has also ordered the establishment of an Environment Protection Authority for Delhi.

These are all landmark judgements but when the governments are unprepared, vested interests active and the public uninformed, the effect of the judgements are at best temporary. As in other megacities, many worlds co-exist in Delhi. The domain of the politicians, the bureaucrats, the captains of industry, merchant princes, traders, manual labour, office staff, teachers or transport workers—all relate to the city in a limited way within the confines of their needs and hopes. Public interest is hard to define and harder to uphold in the best of circumstances.

It is far more so in a fast growing, fiercely competitive city like Delhi. Besides, for centuries its identity has been fractured by the shifting interests of its conquerors. It has been the capital of a democracy for only 50 years now. Yet control and command have been the organising principles for administering the capital city. Despite elaborate and time consuming exercises of planning and legislation, sharing and participation have been waived in terms of patronage and political power rather than of ideas and participation.

A fractured society cannot have a vision. Even if the planners struggle to conceive one, public interest and popular support will be hard to come by. A vision also needs realism. Accepting the limits to growth is a critical component of that realism. The question is whether such realism is within the grasp of any fractured society as in Delhi or other mega cities.

ACKNOWLEDGMENTS

The paper is based on extensive research undertaken by Mr. P. Sisupalan. The Visual material has been prepared by Mr. Arvind Kumar Batt and Mr. Dilip Kumar. Ms. Paramitta Datta and Ms. Sarala Gopinathan helped in getting the manuscript ready.

THE QUESTION OF WATER IN METROPOLITAN BUENOS AIRES

JUAN MANUEL BORTHAGARAY
Universidad de Buenos Ayres, Buenos Aires, Argentina

AVAILABILITY

<u>Fresh Water Resources</u>
The Rio de la Plata is both the supplier of fresh water and the final destination of the waste water discharges of the vast conurbation defined, somewhat arbitrarily, and for census purposes, as the Buenos Aires Metropolitan Area (AMBA). The area houses 11 million people in an extension of 3.800 Km^2.

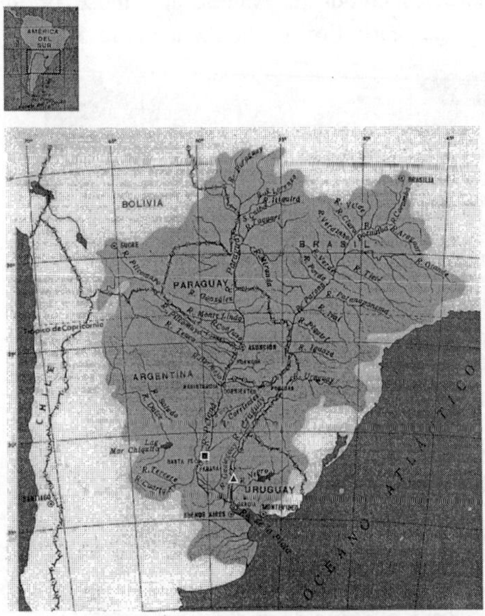

RIO DE LA PLATA RIVER BASIN 3.200.000 KM^2
■ RIO PARANA 15.500 m^3/seg. annual average
▲ RIO URUGUAY 6.000 m^3/seg. annual average

The sovereignty of the river is shared by the coastal states of Argentina and Uruguay, and is ruled by international treaties. As far as fresh water is concerned, the area of the river can be divided into two areas, one that extends from the confluence of the Parana and Uruguay rivers down to an imaginary line that runs between the northern cape of the Samborombón Bay (Punta Rasa) in Argentina to the city of Montevideo in Uruguay.

The area has a length of 180 Km and an average width of 60, therefore it covers around 11.000 Km^2 (11.000 million m^2 with an average depth of 3.5, that is about 40.000 million m^3). It includes both the upper and middle sections of the river. The lower section, going from the far end of Samborombón Bay (Cape San Antonio) to Punta del Este, with 16.000 Km^2 is where the fluvial and oceanic waters gradually mix, with salinity lines drifting subject to tides and winds.

The Parana comes into the Rio de la Plata through an extensive delta, that grows into the river at a ratio of 70m. per year. It carries a sizable amount of clay particles (90% of which originated in the Bermejo (reddish) River, that drains north Argentina and southern Bolivia). These particles colour its waters, and those of the Plata, with a characteristic brown. The average yearly flow of the Parana is 15.500 m^3 per second.

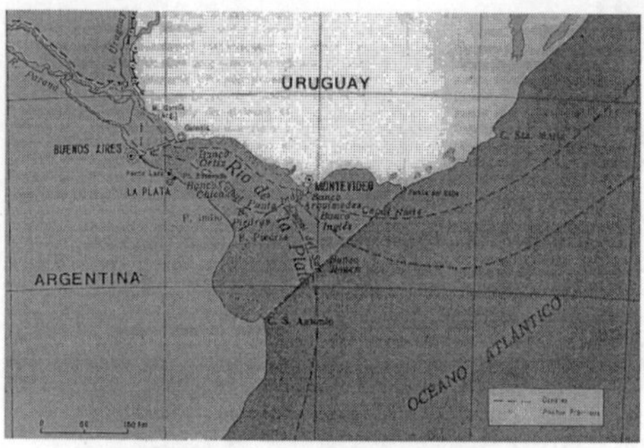

WATER SOURCES OF METROPOLITAN BUENOS AIRES

a) Fluvial
River Plate - Superior and Middle Basins (Softwater)
11.000 Km2, average depth 3.5 - approximate volume 40.000 million m3

Inferior Basin 16.000 Km2 gradual mixing of River and Atlantic Ocean Waters

b) Acquifer
a very extensive and productive one, called Puelchense

The Uruguay runs over a rocky bottom and brings an average yearly **flow of** 6.000 m^3 per second of much clearer water.

Thus, no matter what the effect of the tides of the Atlantic Ocean are on the fresh water reserve of 40.000 million m^3 of Rio de la Plata fresh water mass, 1.857 million additional m^3 pour daily down the rivers Parana and Uruguay.

Treatment of Drinking Water

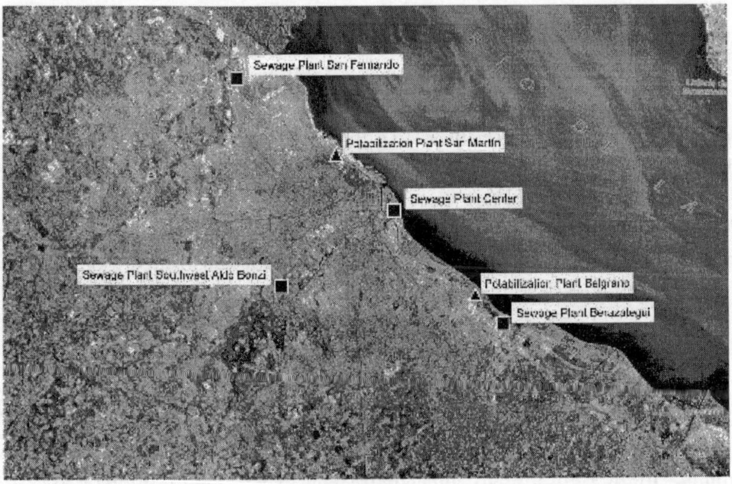

▲ **Potabilization Plant San Martín**	2.400.000 m3/day
▲ **Potabilization Plant Belgrano**	1.600.000 m3/day
	Total: 4.000.000 m3/day

At an estimate average of 500 Hs./person/day they serve 8.000.000 of a total population of 11.000.000 the rest pump from the acquifer.

■ **Sewage Plant Southwest Aldo Bonzi** treats the effluent of 500.000 inh. primary and secondary with percolator and trickling filters.

■ **Sewage Plant San Fernando** treats the effluent of 300.000 inh. primary and secondary with activated sludges

■ **Sewage Plant Center** 20 m3/second - primary and secondary (projected).

■ **Sewage Plant Berazategui** 20 m3/second - primary (partial).

The two plants that serve the Metropolitan Area treat a total of 4 million m^3 per day, and cover 8 million people, with the average of 500 liters per inhabitant.

To this fluvial water, a substantial volume of excellent quality aquifer reserves, called the Puelchense, should be added. Although no reliable estimates are available, it is extensive, and it has, so far, been exploited in areas not yet covered by the networks. Agricultural uses have posed no considerable strain, so far, due to the rainfall averages and distribution, although this may be a growing, whose magnitude is hard to estimate, future concern.

Pollution and Contamination

Upriver from Buenos Aires, the main sources of pollution occur on the right coast of the Parana, where cities, populated by some 2.5 million people are located, of which the most important is Rosario (about 1.5 million). To this may be added industrial installations, of which the oil refineries and metallurgical complexes of Campana, 70 Km upriver from BA, are among the more important.

Minor water currents have been covered by urbanization, all through the urban spread of the metropolis, but the drainage networks that substituted them kept carrying not only the rainwater, but the industrial as well as some clandestine domestic waste.

The two main urban basins, that of the Rivers Reconquista-Luján, and of the Matanza-Riachuelo, still run in the open. Both are heavily contaminated, by waste coming directly or indirectly through the pluvial network that replaced old rivelets.

The worst conditions are located on the Matanza-Riachuelo, a river of very weak output, complicated by the effect of tides, that put it often in negative flow, causing floods, mainly in the right border. This small river, or Riachuelo, drains a very flat area, and had a strongly meandering course, that was subject to rectification. Being the river the border of the jurisdiction of the Federal Capital, the treatment of the deep meanders out of its territory suffered from the poorer budgets of the corresponding municipalities.

Worst still, the old riverbeds were filled with garbage, and later still received dense settlements of population, which have to deal with heavy contaminated soils.

As to the aquifer, the low density peripheral, semirural tissues, extracted their water from wells, and disposed of waste at the superficial water table, through septic chambers. As these areas densified, the system entered crisis stage. Nowadays, the effects of the dense puncturing of the acquifer roof is yet to be measured. The growth of the water supply networks have not been followed by the sewage, thus, larger water disposals have raised the water tables to surface in several, and not small areas.

Overuse and Conservation "Full Cost Pricing"

The water resources of Buenos Aires area are not jeopardized by overuse. Nevertheless the price of the social loss of the amenities of the coastal areas is a heavy figure in the red, and the contamination of the drainage of the urban basins is also a cost that has to be added to the historical non investment in the treatment of waste waters.

Integrated Water Resources Management

Water and sewage systems were developed in Buenos Aires in the aftermath of a series of severe epidemics (yellow fever, cholera) that decimated the population in the 1870s and 80s. The plagues aroused a generation of eminent hygienists, who reacted at the paradox of a city coastal to one of the world´s major software reserves, and yet suffering from deficient sanitation.

A public system, called "Obras Sanitarias de la Nación", was therefore developed, based on the right to water, as much as you needed, called "canilla libre" (that is: free tap). It reached its peak in the early 1930s, by these days it was a world state-of-the-art company, as to the modernity of its installations, the excellence of its technical staff and its own technological developments. This state entreprise had jurisdiction over the national territory.

The system deteriorated gradually. Investment in extension of networks, as well as maintenance itself, was no longer adequately covered by public funding, and tariffs, always little more than symbolic, evaporated through inflation. National coverage exploded into provincial, municipal and even cooperative or individual actions.

By the late 1980s the system had collapsed. The operation of urban networks was open to concessions. The operation of the networks of Buenos Aires Metropolitan Area formerly managed by Obras Sanitarias de la Nación, and two minor systems established by municipalities of the conurbation, were given, after international bidding, in concession to the consortium Aguas Argentinas, led by the multinational Lyonnaise des Eaux.

The concession is supervised by a watchdog called the Ente Tripartito de Obras y Servicios Sanitarios (ETOSS). It is a tri-party outfit because two federal jurisdictions are part of it. One is the Autonomous City of Buenos Aires with its population of 3 million (both the National Capital and a Federal State with all its attributes, after the Constitutional Amendment of 1994) and the other is the Province of Buenos Aires, to which belong 22 municipalities, with 8 milion of the Metropolitan Area. Since more than one Federal Jurisdictions are involved, the partnership of National Government is constitutionally mandatory.

The operation of the networks of the city of La Plata, capital of the Province of Buenos Aires together with the Greater La Plata, was privatized to yet another consortium. Whether La Plata is a separate urban system from Metropolitan Buenos Aires is arguable, because both take water from and drain waste to the same Rio de la Plata. Nevertheless the management of their networks is not integrated.

On the other hand, the operation of the networks and services of most major cities of the province are already privatized or on the way towards concession. A provincial watchdog is meant to supervise the performance of the concessionnaries, as well as the preservation of the subterranean waters, function that is feared to become more virtual than real.

The operation of major urban systems follows this same pattern in the rest of the Argentinian territory, with differing situations that go from extreme abundance, as in Buenos Aires, to worrisome scarcity, compounded with competition for irrigation quotas.

A satisfactory integrated water resources management is yet to be attained.

TREATMENT, STORAGE AND DISTRIBUTION

Current Technology and Cost

The water is taken from the Rio de la Plata at a distance of 2 Km off the shore into two very big Treatment Plants. The older, Planta San Martin dating from the 1910s and 20s but constantly enlarged and kept up to date, is situated in the northern part of the shore of the city of Buenos Aires.

The newer, Planta Belgrano, from the 60s and 70s, is about 45Km downriver from the first. Roughly, they divide the Rio de la Plata shore of the Metropolis into three:

Daily production of Planta San Martin ....2.400.000 m^3
Daily production of Planta Belgrano ....1.600.000 m^3
TOTAL DAILY PRODUCTION...4.000.000 m^3

They can serve 8 million people, at a daily average of 500 lts/person/day. The particles in suspension and the consequent turbidity require a process of coagulation and floculation, carried on by addition of chemicals, sulphates of aluminum, and polyelectrolytes, after what the waters go into fast and slow filters and sedimentation, it is further disinfected by addition of chlore and alkalines.

The cost of the plant itself (excluding land) is estimated at \$50 per m^3 per day.

The operational cost of the plants is \$0.023/m^3 that breaks into.

Personnel.. 30%
Chemicals..45%
Energy... 18%
Other... 7%

Storage

Provisions for storage have been historically overdimensioned. Even if nowadays the capacities do not match what had been initially designed, a reserve for a day's consumption was supposed to be stored at the plants themselves.

A further day's storage should be in huge tanks disguised as modern buildings, each massively covering an entire city block, strategically situated in the highest points of the city. And a further equivalent in the "Rios Subterraneos" system, meant to interconnect the main fake-building reservoirs.

These are the provisions as far as the external system is concerned, because Obras Sanitarias regulations enforced the provision of domestic reservoirs with an additional day's estimated needs. The latter, or domestic system, is still in practice, and proves the most problematic, for its maintenance and periodic disinfection left to the consumers themselves, and is far from being efficiently operated.

Distribution

Water is distributed through the former Obras Sanitarias pipe network, now operated, extended and maintained by Aguas Argentinas.

Inequities in Distribution

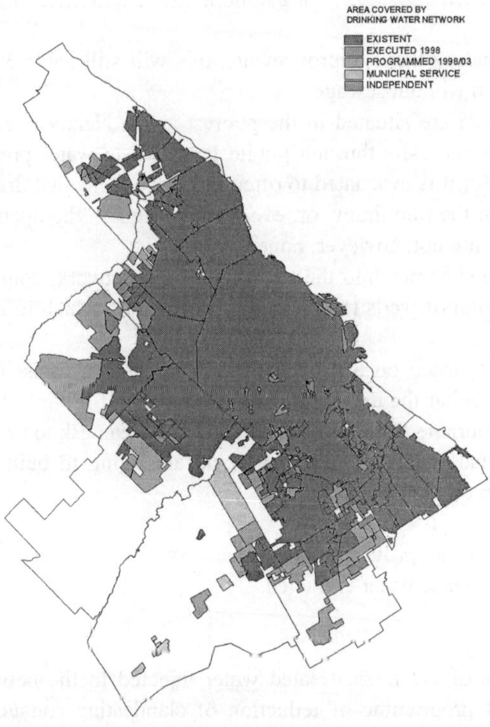

The ideals of the hygienists' generation of the late XIX century could not be totally fulfilled. To begin with, networks could only serve the consolidated urban areas; low density and the fast growth of peripheral areas were sanitation engineers' and planners' constant nightmares.

These problems were compounded with the settlement of the poorer strata in the lowlands of the urban riverlets basins, in which not only waste water could not be carried away by gravity, but the areas were subject also to flooding provoked either by heavy rains or Rio de la Plata extraordinary tides and, worst still, by both effects combined when violent storms coincide with strong southwestern winds, which raise the big river levels.

These areas were the most difficult to serve in state-operation times. They remain the least rewarding in cost-benefit terms in the present private concession system. It is true that a sizable increase in the serviced population has taken place.

YEAR	WATER	WASTE
1993 – start of the concession.........	6 million.....................	5 millon
1999 –	7.5 millon.....................	6 millon
2005 - projected.....................	8.2 millon.....................	7 millon

Notwithstanding these improvements, this will still leave 3 million without safe water and 4 million without sewage

Most of them are situated in the poorest areas. Nearly everybody has access to water, even if that access is through public taps. Waste water presents a more serious sanitary problem, for it is evacuated to often primitive wells that drain to the water table, or are connected to the rain drains or, even worse, runs in the open. Even if these cases are quite rare, they are not, however, nonexistent.

When an area comes into the coverage of the network, connection is mandatory, and further utilization of wells to the acquifer forbidden, which in most cases is a sound policy.

Service is charged on a property's total built area basis, irrespective of actual consumption. But, either the user, or Aguas Argentinas, if it thinks the alternative system might be more favourable from his own point of view, can ask to switch to a 50 % - 50 % system, in which half will continue to be paid according to built area and half to the actual consumption on the basis of:

$0.33/m^3 of water provided and
$0.33/m^3 of waste water extracted.

Dealing with Wastage

An estimated 30% of the fresh, treated water injected to the network is unaccounted. There is a current programme of reduction of clandestine connections and control of filtrations.

Institutional Aspects and Decentralized Systems

As described, practically the whole of the fresh water supply and waste water disposal in the Buenos Aires Metropolitan Area falls into Aguas Argentinas operations. Consumer defence activities and ecological group action are to be channelled through ETOSS, the tri-partite watchdog.

WASTE WATER

Limitations and High Costs of Conventional Systems

The treatment of waste waters is quite recent in the Buenos Aires metropolitan area, because the totality of it was poured raw into Río de la Plata, 50 odd Km south of Riachuelo, the southern limit of the Federal Capital, in the locality of Berazategui, once remote, but nowadays a part of the Metrópolis.

It began with a brick vaulted "cloaca maxima", to which two others were added when the first one proved insufficient. Yet, the system is largely overcharged, and lack of investment resulted in the pouring of waste to the old riverlets and the rain drain piping that substituted them, in clandestine actions, and in the official installation of valves called "espiches" that liberated excessive flow from the main cloacas.

So, at the present stage of improvement of the system, the challenge is twofold: on the one hand, black waters coming through waste piping networks that is poured raw has to be treated, and on the other, black waters that run into rain drain system have to be brought into the corresponding waste water network and also receive proper treatment.

A scheme has been developed to intercept all drains coming to the Rio de la Plata coast north of Riachuelo, and those coming to the northern coast of Riachuelo from Buenos Aires. Only the effluent draining during non-rainy days will be treated, because this is surely not pluvial When it rains, the waters will bypass the plants, their contamination diluted by rainwater. Although not perfect, this system will be a remarkable improvement to the present situation.

The effluent of 11 millon people is the target volume to be treated. At present there are only two small effluent treatment plants.

The older one is located in Aldo Bonzi, called the Southwestern Plant. It handles the effluent of 500.000 inhab/equivalent (250lts/inh/day), with primary treatment consisting of interception of solids and sedimentation and a secondary with percolator and trickling filters.

Its operation cost is $\$0.06/m^3$.

The newer is the San Fernando Plant, recently inaugurated. It processes today the effluent of 150.000 inh/eq. and is being completed for 300.000. It works with a primary treatment of solid interception and sedimentation and a secondary with activated sludges. These are to be recycled for agricultural uses and already a plant is in preliminary operation.

The operation cost is similar, and the cost of construction (land excluded) is of $150 per inh/equ.

The remaining effluent, estimated in 29 m^3/sec, will be treated by two large plants, one located in the southern extreme of the coast of the city of Buenos Aires and the second at the old Berazategui outlet.

As said, all pluvial systems will be collected and poured into the waste water system. This will account for another 10 m^3/sec.

Each new big plant will treat 20 m^3/sec and will cost $400 million. The technology to be employed is primary treatment chemically assisted.

This leaves us with 3.5 million m^3/ day to be treated, with a total cost of \$207.360 per day, and \$6.221.000 per month of operating costs.

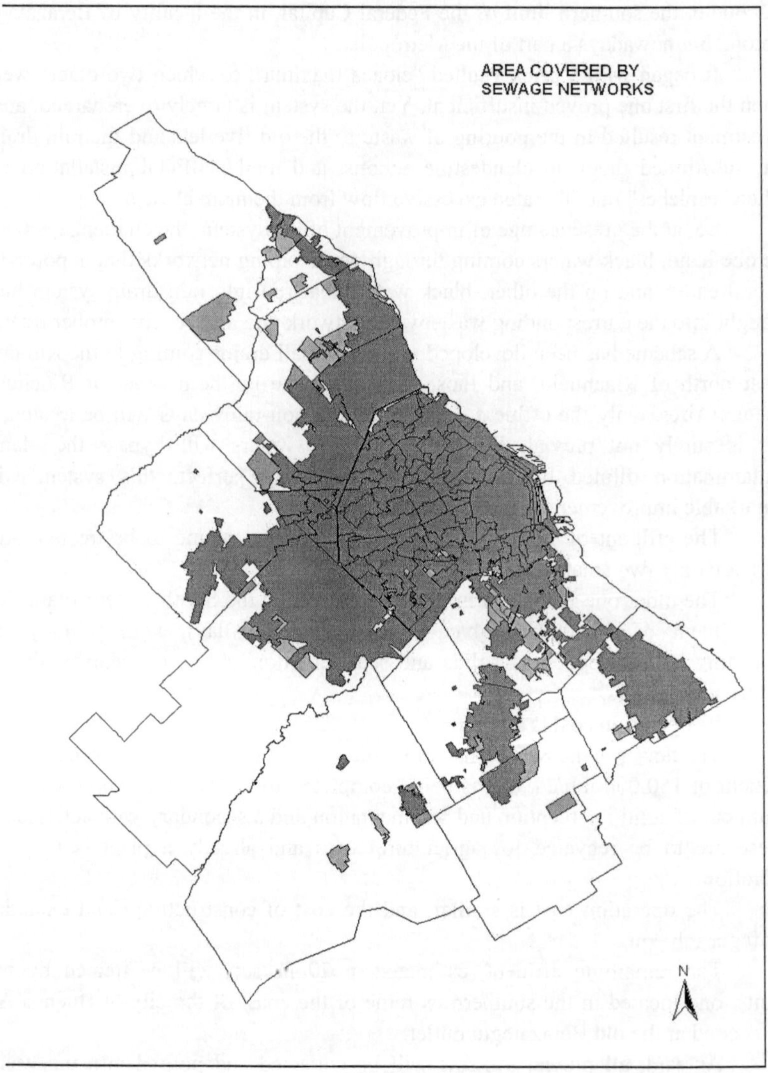

WATER AS AN ELEMENT OF PLANNING AND SHAPING CITIES

Environment and Landscape

In the Buenos Aires Metropolitan Area, the most outstanding presence of water is the waterfront of the Rio de la Plata and its liquid horizon, since the Uruguayan coast is not visible. This particular circumstance has determined that from the first European visits the river was known as the "softwater sea".

The primeval landscape has suffered from the XIX century industrial installations, namely ports, railroad approaches, and powerhouses, the latter, as well as quite a few industries, receiving their fuel, coal, and later on, petrol, through navigation.

The profile of the coast, down to the southern tip of the city founded in 1580, climbs over a 20m "barranca", with a very shallow beach below, of nearly 1km that is left dry in low tide. This led to early fillings as the city grew and land close to the central area could be so easily obtained, determining the loss, therefore, of the shoreline for public amenities. The colonial law, transmitted to the republican statutes, reserved the coastal border for public purposes.

Two "costanera" promenades were developped in early 20th century, but both were not in direct contact with residential areas. In recent times, the northern part of the shoreline has suffered a process of privatization, mostly by marina's development.

The Riachuelo, a riverlet that was established as the limit between the City of Buenos Aires proper, today an accident inmersed in the metropolitan mega-area, marks the limit between two regions, that of the "ondulated pampa", and the "depressed pampa". South of the Riachuelo, the riverfront "barranca" disappears, and a relatively deep fringe of marshes works as an interface between pampa and Rio.

Two other characteristic aquatic landscapes are present in the metropolitan area. One is the Delta of the Parana River, a unique, very extense area of a tree-covered myriad of islands, separated by small, medium sized and majestic branches of the Parana River. Vegetation, flowers, a mild climate and the absence of dangerous species makes the Delta an extraordinary natural reserve.

The "barranca" that borders the "ondulated pampa" is interrupted by the basins of the affluents draining towards the delta, that in times of great rains act as vases of expansion of the otherwise modest currents. This configuration should have, in its turn, determined that of the urban shape, developed in the high lands and preserving the basins of drainage. Instead, the profiles were ignored, filled often with garbage, and still worse, paved, which deprived them of any capacity to retain sudden incomes of water, buying time for slower drain. This should have been a very strong determinant of city-form.

The landscape potential of the acquatic metropolitan system has been mostly sacrificed to "progress", with the exception of the Delta and small traits of costaneras.

Storage and Flooding

Storage has been dealt with before, in point two. Flooding has become a problem since violence has been done to the primeval water behaviour through anthropic action.

Until very recently, it was common wisdom, and as such taught at technical and engineering schools, that industry should be located close to water currents, so as to have easy availability as well as a convenient disposal. This was to bring pollution proportional to the industry's importance and the vulnerability of the river.

But this also brought the settlement of the labor force close to their place of employment. The level of the terrains were often raised through the use of refuses, thus compounding soil and water contamination with flooding caused by the slow flow and the almost inexisting slope of draining towards the Rio de la Plata, once its level has conveniently descended. Because rainstorms are often brought by the southeastern wind, that raises the river level above some of the drain systems that have substituted the old riverlets natural basins.

At the southwestern end of the Capital City, a consistent hydrological management plan has been developed, with regulatory lakes that provide a valuable landscape element.

As a conclusion, the hydraulic structure of the territory should have called for an urban tissue other than the homogeneous, cartesian, carpet-like one, that was developed. Instead, a more irregular one should have been provided, with discontinuitites to allow for water basins, properly forested, with permanent as well as transitory bodies of water, in order to provide green, natural, amenities to the urbanites, that would work at the same time as flood regulators or, better still, take the word flood out of the dictionary.

Taking into account the major basins, this would have produced a star-shaped urbs, with deep wedges of natural land coming very close to the city center, thus facilitating access to greenery traveling short distances perpendicular to the star rays.

Trunk networks work as well as rectangular ones, and drains could have been almost totally saved.

FUNDING FOR RESEARCH AND INNOVATION

At the present state of privatized management, no consistent plan of research and innovation is being carried out locally. No doubt multinationals carry out their own programmes. During the era of state-operated monopolies, the managers were graduates from local universities. As a consequence, allocations were traditionally granted to specific academic institutes. With very few exceptions, these grants have come to a halt, and have not been substituted by others supported by national research budgets.

SÃO PAULO: WATER AS A LIMIT TO DEVELOPMENT

PROF. GERALDO G. SERRA
Universidade de São Paulo, Brazil

INTRODUCTION

São Paulo is a metropolitan area with 39 municipalities. It occupies 8051 km^2, which is 0.1% of Brazil, 4% of São Paulo State and a little smaller than Lebanon (10452 km^2) or Jamaica (10991 km^2). Although the population of the City of São Paulo has stabilized around 10 million inhabitants, the population of the metropolitan area continues growing at an annual geometric rate of 1.51%, and today stands at 17 million inhabitants.

Fig. 1. Paulista Avenue, the central business district of São Paulo.

AVAILABILITY OF FRESH WATER

The Site

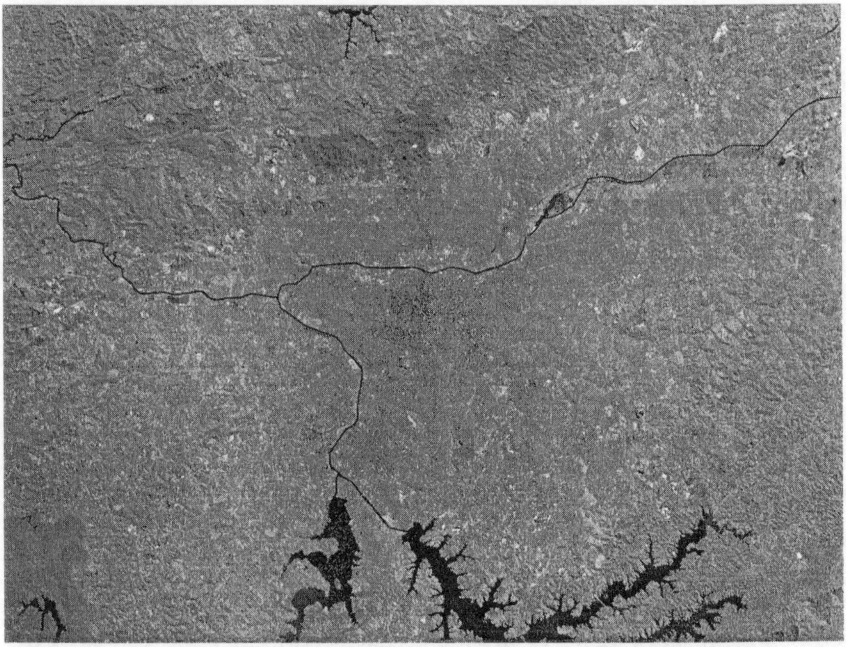

Fig. 2. Satellite image of the São Paulo metropolitan area. Santos harbor is at the bottom right corner and Campinas can be seen in the upper left corner. São Paulo is at 700 m above sea level. With the plateau sloped to the west, rivers run in that direction.

São Paulo was founded in the upper part of the Tietê River basin, near the edge of a 700 m plateau, near the Serra do Mar (Coastal Mountain Range). Several reservoirs were built near the mountains, the location of the main fresh water sources for the megacity. The two largest reservoirs, Billings and Guarapiranga, are both at the south side of São Paulo, and are linked to the Tietê watershed through the Pinheiros River. On the north side there is also the Cantareira system, which is another important source of fresh water to that part of the Metropolitan Region. There is a rich hydrologic system in this area, with more than 1,500 km of streams and rivers. The Tietê watershed area has approximately 5,985 km^2. Considering only the sub-systems that have been exploited and from which water is being withdrawn, we have an available flux of 104 m^3/s. Approximately half of that is still available.

Pluviometry

In São Paulo, the amount of precipitation is between 1500 mm to 2000 mm per year. The maximum rainfall rate or intensity in 24 hours varies from 60 to 100 mm. Although this could seem a very reasonable regimen, most rainfalls actually occurs between December and February, while there is a water shortage between May and September, when less than 20% of the annual amount falls.

The shortage is happening earlier than usual this year, as a consequence of fewer rain days. The reservoirs have insufficient capacity and water must be imported from other watersheds, at ever-increasing costs.

Fig. 3. Guarapiranga reservoir, one of the main sources of fresh water for the south of the city, is shown here during last May, with less than 60% of its capacity.

Social Problems and Pollution

In general, big cities are sources of water pollution and São Paulo is no different. However, where pollution from industrial plants and uncollected sewage is relatively easy to control, water pollution from irregular housing developments and "favelas" is as difficult to resolve as the social problems themselves. For more than 150 years São Paulo was proud to be a land of opportunity for immigrants from many countries and from every Brazilian state. But during the 60s and 70s, the geometric rate of population growth, mostly caused by immigration, was too much for the metropolitan economy to absorb and federal and local governments approved very strict rules about new developments. Both factors put upward pressure on land prices and led to irregular developments and "favelas".

GEOMETRIC RATE OF POPULATION GROWTH			
1970/1960	**1980/1970**	**1991/1980**	**1996/1991**
5,44	4,46	1,88	1,46

In an attempt to avoid the occupation of the areas adjacent to reservoirs, São Paulo state government approved laws creating headwater protection areas around them, including their entire watersheds. These laws made most economic activities in these areas illegal, forcing a strong decline in land prices. As a perverse consequence of such well-intentioned regulations, these unoccupied areas become the preferred location for irregular developments, marginal occupation and “favelas”.

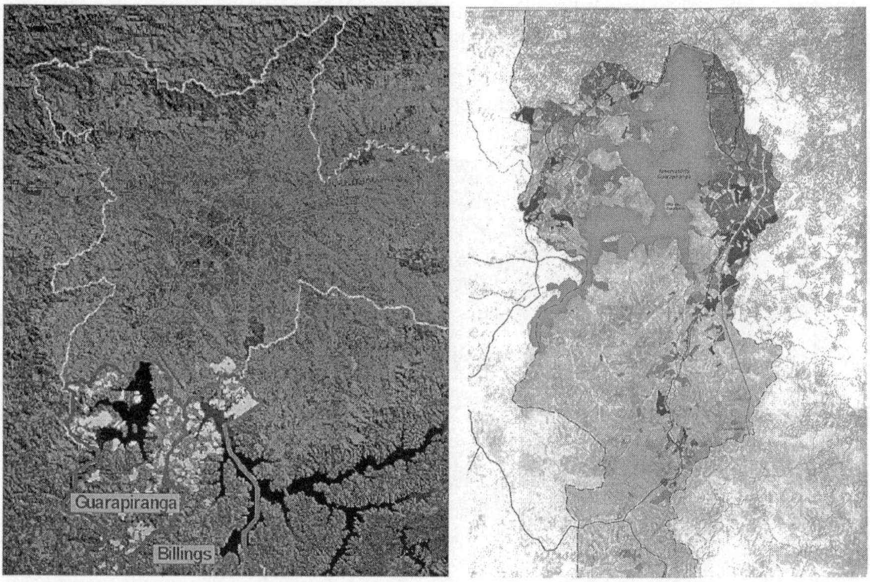

Fig. 4. Irregular developments around the reservoir and high-density occupation in Guarapiranga.

In recent years the occupation has become so intense that the state and municipal governments have mobilized to fight it and have created programs to bring municipal infrastructure to these areas, including sewage collection and treatment, to prevent pollution of the reservoir.

To clean the rivers, very heavy investments are made in sewage treatment plants and stopping industrial sewage discharge. Water quality is expected to improve over the next ten years.

DRINKING WATER

Demand and Availability

The system of drinking water treatment and distribution now reaches almost 100% of the metropolitan population. That system has more than 30,000 km of pipes. The water company is owned by São Paulo State, who does not seem to want to privatize it and announced in 1998 that enough investments had being made in the system to attend the demand without any rationing, only to announce now – two years later – that there will be strict rationing in half of the city, as a consequence of unusually low precipitation. The general perception now is that we are working at the limit of fresh water resources and that heavier investments should be made over the next few years to avoid disaster. A new canal is being built linking the Billings and Guarapiranga reservoirs, because the level of the latter is very low. Projects to bring water from other watersheds are under consideration. But rationing will likely be maintained until October or November, when the rainy season begins.

The average production of treated water is 65 m^3/s, but average consumption is only 52 m^3/s, implying a leakage of 19%, which is reasonable. Average production per inhabitant is 317 l/day and consumption is 254 l/day. But at peak hours demand increases to 75.5 m^3/s. The total consumption of water in the watershed is more than 80 m^3/s, with 61.1 m^3/s being domestic and 16.5 m^3/s industrial. An additional 12.6 m^3/s is used for irrigation.

Water shortages occur between June and August, because during these months the amount of precipitation is very low and reservoir capacity is insufficient in especially dry years. The water flux available during these months is around 17 m^3/s. To operate, the system needs to pump water from outside the Tietê watershed.

To cope with this situation, new reservoirs are being planned. Indeed, if the total amount of annual precipitation is more than the demand of municipal water, more reservation is needed, considering the seasonal differences. Reservation could also be an answer to annual floods and could be an element of landscape design to improve urban form and environment.

SEWAGE SYSTEM AND FINAL DISPOSAL

Collection System

About 33 m^3/s of the domestic and industrial sewage in São Paulo is collected, which is 80% of the total produced. The state owned water utility has plans to increase the percentage of sewage collected to 90%, within three years, and is implementing a plan to treat all that material. Although the capacity of the treatment plants is around 18 m^3/s, currently only half of that capacity is being used, which means less than 30% of the sewage is being treated. This is because the sewage system is not completed. However, effluent without treatment is currently being discharged into in the Tietê River.

Treatment and Final Disposal

Nowadays sewage treatment is done in 5 large plants. Domestic sewage is responsible for most of the organic pollutants discharged in the Tietê watershed, totaling approximately 570 tons of BOD[1]/day. Industries discharge only 120 tons of BOD/day, but are also responsible for the toxic material and metals. However, industrial wastes can be dealt

---

[1] BOD stands for biochemical oxygen demand, or the amount of oxygen used by microorganisms in the process of breaking down organic matter in the water.

with more easily, because they can be collected and treated at the industrial site. Indeed, of the 1,250 worst cases, 95% are now carrying out such treatment.

The solution being implemented for domestic sewage consists of plants that make pellets dried by thermal treatment using natural gas and methane produced by the system itself. The dried solid part is disposed of in special landfills.

Conclusion

The situation of the sewage system of collection, treatment and disposal is very bad, especially when compared with the drinking water system. If, in one hand, 80 to 90% of all the sewage produced is being collected, less than 20% is being treated. This situation, of course, has heavy consequences for the environment and for the availability of fresh water itself, without mentioning the landscape and urban design.

FLOODING AND DRAINAGE

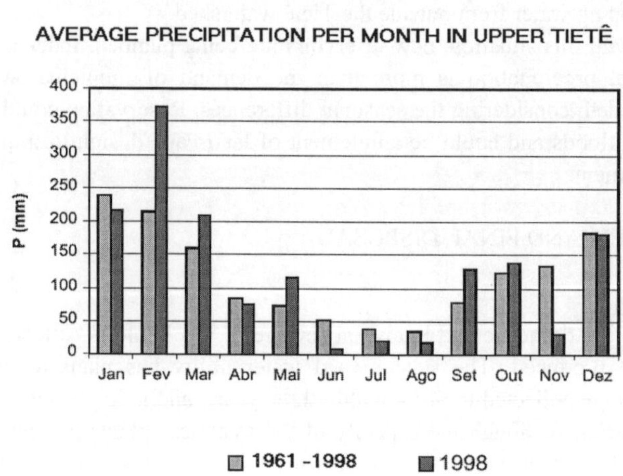

Fig. 5. Most rainfall occurs between December and March, but in 1998 the rainfalls were concentrated in an even shorter period than usual.

The metropolitan region of São Paulo covers an area of approximately 8,000 km^2, of which almost 6,000 km^2 lie within the Tietê watershed. The total annual precipitation is sufficient to provide for all the city's needs, but the distribution of this precipitation throughout the year is very irregular. Indeed, more than 50% of the annual rainfall occurs during only 4 months in the summer – from December to March - which is the rainy season.

The maximum rainfall rate or intensity within 24 hours ranges from 60 to 100 mm. However, as Figure 6 shows, the first two days of February 1998 had much higher precipitation, with some regions receiving 200 mm in a period of 24 hours. Of course, under such conditions it is difficult for the drainage system to handle the runoff, in this case extensive flooding occurred.

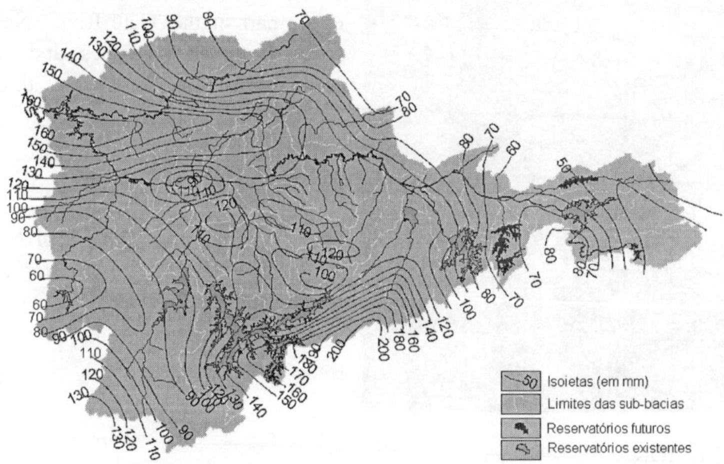

Fig. 6. February 1 and 2, 1998. Isohyets of the rainfall.

Two sorts of approaches have been adopted to prevent flooding. The first is to retain water in dams built upstream and in big concrete pools, in order to delay the natural drainage system flux. The second approach is to accelerate the water flux downstream, in order to drain away the water as quickly as possible. The City of São Paulo City has built several pools in regions plagued by frequent floods. The State of São Paulo is building dams upstream and deepening the Tietê River bed, which is, of course, a very expensive solution.

<u>Drainage Macro System Project</u>

In 1894, the water flow of the Tietê River through the city was estimated to be 174 m^3/s, but it has grown steadily and proportionately to the extension of the paved area, and is today estimated at more than 800 m^3/s. Simulations of water flow in 2020 have proposed estimates of more than 1300 m^3/s. As a result of the almost annual floods in several parts of the Metropolitan Area, the State of São Paulo developed a macro drainage system plan. The main feature of this plan is the deepening of the Tietê River bed. That solution is proposed because the main highway systems were built along the riversides and did not leave space to widen the channel. Figure 7 shows an example of a deepened section.

In any case, it is absolutely vital to avoid new urban developments on the upper part of the Tietê River. New paved areas along the marshes and riversides could double the water flow through the urban section of the river.

Fig. 7. Flood near Tietê River borders. In 1998, São Paulo experienced what is probably the worst flood.

CONCLUSION

For many years all studies and projects about water in São Paulo were much more concerned with floods than with water supply and sewage disposal. A favorable precipitation regime, along with a very extensive hydrologic system was sufficient to avoid water shortages and to carry sewage easily downstream. The tremendous growth in population was accompanied by ever-larger paved areas, increasing runoff and making floods an annual event. But it brought also environment concern about pollution and about water scarcity.

Although the deepening of the Tietê River bed follows the old style idea of draining the water away as quickly as possible, the building of new dams upstream and underground pools point to a new direction, that is to retain water upstream and to reduce its velocity.

There is also a strong movement and consciousness against pollution and about the urgent need for improvements in sanitation. Consequently, millions of dollars have been invested each year in new dams upstream, new water treatment and new sewage treatment plants.

The effort to control urbanization through legal instruments has been a disappointment, having failed under the pressure of migration and social marginality. Nowadays those irregular developments, "favelas" and other illegal occupation of legally protected land around water resources cannot be ignored or eliminated, but now must be accepted as part of the social and urban reality. Thus the solution is more investment in sewage, drainage and bringing urban infrastructure to all these areas.

In spite of all these problems and the investment required to address them, it is not accurate to say that water scarcity, pollution, sewage or flood is a severe deterrent to new developments. On the contrary, there have been significant improvements in all these aspects during the past few years. Migration has slowed and there is a corresponding reduction in the rate of population growth. This seems to be a consequence of locational decisions by industries that prefer to invest in other regions, like the hinterland of São Paulo and other Brazilian states, bringing new jobs to these regions. Indeed, most new jobs in the metropolitan region are in managerial and financial areas, reflecting a strong change in the economic structure of the city.

REFERENCES

São Paulo State, DAEE. "*Plano Diretor de Macrodrenagem da Bacia Hidrográfica do Alto Tietê*". http://www.daee.sp.gov.br/

São Paulo State, CETESB. "*Ciência e Tecnologia a Serviço do Meio Ambiente*". http://www.cetesb.br/index.htm

Marcos Carrilho, Arquitetos S/C. "*Guarapiranga*". PMSP, São Paulo, 1998.

São Paulo State. "*Ações do Governo*" http://www.saopaulo.sp.gov.br/acoes/saneamen/index.htm

Emplasa. "*Metrópoles em dados*". http://www.emplasa.sp.gov.br/metrodados.htm

9. MISSIBLE PROLIFERATION AND DEFENSE — INFORMATION SECURITY

INFORMATION CHALLENGES TO SECURITY

VITALI TSIGICHKO
Institute for systems Analysis, Russian Academy of Sciences, Moscow, Russia

New information and network telecommunicative technologies are a powerful influence in all spheres of life: politics, economy, culture, international relations and the sphere of national and international security. Modification of the world's information area appears to be a global factor of development and determine the main directions of social progress.

Moreover, development of the world community informatisation process creates a whole complex of negative geopolitical consequences. First of all it is speeding up polarisation of the world, increasing the gap between rich and poor, technically backward and advanced countries.

In this way the informational revolution does not only increase the progress of civilisation, but even creates new threats to national, regional and global security mainly to the developing countries.

High technological complexity of all systems that are the basis for the world, regional and national information areas and vulnerability of their infrastructure, present a number of complex problems for members of the world community that require immediate and efficient decisions. First of all these decisions are connected with the appearance of information weapons.

As INFORMATION WEAPONS we mean means and methods that are employed with the aim of inflicting damage to information resources, processes and systems of the state, informational effect on defence, management, political, social, economic and other systems of the state, in psychological treatment of the population with the aim of destabilising policy and economy situation.

The accumulated information about potential uses of information technologies as a means of armed struggle allows us to offer a tentative classification of the information weapons based on their intended use and principle of functioning. The information weapons of direct military application allow the following military function:

- conventional ammunition siting the targets identified by radio and electronic reconnaissance using electronic homing devices;
- high precision ammunition of the new generation, so-called intellectual ammunition, capable of independent search, selection and targeting of the most vulnerable elements;

- masking ECM jamming;
- imitation of electronic devices of communication, command and control, and data processing with powerful electromagnetic pulses and high levels of ionic radiation;
- power impact by high-voltage pulses through electric power lines;
- damaging the medium of radio waves propagation to disrupt radio communication;
- enemy disinformation by invading communication channels and using the means of human voice generation of concrete people (political leaders, commanding generals, field commanders).

The second form of information weapons are the means employed to destroy, distort or steal information resources, extract information, break protection mechanisms, disorganize the smooth function of technical means, destroy data banks, software, telecommunication systems, computer systems, energy blocks, and system of state administration: in short, the entire range of high-tech support of society's existence and state functioning.

These are the types of information weapons used to attack computer and telecommunication systems and networks:

- computer viruses able to multiply, attach themselves to software, travel along the communication lines and information transmission networks, penetrate electronic telephone stations and management systems and cripple them;
- logic bombs introduced in advance into the information and commanding centers of military and civilian infrastructure.

Activised either with a signal or at set time, they destroy or distort information and disorganize the software and hardware functioning.

- means of suppressing information exchange in telecommunication networks, its falsification and transmission along the channels of state administration and military command and along the media channels;
- methods and means used to introduce computer viruses and logic bombs into state and corporation information and to manage them at a distance (ranging from microprocessors and other components introduced into electronic devices sold on the world market, to setting up internal information networks and systems).

The appearance of information weapons sets the task of information security at the same level as the spread of nuclear, chemical and bacteriological weapons, international terrorism, dissemination of narcotics and others. All these problems are linked by their global character and impossibility to solve them within the framework of one or several countries.

Nowadays there are three main spheres for using information weapons, that is *information warfare, information terrorism and information crime.*

By INFORMATION WARFARE we mean actions for achievement of information superiority by means of inflicting damage to information, processes which are based on information, and enemy information systems with simultaneous defence of the native information, processes which are based on information, and information systems.

Changes under the influence of new information technologies in the military sphere become today the most radical and dangerous ones. Information and technological revolution leads to a drastic increase of combat capabilities of military forces. It changes not only forms and methods of leading military operations of different scales, but also changes the whole traditional operational paradigm from a tactical to a strategic level.

The greatest damage is done when information weapons are applied against military and civilian objects that should function uninterruptedly and on-line-government information systems, systems for the command and control of strategic missile forces and systems for management of transportation, power engineering, especially atomic power stations, industry, credit and financial structures.

The result may be catastrophic and comparable with those produced by mass-destruction weapons.

In principle, development of information weapons has changed the scheme of escalating armed conflict. According to views of American experts, even selective employment of information weapons against targets of military and civilian information infrastructure would terminate the conflict at an early stage, that is before the start of active military operations of both sides, because the threat of information assault escalation could have disastrous results for the object of the information attack.

So possessing information weapons, as well as nuclear ones, provides for overwhelming military superiority over countries which do not possess them.

Relative accessibility and cheapness of information weapons, together with the possibility of their secret development, accumulation and introduction, their extraterritorial and anonymous influence constitute important features of information weapons. All this makes extremely dangerous their non-controlled distribution. According to information from the U.S. financial control department, about 120 countries are engaged in or have completed elaboration of possible information-computer impact on potential enemy's information resources.

Employment of new information technologies for improvement of weapons and military technology together create new types of weapons of mass destruction. Adapting these means as weapons and their distribution provides a powerful destabilising factor, which violates the set military and strategic equilibrium, regional and global balances of forces, increases the potential threat in new points of instability and growing military conflicts. Many international agreements, that support strategic stability, lose their significance.

A real threat to national security is presented today by using information weapons such as *INFORMATION CRIME AND INFORMATION TERRORISM.*

Information crime comprises the actions of separate persons or groups, aimed at breaking protection systems, misappropriation or destruction of information for profit or ruffian aims. The so-called "hackers" and computers thiefs are the typical representatives of an information criminal. Criminal actions as a rule constitute a single crime against a definite object. The number of computer crimes is doubled each year. There are hundreds of thousands of registered computer-associated crimes all over the world.

In fact, a new type of international organised crime exists already, as information attack from abroad, aimed at opening codes, passwords and other means of computer system protection at defence, credit and finance institutions. It creates the means for economic and politic blackmail on the part of the criminal.

Information terrorism differs by its aims, that remain peculiar to political terrorism as a whole. The means for realisation of information terrorism actions may be varied to a wide extent and include all types of modern information weapons. At the same time the tactics and methods of its employment differ greatly from the tactics of information warfare and methods of information crime.

Simplicity and cheapness in the realisation of the approach to information infrastructure serves as an important issue for warfare and fast dissemination of information criminal and terrorism. This is facilitated by blurred borders in international infrastructure, blurred distinctions in geographic, bureaucratic, judicial and even conceptual borders that are traditionally connected with national security. As a consequence, there may be some impossibility to clearly distinguish between internal and foreign sources of danger for the country's security and between various forms of action against the state (from ordinary criminal activity up to military operations).

The Internet network is an important site to perform criminal and terrorist actions. It possesses a branched, worked out structure, along the communication channels of which are circulating considerable volumes of information with scientific and technical, economic and political character. There is a possibility for misappropriation of closed information, its destruction or distortion through the Internet. Propaganda materials from criminal organisations and formulas for manufacture of terrorist weapons can be circulated through the net, starting with trivial bombs up to refined algorithms for code deciphering, misinformation and political slogans, capable of provoking situations of social risk.

All of the above mentioned allows the formulation of a list of principal threats in the field of information security:

- creating and using the means of influence and inflicting damage upon information resources and systems of another state with simultaneous defence of native infrastructure;
- goal-oriented information influence on defence and other critically important structures of another state;
- information influences aimed at undermining the political and social systems in a state, psychological brainwashing of the population with the purpose of destabilising the society;

- actions of states that lead to their domination and control in information space, counteracting the access to the latest information technologies, creating conditions for technological dependence in the sphere of informatisation;
- actions of international terrorists, extremist and criminal societies, organisations, groups and separate law offenders, presenting a threat to information resources and important state structures;
- development and adoption on the part of states for plans and doctrines, that foresee the possibility of conducting information warfare and capable of provoking an arms race in the sphere of information;
- threat of using information technologies and means to the detriment of the principal rights and freedom of men, realised in the sphere of information;
- threat of uncontrolled transborder dissemination of information, contradicting the principles and norms of international law and also to the home legislation of concrete countries;
- danger of manipulation of information streams, misinformation and concealment of information with the aim to influence the psychological medium of society, for erosion of traditional cultural, moral, ethical and aesthetic values and norms.

Certainly, it could serve as an illusion of some sort to lay obstacles to scientific and technical progress, including that in the field of defence. Still, the world community has already accumulated a substantial experience in struggle with various types of weapons of mass destruction in order to realise the future menace by the spread of information weapons and to take the necessary measures of international character, placing this process under strict national and international control. In this connection, there appears an evident necessity in the international legal regulation of world procedures for civil and defence informatisation, for the development of an agreed platform for the problem of information security.

INTERNATIONAL INFORMATION SECURITY CHALLENGES FOR MANKIND IN THE XXI CENTURY

ANDREI KROUTSKIKH
Head Directorate of the Department on Security and Disarmament of the Ministry of Foreign Affairs of Russia, Member International Informatization and Telecommunications Academies, Moscow, Russia

Initiating in 1998 a discussion in the United Nations on international information security, Russian foreign minister Mr. I. Ivanov noted in his letter to the UN Secretary General that mankind is witnessing now the formation of a truly global information society in which information is acquiring a new, revolutionary quality, significance and influence, both nationally and universally.

A single technological line is formed by computers, telephone, radio/TV, and space-based communication systems. Society today greatly depends on its smooth functioning. In fact, it is experiencing a land-slide with the worldwide introduction of hi-tech telecommunication and cybernetic means. Local and global networks have created a new quality of transborder information exchange. All this directly affects politics, economy, culture, international relations, national, regional, and international security.

A single worldwide information space is emerging as a global development factor and, as such, is determining the main trends of social progress. Information is becoming the states' major strategic resource.

The global information-technological revolution we are living through today has brought obvious boons and is promising more. At the same time it has created new fundamental threats. The scientific and technological achievements can be abused to reach aims that have nothing in common with international peace, stability and security, rejection of the use of force, non-interference in domestic affairs of other states, and respect for human rights. No wisdom is needed to predict the information-technological threats evolving into serious challenges to twenty-first century international security.

A new type of extremely destructive weapons can be developed—the information ones. The pace of their development and growing interdependency of national and international information infrastructures leaves, as it seems, no chance to any state to be immune to possible hostile transborder actions either with the use of information technologies or against critical information resources.

So far, the term "information weapon" has not yet received an exact definition. It was first used by the American military in 1991 after the Gulf War. It is hard to define it because the bulk of information technologies are of dual or non-military application. But whatever the terms are, the huge potential of the information-computer technologies can be used to ensure military-political domination, the use of force and blackmail and cannot exclude a possibility that in the foreseeable future punitive expeditions against international outcasts will use information weapons rather than cruise missiles and bombs. This will turn conflicts into information warfare.

Sophisticated scenarios of information warfare took our breath away in horror films of the past: computer viruses secretly introduced into electronic systems of a state and its economic administration, and military command coming to life and paralyzing them; acting at long distances, scoundrels used electronic devices to remove funds from the adversary's bank accounts and cripple industry, communication, power production, transport, municipal services, ecological monitoring, atomic power stations, airports, and strategic forces' command points; powerful generators of electromagnetic impulses destroyed software and deleted vitally important data bases of protected computer systems. All this created panic among civilian population and deprived the state leaders of correct information.

Little by little alike TV horrors reached real life. Back in the mid-eighties the United States embargoed Iranian bank deposits during the American-Iranian hostages crisis by using a computer program. Military experts called the Desert Storm punitive operation that efficiently employed radioelectronic means of warfare the first "information Hiroshima."

Deliberate information impact on the enemy has a history that is as old as the world itself. Today, due to the latest technologies, it is developing from scattered acts of information sabotage and disinformation into a fully-fledged method of international policy applied on a mass scale.

Information weapons are used to: achieve information superiority, damage information, information resources, processes, and systems; improve traditional and create new types of armaments and military technology aimed at a further direct armed impact on the enemy; put out of order civilian objects and life-support systems; disorganize state administration; introduce economic chaos and sabotage; damage national financial systems based on information-computer networks; psychological brainwashing of the population to achieve social disorganization. Any of the above technologies when used by one state against another can be called information warfare or war.

The greatest damage is done when information weapons are applied against military and civilian objects that should function uninterruptedly and online (early warning systems; anti-air and anti-missile defense systems: power production complexes, especially atomic power stations; industry). The results may be catastrophic and comparable with those produced by mass-destruction weapons.

Information weapons are qualitatively universal, highly efficient, and easily accessible. They offer a wide choice of time and place of use; they do not require large armies, which makes information warfare relatively cheap. Their application can easily pass for routine action, at the same time it is hard to pin it on any particular state. Information weapons are indifferent to long distances and state frontiers.

The weapons can be used without a declaration of war; they do not need large-scale and obvious preparations. Sometimes a victim remains unaware that information impact is applied to it. It is much harder to respond to information aggression because there are no systems and methods to assess the threat of attack and warn about it.

The information weapon has produced a revolution in warfare. Many traditional military concepts such as "defense" or "assault" have been transformed. In local clashes there is no longer need to seize territories or take POWs; it has become possible to reduce loss of life of one's own army and to entrust initiative in combat assignments to pilotless means.

The sham humanitarian nature of information weapons should be also emphasized. Many methods of information warfare such as crippling telecommunication systems, virus programs, jamming, blocking communication systems, etc. while dealing a heavy blow at economy do not cause directly bloodshed, loss of life and visible destruction common in conventional warfare. As a result none is deprived of food, dwelling, etc., needed to maintain life. There will be probably no refugee problem. This may lower the moral threshold of political decision-making. All talk about the humanitarian nature of the information-cybernetic means and methods of military-political impact may produce dangerous light-heartedness and tolerance where their use is concerned. There may be a tendency to excuse their use as unilateral sanctions on the ground that no blood was spilt.

Man-in-the-street may approve such means and methods because they do not require a build-up of the armed forces; they even lead to their shrinking. Development of military-information potential camouflages itself as part of technological progress. Budget allocations for military purposes can be easily passed for and realized as spendings for large-scale peaceful programs. The miracle weapon looks tempting. Indeed, according to information of a U.S. financial control department, about 120 countries are engaged in or have completed elaboration of possible information-computer impact on a potential enemy's information resources.

Further development of the information weapons and progress in the use of civilian information-cybernetic networks and means for military purposes may let out a "technological genie" of a new generation to supplant the nuclear one.

There are practically no international laws to regulate the use of information weapons, to limit them as this is done under treaties with other weapon types and military activities. This cannot but aggravate the situation.

The emergence and proliferation of this weapon, and militarization of peaceful information technologies are a powerful destabilizing factor in international relations. The present military strategic balance, the local and global balance of forces, and greater risks of an attack or blackmail are the price for the new technological experiment. The entire system of international agreements on maintaining strategic stability, curbing arms race, at the regional level as well, will be put to a serious test.

It never rains but pours: information-cybernetic technologies can be used by criminals and terrorists. On the one hand, the technologies are easy to use, an access to the means of communication and data transmission means is cheap, global information networks are cosmopolitan. On the other, the world-wide information resources and infrastructure are vulnerable.

Individuals and groups acting towards unsanctioned penetration into information-cybernetic systems irrespective of their affiliation, breaking protection systems, stealing or destroying information for mercenary reasons or out of hooliganism are criminals. Computer thieves, or hackers, are also criminals. There are hundreds of thousands of registered computer-associated crimes all over the world and they double every year. According to the Pentagon figures, in 1995 alone hackers penetrated the U.S. Ministry of Defense computers through Internet over 165 thousand times. Criminologists believe that there is a new type of organized crime in the world specializing in overcoming computer protection of military departments or objects, credit and finance organizations and in stealing secrets and money.

Information terrorism differs from information warfare and crime not so much by its method as by its aims typical of terrorism, and the tactics employed. Relying on the same technological foundation, the three types of menaces make the problem of information security as topical as other global problems: nonproliferation and liquidation of nuclear missile, and chemical weapons, banning bacteriological weapons, anti-terrorist and anti-drug struggle, etc. They are more or less the same in scope and are part and parcel of international relations. One or several countries united in a bloc cannot deal with them. It is equally impossible to cope with these global problems on the "every man for himself" principle. The world information space cannot be divided while the information systems are interconnected.

The world community should not and cannot afford to permit itself to be involved in a new—information technology this time—area of confrontation, to face a possible escalation of the arms race in this field and an endless chase of countermeasures for offensive inventions as was typical for nuclear age.

There is an objective need to legally regulate the world-wide processes in civilian and military informatization and to create a concerted international platform of information security. The UN members have to formulate their own assessments of such threats and provide related basic definitions, including those of "unsanctioned interference" and "illegal use of information systems and resources".

There is no doubt that the complexity and the extent of security problems in the realm of global information community are enormous. It involves legal, economic, ethic, political, military, technological and other aspects making the issue sound at both national and international levels. But at the same time, this complexity makes it even more convincing—it is indeed necessary to address the global problem at the global level. Therefore, for a general consideration or guidance, at least at this first initial stage, it needs an appropriate universal approach facilitated by the forum of the UN General Assembly. It is also essential that such a consideration be carried on a maximum wide, joint basis to identify all national existing approaches, positions, views or concerns and then accumulating them in a sort of international concept.

While those trends and approaches are identified they can be initially summarized in a document such as "general principles" to form the basis for a future internationally-adopted regime or code of conduct for States, which would be aimed at confining the emerging threats and to strengthen overall international information security. Practically, those principles can be incorporated in a multilateral declaration or further enhanced in an appropriate international legal instrument of an "umbrella" type.

One can try to visualize a possible subject of international negotiations on the information-related aspects even if it is impossible to predict how far the world community is ready to go to assume specific obligations to limit current threats to international security in international legal anti-military spheres.

A ban on or voluntary rejection of elaboration, production and application of especially dangerous information-cybernetic technologies and methods of their use can be high on the list of priorities. The same can be said about possible exclusion of the information means and methods of destructive impact on man and the human organism. It is equally important to introduce a non-proliferation regime of military informational technologies, means and programs. There is a need to coordinate, on a world scale, anti-terrorist and anti-crime efforts; it is advisable to create international norms and legal acts and, probably, general technical measures to protect national information resources, regulate transborder information exchange, minimize negative information impact on mass consciousness, ban the use of information-cybernetic technologies for aggressive aims against certain objects.

This work may draw on the rich experience international diplomacy has accumulated when elaborating maritime law, the legal regime of the use of space, wide-scale international conventions and agreements.

These efforts can go side by side with coordinating national legislations în information activity through parliaments. To extend the efforts to maintain information security it would be wise to set up a permanent international monitoring of information threats; centers of information technological aid to the countries that fell victim to information aggression or any other illegal use of information means; international groups of experts to promptly react to all cases of information terrorist threats.

As for the fora or other international bodies to start practical work in this direction, it can be a UN group of governmental experts reporting to the Geneva Disarmament Conference or to the UN Secretary General and the General Assembly. These efforts can also be supported by other UN and multilateral entities dealing with related but more specific issues like the International Telecommunication Union, UNIDIR or International Institute for System Analysis.

To summarize:

- It is obvious that the matters of information security cause concern for the whole international community;
- The issue is quite versatile and at the same time involves interrelated problems;
- So far there are no necessary political or legal means to regulate the problems manifested;

- The UN General Assembly can give general guidance for further consideration of the issue;
- To deal with the matter in an applied, systematic way, proper international fora are to be identified;
- There is a need for a thorough expert review of all the issues involved;
- This review is to result in a kind of concept to form a basis for a future international legal regime on information security.

THREATS TO INFORMATION SECURITY BY COMPUTER-BASED INFORMATION HIDING TECHNIQUES

AXEL LEHMANN
Institut für Technische Informatik, Universität der Bundeswehr München, Neubiberg, Germany

ABSTRACT

Along with rapid innovations and dramatic distribution of computers, information and telecommunication technologies, and with their ubiquitous usage in our public and private life, new risks are arising from the misuse and from information warfare which need to be addressed. As far as information warfare is concerned – "malicious attacks against an organization's information base through electronic means" – we have to distinguish between attacks from outsiders over networks and potential invasions of one's system enabled by insider turned foes.

At first, this presentation reports on political and economical risks and damages caused by information warfare attacks, in general. I will classify attacks to electronically stored data and information according to active and passive attacks. More in detail, we will specifically focus on passive intrusion attacks based on information hiding techniques, especially on steganography. Such passive intrusion techniques can be implemented by insiders as part of standard COTS-software products (COTS: commercially-off-the-shelf software) bought by users or by usage of copied application software. Besides an overview of different approaches for application of information hiding techniques, this presentation will also address possibilities to prevent passive intrusion attacks on networked computers by means of intrusion detection approaches.

The primary focus of this presentation is directed towards awareness and sensitivity of the audience with respect to the potential risks for governments, industry, private and public life arising from passive information warfare.

NEW STRATEGIC EVIRONMENT AND RUSSIAN MILITARY DOCTRINE

ANDREI PIONTKOVSKY, VITALI TSIGICHKO
Institute for Systems Analysis, Russian Academy of Sciences, Moscow, Russia

The Military Doctrine of the Russian Federation was approved by presidential decree of April 21, 2000. Point 8 of this doctrine, concerning the use of nuclear weapons, is the most debated element in Russia and abroad. It says: "The Russian Federation retains the right to use nuclear weapons in reply to the use of nuclear and other mass destruction weapons against it and/or its allies, as well as in reply to a large-scale aggression with the use of conventional weapons in situations critical to the national security of the Russian Federation." The ado created by that phrase was partially unexpected. Firstly, it contains nothing new from the viewpoint of Russia's official military strategy. A similar thesis replaced the traditional Soviet pledge not to be the first to use nuclear weapons in "The Basic Provisions of the Military Doctrine of the Russian Federation," approved by presidential decree on November 2, 1993. And secondly, this is a standard provision for Western military doctrines, including the NATO strategy concept. Russia has lost its superiority in conventional weapons, and countries and military blocs have appeared close to Russia, which are superior to it in this sphere. The new Russian military doctrine stipulates that nuclear weapons have become the main deterrence factor in this situation. Consequently, it would be inexpedient to make a no-first-use pledge now. As long as Russia feels a potential non-nuclear threat, it will pursue the logical strategy of "defence in all directions," with the possession of nuclear weapons being a substantial element thereof. This strategy, which was clearly and openly formulated in the new military doctrine (although it contains some confrontation and rhetorical passages), is apparently defensive and should not worry our neighbours and partners. As for Russia's security, although the nuclear factor is a vital element, we should not overrate the doctrine's stake on the nuclear umbrella. We offer you the analysis of the situation in specific directions to prove the point.

Most experts agree that the 1999 Kosovo crisis was the pivotal element in Russia-NATO relations. But neither Brussels, nor Moscow drew the proper conclusions from it, for they would be embarrassing to both sides. The NATO operation in Yugoslavia reached a dead-end in mid-May 1999. NATO was tottering on the brink of a split over two key questions: the continuation of the air strikes, and the possibility of a ground operation. The bombing raids, however pinpoint they were, still increased "collateral damage," meaning the death of peaceful civilians and the destruction of Yugoslavia's infrastructure, which dramatically undermined the European public support for the operation. Democratic countries cannot wage wars without public support. Greece virtually spoke up against the military operation. The governments of Italy and Germany were on the verge of getting a no-confidence vote from their parliaments. Besides, it turned out that the bombing raids alone could not force the Yugoslav army to leave Kosovo. NATO has admitted today, a year after the Kosovo operation, that the air operation against the Yugoslav army brought minuscule results. The Western society today is not prepared to sustain military losses, at least in a war that does not threaten its existence. This is the effect of the Mogadishu criterion: The war ends if five servicemen are killed and their dead bodies are shown on TV. NATO was not ready to wage a ground operation in Yugoslavia, even under the threat of public humiliation and a review of the Cold War results. We must agree with the opinion of General Klaus Naumann, the recently retired chairman of the NATO Military Committee, who said that NATO was saved by a miracle in Kosovo. That miracle has a name - Viktor Chernomyrdin. We are not going to criticise Moscow's position at the final stage of the Kosovo conflict. Despite the attractive elements of the confrontation scenario, which gave us a chance "to put NATO in its place," it would be counterproductive in the long-term perspective from the viewpoint of Russia's interests. The new strategic concept, which was approved at the NATO jubilee session and which provides for humanitarian interventions beyond the framework of Article 5 of the NATO Charter, was stillborn. Kosovo was the first and only instance when it was applied. This is the main lesson of the Kosovo conflict. The NATO military experts are perfectly aware of this, but prefer not to speak about it–for understandable reasons. Our military experts are top-class professionals too, but it was not in their interests to hinder the anti-Western and anti-NATO hysteria, which swept our politicians off their feet. "Yugoslavia yesterday, Russia tomorrow." This slogan is still popular. It is much easier to lobby for larger military allocations in this atmosphere. On the other hand, this noble cause must in no way prevent us from adequately evaluating the world around us. We can dislike the West for the very fact of its existence. Or we can regard it as an economic, information and spiritual challenge. Or we can believe in the imminent hostility and evil intentions of the West with regard to Russia - if we like this and think that this flatters our vanity. But we must admit that the modern democratic West, with its highly vulnerable infrastructure, does not present a military threat to Russia. As for nuclear weapons, their possession is a major political and psychological factor in Russia-West relations. But the practical scenario stipulated in Point 8 of the Russian military doctrine - the use of tactical nuclear weapons to deter or repel an aggression, is not effective in the Western direction. The nuclear factor is excessive in deterring a potential aggression in the Western direction.

Let's look at the southern direction. The nature of threats coming from it is connected above all with the possible involvement of Russia in local conflicts close to the state borders of Russia and its allies. No matter what political or legal description we provide for the Chechen conflict, it is military-wise a guerrilla war with separatists in a border region of Russia. It is clear that nuclear weapons cannot be used in such conflicts as a means of deterrence, let alone as a means of warfare. We should train special professional units. But it is even more important to use political methods to preclude the involvement of Russia in local conflicts on its southern borders. To make these methods effective, we should have an adequate understanding of the neighbouring culture. This means above all Islamic countries in the south. The Islamic world is highly unstable and plagued by social and ethnic conflicts, which sometimes result in the creation of extremist groups. Initially, these groups were not hostile to Russia. Moreover, the attitude of the Islamic world to Moscow had been mostly positive. But we seem to be doing everything possible to ruin our relations with the Islamic world and to incur the wrath of its most radical factions. Do our strategists ever think about this? Especially when they encourage our leaders, who visit Western capitals, to laud Russia as the shield protecting the Western civilisation from Islamic extremism? These propaganda efforts would have been laughable, if they were not dangerous. The recent statement made by our high-ranking officials on Russia's intention to deliver strikes at the fighters' bases in Afghanistan was particularly irresponsible. It is apparent that it will be used as a weighty argument by the Islamic extremists and attract thousands of fanatics into the terrorist groups. We cannot understand the military goal of the suggested plan. The bombing of terrorist training camps is an unrealistic task, since it is virtually impossible to organise effective reconnaissance of the Afghan territory. Strikes can be delivered only at large stationary objects, such as cities, military bases of government troops, and airfields, which would be tantamount to beginning a new Afghan war, with catastrophic consequences for Russia. Maybe we want to scare the Taliban, who have been waging a war in their country for more than ten years now? It is time to see that deterrence does not mean anything to ideological fanatics.

As for the Far Eastern direction, we have developed a strange tradition of avoiding a comparative analysis of the Russian and the Chinese armed forces, although we analyse possible—and frequently speculative—scenarios of potential conflicts between Russia and the USA or Russia and NATO. This professionally aloof analysis is an obligatory element of creating a stability system and has nothing in common with fostering hostility. Well, if we regard Russia and China as a couple of states with their military capabilities, we might think that this is the classical case when the conventional superiority of one country (China) is counterbalanced by the threat of the first use of nuclear weapons by the other country (Russia). But such analysis disregards such a vital parameter of military strategy as unacceptable damage. Since nuclear strategy is in fact psychology, the advantage in this psychological duel can be snatched not by a country that has more sophisticated nuclear weapons, but by a country whose culture is more tolerant of human losses. If we regard the potential Russia-China conflict from this viewpoint, we will have to drop the illusion that the threat of tactical nuclear weapons will deter the opponent. The readiness to lose human lives will allow China to up the stakes in this nuclear gamble. If China becomes our military opponent, it will be a superior opponent at all stages of the escalation of the conflict. With the exception of the last stage—an all-out nuclear war, in which we are assured a draw with total destruction of each other. China is moving, although slowly, in the same direction as the bulk of civilised countries. This is why the best guarantee of Russia's security is the political and ideological evolution of China towards the Western values, above all the fundamental value of human life. So, despite its seeming logic and attractiveness, the concept of reliance on the nuclear factor is vulnerable. It is excessive in the first, senseless in the second, and dangerous and counter-productive in the third strategic direction. Even a cursory analysis of the nature of threats in each of these directions shows that Russia's security cannot now be guaranteed only by a package of military means. Political and civilisation factors, and their adequate understanding and use, have a no less important part to play in relation to our neighbours and partners.

MISSILE DEFENSE AND PROLIFERATION

GREGORY H. CANAVAN
Los Alamos National Laboratory, Los Alamos, New Mexico, USA

INTRODUCTION

Missile defense and proliferation are closely coupled. Proliferators want missiles and weapons of mass destruction because of their perceived value in deterring, coercing, or striking others quickly and certainly from long range. Missile defenses are intended to destroy missiles, which would eliminate the attributes proliferators want. The two are antithetical. The better the defenses, the less valuable the missiles, and hence the less incentive for proliferation. The present situation is essentially the converse: even major powers have little or no defense, so a few missiles pose a threat to them. That gives missiles significant value to states of concern, so the incentive for proliferation is currently high.[1]

The obvious answer would seem to be to build defenses to reduce that incentive. However, there are other considerations, the dominant one being stability. As a country builds missile defenses, their impact on stability depends on how it intends to use them. If it intends to withdraw behind a missile shield—possibly reducing or eliminating its offensive forces in the process—that should not be threatening. However, if it wishes to retain the ability to deter or damage an adversary, defenses can be viewed as a shield against second strikes that could make its first strikes more effective. Intent clearly differentiates between the two cases, but one's intent is difficult for another to judge, and the history of the last five decades and the large number of offensive weapons extant do not give particularly positive indications of benign intent.

Thus, the question whether to build missile defenses is not determined solely by their effectiveness and impact on proliferation. It also depends also on the tradeoff between the benefits of reduced proliferation and the loss of stability that could occur. As both effectiveness and stability impact depend on the nature of the defenses deployed, the discussion below reviews the defenses currently under discussion in enough depth to estimate their effectiveness. It then discusses the impact of that level of effectiveness on stability, which provides the inputs needed to trade defenses against proliferation.

BACKGROUND

Defenses are named according to the portion of flight in which the weapon is engaged. Boost phase refers to intercepts before the booster burns out; midcourse to intercepts in the long portion of the trajectory between booster burnout and reentry into the atmosphere; terminal to intercept in the atmosphere. Each has strengths and weakness. In boost, missiles are bright, and there are no credible decoys, but the engagement opportunity is short. Midcourse engagement times are much longer, but since all objects follow ballistic trajectories, it is difficult to discriminate decoys. In terminal, air resistance aids discrimination, but interceptor footprints are limited, so the defense of population would require very many of them.[2]

Historically, development went from terminal to midcourse to boost. Terminal was first because it could be attempted with small interceptors. It was abandoned when it was shown that the limited atmospheric battlespace could not be used to defend military targets, let alone population. For that reason, it will not be discussed further below.

Midcourse was attempted when larger interceptors became available. The U.S. developed the Safeguard system in the 1070s around the Spartan exo-atmospheric interceptor, but abandoned it when it was shown that the Perimeter Acquisition Radars that controlled the nuclear Spartan interceptors could not track targets in the disturbed environment from attacks at the level expected from China. Russia still maintains such a system of 100 nuclear-tipped interceptors.[3]

The boost phase was studied after about 1975 as a way to avoid the decoys and nuclear effects of midcourse. The U.S. Strategic Defense Initiative (SDI) got mid-way through the development of boost phase interceptors before it was cancelled in 1993. After the Rumsfeld Commission assessment of the emerging "rogue nation" (now state of concern) threat, missile defense development was restored in 1997, but redirected to a midcourse defense based on the hit-to-kill underlay of the earlier SDI program.[4]

MIDCOURSE DEFENSES

Although much current discussion centers on performance in recent tests, there is little doubt that midcourse hit-to-kill interceptors can be made to work.[5] Neither ranges nor response times are particularly stressing, and the engagement of at most a few dozen missiles should not degrade the performance of current radar and infrared sensors. There are, however, concerns about the level of performance required.[6] $I = 80$ independent interceptors of kill probability of $p = 0.9$ could reduce leakage of $W = 20$ weapons to $\sim (1 - p)^{I/W} = (1 - 0.9)^{80/20} = 0.01\%$, or $\approx 10^{-4} x 20 \approx 2 x 10^{-3}$ penetrating weapons. That is roughly the goal of the current U.S. NMD system, i.e., very low leakage from a few tens of weapons with about 100 interceptors.

The deeper concern in midcourse is the discrimination of the countermeasures and decoys now available even to states of concern through alliances or trade. The amount of leakage depends sensitively on the degree of discrimination. If it is very good, as should be the case for expected threats containing decoys already encountered, performance should approach that estimated above. However, new or unexpected countermeasues could cause more rapid degradation.[7] For D non-discriminable objects per RV, the defender is forced to intercept (1 + D)W objects, for which the leakage is $(1 - p)^{I/(1+D)W}$. While D = 0 recovers the previous result, D = 1 gives leakage $(1 - p)^{80/2x20} = (1 - p)^2 =$ 1%, and D = 3 gives 1 – p = 10%, which is probably unacceptable for defense of value.

These fundamental limitations on midcourse defenses are not new; they were major factors in terminating the Safeguard system in the 1970s. The problem is just in modern guise. The defense now has much better sensors and discriminants, but the attacker has several more decades of countermeasure development to draw on, much of which is now more readily available. Which will ultimately prevail is not known.

BOOST-PHASE DEFENSES

The sensitivity of midcourse defenses to countermeasures and disturbed environments was the basis for the decisions in the 1980s and 1990s that the main element of the SDI would be boost-phase interceptors backed up by a modest midcourse underlay. Important to that decision was the recognition that a boost phase defense could produce the low and reliable leakage levels needed to protect populations against attacks.[8] SDI concentrated on boost-phase defense from space, which was the only basing mode that would allow interceptors to survive Soviet suppression long enough to function. With the end of the Cold War, and the shift of emphasis to states of concern, interceptors can be based nearby while remaining sufficiently survivable.

Surface-based boost-phase defenses

Proposals have been advanced for both ship- and ground-based missile defenses from missile threats of certain known states of concern, such as North Korea, where geography is favorable. The interceptors could be derived from developed ground- and sea-based interceptors for midcourse, alerted by existing satellite warning systems, and controlled by land- or sea-based radars. Their scaling can be illustrated algebraically. Current liquid-fueled ICBMs accelerate to intercontinental velocities $V \approx 7$ km/s in times $T \approx 300$ s. An interceptor launched at the same time from below its path to $v \approx 7$ km/s could reach it by burnout from a range roughly equal to the sum of their average velocities times the burn time, or $T(V + v)/2 \approx 2{,}000$ km. The actual range is reduced by intercept altitude and delays. The results of more careful calculations are shown in Figure 1 for typical 6g (net average acceleration 6 times gravity) interceptors with maximum speeds of 4 and 7 km/s. The 7 km/s curves reproduce the above estimates for a typical range of warning delays. The 4 km/s curves show that even current interceptors could have useful capability[9].

With ranges this large, interceptors could be based on sea or land, depending on their cost and other advantages. An ICBM from North Korea to the U.S. could arguably be intercepted from a naval ship in the Sea of Japan, a land base in northern Japan, or in a shared facility in Vladivostok, which is further north but closer to the ICBM's track. Decoys are probably are not an issue, as it essentially takes an ICBM to decoy an ICBM, and their simultaneous, credible launch would be difficult and expensive.[10]

MIXTURES

Absent decoys, leakage could be reduced to the desired level by committing more interceptors. Two interceptors would produce ≈ 1% leakage. Whether additional attrition should be supplied by boost-phase or midcourse interceptors is a matter of relative cost and preference. Insensitivity to decoys reduces the cost and risk of surface boost-phase systems, but a mix would minimize the risk of unexpected degradation of either layer. An additional factor in the decision is that surface-based boost-phase defenses only provide defense against a specific country, while midcourse defenses provide some defenses against all possible launch areas. Selecting a midcourse underlay to surface-based boost-phase defenses against a few countries could provide useful protection against launches from other states that the surface-based boost-phase defenses would not provide if deployed alone.

Surface-based boost-phase systems are plausible and efficient solutions to individual countries of concern, but they are country—and target—specific because of the specific trajectories they must lie along and the short alert and reaction times they must support, which preclude intercepts of distant missile launches. While surface-based boost-phase systems are probably the lowest cost for a single country and target, costs increase rapidly as the number of states of concern and targets increases.

Space-based boost-phase defenses. While ground-basing affords a number of basing options for North Korea, others of concern such as Iraq, Iran, or Libya are more challenging. For them, other, and new threats, space-based boost-phase interceptors offer direct access to launch as well as automatic global coverage. The interceptors required are essentially miniaturized versions of the ground-based interceptors for midcourse defense. They are pre-placed in orbit so that one or more of them can reach a missile launched anywhere on the surface of the Earth before it burns out, which makes it very difficult for the attacker to use decoys effectively.

While the detailed analysis of boost phase systems is complex, their essential scaling is algebraic. If a low-altitude interceptor has a maximum divert velocity $v \approx 6$ km/s, which is within the state of the art for space propulsion, and the missiles have burn times $T \approx 300$ s, which is about that of current liquid fueled rockets, the interceptor can reach missiles from an area below of $\approx \pi(vt)^2 \approx \pi(6 \text{ km/s} \times 300 \text{ s})^2 \approx 10^7 \text{ km}^2$. The surface of the Earth has area $\approx 5\text{x}10^8 \text{ km}^2$, so complete single coverage requires $\approx 5\text{x}10^8 \text{ km}^2/10^7 \text{ km}^2 \approx 50$ interceptors. That number could be reduced by a factor of 2 to 3 if coverage was restricted to the "SCUD belt" to 30 degrees latitude, which would also minimize impact on strategic missiles. Conversely, the attacker could increase the number required by radically shortening the boost time or placing many missiles in the area covered by each interceptor.[11]

While it is difficult for the attacker to use decoys in boost, leakage can still result from imperfect performance. For a single layer of $p = 0.9$ interceptors, a booster would have a probability of 10% of penetration. If the defensive constellation were doubled to 100 interceptors, the missile would still have a 1% probability of penetration. Whether additional attrition should be supplied by space- or ground-based interceptors is largely a matter of cost. In the past, cost trades have favored space-based systems, which have low investment and operational costs, but actual decisions have favored mixes, which provide the maximum insurance against the unexpected degradation of either layer.[12]

Interceptor autonomy and hardening were critical issues during the Cold War, as it was necessary to insure that they were not vulnerable to preemption or dependent on information that could be denied to them by the destruction of missile launch warning systems. That led to a particular embodiment of the interceptor known as the "Brilliant Pebble," which was extensively developed by SDI and the multilateral Global Protection Against Limited Strikes (GPALS) before its cancellation in 1993.[13] With the end of the Cold War and the shift of emphasis to other states of concern, these features are no longer as critical, so a simpler version called the "Burro was discussed," which used external sensor data and was not designed to survive preemption.[14] The distinction between them for current applications is not just a matter of cost or performance. Burros' potential for international construction and control must be weighed against the greater capability and survivability of the Brilliant Pebble, which in time could become an important factor for advanced threats.[15]

It has previously been demonstrated that both systems are amenable to human control and that accidental activation or intercept are not issues even for short reaction times. Modest inspection procedures could reliably prevent accidental use against peaceful space launches. They are needed to prevent the misuse space in any case.

AREAS FOR POSSIBLE COOPERATION

While U.S. and Russian missile defense programs have generally proceeded independently, there have been areas of cooperation. Some involved treaties that controlled competition by limiting numbers or types of defenses; others have explored areas for possible cooperation on technology or deployment. Some possibilities largely lost after 1993 may be resurfacing.[16] The U.S. has expressed its disinclination to protect only itself, and Russia has put forward some interesting initiatives for the multilateral defenses. U.S. initiatives have been confined to midcourse defenses to date, while President Putins' have focused on boost-phase, but the two sets of suggestions have gone a long way towards legitimizing defense schemes in the minds of thoughtful international analysts. A number of scholars have argued that the boost phase concepts noted above—particularly those like surface-based or low-latitude space-based defenses that can separate coverage of state and strategic missiles—need not raise ABM Treaty issues. President Putin appears to agree, at least on the essentials.

While current policy is to continue the deployment of midcourse defenses, others propose to supplement them with more robust alternatives. There are proposals to add more ground-based interceptors or sites or sea-based defenses, but that would only add interceptors with the same fundamental limitations as current midcourse interceptors.

There are also a number of advocates for boost-phase defenses, including two former Clinton Administration Directors of Central Intelligence, two former Clinton

Administration Deputy Secretaries of Defense, and the Carter Administration Secretary of Defense.[17,18] Moves towards the boost phase, robust multi-layer defenses, and international cooperation are resisted for ideological reasons, but the U.S. academic and policy communities realize that such a shift is overdue.

At first glance these choices appear mutually exclusive; on inspection they are not. Although ground-based midcourse interceptors have received considerable emphasis in the last few years, those for ground-based boost-phase could be derived from them in roughly the time for the deployment of the former. While it would be necessary to make up for the lost eight years since the 1993 cancellation of the Bush Administration's "first to deploy" boost-phase defensive program, it had proceeded far enough in development to provide a basis for recovery on roughly the same time scale.

It would not be necessary to choose immediately which of the boost phase options to emphasize, as some work would have to be done on each. Given current development in ground-based interceptors, ground-based boost-phase defenses could develop first. If so, they could be deployed to address the limited number of states of concern for which they are the logical, low-cost boost-phase option. Such an option could be implemented at no additional cost by modestly scaling back presently-planned midcourse defensive expenditures, while realizing large gains in overall defensive effectiveness arising from a multi-layered defense.

If states of concern increase in number, shift to areas less suited to surface basing, or if it is decided to provide insurance against launches from anywhere on the globe to any other state, space basing is appropriate. Given modest funding, it could be available when needed. The Burro should involve a simpler level of technology that could be subject to international deployment and control. The more capable Pebble could be developed as a reserve. The technology and control of such systems could realistically be shared with allies and other nations. Those were key factors that stimulated enthusiasm for international cooperation in the past, and that could do so again.[19]

EXCHANGE ILLUSTRATIONS

Even unilateral deployment of defenses would impact both sides. Russia's stated concern is that U.S. defenses would reduce stability by reducing its second strike. The U.S. response is that its defenses could only defend against a few tens of long-range rogue missiles, so they could not threaten Russia's strategic deterrent even at START III levels. The arguments can be illustrated with equal 1,000 ICBM forces, to which side U adds 100 interceptors and P does not. The labels U and P are arbitrary; they are chosen to make the argument country neutral.

It is shown above that for interceptors with kill probability 0.9, to achieve 0.01% leakage would take 4 interceptors per weapon, so 100 interceptors could address about 25 weapons. However, in defending against a 1,000-weapon attack, it would be appropriate to commit one interceptor per weapon. Then 100 interceptors would negate $\approx 0.9 \times 100$ missiles and save ≈ 90 targets, which could constitute an adequate second strike. However, a first strike by U missiles with kill probability 0.9 would leave $\approx 0.1 \times 1000 \approx$

100 survivors, which is about the number of missiles U's 100 interceptors could engage, so P's second strike would be reduced to leakage.

If P struck first with 1,000 missiles, about 100 U missiles would survive by leakage, and another $\approx 0.9 \times 100 \approx 90$ would be protected by U's 100 interceptors, so $\approx$190 U missiles would survive. As all would penetrate, U would still be in a position to execute a counterforce strategy, while P would not. That was the origin of former Pr. Gorbachev's statement that "the U.S. cannot develop defenses that can handle our first strike, but it could develop defenses strong enough to handle our ragged retaliation."

STABILITY IMPACT

Stability impact varies with the specific form of the defenses. For instance, the surface-based boost-phase systems discussed above, whether ground- or sea-based, should have little impact on strategic forces because their range is limited. While they have a range $\approx$1,800 km against missiles coming towards them, they would have a range of 900km or less against strategic missiles accelerating away. Moreover, ground-based interceptors would be vulnerable and could be suppressed before the launch of the force. The Burro, although a space-based boost-phase system, would have little impact on stability because it is not designed for survivability. Even Pebbles would have little impact if deployed at low latitudes for suppression of the SCUD belt.

The two deployments that do offer concerns are space based interceptors (SBIs) deployed at the latitudes of the strategic missile fields and midcourse defenses. Midcourse defenses should be a limited concern, because they could be suppressed by large attacks. However, midcourse defenses designed to work against all azimuths will have some limited impact on strategic forces. As illustrated above, that impact can be awkward. Thus, the discussion below only has to cover the stability impacts of SBIs over strategic missile fields and midcourse defenses.

CRISIS STABILITY

The incentives for strategic exchanges can be explored with game theory, which is discussed further in the Appendix. Crisis stability, the incentive to strike first in a crisis, can be evaluated with a game in extensive form that has conventional tree structure, nodes, order of decision making, and payoffs. [20] The only modification needed is to recognize that the costs of striking first or second (C_1 or C_2), which can be evaluated with aggregate models to the level required for both ground- and space-based defenses, are appropriate payoffs for the game. [21] The solutions are Nash equilibria that give both sides' optimal decisions for any given combination of forces,[22] which bypass the ambiguities of predictions from the conventional metric of the ratio of first to second strike costs.[23] The two sides are identified only as U and P, in accord with the symbols for their forces and costs (which are unprimed or primed, respectively). That avoids identification with specific countries, which is essential, as it is shown below that damage objectives are more important than specific forces or strikes.

Ground based defenses

Figure 2 shows U's first and second strike costs as U's defenses increase from D = 0 to 1,200 ideal ground-based interceptors.[24] As D increases, U's first and second strike costs decrease, and P's increase. The two top curves show that P's first strike cost C_1' is always greater than its second strike C_2'. U's first strike cost C_1 falls below C_2 for $D > 500$. That would indicate instability under the cost ratio stability metric, but does not in the game metric, where the critical defense size is where C_1 falls below U's damage objective, which is about 0.23 for conventional assumptions about the two sides' relative damage preferences (Appendix).

Figure 3 shows the costs to both sides of U deploying ground-based defenses. The costs are independent of D less than 900, because smaller defenses produce no incentives for exchanges, do not alter costs, and hence have no impact on stability. However, U can reduce its cost below its damage objective by deploying $D > 900$. The resulting first strike would reduce U's costs only marginally, but would greatly increase P's. For these conditions, P never sees an incentive to strike.

Space-based defenses

Figure 5 shows the first and second costs for ideal space-based interceptors.[25] For $D > 100$, $C_1 < C_2$, and for $D > 225$, $C_1' < C_2'$, each of which indicates instability in a cost-ratio metric. Figure 6 shows that U and P's game theoretic costs are independent of $D < 200$. For $200 < D < 275$, U's costs are *greater* than its damage objectives, which acts as a barrier to deployment of the larger defenses that could lead to strikes. The barrier results from P, anticipating U's incentive to strike at higher levels, seeing an incentive to preempt, which gives U a disincentive to proceed to such levels. The cost to P for striking would be large, but smaller than for allowing itself to be struck first by U.

Figure 7 shows U's cost as a function of D and u, the probability that U will strike first in a crisis, which was introduced into game theory by Schilling.[26] The peak cost to U is greatest if u = 0, i.e., if U would never strike first. It falls monotonically as u increases. For $u > 0.3$, U no longer sees a barrier to deployment. Thus, Pr. Gorbachev's statement is still reflected in the optimal solutions in that "you cannot develop a defense good enough to negate our first strike [due to the large barrier at u = 0], but can develop one good enough to mop up our ragged retaliation [because the barrier falls at large u]."

As long as U is viewed as less than half as likely to strike as P, space-based defenses are less threatening than ground-based defenses because the barrier discourages the deployment of defenses that could cause strikes. Otherwise, space-based interceptors differ from ground-based ones only in requiring fewer interceptors locally due to their greater efficacy against multiple-weapon missiles. However, that is offset by their absenteeism, which increases the number of interceptors in their overall constellation to about that for ground-based defenses.

Although moderate space-based defenses do not impact stability, and intermediate ones have a barrier to destabilizing deployments, large deployments exhibit progressive decreases in cost that could be provocative. That is because for any u, U can deploy some D for which C_1 drops below U's damage objectives. To deploy large defenses without crossing this threshold, U must decrease its damage objectives as it increases its defenses--and demonstrate to P that it has done so. Thus, a key element of deploying large defenses is the transmission of one's current objectives. Cooperation on defenses could be a fruitful avenue for doing so.

PROLIFERATION

The sections above have shown that defensive concepts have various advantages and disadvantages in effectiveness against attacks by states of concern, proliferation, effectiveness against strategic systems, stability, and cooperation. The table below summarizes those results for each systems discussed.

	St Concern	Proliferation	Strategic	Stability	Coop
Midcourse	M-H	H	L	L-M	L
Boost-surface	H	H	L	L	M
Boost-Burro	H	H	L	L	H
Boost-SBI	H	H	H	L-H	M

Midcourse systems provide a rough baseline. They should have medium (M) to high (H) effectiveness against states of concern, depending somewhat on performance and more on the level of discrimination needed and possible, which are unresolved. Thus, they should have high effectiveness in discouraging proliferation. They would have low (L) capability against large strategic systems, although they should have good performance against accidental or unauthorized launch of a few missiles. Their impact on stability should be low to medium, depending on the number of interceptors and the perception of how they would be used. They would have limited potential for international cooperation—particularly with Russia—as currently configured.

Surface-based systems provide a rough baseline for boost-phase systems. They should have medium to high effectiveness against the particular state of concern, because they are not sensitive to decoys and discrimination. They should have high effectiveness in discouraging proliferation. They would have little capability against large strategic systems and none against accidental or unauthorized launches. Thus, their impact on stability would be low, independent of the number of interceptors used. They would have moderate potential for international cooperation—even with Russia—depending on political and military acceptance.

Burros are essentially the space-based equivalents of ground-based boost-phase systems, except that they provide such limited protection globally. They should have high effectiveness against any state of concern, identified or not. They are not sensitive to decoys and discrimination. They should have high effectiveness in discouraging proliferation. They should have some effectiveness against accidental or unauthorized launches, but little against large strategic systems. Thus, their impact on stability should be low, independent of constellation size. They would have high potential for international cooperation, given political acceptance.

SBIs are the limit of space-based boost-phase technology and capability that would provide robust defense globally. They would have high effectiveness against any state of concern, identified or not, and are not sensitive to decoys and discrimination. They should have high effectiveness in discouraging proliferation. They should have high effectiveness against large accidental, unauthorized, and strategic launches. Their impact on stability could be low to high, depending on the size of their constellation. However, at any given number of interceptors, it appears to be less than or equal to that of midcourse defenses. They would have moderate potential for international cooperation with Russia, depending on political acceptance of technology transfer and sharing of control.

Viewed another way, all of the boost phase defenses have the traditional advantage over midcourse defenses that must discriminate. Depending on how the threat and technology evolve, that may or may not be a significant advantage. All of the concepts should be excellent in preventing proliferation. Only the SBI would have any significant capability against a large, intentional Russian attack. The surface and Burro boost-phase defenses have the least impact on stability, because they don't have enough survivability to impact large, intentional launches. The impact of small midcourse defenses should be low, but the effect and perception of larger deployments could be disproportionate. SBI impact would increase with constellation size.

Given that effectiveness against states of concern and proliferation is good, impact on strategic forces and stability is not, and potential for cooperation is a bonus, there is no clearly best system for all these criteria. There are, however, some combinations that appear useful. A moderate midcourse defense is a useful underlay that could provide early global coverage. Discrimination issues could be offset by using surface-based boost-phase defenses to provide the first few orders of magnitude of attrition at any given state of concern. If the states became awkward in number or location, Burros could be added to give global boost-phase coverage. That combination would give high effectiveness and deterrence of proliferation, while minimizing impact on strategic systems and stability and retaining significant options for cooperation. SBIs could be added as concern for improved state, accidental, or unauthorized launches required enhanced capability, warranting their impact on strategic systems and stability.

CONCLUSION

Effective defenses are needed for missile threats that are developing in unpredictable ways at uncontrollable rates, driven by missiles' ability to deter, dissuade, and destroy. Due to past decisions, ground-based defenses are the earliest and apparently appropriate way to start. However, their intrinsic sensitivity to countermeasures could be exploited by sophisticated threats, even from unsophisticated states. Ground-based, boost-phase defenses could blunt threats from specific countries and discourage proliferation with little impact on stability. If the number of states grew to levels and locations inappropriate for ground-based defenses, space-based boost-phase Burros could provide global coverage and dissuasion of proliferation with little impact on strategic forces or stability. A combination of the three would have the advantages of each—global coverage from midcourse, robust local defense from surface-based boost phase, and global coverage from the space-based Burro—without significantly impacting stability. More capable SBIs could be added as the threat required and justified impact on stability.

If coverage was restricted to low latitudes, it should be possible to develop a such a defense in concert with like-minded nations without fundamentally altering treaty obligations. The attempt to do so could be as important as the product. The greatest impediment to deploying defenses may be each side's ignorance and mistrust of the others' intentions. Communicating intentions is a difficult task. It would be difficult and dangerous to try to transmit them through offensive force reductions alone. Cooperation on the development, deployment, and control of global defenses could be a step towards eliminating decades-old suspicions.

APPENDIX: GAME THEORY

Decision analysis is defined by a graph in extended form, in which decisions are formulated in a forward manner and solved with a backward sweep. Decision nodes are labeled from right to left and top to bottom in accord with the solution process. At node 8, the two sides decide whether to compete. If they choose not to, the engagement terminates with expected costs to U and P of (x, ,x'). If they do compete, each side deploys m (=m') weapons on each vulnerable missile to and the engagement moves to node 7. There *nature* (N) decides whether U or P has the opportunity to strike first in a crisis, choosing U, the upper branch, with probability u, and P, the lower, with probability 1 – u.

If N chooses U, the engagement moves to node 5, where U can strike P (upper branch) or not (lower). If U strikes, the engagement moves to node 1, where P can strike back (upper) or not (lower), which terminates that path. If U does not strike at node 5, the engagement moves to node 2, where P can strike (upper) or not (lower), after which U can retaliate or not. If N selects P at node 7, the engagement moves to node 6, where P can strike or not. If he does, the engagement moves to node 3 where U can strike back or not. If P does not strike at node 6, the engagement moves to node 4 where U can strike or not. This constructive solution produces a Nash equilibrium.

Both sides are assumed to have a nominal damage preference L = 0.3, which implies damage objectives for each of $L/(1 + L) \sim 0.23$.

REFERENCES

1 Report of the Rumsfeld Commission (U.S. Government, 1997)

2 G. Canavan, "Missile Defense in Modern War," *American Physical Society Forum on Physics and Society*, July 1999; Los Alamos National Laboratory Report LA-UR-99-2230, April 1999.

3 R. Garwin and H. Bethe, *Scientific American*, 1972.

4 H. Cooper, letter to Chairman, Senate Armed Services Committee, 31 July 2000.

5 L. Welch, Independent Review Team Report on National Missile Defense, May 2000.

6 B. Richter, "It Doesn't Take Rocket Science," Washington Post, 23 July 2000, p. B2.

7 A. Sessler, et al, *Countermeasures* (Union of Concerned Scientists, Cambridge Mass, 2000).

8 H. Cooper op cit

9 G. Canavan, "Missile Defense in Modern War," op. cit.

10 R. Garwin, "Technical Aspects of Ballistic Missile Defense," *American Physical Society Forum on Physics and Society*, July 1999.

11 G. Canavan, "Missile Defense in Modern War," op. cit.

12 H. Cooper, op cit

13 G. Canavan, and E. Teller, "Strategic Defence for the 1990s," *Nature*, 1990, 344(April), p. 699-704.

14 G. Canavan, "Burros: Simple, Affordable, Effective Space Transportation," Los Alamos Manuscript Report LA-12197-MS, May 1992.

15 R. Woolsey, "The Way to Missile Defense," National Review, 19 June 2000, 36-41.

16 K.Payne, ed, *Proliferation, Counterproliferation and Missile Defense, Task IV: U.S.-Russian Mutual Accommodation* (National Institute for Public Policy, Aug 1997).

17 J.Deutch, H. Brown, and D. White, "National Missile Defense: Is There Another Way?" Foreign Policy, Summer 2000.

18 R. Woolsey, "The Way to Missile Defense," op. cit

19 K.Payne, ed, *Proliferation, Counterproliferation and Missile Defense, Task IV: U.S.-Russian Mutual Accommodation*, op cit.

20 R. Powell, *Nuclear Deterrence Theory* (Cambridge, University Press, 1990).

21 G. Canavan, "Crisis Stability and Strategic Defense" *Proceedings of the Military Modeling and Management Session of the ORSA/TIMS National Meeting, November 12–14*, S. Erickson, Ed. (Operations Research Society of America: Washington, 1991).

22 G. Canavan and J. Immele, "Stability Against Strategic Reconstitution by Transparency," STRATCOM Stability Workshop, July 1999; Los Alamos National Lab LA-UR-2636, August 1999

23 G. Kent and R. DeValk, "Strategic Defenses and the Transition to Assured Survival," RAND Report R-3369-AF, October, 1986.

[24] G. Canavan, "Stability at START III Levels with Midcourse & Terminal Defenses," Los Alamos National Lab LA-UR-00-1734, April 2000.

[25] G. Canavan, "Stability with Space-Based Defenses," Los Alamos LA-UR, February 2000.

[26] R. Powell, *Nuclear Deterrence Theory,* op cit.

10. COSMIC OBJECTS

NEOs: PHYSICAL PROPERTIES

W. F. HUEBNER
Southwest Research Institute, San Antonio, TX 78228-0510, USA
A. CELLINO
Osservatorio Astronomico di Torino, 10025 Pino Torinese (TO), Italy
A. F. CHENG
Johns Hopkins, Applied Physics Laboratory, Laurel, MD 20723, USA
J. M. GREENBERG
University of Leiden, Huygens Laboratory, 2300 RA Leiden, The Netherlands

Space missions over the past three decades have established the importance of cosmic impacts by asteroids and comets in shaping the surfaces of the Moon and the inner planets. Asteroids and comets, which are the debris from aborted planetary accretion, still collide with the planets. Impacts that would threaten civilization, corresponding to the collision with an asteroid of at least 1 km in diameter, occur roughly once in a million years. These are random events. No asteroid is now known to be on a collision course with Earth and we do not know when the next catastrophic impact might occur.

According to the most recent estimates, there are at least 500 and maybe as many as 1,100 near-Earth objects (NEOs: asteroids and comets) larger than 1 km in diameter that can cause catastrophic global effects in a collision with Earth. Using a common power-law extrapolation, one may predict 10,000 to 25,000 NEOs that are larger than 200 m in diameter. NEOs as small as 200 m can cause catastrophic regional or local effects including tidal waves in case of an ocean impact. Detection of NEOs is well in progress excepting (1) faint NEOs (objects less than 1 km in diameter), (2) Atens, (3) objects with orbits completely interior to that of the Earth (IEOs), (4) small comets, and (5) unpredictable long-period comets.

To deflect or destroy potentially hazardous objects (PHOs) effectively, we must know their internal structure and bulk material properties, particularly their material strengths. Physical characterization of NEOs is usually only carried out by remote sensing. Remote sensing includes crucially important determinations of albedo, size, shape, spin state, mass, and various inferences regarding composition and topography of the *surface* of an asteroid, but it does *not* include the most important determination of an NEO's *internal* structure or material strengths. No program exists to determine internal structure and material strengths of NEOs and even remote sensing increasingly lags the rate of NEO discoveries.

With the Near Earth Asteroid Rendezvous (NEAR) mission to Asteroid 433 Eros, a start has been made in the detailed study of objects that some day may impact the Earth. We discuss the status and needs of NEO research and outline science highlights from the NEAR mission, but concentrate on methods for determining the bulk properties and geologic structures of NEOs. The need of a planned program to determine physical properties and establish a database to record these properties is outlined.

INTRODUCTION

Vast numbers of asteroids and comet nuclei orbit the Sun. Because of planetary perturbations (mostly with Jupiter), a fraction of them is continuously removed from their original orbital locations, and is injected into the inner regions of the solar system. Such objects have collided with the terrestrial planets including the Earth since early phases in solar system history. They may have brought water and prebiotic materials to the Earth to support the origins of life.[1] They also have left their marks of devastation in the form of giant impact craters on Earth's surface, similar to the craters easily identified on the Moon, Mars, and other planets. The threat to Earth is real as evidenced by the impact in Tunguska, Siberia, in 1908 and the dramatic collisions of about 26 large fragments of Comet Shoemaker-Levy 9 with Jupiter in 1994. Many of these collisions with Jupiter caused marks in atmospheric layers that were as large as the Earth itself.

Objects from space hit the Earth all the time at speeds of more than 15 km/s. Some objects in retrograde orbits even have speeds relative to the Earth of about 70 km/s. Most small objects burn up harmlessly, as "shooting stars." Larger objects reach the ground as meteorites. However, the threat and consequences of large near-Earth objects (NEOs) colliding with Earth have only recently been recognized.[2] Such impacts represent a significant peril to human and other forms of life. For the first time in man's history, the means exist to mitigate against an NEO collision with Earth.

Several meetings and workshops have been held about the potential danger of NEOs colliding with the Earth starting at Aspen, Colorado, in June of 1981. However, it was not until a decade later in 1991, at San Juan Capistrano, California, and the Spaceguard Survey report[2] that the subject matter was addressed more aggressively. In 1992, the NEO interception workshop[3] was held in Los Alamos. A summary of this workshop was published by Rather et al.[4] These two reports are fraught with disagreements (see Appendices of Los Alamos report[3]). A follow-up workshop was held at Livermore: The Planetary Defense Workshop.[5] Perhaps because of the disagreements at the Los Alamos workshop, there are no editors for this report! A more balanced picture emerges from the book *Hazards Due to Comets and Asteroids*,[6] which is the result of a conference held at Tucson, Arizona, in 1993. All topics of relevance are described to varying degrees of completeness. This was followed by the Chelyabinsk-70 meeting in 1994 in Russia[7] and most recently, in 1999, the IMPACT Workshop in Torino, Italy.

Many methods are discussed for impact mitigation procedures, including kinetic energy devices, lasers, and atomic devices. The use of atomic weapons is emphasized in particular for the larger objects. When an explosion occurs near a consolidated object, a shock wave travels through the object. When the shock wave arrives at the far end of the object, a layer of material receives such a large impact that it is knocked loose; i.e., the spalled material carries momentum with it. Spall has not received much attention, but can substantially decrease the amount an NEO is nudged out of its orbit.

Types of NEOs

NEOs are not only visually but also geologically diverse objects composed of iron-nickel, stony materials, carbonaceous materials, or ice-and-dust mixtures (comet nuclei). They may also be transition objects; i.e., inactive comet nuclei in their end stage of evolution with most of their ices evaporated, giving them the appearance of an asteroid. Comet 107P/Wilson-Harrington is such an object. In 1992, the Spacewatch telescope recovered the dark Amor asteroid 1979 VA. This led to a pre-discovery search[8] that showed that the object is the same as the Comet 107P/Wilson-Harrington, which had a tail in two plates taken in 1949.[9,10] In forty years, this comet had evolved into a dormant body two magnitudes dimmer.

The structure of NEOs can be monolithic, porous and fluffy, fragmented aggregates (rock masses disconnected by faults and fissures held together only by their own gravitational attraction), or rubble piles. Each combination of composition and structure requires different technologies for hazard mitigation or resources exploitation. Beyond enabling reliable hazard mitigation decisions, knowledge of NEO internal structures and their response to impact also bears directly on resources exploitation and understanding the formation and collisional evolution of planetesimals. Physical characterization of NEOs has only been carried out by remote sensing of the *surfaces*. It does *not* include determination of internal structure or material strengths. Furthermore, remote-sensing characterizations cannot keep up with the rapidly increasing rate of discovery. This leads to a marked decline in the percentage rate of NEO size-albedo determinations.

The perihelion and aphelion distances of the Earth are 0.983 AU and 1.017 AU, respectively. These values define three subgroups of near-Earth asteroids (NEAs) that can intersect the capture cross section of the Earth: Atens (about 6% of the three groups of NEAs) have aphelia $Q > 0.983$ AU and semi-major axis $a < 1$ AU, Apollos (about 65% of the NEAs) have perihelia between $0.983 < q < 1.017$ AU, and Earth-crossing Amors (about 29% of the NEAs) have perihelia $1.017 < q < 1.3$ AU.[11] Atens orbit mostly in a region interior to the Earth's orbit, but have eccentricities sufficient to allow them to cross Earth's orbit near aphelion. The Amors may traverse the Earth's capture cross section since their orbits evolve because of long-range planetary perturbations over tens of thousands of years. No currently known asteroids are on a collision path with Earth in the near future, but we do not know the state of orbital evolution of still undiscovered asteroids. NEOs also include short-period comets (with a period $P < 200$ yr) and long-period comets ($P > 200$ yr). Finally, as already mentioned, some objects cannot be identified as asteroids or comets. These transition objects may be extinct comet nuclei. Many of the Atens, Apollos, and Amors may be transition objects. Structurally and compositionally, we know very little about them.

The rate of discovery of NEOs has dramatically increased. We now know orbits for about 50% of the expected NEOs with diameters larger than 1 km, but we know only a very small fraction of the orbits of the expected number of smaller NEOs. *However, high discovery rates alone will not solve the hazard problem of collisions with Earth.* Knowing how to protect ourselves from impacts by potentially hazardous objects (PHOs*) is just as important as finding them! Even before we can make plans for scenarios how to mitigate the danger of collisions with Earth, we must understand the physical bulk properties and spin states and PHOs and know the true inventory of the small-sized objects. A nickel-iron object or a stony object will require different techniques of mitigation than a porous carbonaceous object or an ice-and-dust body like a comet nucleus. However, not only material composition and strength matter, but knowing the geologic structure, location of the center of mass, distribution of mass in the object, and the moments of inertia of the object is equally important. Deflecting or destroying a rubble pile or a fragmented aggregate requires a very different approach than deflecting or destroying a strong monolithic object or a large but fragile object. For example, Melosh et al.[12] have suggested several approaches of applying impulses to a PHO to nudge it out of its collision path. Also, forces must be applied close through an object's center of mass, or the energy will be wasted into spinning the object instead of nudging it in its orbit.[13] The true inventory of relatively small-sized NEOs can be obtained only by extrapolating their size-frequency relation, which is still only very poorly known. There also are many transition objects with properties between those of asteroids and comets. They have material bulk properties and internal structure about which we know even less.

The PHOs are the group of objects for which mitigation procedures for collision avoidance with Earth must be developed. In addition, local measurements of surface properties must be carried out and linked in a database to the physical characterization from remote sensing and to the bulk properties. Not all objects can be investigated in all details. A database, linking physical characterizations of surfaces to likely internal properties, will serve as a very useful tool for guidance. These matters are at the heart of the present discussions.

---

* A sub-class of NEOs that passes the Earth's orbit within 0.05 AU (about 20 times the Earth – Moon distance).

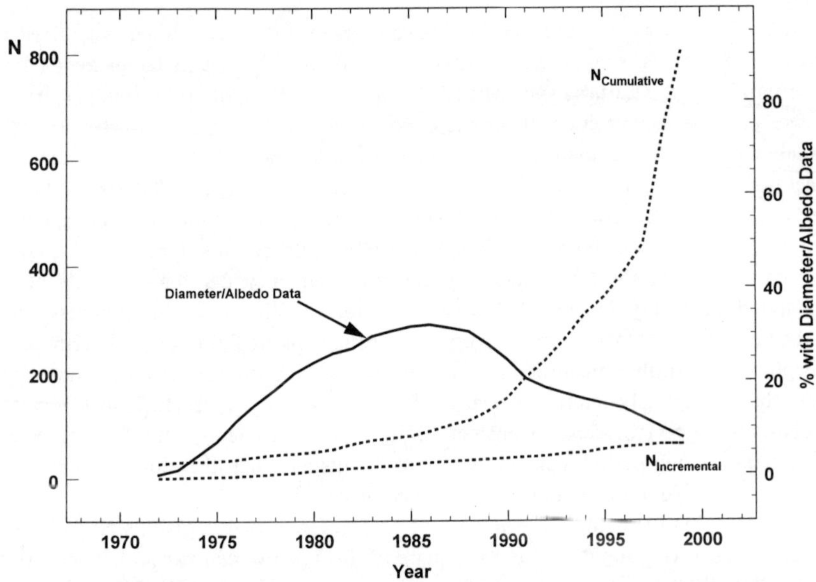

Fig. 1. Cumulative and incremental discovery rates of NEOs (dotted lines with scale on the left). Diameter and albedo determination of NEOs (solid line) expressed as a percentage of known NEOs (scale on the right). Courtesy Tedesco et al.[14]

The threat from NEOs raises major issues: Inadequate current knowledge, confirmation of a potential hazard after initial observations, reliable communication with the public, disaster management (in case of an impending impact), and, most important of all, methods for collision mitigation. The largest uncertainty in risk analysis arises from our incomplete knowledge of NEOs. With more data about their structure, mass, and physical strength, better plans for collision avoidance can be made.

One of the major problems, as was already pointed out by Tedesco et al.,[14] is that even physical characterization by remote sensing is drastically lagging behind the rate of new discoveries. Figure 1 illustrates this point.

Characterization by remote sensing is of great important. It provides size and albedo measurements from so-called radiometric techniques, spin rates from photometry, and overall surface compositions from visible and near-IR reflectance spectroscopy. While such techniques are well developed, the rate of property determinations significantly lags the discovery rate and falls further behind with each new discovery.

Below we list problem areas of NEO research and technology issues ranging from NEO detection to development of mitigation techniques.

Discovery Rate of NEOs

The rate of discovery of NEOs has dramatically increased. New estimates[11,15] based on recent NEO surveys, suggest that there may be 500 to 1100 objects larger than 1 km in the proximity of the Earth's orbit. The size-frequency relation of NEOs follows generally a power-law trend, with a characteristic exponent that is still poorly constrained by the available observational evidence. At sizes of the order of 200 m, that can cause catastrophic regional or local effects, the NEO population includes 10,000 objects in the most optimistic case. The objects might be several times more numerous. Thus, we have here one of the most important problems that NEO science must face, namely the very poor knowledge of the NEO inventory and size-frequency distribution, even at sizes corresponding to very dangerous objects. Detection of NEOs is well in progress, mostly for the objects larger than 1 km, but very little is made in the field of size determinations. This problem is further magnified by the existence of NEO subclasses (Atens and objects with orbits completely interior to that of the Earth) that are most difficult to detect by ground-based observatories. Considering the above problems, and the presence of unpredictable long-period comets, we must conclude that the current impact risk estimates available in the literature are still very uncertain.

Assuming the cumulative number of objects larger than a given size varies (very conservatively) with about the inverse square of the size, we can expect between 10,000 and 25,000 NEOs larger than about 200 m in diameter. When a 200 m NEO impacts in an ocean, it creates a tidal wave (tsunami) that can wipe out coastal cities along the waterfronts.[16] Considering that 70% of the Earth surface is covered by oceans and that some of the largest cities are along coasts, these objects are extremely dangerous. Analysis of events of very low probability but with devastating consequences shows that 200 m diameter asteroids may be the greatest threat to society.[17] NEOs in this size range are only beginning to be observed.[15,18,19]

International coordination of astronomical observations is a necessity. This is often ignored because much progress is being made in finding the larger objects. However, the smaller objects in the 200 m – 1 km size range, are much fainter, harder to discover, and much more numerous.

As mentioned earlier, the NEO population includes groups of objects having different orbital parameters. Among them, the so-called Aten asteroids orbit mostly in a region interior to the Earth's orbit, but have eccentricities sufficient to allow them to cross Earth's orbit near aphelion. They can be observed at opposition, but this is rare. They are most of the time at small heliocentric distances and are visible when located at small angular distances from the Sun. This makes ground-based observations very difficult.

In addition to the Atens, another class of objects with orbit completely "interior to Earth's orbit" (IEOs) have been postulated to exist.[20] The postulate is the result of numerical integration of the orbits of all known classes of NEAs. It was found that many objects might spend a significant fraction of their lifetime as IEOs. The IEO abundance should be about one half the abundance of the Atens. It is therefore an important but extremely difficult to observe population, since IEOs never reach large solar elongations.

The problems of detecting short-period and long-period comets have been discussed by Shoemaker et al.[21] and by Marsden and Steel,[22] respectively. The Spaceguard search region was selected to cover 60° in longitude along the ecliptic and ±60° in ecliptic latitude. Since most short-period comets spend a large fraction of their time in the neighborhood of Jupiter, the selected region should find most asteroids and short-period comets if the limiting magnitude of the telescopes is about 22. Some Halley-family (short-period) comets are at higher ecliptic latitudes and larger aphelia and therefore more difficult to discover. There is no reliable approach to detect long-period comets, i.e., comets with a period of more than 200 years that can be as long as 2 million years if they come directly from the Oort cloud.

That comets can be potentially hazardous objects on a collision course with Earth is demonstrated by the existence of meteor showers. Meteor showers result when the Earth passes through a meteoroid stream and the particles, traveling at high speeds through the atmosphere, reveal themselves as luminous streaks in the sky as they heat up from the friction with Earth's atmosphere. A dust trail is produced when a comet nucleus sheds dust in its orbit around the Sun. The dust trail remains in the orbit of the comet. Meteoroid streams are extremely large structures consisting of material spread over an entire comet orbit. Comet trails consist of large, millimeter to centimeter sized particle aggregates that extend over only small portions of a comet orbit and are less than hundreds of years in age (compared to thousands of years for meteoroid streams). They are ejected at speeds of a few m/s, represent the beginning of meteoroid stream formation, and appear to be preferentially associated with short-period comets that have the smallest perihelion distances.

The stronger the meteor shower, the closer is the intersection of the two orbits and the more recently has the comet passed. A collision or close encounter between Earth and the comet is avoided because the comet is not at the crossing point when the Earth is there. Some suspected sources of known meteoroid streams are the Comets C/1861 G1, 1P/Halley, 109P/Swift-Tuttle, 21/PGiacobini-Zinner, 2P/Encke, 55P/Tempel-Tuttle, 3D/Biela, and 8P/Tuttle. A recent analysis[23] of comet dust trails list eight trails associated with the short-period Comets 67P/Churyumov-Gerasimenko, 2P/Encke, 65P/Gunn, 22P/Kopff, 7P/Pons-Winecke, 29P/Schwassmann-Wachmann 1, 9P/Tempel 1, and 10P/Tempel 2. The Asteroid 3200 Phaethon is in the same orbit as the Geminid meteoroid stream; it may be an extinct comet nucleus. The near-Earth Asteroids 2101 Adonis and 2201 Oljato have orbits similar to those of known meteor showers. These objects are difficult to observe, but they may be extinct comets, i.e., transition objects.

In several cases, the source of a meteoroid stream is not known. These are likely disintegrated short-period comets (periods less than 200 years), however, the Quadrantids shower corresponds to a period of almost 200 years and the source is therefore difficult to trace. It could be a long-period comet. The most famous of the meteor showers, a periodic episode of meteor activity, is the Leonid shower, which occurs middle of November and is particularly strong every 33 years (which is the period of the comet's orbit around the Sun). Figure 2 shows the Earth's path (from bottom right to top left) with respect to the Leonid meteoroid streams. The ovals indicate the positions of the streams from previous passages of Comet 55P/Tempel-Tuttle. There are several reasons why the streams are displaced for each period: In 1883 Comet Tempel-Tuttle came close to Jupiter; its gravitational forces changed the comet's orbit. This caused the jump in the positions of the trails toward the bottom right. Within each of the two groups, nongravitational forces change the orbit of the comet and therefore the dust trails. The meteor showers are not of direct interest here. Interesting is that the comet causing the showers shifts its orbit because of nongravitational forces due to outgassing and that the comet can be very close to Earth. Nongravitational sources are the result of outgassing of the comet when it is close to the Sun. Gases evaporate on the sunlit side of the nucleus, but since the nucleus spins, the recoil effect is not exactly in the antisunward direction. This makes prediction of the orbit of the comet nucleus more difficult.

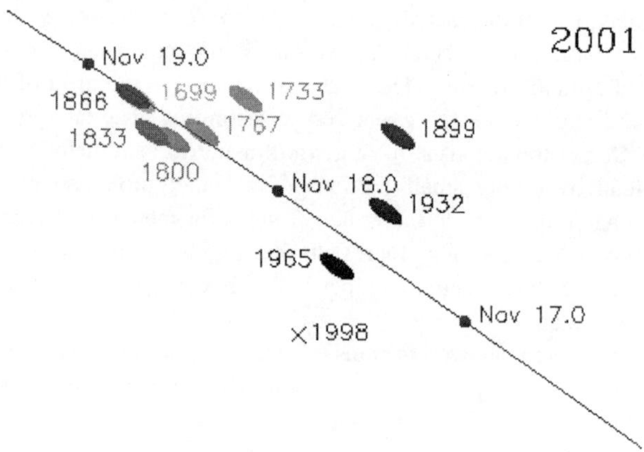

Fig. 2. Cross sections (ovals) of meteoroid streams from various passages of Comet 55P/Tempel-Tuttle. Note the almost exact superposition of the dust trails of 1699 and 1866. The two groupings of cross sections (upper left and lower right) are the result of a change in the comet's orbit caused by a gravitational encounter with Jupiter. The scatter within each of the two groups, e.g., the trails that were created in 1899, 1932, and 1965, are caused by changes in the comet's orbit because of nongravitional forces. The positions at any time are critically dependent on where the comet was in previous years. The comet's motion is retrograde, but about 17° out of the ecliptic. The retrograde motion means high relative velocity with respect to the Earth. The comet's 1998 crossing of the ecliptic is indicated by a cross. (With permission from David Asher, Armagh Observatory).

Follow-up Observations and Accurate Orbit Determination

Follow-up observations are important for orbit determination. However, presently, there are insufficient follow-up observations. In particular, the Southern Hemisphere presents a huge gap in the fraction of celestial sphere coverage. Many recently discovered NEOs have been lost again. Most observers currently involved in follow-up activities are amateurs who do not have sufficiently large telescopes to track the faintest objects. Dedicated telescopes for follow-up observations of NEOs are needed. Nongravitational forces play an important role in tracking and determination of orbit uncertainties of potentially hazardous comets. We return to the discussion how to improve follow-up observations of PHOs briefly in the section of Penetrator Probes.

Cataloging

Excellent progress has been made in cataloging NEOs and their orbits. NEOs are a worldwide problem. No nation is too poor not to contribute in some form to the solution of this problem. The Minor Planet Center tracks NEO discoveries and maintains the web site: http://cfa-www.harvard.edu/iau/lists/Unusual.html and the EARN web site lists measured albedo-diameters at http://129.247.214.46/nea/.

Other important NEO web sites are: http://neo.jpl.nasa.gov, http://impact.arc.nasa.gov, http://newton.dm.unipi.it/neodys, and the Torino scale can be found at: http://impact.arc.nasa.gov/torino/index.html.

Size-Frequency Determinations

The determination of NEO sizes and size-frequency relation is one of the most urgent tasks that wait to be accomplished in the framework of NEO research. The reason is that knowledge of sizes and their distribution is critically needed to assess the number of existing objects corresponding to different kinds of threats. The size is also needed for determining the mass, which is critically important in order to develop mitigation options. However, sizes of small solar system bodies are extraordinarily hard to measure. Direct measurements by purely imaging techniques are ruled out by the exceedingly small apparent angular diameters of the objects. This means that indirect techniques are needed. One possibility is offered by polarimetry, through an analysis of the variation of the degree of linear polarization as a function of varying phase angle (the Sun - NEO - Earth angle). This technique, however, is demanding in terms of telescope size (since polarimetry always requires splitting of the incoming light beam) and in terms of time, since each single object must be observed over periods of the order of weeks. The only viable option for a quick and efficient survey aimed at obtaining NEO sizes is provided by radiometry, a technique that is based on the simultaneous measurement of the visible scattered radiation and the thermal flux. However, this technique needs observations at mid-IR wavelengths[24] (around 10 μm) that can hardly be performed from Earth because of the absorption and emission properties of the atmosphere at these wavelengths. For this reason, the vast majority of available information on asteroid sizes and albedos has been obtained by means of IR satellites, such as IRAS and MSX. It is straightforward to conclude that a major step forward for NEO science would be the development of a dedicated space-based observatory (satellite) equipped with a modest-sized telescope, and both a visible CCD and an IR array.[25]

Composition and Bulk Properties

Much new information has been gained about asteroids from recent ground-based observations and from spacecraft missions. Unexpectedly low densities have been determined for a few objects. For example, the density of Asteroid 253 Mathilde was determined to be only 1300 ± 300 kg/m^3 from data obtained during the flyby of the NEAR spacecraft.[26] A similarly low density of about 1200 kg/m^3 was determined for asteroid 45 Eugenia from the orbit of its newly discovered moon.[27] The density of the S-type Asteroid 243 Ida is 2500 kg/m^3, as determined from the orbit of its moon Dactyl and other constraints. It also is less dense than its stony structure suggested.[28] The NEAR mission to Asteroid 433 Eros will determine the mass and volume (i.e., the density) of the asteroid, its spin state, and higher moments of its gravity field very accurately. From the 1999 flyby of Asteroid 433 Eros we already have an estimate of the density – it is 'Ida-like' at 2500 kg/m^3.[29] These lower than anticipated densities reflect the uncertainties in our knowledge about the internal structure of these objects.

Determination of whole-body properties is poorly funded, but crucial for the development of impact mitigation techniques. It suffers from a minimum in progress. We must develop and launch a number of coordinated multiple rendezvous space missions, possibly based on relatively inexpensive microsatellite technology, to visit different types of NEOs to establish their detailed structure and physical properties.

Databases for Surface and Bulk Properties

It will be impossible to send spacecraft missions to all NEOs. However, we must collect a meaningful statistical sample of NEO properties to extrapolate the data to thousands of objects. Even in an ideal case, a limited sample would require missions to 100 to 150 NEOs. Physical characterization from ground-based observations or from flyby missions will be helpful. However, the data from such observations must be linked to more detailed data of internal structure and whole-body physical properties to maximize our understanding and interpretation. For this purpose, a database will be needed.

Development of Mitigation Techniques

Orbits of long-period comets are poorly known. Their periods (P > 200 years and about two million years for an Oort cloud comet) are too long to predict their arrival. They are usually not discovered until a few years before their entry into the inner solar system. If such a comet is threatening Earth, insufficient time may be available to nudge it out of the way gradually, and a nuclear explosion to nudge it out of the way may be the only solution. However, for a known PHO with a well determined orbit the best solution may not be to send a fleet of rockets carrying nuclear bombs and detonate them a kilometer from the object.

Various procedures have been proposed, but they depend on whole-body properties that are very poorly understood. We must set up studies to look into various alternatives for mitigating collisions with Earth. Some organizations like COSPAR have recognized the importance for determining material properties as the next most important step in this direction.

We know very little about the internal structure of NEOs. Any further progress in this field will strongly depend on advances in understanding the internal structure of NEOs. This is particularly true for the transition objects. An investigation of internal properties of NEOs has never been carried out. The information is not only needed for mitigation of collisions with Earth; it is also useful for exploration of natural resources. Low gravity and unknown surface properties of comet nuclei are an especially troublesome problem for landing and anchoring of spacecraft.

Several methods exist for determining interior properties. Radio tomography experiments can determine electrical properties, in particular the complex electric permittivity, which is related to the complex refractive index. Changes in the index can then be used to infer internal structures. Radio tomography will be carried out on the Rosetta mission. Ground penetrating radar can "see" several meters below the surface. Artificially activated seismic experiments can yield information about bulk material strengths by measuring seismic wave propagation properties from which material strengths can be extracted. Among artificial seismic sources applicable to NEO probing are impacts or impact-surviving explosive charges equipped with time-delayed or remotely triggered firing devices. Seismic experiments have been carried out on the Moon and are proposed for Mars missions. Penetrator experiments can be used for artificial seismic activation as well as for delivery of instruments. Gravity mapping has been carried out on several space missions. *In situ* experiments, such as drilling, have been proposed for the Rosetta mission. Such experiments will sample to a depth of about a meter below the surface. Finally, there are laboratory and simulation experiments. One series of simulation experiments was carried out in a large space chamber at the DLR in Cologne, Germany. Another set of experiments relevant for asteroids will be carried out in the Arkansas-Oklahoma Center for Space and Planetary Science. Other simulation experiments could be carried out on the Space Shuttle and the International Space Station. Below we discuss these methods in more detail.

COMET NUCLEI

In Table 1 we list approved missions to comets. Only the Rosetta mission and the Deep Impact mission promise to reveal some detailed information about internal material properties and structure of comet nuclei.

Table 1. Approved missions to comets.

Mission	Launch	Target Nr. 1	Flyby (F) or Impact (I)	Target Nr. 2	Flyby	Target Nr. 3	Rendezvous (R) or Flyby (F)
DS1	25-Oct-98	1992 KD	F-28-Jul-99	W-H	15-Jan-00	Borrelly	F-20-Sep.-01
Stardust	12-Feb-99	Wild 2	F-2-Jan-04	-	-	-	-
Rosetta	20-Jan-03	Mimistrobell	F-2006	Rodan	2008	Wirtanen	R-2012
CONTOUR	04-Jul-02	Encke	F-Nov-03	S-W 3	18-Jun-06	d'Arrest	F-16-Aug-08
Deep Impact	?-Jan-04	Tempel 1	I-Jul-4-05	-	-	-	-

S-W 3 = Schwassmann-Wachmann 3

W-H = Wilson-Harrington

1992 KD, Mimistrobell, and Rodan are asteroids.

The stardust mission will return a dust sample on 15 January 2006. The end of the mission for Rosetta is July 2013.

Dangers that Comet Nuclei Present

The hazard from a collision of a comet nucleus with Earth is between 10 and 30% of that from asteroids. The biggest risk for a collision with Earth is from long-period comets that have periods of two hundred to two millions of years. Coming from large distances in the solar system, their velocities and therefore their kinetic energies are much greater than those of asteroids. In addition, they can be in retrograde orbits, increasing their velocity relative to the Earth to about 70 km/s. The Comet Halley flybys in 1986 were at relative speeds of more than 65 km/s. The advance-warning period for a potential impact from a long-period comet may be as short as a few months compared to decades or centuries for asteroids. For example, Comet Hyakutake was only discovered a few months before it passed the Earth at relatively close distance. Furthermore, it was discovered by an amateur astronomer, not by a telescope dedicated to NEO searches.

Detection of inbound Oort cloud comets at large heliocentric distances is extremely difficult. The two recent bright Comets Hale-Bopp (C/1995 O1) and Hyakutake (1996 B2) were not found by NEO search telescopes, they were found by amateur astronomers. This is not intended to be a criticism of the NEO program. NEO search telescopes do find comets, but the NEO program is concentrating on the most promising regions of the sky to find NEOs. They did not search the regions from which Comets Hale-Bopp and Hyakutake came. Comet IRAS-Araki-Alkock (1983 II1) was discovered only 15 days before it passed the Earth at a distance of only 0.031 AU. It was not even recognized as a comet until only 8 days before closest approach! With more telescopes coming on line, this situation may be resolved soon.

Brandt et al.[30] argue that there also must be a large population of small comets that escape detection. As examples of small comets they cite Comets 41P/Tuttle-Giacobini-Kresák, 7P/Pons-Winnecke, Sugano-Saigusa-Fujikawa (C/1983 J1), 45P/Honda-Mrkos-Pajdušáková, and 46P/Wirtanen. Small comets are intrinsically faint.

Their detection requires an optimized approach based on unique cometary features: a diffuse source of reflected light from the coma (rather than from the nucleus), motion with respect to background stars, and characteristic fluorescence spectra (e.g., CN).

ASTEROIDS

Asteroids are classified according to their visible reflectance spectra. Generally recognized asteroid spectral types include the S (stony) asteroids, the C (carbonaceous) asteroids, the M (metal) asteroids,[†] and the D asteroids (primitive and rich in organics like the C types, but with redder spectra). Several additional types have been recognized.[31] The S asteroids belong to the most numerous type in the inner solar system. The two asteroids visited by the Galileo spacecraft and the NEAR rendezvous target (Asteroid 433 Eros), are all S asteroids. These objects contain the silicate minerals pyroxene [(Fe, Mg)SiO_3] and olivine [$(Mg, Fe)_2SiO_4$] as well as metallic iron. Most of the current knowledge on the surface composition of asteroids and NEOs come from remote sensing observations, and comparisons with reflectance spectra of different classes of meteorites. Only recently, we have had the beginning of a new era of *in situ* exploration by means of space probes.

The NEAR-Shoemaker Mission

The Near Earth Asteroid Rendezvous (NEAR) is a mission of the NASA Discovery Program to rendezvous with the near-Earth Asteroid 433 Eros for a yearlong comprehensive scientific study. Asteroid Eros, whose maximum dimension is 32.7 km, is one of the largest near-Earth asteroids. The NEAR-Shoemaker spacecraft, named after Eugene M. Shoemaker (1928-1997), is three-axis stabilized and carries a payload of five scientific instruments: a multispectral imager[32] (MSI), a near-infrared spectrometer[33] (NIS), an x-ray/gamma-ray spectrometer,[34] a laser rangefinder,[35] and a magnetometer.[36] The scientific investigations on NEAR,[37,38,39,40,41,42] which include a radio science investigation using the spacecraft coherent X-band telemetry system, address key issues of the asteroid's surface morphology, surface composition, and interior structure.

NEAR was launched on February 17, 1996 and executed a flyby of the main belt Asteroid 253 Mathilde on June 27, 1997. This was the first spacecraft encounter with a C-type asteroid, and it was the first science return from this NASA Discovery Program. Initial reports were published in a special issue of Science.[43,44] Subsequently, NEAR flew by Earth again on January 23, 1998, receiving a gravity assist that targeted the spacecraft to its rendezvous with Asteroid 433 Eros.

On December 20, 1998, NEAR-Shoemaker was scheduled to begin its rendezvous with Asteroid Eros, but the first rendezvous burn was aborted, and contact with the spacecraft was lost for 27 hours. After recovery of communications, the NEAR-

† Recent spectroscopic investigations at 3 μm has shown the presence of water of hydration in about one third of the cases. Thus, it is likely that at least one third of M-type asteroids are not at all metal-rich. For them, a new taxonomic class (W, for "wet") has been proposed.

Shoemaker spacecraft executed a flyby of Asteroid 433 Eros on December 23, 1998. The rendezvous burn was executed successfully on January 3, 1999, targeting NEAR-Shoemaker for a return to Eros in February 2000. The 1998 Asteroid Eros flyby yielded important first measurements[45,46] of the mass and shape of the asteroid, which reduced risk for the later orbital operations at Asteroid Eros.

On February 14, 2000, the NEAR-Shoemaker spacecraft passed directly between the Sun and Asteroid Eros, and the highest priority near-infrared spectral maps were obtained successfully at low phase angles (when shadows on the surface are minimized). Later that same day, orbit insertion around Asteroid Eros was accomplished successfully. Since then, the NEAR-Shoemaker spacecraft has operated in nearly circular orbits at radii as small as 35 km. On October 26, 2000 the NEAR-Shoemaker spacecraft executed a low altitude fly-over at a minimum altitude of 5.3 km from the surface.

The NEAR flyby of Asteroid Mathilde provided important new scientific results from the first close-up look at a C asteroid (albeit an unusual one, see below). The NEAR rendezvous with Asteroid Eros obtained the first x-ray spectra and the first laser ranging measurements of an asteroid.

Asteroid 253 Mathilde Flyby

The significance of the NEAR encounter with Asteroid 253 Mathilde is that it provided the first close-up look at a completely different type of object from the S asteroids explored by Galileo (and to be explored by NEAR). Asteroid Mathilde belongs to the C taxonomic type that predominates in the central portion of the main belt of asteroids between Mars and Jupiter. The carbonaceous composition is an inference, never confirmed by direct observation, based on the idea that most meteorites must be fragments of asteroids and on the spectral similarity of C asteroids and the carbonaceous chondrite meteorites. The nature and origins of the primitive asteroid types (including the C types) and their relationships to comets and dark objects in the satellite systems of the outer planets are among the most important unresolved issues in solar system exploration.

Apart from its importance as the first example of the C asteroids to be explored, Asteroid Mathilde was discovered to be extremely slowly rotating.[47] Its 17.4-day period is the third longest known and is at least an order of magnitude longer than that of typical asteroids. The origin of these very slow rotation states is puzzling.[48]

The NEAR flyby obtained the first direct mass determination of an asteroid: 44. The measured mass of 1.03×10^{17} kg and estimated volume of 78,000 km^3 imply a density of 1300 ± 300 kg m^{-3}. The volume must be estimated because only one face of Asteroid Mathilde could be imaged during the 25-minute NEAR flyby. The inferred density44 was unexpectedly low, half or less than that of carbonaceous chondrite meteorites, which are the closest spectral analogs. It implies a high porosity of 50% or more.[43] No natural satellite of Asteroid Mathilde was found,[43] although a few main belt asteroids, such as 243 Ida, 45 Eugenia, and 762 Pulcova are known to have satellites. The surface of Asteroid Mathilde is heavily cratered, with at least five giant craters whose diameters are comparable to the 26.5 km mean radius of the asteroid itself. The areal density of smaller craters less than 3 km diameter is approximately at equilibrium, similar to that of 243 Ida.[43] However, the presence of the five giant craters was a surprise, because impacts of the magnitude required to make such large craters are believed to be close to that for complete disruption of the asteroid (i.e., giant craters of diameter about equal to the radius of the asteroid are the largest that can be created without destroying the target). It was therefore remarkable that Asteroid Mathilde survived at least five giant impacts. Finally, Asteroid Mathilde proved to be remarkably uniform in both color and albedo.[43] The asteroid was known from ground-based observations to be a C-type asteroid and therefore dark and spectrally neutral, but the ground observations could not rule out the possibility of small bright patches (e.g., of ice) or spectrally distinct regions. The NEAR observations revealed no evidence of any albedo or spectral variations, implying a homogeneous composition. The measured albedo of 0.035 to 0.05 was consistent with telescopic observations.[43]

Asteroid 253 Mathilde Flyby Implications.

At this time, the results of the NEAR flyby of Asteroid Mathilde have yet to be fully understood. The low density suggests that Asteroid Mathilde be composed of high porosity, unconsolidated rock. In this sense, it can be said that Asteroid Mathilde is a "rubble pile". This is clearly a significant result for Asteroid Mathilde's geologic history, but the implications are unclear.[49] To infer how the asteroid came to be a porous body, we must know whether the porosity is microscopic, and what is the nature and distribution of voids within the interior - that is, we must probe Asteroid Mathilde's interior structure. Although it is unlikely, the porosity of the asteroid may be primordial, i.e., it may have originally accreted as a porous structure and survived as such to the present. This picture would suggest microscopic porosity similar to that of interplanetary dust particles. Alternatively, the structure may be an agglomerate of fragments of diverse bodies, subsequently accreted to form Asteroid Mathilde. In this case, macroscopic voids would be expected, possibly in addition to microscopic porosity. Another possibility is that the asteroid was thoroughly fractured by impacts but not dispersed, so it was ground into rubble; in this case macroscopic voids might preserve some spatial correlation to impact craters.

The NEAR observations provide important clues to the nature of Asteroid Mathilde. There is no evidence of any layered structure or of any compositional heterogeneity, despite the presence of giant craters that probed kilometers below the surface of Asteroid Mathilde. If the asteroid accreted fragments of diverse parent bodies, these must have had remarkably uniform albedos and colors, or else the fragments must be smaller than about 500 m and therefore not spatially resolvable in the NEAR images.

The giant craters provide additional clues to Asteroid Mathilde's history and nature. The asteroid's porosity makes it more difficult to crater and enhances the likelihood of survival of giant impacts.[49,50] Moreover, effects of oblique impacts need to be considered.[49] Roughly half of all impacts are more oblique than 45°. An oblique impact generates lower peak pressure and lower peak strain rates than a normal impact that creates the same sized crater, so Asteroid Mathilde is more likely to have survived *oblique* giant impacts.[49] Moreover, oblique impacts most often do not create elongated craters (none have been found on Asteroid Mathilde). Curiously, no ejecta blankets, and no ejecta blocks, have been identified on Asteroid Mathilde.[43] If account is taken of oblique impacts and of the asteroid's porosity, then the probability of making a giant crater is 2.1 to 2.6 times more than the probability of disruption.[49]

Fig. 3. Clockwise from far left: Asteroids 253 Mathilde, 433 Eros, 243 Ida, and 951 Gaspra are shown to correct relative scale, but Asteroid Mathilde is much darker than the other three asteroids. The 5-km crater Psyche (proposed name) is clearly evident on Asteroid Eros.

<u>NEAR at Asteroid 433 Eros</u>

The NEAR mission addresses two fundamental questions about asteroids. The first is whether some S asteroids are examples of primitive asteroids that have not melted or differentiated. There are distinct sub-types of S asteroids,[31] some of which are believed to represent differentiated asteroids (including 951 Gaspra). However, the sub-type to which both Asteroids 433 Eros and 243 Ida belong may be primitive. A primary objective of the NEAR-Shoemaker mission is to determine whether Asteroid Eros, and by inference similar S asteroids, are primitive or evolved objects. The NEAR-Shoemaker x-ray and gamma ray spectrometers measure the abundance of key elements such as Si, Mg, and Fe to make this determination. In addition, NEAR-Shoemaker infers the silicate mineralogy of the surface from visible and near infrared spectra, and constrains the thermal history by searching for a magnetic field of Asteroid Eros.

The second fundamental question deals with the collisional history of small bodies in the early solar system, when the terrestrial planets formed. An essential issue is the competition between disruption by violent impacts and accretion from gentler collisions. The question is whether Asteroid Eros has been battered into a loose agglomeration of much smaller bodies (that is, a so-called rubble pile), or is instead an intact collisional fragment from a larger parent body.

First results from the Asteroid Eros orbit have been published in *Science*.[51,52,53,54] Asteroid 433 Eros is a primitive, undifferentiated asteroid and a consolidated object, not a rubble pile.[53,54] The bulk elemental composition of Asteroid Eros is consistent with that of ordinary chondrites based on the areas so far analyzed,[53] but a primitive achondrite-like composition is not ruled out. The silicate mineralogy of Asteroid 433 Eros is also inferred from visible and near infrared spectra to be consistent with low-iron ordinary chondrites.[54] Asteroid Eros has not melted nor differentiated fully, but some degree of partial melting or differentiation is possible. No evidence for intrinsic magnetization of Asteroid 433 Eros has yet been found (B. Anderson, private communication). The absence of magnetization may also be consistent with a thermal history in which Asteroid Eros was never heated to melting. Also, there are subtle variations in spectral properties across the surface, but no firm evidence for compositional heterogeneity has been found.

The average density of Asteroid 433 Eros is about that of Earth's crust,[46] as first found in NEAR's December 1998 flyby and since confirmed in orbit. This average density of 2700 kg/m^3 is less than the average bulk density of ordinary chondrite meteorites as measured in the laboratory. This suggests that the bulk of Asteroid Eros is significantly porous and/or fractured, but not to the same extent as the >50% porous Asteroid Mathilde.[54] The interior of Asteroid Eros is nearly uniform in density, as inferred from its gravity field, which is very near to that which would be expected from a uniform density object of the same shape.[46,52] There is a small center of mass offset from the center of the figure that may be consistent with a regolith layer of up to 100 m depth.[52]

The NEAR-Shoemaker mission has shown that Asteroid Eros is a consolidated body, not an agglomeration of smaller component bodies bound mostly by gravity. There is a pervasive global fabric consisting of a variety of ridges, grooves, and chains of pits or craters. This can be seen in Figure 4 in the linear features in Himeros and Shoemaker Regio and the crosscutting grooves and ridges near Selene. Coherent systems of linear features extend globally across Asteroid Eros. Many craters appear to be jointed and/or structurally controlled (e.g., small craters near Selene in Fig. 4). Steep slopes are found, well above expected angles of repose, which indicate the presence of a consolidated substrate.[52] Tectonic features are found, including one ridge that extends over 15 km across the surface.[54] These findings suggest that Asteroid Eros is a collisional fragment from a larger undifferentiated parent body.

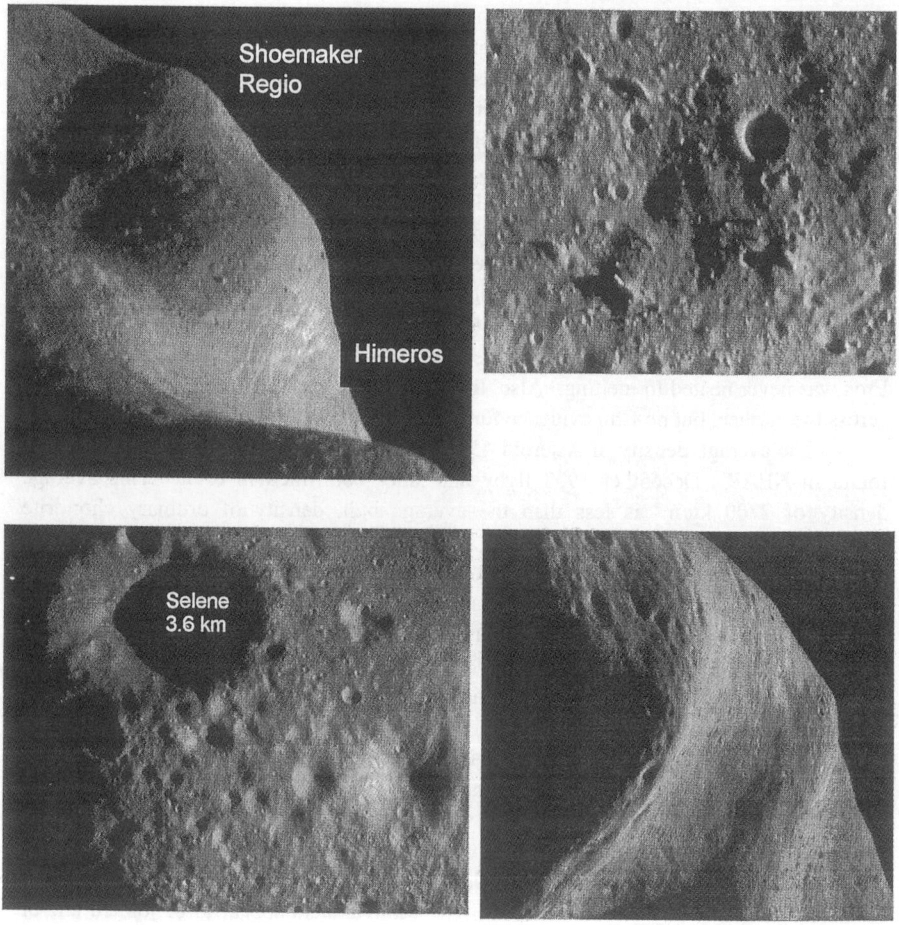

Fig. 4. Counterclockwise from top left: Himeros and Shoemaker Regio; Selene crater; east and west rims of Himeros; Shoemaker Regio (proposed names).

Most of Asteroid Eros's surface is old and close to equilibrium crater saturation, but some regions appear to be relatively young and extensively resurfaced.[54] Blocks and boulders are ubiquitous but are not confined to gravitational lows.[54] The surface of Asteroid Eros is extremely rough and exhibits a fractal structure from scales of a few meters up to more than a kilometer.[55] Peak-to-trough amplitudes of ridges and grooves southwest of the 5 km crater Psyche (proposed name; see Fig. 3) exceed 100 m, but are under 40 m southeast of the saddle-shaped depression Himeros (proposed name) on the opposite face of Asteroid Eros. Examples of downslope motion have been found on

Asteroid Eros,[55] associated with steep slopes in crater walls. These data suggest depths of unconsolidated regolith a few tens of meters thick at widely separated locations.[55]

Objects with orbit completely interior to Earth's orbit (IEOs)

Apart from Asteroid Eros, what we know about NEOs comes from remote-sensing observations, including some spectacular radar "images" obtained by S. Ostro and co-workers, who have also discovered the presence of a number of contact binary systems among these objects. NEOs are among the best possible targets for remote radar experiments, since they can approach the Earth to relatively small distances. In spite of being intrinsically very powerful, radar techniques suffer from the inverse fourth-power dependence on the distance for the received signal. However, the conventional observing techniques based on ground-based telescopes, for both discovery and physical characterization, also suffer from some intrinsic and unavoidable limitations that put some serious limits in their performances. First, it is obvious that any ground-based telescope cannot monitor the entire celestial sphere. This means that different instruments, located at several locations in latitude and longitude, are needed to ensure a satisfactory coverage of the sky. This is a current problem encountered in follow-up activities, which suffer from a poor coverage of the Southern Hemisphere. The situation is even worse for physical characterization. For example, mid-IR observations can be made only in a few places. Another problem that can hardly be solved is caused by the presence of objects that orbit partly (Atens) or entirely (IEOs) inside Earth's orbit.

Numerical integrations of the orbits of known NEOs belonging to all the orbital subclasses (Atens, Apollos, Amors) show that a population of IEOs must exist, and the estimated abundance of these objects is not negligible.[20] IEOs should account for between 0.6 and 0.7 of the total number of existing Atens. In turn, Atens account for 6 to 7% of the Aten plus Apollo plus Amor population, and 13% of the population of Earth-crossing asteroids. These numbers, however, are only a lower limit, since it is known that the discovery of Atens is made difficult by the fact that they spend most of the time at small elongations from the Sun. For this reason, Atens are expected to be actually more abundant than the current observational evidence seems to indicate. In the case of IEOs, the situation is even worse, in the sense that these objects are never visible at opposition, and are always located at small solar elongations. For this reason, we are still waiting for the discovery of the first object of this important class. Any ground-based telescope will always be strongly limited in its capability of detecting these objects. They can be visible only for brief periods during dusk, low above the horizon. For this reason, it is easy to predict that any significant progress in the discovery and physical characterization of IEOs (and of Atens) cannot be made without the help of dedicated space-based observatories. Space-based telescopes could observe a much larger fraction of the celestial sphere, including the regions close to the Sun, much more efficiently than any ground-based facility.[14,25]

BULK PROPERTIES OF NEOS

Material strengths, mass, density, moments of inertia, center of mass (higher moments of the gravitational field), internal structure (monolithic, rubble pile, etc.) must be determined. Such properties can be determined from properties of wave propagation through the interior of NEOs. In the next sections, we describe some experiments based on electromagnetic and sound wave propagation.

Radio Tomography Experiments

Electromagnetic transmission and reflection tomography can be used to explore electric properties (electric permittivity) of the interior. Changes in the complex permittivity result in attenuation, changes in direction of propagation, and reflection of signals. First, we discuss transmission tomography.

Electromagnetic wave transmission and absorption (attenuation) measurements (similar to the CONSERT experiment on the Rosetta mission to Comet 46P/Wirtanen[56]) are a function of the complex permittivity of the materials in the NEO. Electromagnetic radiation is strongly attenuated and reflected by conductors. Therefore, radio tomography will be most effective for nonmetallic objects such as comet nuclei and carbonaceous asteroids. It is an important complementary tool to seismology for investigating stony asteroids. However, the presence of small concentrations of trivalent cations can significantly increase the conductivity of orthopyroxene (orthorhombic $FeSiO_3$) and greatly change the conductivity of olivine $[(Fe, Mg)_2SiO_4]$ as a function of Mg/(Mg + Fe).

A signal is transmitted from the orbiting spacecraft through the body of the NEO and then detected by receivers on the surface of the NEO. The signal is re-transmitted to the orbiter where the roundtrip time delay (phase shift) and attenuation can be determined. Plane changes of the orbiting spacecraft provide planar "cuts" of the data through the NEO. The more orbit planes can be scanned and the more transmitters there are on the surface, the more unique and detailed will be the interpretation of the data.

Solving the scalar Helmholtz equation

$$\nabla^2 E + k^2 \varepsilon(r) E = 0\,, \qquad (1)$$

with the appropriate boundary conditions allows evaluation of the complex permittivity, $\varepsilon(r)$. In Eq. (1), k is the wave number for a given frequency. The inversion problem of extracting the complex permittivity uniquely throughout the body is complicated. The Moon is the only extraterrestrial body whose response to electromagnetic disturbances has been studied in detail.

Kofman et al.[56] have constructed an electromagnetic model for the interior of a comet for application to the CONSERT experiment on the Rosetta mission. Surfaces scatter the radiation. Internal scattering is treated by more complex treatments of the ray paths and scattering properties of the body. The signals are sensitive to variations in the permittivity of the material. These variations can be introduced by fissures or strains in the material, changes in composition, or changes in porosity. Changes of the permittivity detected and recorded at various positions of the spacecraft in each orbital plane reveal inhomogeneities that can be interpreted as fracture geometries, compositional or structural boundaries, voids, or mineralogical inclusions in the object.

Alternatively, radio reflection tomography can be applied.[57] In this procedure a radio signal illuminates the NEO and records the reflected signals from both the surface and interior scattering regions. The reflected signals are sensitive to variations in the complex permittivity of the material and are analyzed to achieve an image of the interior.

Reflection tomography offers several important advantages over transmission tomography when the objective is to image the interiors of moderately lossy objects like an asteroid, whose absorption may be much larger than that of a comet. Reflection tomography avoids the inherent risk and cost associated with receivers and transmitters that must be landed on the NEO surface and the risk that the signal may not fully penetrate through the object. In addition, since reflection tomography gathers its projection data from one side only, the technique returns useful data even with partial penetration and partial coverage.

Radar reflection measurements at different frequencies can be used to test the structure to various depths. Low-frequency radar penetrates deep into an NEO and is useful for examining the interior structure. High-frequency radar reflection tomography can determine shallow surface characteristics such as depth of regolith on asteroids and comets and the geology beneath and around craters. Thus, multifrequency radar experiments result in complementary information.

Measuring Artificially Induced Seismic Activity

Seismology will be most effective for compact objects such as metallic and stony asteroids. It will be less effective for fluffy objects such as comet nuclei and rubble pile asteroids because such materials will display high attenuation and scattering. Thus seismology experiments are complementary to radio tomography experiments.

The only extraterrestrial object investigated seismologically is the Moon. These investigations started with the Apollo 11 mission in July 1969. More advanced seismometers were deployed at the landing sites of Apollo 12, 14, 15, and 16. By September 1977, over 12,000 seismic events (including impacts from over 1700 large meteoroids) were recorded with these instruments and their data transmitted to Earth.

Moonquakes occurring naturally are scientifically interesting. They reveal magnitude and frequency of meteoroid impacts on the lunar surface or stresses induced by lunar tides as the Moon orbits Earth. However, characterizing the response to man-made seismic events, such as jolting impacts, explosives, or vibrators, is more useful since the place and time of initiation of the event can be controlled and determined precisely. Such events were achieved by deliberately crashing rocket stages and ascent stages of lunar modules onto the surface. Crashes with masses of 0.5 to 5000 kg and other active experiments with explosives (e.g., during the Apollo 17 mission in December 1972) created moonquakes at known times and locations. Structural data and material properties for the upper kilometer of the lunar crust were obtained from these measurements. It was found that the seismic P-wave velocity is between 100 and 300 m/s, much lower than for solid rock on Earth. These velocities are consistent with highly fractured (brecciated) material produced by prolonged meteoritic bombardment of the Moon. Although the Moon was the only extraterrestrial body investigated seismically, this situation may change soon with the intensifying exploration of Mars. Similar results as for the Moon can be expected for artificially induced seismic activity on NEOs.

The most useful seismic phenomena for determining interior properties are those artificially induced and associated with the whole body response of an NEO. These phenomena include the circumferential surface (Rayleigh) wave travel-times and the modal resonances of the NEO as an irregularly shaped "geoid." In consolidated materials, one can assume constant seismic velocity of the waves. Since pressures within an NEO are low, compression of materials will not occur and if no physical or chemical discontinuities or gradients exist, one can expect seismic signals to travel approximately in straight lines through the interior.

Several seismic effects can occur on an NEO with the potential for revealing useful information on structural and bulk physical properties. For example, the propagation velocity of direct-path compressional and shear waves traveling through an NEO will be accurately indicative of a broad class of possible materials and their strengths. In its simplest form (purely elastic, isotropic, homogeneous medium), the speed of longitudinal pressure (P) waves, v_P, and of transverse shear (S) waves, v_S, are related through thermodynamic properties of the medium to the adiabatic bulk modulus of compressibility

$$\kappa = \rho\,(\partial P/\partial \rho)_S\,, \tag{2}$$

and the shear modulus, μ, of the body by

$$v_P^2 = (\kappa + 4\mu/3)/\rho\,, \tag{3}$$

$$v_S^2 = \mu/\rho\,. \tag{4}$$

Here ρ is the density of the material in its undisturbed state as determined, e.g., by independent mass (from gravity) and volume determinations and S in Eq. (2) is the entropy. Thus, determining propagation velocities is an important objective of seismic experiments.

Reflected and refracted waves within the body, recognized by comparing seismic event travel times with those from directly arriving waves, will indicate the presence of structural inhomogeneities and discontinuities in the NEO. Multi-path reverberations (scattering) and anelastic attenuation will reveal the general rigidity and inelastic energy absorbing characteristics of the NEO. Information on regolith thickness from imaging, radar, and penetrator probes may complement the seismic observations. Modal resonances of the body are unique to the NEO size and shape. They will exhibit selective responses dependent on the location of the activating source on the NEO surface and the dominant wave types (compressional or shear) active in each mode.

Surface (Rayleigh) waves are two-dimensional waves bound to the surface of the NEO and can be readily generated by a surface detonation or impact. Such waves typically travel at about 85-90% of the shear wave velocity. The higher frequency spectral components, traveling as surface waves, will be governed largely by the seismic bulk properties of the materials located within the top 200 m of the NEO surface.

Although many modal resonances are possible in arbitrarily shaped objects, the fundamental resonance frequency of these modes generally requires the controlling dimensions of the resonant body to be approximately one-half of a wavelength. For consolidated rocky materials such as may be found in stony NEOs, compressional wave velocities will be in the range of 3,000 – 9,000 m/s and shear waves will be in the range of 1,000 – 5,000 m/s. Thus, the spectral range of 0.1 – 100 Hz will suffice to excite most of the modal resonances in consolidated rock for NEOs in the size range from 200 m to a few kilometers in diameter. Seismic data (modal resonance frequencies, pulse shapes, `ring-down' decay times, and encircling-time delays of surface-waves), when combined with other independently observed or derived parameters (e.g., mass, size, and shape of the body), lead to detailed whole-body models of the NEO.

Seismic wave propagation velocities span a wide range of values depending on the elastic moduli (strength parameters), density, and the seismic wave particle motions (compressional or shear) involved. Velocities in soft or unconsolidated materials are slower than velocities in hard solid materials and, in general, low-velocity materials are more dissipative than high velocity materials. Surface waves and shear waves travel at slower velocities than compressional waves, while surface waves are inherently always slower than shear waves by about 5 - 15%. In consolidated materials, shear waves may be as high as about 60 – 70% of the compressional wave velocity. In unconsolidated soils (e.g., regolith), the shear wave velocity may be only about 20 - 30% of the compressional wave velocity in the same material. These relationships can provide practical guidelines for interpreting NEO whole-body seismic responses.

Vibrational damping and seismic wave attenuation, expressed in terms of the dimensionless parameter Q, will determine the decay time of shock-excited NEO resonances. The parameter Q is a measure of the peak elastic energy stored in an oscillation relative to the energy dissipated over one complete cycle of the oscillation. NEOs composed of consolidated high-strength materials are anticipated to have long resonance `ring-down' times; typically in the range of 30 - 60 seconds and possibly longer for Earth-like values of $Q < 100$. For a seismic environment with Q values similar to those within the Moon ($Q > 5000$), such "ringing" (when measured near 0.1 Hz) could last for more than one hour.

NEOs composed of weak or unconsolidated materials will have much shorter `ring-down' times; generally about 10 - 20 seconds. Thus, by recording the NEO whole-body resonance for 20 seconds or longer, the resonance damping time constant can be measured and possibly several circumferential passes of encircling surface wave events can be captured.

For very fluffy comets or "rubble pile" asteroids, we must anticipate very low sound speeds and anomalous wave propagation. When the medium is a collected body of stony fragments or rubble held together only by gravity, the concepts of conventional seismic propagation are not valid. In this case, seismic waves will be multiply scattered. This will result in prolonged seismic signals, the so-called "seismic codas," as, e.g., observed on the Moon. In an extreme case, seismic wave propagation can be described by a diffusion-like process, as has been demonstrated for the strongly scattering lunar surface layers. Thus, in comets or rubble pile asteroids, carefully selected placement of seismometers and explosives or impacts to activate seismic events is very important to maximize useful data return.

Excellent progress has been made in developing and testing rugged, high-sensitivity micro-electromechanical systems (MEMS) acceleration sensors for use in geophysical exploration applications.[58,59] The MEMS technology designs provide a three-axis accelerometer in a single package. It has been tested in prototype form to demonstrate that it is equally sensitive, has a greater dynamic range, and a wider seismic signal bandwidth than conventional high-quality geophones. This sensor design achieves the indicated superior performance capabilities by operating as a force-balance sensor system. Most commercial geophones have a low-frequency response cut-off limited at about 5 – 10 Hz because of the size of the spring-mass moving coil design. The dynamic feedback technique of the micro-miniature sensor system, on the other hand, has a low-frequency response down to about 1 Hz and a high-frequency response up to 250 Hz. In principle, the force-balance sensor technology is not physically constraint to the low-frequency limit. With some design modifications, the MEMS sensor can be made to have a frequency response range down to about 0.1 Hz. Such low frequency response may be useful for applications on NEO probing missions. Since the data analysis process can translate recorded data to any desired triaxial orientation, the integrated three-component accelerometer design eliminates the need for the sensors to be physically oriented in a prescribed way at the point of their deployment.

Penetrator Probes

Penetrators can be used for local measurements of composition and of material properties in general. For example, accelerometers can determine the local resistance encountered by a penetrator as it enters the target body and thus determine the local material strength. Launching and implanting a swarm of mini-penetrators can reveal material properties over a large area of an NEO. However, even without implanted seismometers, some seismic measurements may be possible since there is no atmosphere and negligible gravitation on an NEO. For example, dust can be lofted by a seismic wave traveling through or on the surface of an NEO. Information can then be gathered from the propagation speed and the decay of the lofted dust with distance from the activating source.

Another useful measurement is the composition below the surface. It can be measured, e.g., with miniaturized mass spectrometers and complementary instruments implanted by penetrators.

Important global measurements can be made with seismometers delivered with penetrators. MEMS seismometers can withstand high-*g* decelerations for short times. Seismic activity can be induced by impacts of penetrators. The kinetic energy of a passive 1-kg mass projectile impacting at a speed of about 3.2 km/s or greater exceeds the sensible heat energy of the same mass of high explosive material detonated at rest on an NEO. Thus, explosive supplementation of a passive impact seismic source is only significant when the impact velocity is relatively small.

Other instruments such as radio beacons can be delivered. Radio beacons can be used to measure Doppler shifts of the carrier frequency caused by the motion of the NEO relative to Earth. Determining the velocity of an NEO relative to Earth would supplement ground-based observations of the positions of an NEO on the sky and thus double the amount of data for determining the NEO's orbit. In addition, it would reveal the spin rate of the NEO. The beacon only needs to be interrogated once or twice a year; thus, power requirements might be overcome relatively easily.

Gravity Measurements

Unpowered low velocity passes close to the surface of the NEO cause gravitational forces to produce small changes in the trajectory of the spacecraft. From measurements of these changes, the mass of the NEO can be determined and its gravitational field can be mapped. Combining the mass with the volume of the NEO obtained from imaging allows determination of the bulk material density. The center of mass of the NEO and its moments of inertia can be determined from the higher orders of gravitational moments. Mass determinations have been carried out on several asteroid flyby missions but measuring the higher gravitational moments was perfected during rendezvous of the NEAR-Shoemaker spacecraft to Asteroid 433 Eros.

In Situ Experiments and Sample Returns

In situ surface examination and sample collection are already planned for some missions, e.g., the Japanese mission Muses C to Asteroid 4660 Nereus. A sample will be returned to Earth for laboratory analysis. Drilling and *in situ* subsurface examinations are planned for the Rosetta mission to Comet 46P/Wirtanen. Samples will be extracted from the comet nucleus from depths up to about 1 m and analyzed *in situ*. A deep hole will be blasted by a 500-kg inactive kinetic energy projectile in the Deep Impact mission to Comet 9P/Tempel 1 and the ejecta materials will be examined remotely. Such tests can be extended by active (explosive driven) impacts to examine subsurface materials.

Using mirrors to focus and concentrate the Sun's radiation on selected spots on the surface of a comet nucleus can be used to "excavate" surface materials. For example, at 1 AU heliocentric distance, the Sun's energy at the subsolar point vaporizes ~10^{22} H_2O molecules/(m^2 s). For a mean density of 300 kg/m^3, this corresponds to a depth of 1 μm/s. In 10^4 s (about 3 hours of sunshine) the excavated depth is 1 cm. With a mirror of 10 m^2 size (less than 2 m radius), a hole with a cross section of 1 m^2 can be excavated to a depth of 1 m in 30 hours of sunshine.

Simulation Experiments

Gravity on an asteroid or a comet nucleus with average radius of about 1 km is typically 10^{-4} to 10^{-5} times that on Earth. These values are in the range of the residual acceleration on the Shuttle and Space Station. Microgravity experiments are a natural consequence of research on small solar system bodies such as asteroids and comets. In general, simulation experiments of various physical and chemical properties of surface materials on small solar system bodies should be carried out. As the Near-Earth Objects program increases in significance and the need for physical characterization of NEOs grows, so will the importance of microgravity experiments. Cratering experiments are important for determining internal material strengths. Such experiments are useful for collision mitigation of a potentially hazardous object (PHO) with Earth and for mining of asteroids (and perhaps even of comets) for *in situ* resource utilization.[60]

The stability of slopes of loose (non-cohesive), granular material (soil) depends on the angle of repose and the extent to which the soil is disturbed. The angle of repose (the angle between a horizontal plane and the maximum slope at which loose soil is stable) is a recognized characteristic of non-cohesive materials. It is independent of the height of the slope.

Asteroid regolith results from the continuous impact of large and small meteoroids and the steady bombardment of solar wind and cosmic ray particles. Cometary regolith forms by similar processes on mantled surface areas of the nucleus and on active areas, when evaporation of ices expose dust particle aggregates that are inherent in the mixture of ice and dust of which comets are composed. Most of these bodies are too small to be formed into spherical objects under their own gravity. They not only have an irregular shape, they also have hills and valleys - besides craters.

Of particular interest are non-cohesive granular materials on asteroids and comet nuclei. More specifically, conditions may exist where, under microgravity conditions, very small cohesive forces (e.g., van der Waals forces, negligible under normal conditions) permit an essentially non-cohesive granular material to acquire a slope steeper than the angle of repose. A minor perturbation may break the weakest link in the cohesive forces and initiate an avalanche. In an avalanche material slides down a slope picking up more material during the slide until a more level surface is reached with a slope smaller than the angle of repose. A slope close to the angle of repose will be established. The angle of repose is being investigated by the NEAR mission on Asteroid Eros (see, e.g., the section on NEAR at Asteroid 433 Eros). The question that needs to be answered is if avalanches can occur on small, low-gravity, solar system bodies. A meteoroid impact on an asteroid or a fluctuation of the outgassing on the surface of a comet nucleus can act as the trigger for an avalanche.

The Cosmic Dust Aggregation Experiments: CODAG and PROGRA, in which the growth of dust particles is examined experimentally, are well in progress. The experiment is designed for Space Shuttle missions to provide basic knowledge about the physics of aggregation under realistic environmental conditions. Optical properties of the growing particles will also be studied by light scattering.

CONCLUSIONS

A short-term goal of NEO research is to discover 90% of NEOs larger than 1 km in mean diameter, follow them up with telescopic observations (for orbit determination), characterize their physical properties, and catalog them. However, NEOs larger than 200 m in mean diameter are perhaps even more dangerous than the 1 km sized objects because they are much more numerous. These smaller objects can cause tidal waves (tsunamis) when they fall into an ocean, wiping out coastal cities on its peripheries where most of the world's population lives. Ocean impacts are more likely than land impacts since 70% of Earth's surface is covered by oceans. Most of the world's population lives in cities along the oceans. It is thus imperative that we prepare to find and examine smaller NEOs. The properties of smaller NEOs may be different than those of large NEOs. Some small NEOs have been observed to spin faster than their tidal disruption would allow. Thus, these small NEOs are more likely monolithic in structure. Dedicated telescopes are needed for follow-up observations, while larger NEOs may be fragmented aggregates or rubble piles, requiring very different collision mitigation procedures to nudge them out of their orbit.

It is evident that the existence of objects like Atens, IEOs, and small comets is a strong argument in favor of dedicated space-borne observations of NEOs, since a satellite should be able to observe at small solar elongation angles. At the same time, a space-based observatory should be able to efficiently perform the tasks for physical characterization of NEOs, with good cost effectiveness.

It will take a long time to build a materials properties database for NEOs. Ideally, to obtain a statistically meaningful sample, we must examine about 100 to 150 NEOs. Rendezvous and lander missions are expensive; thus, we should develop techniques to gather relevant materials properties data from flyby missions. We also must develop a long-term plan for sample returns.

In spite of all the current activities, we do very little about physical properties of NEOs. NEO physical characterization is currently losing the race against NEO discoveries. This constitutes a serious problem, since physical characterization is critically important in order to assess the NEO impact risk, and to solve the theoretical open problems about the sources and evolution of the NEO population.

Determination of whole-body NEO properties is poorly funded, but crucial for development of impact mitigation techniques. It shows only a minimum of progress. We must develop and launch a number of coordinated multiple rendezvous space missions, possibly based on relatively inexpensive microsatellite technology, to visit different types of NEOs to establish their detailed structural and physical properties.

ACKNOWLEDGMENTS

W. F. H. gratefully acknowledges support from NASA grant NAG5-6785. A. F. C. acknowledges support from NASA under the NEAR Project.

REFERENCES

1 Oró, J., Comets and the formation of biochemical compounds on the primitive Earth. *Nature* **190**, 389, 1961.

2 Morrison, D., The Spaceguard Survey: Report of the NASA International Near-Earth-Object Detection Workshop. JPL/Cal/Tech report, Pasadena 1992.

3 Canavan, G. H., Solem, J. C., Rather, J. D. G. (eds.) *Proceedings of the Near-Earth-Object Interception Workshop.* Los Alamos National Laboratory report LA-12476-C, 1993.

4 Rather, J. D. G., Rahe, J. H., Canavan, G. *Summary Report of the Near-Earth-Object Interception Workshop.* NASA, Washington, DC, 1992.

5 *Planetary Defense Workshop.* Lawrence Livermore National Laboratory report CONF-9505266, 1995.

6 Gehrels, T. (ed.) *Hazards Due to Comets and Asteroids.* The University of Arizona Press, Tucson, London, 1994

7 Chelyabinsk-70. Space Protection of the Earth against Near-Earth Objects. Organized by VNIITF, Russian Federal Nuclear Center, 1994.

8 Bowell, E., West, R. M., Heyer, H.-H., Quebatte, J., Cunningham, L. E., Bus, S. J., Harris, A. W., Millis, R. L., Marsden, B. G., (4015) 1979 VA = Comet Wilson-Harrington (1949 III). *IAU Circ.* **5585**, 1992.

9 Cunningham, L. E. Comet Wilson-Harrington (1949g). IAU Circ. **1248**, 1. 1949.

10 Fernandez, Y. R., McFadden, L. A., Lisse, C. M., Helin, E. F., Chamberlin, A. B. Analysis of POSS images of comet-asteroid transition object 107P/1949 W1 (Wilson-Harrington). *Icarus* **128**, 114-126, 1997.

11 Bottke Jr., W. F., Jedicke, R., Morbidelli, A., Petit, J.-M., Gladman, B. Understanding the distribution of near-Earth asteroids. *Science* **288**, 2190-2194, 2000.

12 Melosh, H. J., Nemchinov, I. V., Zetzer, Yu. I. `Non-nuclear strategies for deflecting comets and asteroids,' in Hazards Due to Comets and Asteroids. (T. Gehrels and M. S. Matthews, eds.) University of Arizona Press, Tucson, p. 1111-1132, 1994.

13 Huebner, W. F., `Physical and chemical properties of comet nuclei.' In International Seminar on Nuclear War and Planetary Emergencies, 23rd Session, K. Goebel, K. (ed.) p. 169-179, 1999.

14 Tedesco, E. F., Muinonen, K., Price, S. D. Space-based infrared near-Earth asteroid survey simulation. *Planet. Space Sci.*, in press, 2000.

15 Rabinowitz, D., Helin, E., Lawrence, K., Pravdo, S. A reduced estimate of the number of kilometre-sized near-Earth asteroids. *Nature* **403**, 165-166, 2000.

16 Hills, J. G., Mader, C. L. `Near-Earth Objects' in Annals New York Acad. Sci. **822**, p. 381, 1997.

17 Ward, S. N., Asphaug, E. Asteroid impact tsunami: A probabilistic hazard assessment. *Icarus* **145**, 64-78, 2000.

18 Binzel, R. P., Bus, S. J., Burbine, T. H. Size dependence of asteroid spectral properties: SMASS results for near-Earth and main-belt asteroids. *LPSC* XXIX, abstract no. 1222, 1998.

19 Pravec, P., Sarounová, L., Benner, L. A. M., Ostro, S., J., Hicks, M., D., Jurgens, R. F., Giorgini, J. D., Slade, M. A., Yeomans, D. K., Rabinowitz, D. L., Krugly, Y. N., Wolf, M. Slowly rotating Asteroid 1999 GU_3. *Icarus*, submitted, 2000.

20 Michel, P., Zappalà, V., Cellino, A., Tanga, P. Estimated abundance of Atens and asteroids evolving on orbits between Earth and Sun. *Icarus* **143**, 421-424, 2000.

21 Shoemaker, E. M., Weissman, P. R., Shoemaker, C. S. The flux of periodic comets near Earth. In Hazards due to Comets and Asteroids. (T. Gehrels and M. S. Matthews, eds.) University of Arizona Press, Tucson, p. 313-335, 1994.

22 Marsden, B. G., Steel, D. I. `Warning times and impact probabilities for long-period comets.' In Hazards due to Comets and Asteroids. (T. Gehrels and M. S. Matthews, eds.) University of Arizona Press, Tucson, p. 221-239, 1994.

23 Sykes, M. V., Walker, R. G. Cometary Dust Trails. I. Survey. *Icarus* **95**, 180-210, 1992.

24 Harris, A. W., Davies, J. K. Physical characteristics of near-Earth asteroids from thermal infrared spectrophotometry. *Icarus* **142**, 464-475, 1999.

25 Cellino, A. Physical properties of near-Earth objects: Open problems. *Adv. Space Res.* in press, 2000.

26 Veverka, J., Thomas, P., Harch, A., Clark B., Bell III, J. F., Carcich, B., Joseph, J., Murchie, S., Izenberg, N., Chapman, C., Merline, W., Malin, M., McFadden, L., Robinson, M. NEAR encounter with asteroid 253 Mathilde: Overview. *Icarus* **140**, 3-16, 1999.

27 Merline, W. J., Close, L. M., Dumas, C., Chapman, C. R., Roddier, F., Ménard, F., Slater, D. C., Duvert, G., Shelton, C., Morgan, T. Discovery of a moon orbiting the asteroid 45 Eugenia. *Nature* **401** 565, 1999.

28 Belton, M. J. S., Chapman, C., Thomas, P., Davies, M., Greenberg, R., Klaasen, K., Byrnes, D., D'Amario, L., Synnott, S., Merline, W., Petit, J.-M., Storrs, A., Zellner, B. Bulk density of Asteroid 243 Ida from the orbit of its satellite Dactyl. *Nature* **374**, 785-788, 1995.

29 Cheng, A. F. The NEAR Mission: Results to date. *BAAS* **31**, 1071, 1999.

30 Brandt, J. C., A'Hearn, M. F., Randall, C. E., Schleicher, D. G., Shoemaker, E. M., Stewart, A. I. F. On the existence of small comets and their interactions with planets. Earth Moon Planets **72**, 243-249, 1996.

31 Gaffey, M. J., Burbine, T. H., Binzel, R. P., Asteroid spectroscopy – Progress and perspectives. *Meteoritics* **28**, 161-187, 1993.

32 Hawkins, S. E., Darlington, E. H., Murchie, S. L., Peakock, K., Harris, T. J., Hersman, C. B., Elko, M. J., Prendergast, D. T., Ballard, B. W., Gold, R., Veverka, J., Robinson, M. S. Multi-spectral imager on the Near Earth Asteroid Rendezvous mission. *Space Sci. Revs.* **82**, 31-100, 1997.

[33] Warren, J., Peakock, K., Darlington, E. H., Murchie, S. L., Oden, S. F., Hayes, J. R., Bell III, J. F., Krein, S. J., Mastandrea, A. Near infrared spectrometer for the Near Earth Asteroid Rendezvous mission. *Space Sci. Revs.* **82**, 101-167, 1997.

[34] Goldsten, J. O., McNutt Jr., R. L., Gold, R. E., Gary, S. A., Fiore, E., Schneider, S. E., Hayes, J. R., Trombka, J. I., Floyd, S. R., Boynton, W. V., Bailey, S., Brückner, J., Squyres, S. W., Evans, L. G., Clark, P. E., Starr, R. The x-ray/gamma-ray spectrometer on the Near Earth Asteroid Rendezvous mission. *Space Sci. Revs.* **82**, 169-216, 1997.

[35] Cole, T. D., Boies, M. T., El-Dinary, A. S., Cheng, A., Zuber, M. T., Smith, D. E. The Near Earth Asteroid Rendezvous laser altimeter. *Space Sci. Revs.* **82**, 217-253, 1997.

[36] Lohr, D., Zanetti, L. J., Anderson, B. J., Potemra, T. A., Hayes, J. R., Gold, R. E., Henshaw, R. M., Mobley, F. F., Holland, D. B., Acuña, M. H., Scheifele, J. L. NEAR magnetic field investigation, instrumentation, spacecraft magnetics and data access. *Space Sci. Revs.* **82**, 255-281, 1997.

[37] Cheng, A. F., Santo, A. G., Heeres, K. J., Landshof, J. A., Farquhar, R. W., Gold, R. E., Lee, S. C. Near-Earth Asteroid Rendezvous: Mission overview. *J. Geophys. Res.* **102**, 23695-23708, 1997.

[38] Veverka, J., Bell III, J. F., Thomas, P., Harch, A., Murchie, S., Hawkins III, S. E., Warren, J. W., Darlington, H., Peakock, K., Chapman, C. R., McFadden, L. A., Malin, M. C., Robinson, M. S. An overview of the NEAR multispectral imager-near-infrared spectrometer investigation. *J. Geophys. Res.* **102**, 23709-23727, 1997.

[39] Acuña, M. H., Russell, C. T., Zanetti, L. J., Anderson, B. J. The NEAR magnetic field investigation: Science objectives at asteroid Eros 433 and experimental approach. *J. Geophys. Res.* **102**, 23751-23759, 1997.

[40] Trombka, J. I., Floyd, S. R., Boynton, W. V., Bailey, S., Brückner, J., Squyres, S. W., Evans, L. G, Clark, P. E., Starr, R., Fiore, E., Gold, R., Goldsten, J., McNutt, R. Compositional mapping with the NEAR x ray/gamma ray spectrometer. *J. Geophys. Res.* **102**, 23729-23750, 1997.

[41] Zuber, M. T., Smith, D. E., Cheng, A. F., Cole, T. D. The NEAR laser ranging investigation. *J. Geophys. Res.* **102**, 23761-23773, 1997.

[42] Yeomans, D. K., Konopliv, A. S., Barriot, J. P. The NEAR radio science investigation. *J. Geophys. Res.* **102**, 23775-23780, 1997.

[43] Veverka, J., Thomas, P., Harch, A., Clark, B., Bell III, J. F., Carcich, B., Joseph, J., Champman, C., Merline, W., Robinson, M., Malin, M., McFadden, L., Murchie, S., Hawkins III, S. E, Farquhar, R., Izenberg, N., Cheng, A., NEAR's flyby of 253 Mathilde: Images of a C asteroid. *Science* **278**, 2109-2114, 1997.

[44] Yeomans, D. K., Barriot, J.-P., Dunham, D. W., Farquhar, R. W., Giorgini, J. D., Helfrich, C. E., Konopliv, A. S., McAdams, J. V., Miller, J K., Owen Jr., W. M., Scheeres, D. J., Synnott, S. F., Williams, B. G. Estimating the mass of Asteroid

253 Mathilde from tracking data during the NEAR flyby. *Science* **278**, 2106-2109, 1997.

45 Veverka, J. Thomas, P. C., Bell III, J. F., Bell, M., Carcich, B., Clark, B., Harch, A., Joseph, J., Martin, P., Robinson, M., Murchie, S., Izenberg, N., Hawkins, E., Warren, J., Farquhar, R., Cheng, A., Dunham, D., Chapman, C., Merline, W. J., McFadden, L., Wellnitz, D., Malin, M., Owen Jr., W. M., Miller, J. K., Williams, B. G., Yeomans, D. K. Imaging of Asteroid 433 Eros during NEAR's flyby reconnaissance. *Science* **285**, 562-564, 1999.

46 Yeomans, D. K., Antreasian, P. G., Cheng, A., Dunham, D. W., Farquhar, R. W., Gaskell, R. W., Giorgini, J. D., Helfrich, C. E., Konopliv, A. S., McAdams, J. V., Miller, J. K., Owen Jr., W. M., Thomas, P. C., Veverka, J., Williams, B. G. Estimating the mass of Asteroid 433 Eros during the NEAR spacecraft flyby. *Science* **285**, 560-561, 1999.

47 Mottola, S., Sears, W. D., Erikson, A., Harris, A. W., Young, J. W., Hahn, G., Dahlgren, M., Mueller, B. E. A., Owen, B., Gil-Hutton, R., Licandro, J., Barucci, M. A., Angeli, C., Neukum, G., Lagerkvist, C.-I., Lahulla, J. F. The slow rotation of 253 Mathilde. *Plan. Space Sci.* **43**, 1609-1613, 1995.

48 Harris, A. W., Tumbling Asteroids. *Icarus* **107**, 209-211, 1994.

49 Cheng, A. F., Barnouin-Jha, O. S. Giant craters on Mathilde. *Icarus* **140**, 34-48, 1999.

50 Asphaug, E., Ostro, S. J., Hudson, R. S., Scheeres, D. J., Benz, W. Disruption of kilometre-sized asteroids by energetic collisions. *Nature* **393**, 437-440, 1998.

51 Yeomans, D. K., Antreasian, P. G., Barriot, J.-P., Chesley, S. R., Dunham, D. W., Farquhar, R. W., Giorgini, J. D., Helfrich, C. E., Konopliv, A. S., McAdams, J. V., Miller, J. K., Owen Jr., W. M., Scheeres, D. J., Thomas, P. G., Veverka, J., Williams, B. G., Radio science results during the NEAR-Shoemaker spacecraft rendezvous with Eros. *Science* **289**, 2085-2088, 2000.

52 Zuber, M. T., Smith, D. E., Cheng, A. F., Garvin, J. B., Aharonson, O., Cole, T. D., Dunn, P. J., Guo, Y., Lemoine, F. G., Neumann, G. A., Rowlands, D. D., Torrence, M. H. The shape of 433 Eros from the NEAR-Shoemaker laser rangefinder. *Science* **289**, 2097-2101, 2000.

53 Trombka, J. I., Squyres, S. W., Brückner, J., Boynton, W. V., Reedy, R. C., McCoy, T. J., Gorenstein, P., Evans, L. G., Arnold, J. R., Starr, R. D., Nittler, L. R., Murphy, M. E., Miheeva, I., McNutt Jr., R. L., McClanahan, T. P., McCartney, E., Goldsten, J. O., Gold, R. E., Floyd, S. R., Clark, P. E., Burbine, T. H., Bhangoo, J. S., Bailey, S. H., Petaev, M. The elemental composition of Asteroid 433 Eros: Results of the NEAR-Shoemaker x-ray spectrometer. *Science* **289**, 2101-2105, 2000.

54 Veverka, J., Robinson, M., Thomas, P., Murchie, S., Bell III, J. F., Izenberg, N., Chapman, C., Harch, A., Bell, M., Carcich, B., Cheng, A., Clark, B., Domingue, D., Dunham, D., Farquhar, R., Gaffey, M. J., Hawkins, E., Joseph, J., Kirk, R., Li, H., Lucey, P., Malin, M., Martin, P, McFadden, L., Merline, W. J., Miller, J. K.,

Owen Jr., W. M., Peterson, C., Prockter, L., Warren, J., Wellnitz, D., Williams, B. G., Yeomans, D. K. NEAR at Eros: Imaging and spectral results. *Science* **289**, 2088-2097, 2000.

55 Cheng, A. F., Barnouin-Jha, O., Prockter, L., Cole, T., Guo, Y., Zuber, M. T., Neumann, G., Smith, D. E., Garvin, J., Robinson, M., Veverka, J., Thomas, P. Small scale topography of 433 Eros from laser altimeter and imaging. *BAAS* **32**, 994, 2000.

56 Kofman, W., Barbin, Y., Klinger, J., Levasseur-Regourd, A.-C., Barriot, J.-P., Hérique, A., Hagfors, T., Nielsen, E., Grün, E., Edenhofer, P., Kochan, H., Jpicardi, G., Seu, R., van Zyl, J., Elachi, C., Melosh, H. J., Veverka, J., Weissman, P., Svedhem, L. H., Hamran, S. E., Williams, P. I. Comet nucleus sounding experiment by radio transmission. *Adv. Space Res.* **21**, 1589-1598, 1998.

57 Kak, A. C. Principles of Computerized Tomographic Imaging. IEEE Press, New York, 1988.

58 Maxwell, P. W., Cain, B., Roche, S. L. Field test of a micro-mechanical, electro-mechanical digital seismic sensor. *Soc. Expl. Geophysicists*, 69th Annual Meeting, Oct. 31-Nov. 5th, Houston, Texas, 1999.

59 Gannon, J. C., McMahon, M. G., Pham, H. T., Speller, K. E. A seismic test facility. Soc. Expl. Geophysicists, 69th Annual Meeting, Oct. 31-Nov. 5th, Houston, Texas, 1999.

60 Huebner, W. F., Greenberg, J. M. Needs for determining material strengths and bulk properties of NEOs. *Planet Space Sci.* **48**, 797-799, 2000.

11. DESERTIFICATION, CARBON SEQUESTRATION AND SUSTAINABILITY

STORING CARBON IN AGRICULTURAL SOILS TO HELP HEAD-OFF GLOBAL WARMING AND TO COMBAT DESERTIFICATION

NORMAN J. ROSENBERG AND ROBERTO C. IZAURRALDE
Pacific Northwest National Laboratory, 901 D Street SW, Washington, DC, USA

We know for sure that addition of organic matter to soil increases water-holding capacity, imparts fertility with the addition of nutrients, increases soil aggregation and improves tilth. Depending on its type–humus, manure, stubble or litter–organic matter contains between 40 and 60 % carbon. We also know that carbon (C, hereafter), in the form of carbon dioxide (CO_2), is currently accumulating in the atmosphere as the result of fossil fuel combustion, land use change and tropical deforestation (Table 1). The atmospheric concentration of carbon dioxide has increased by ~32%, from about 280 ppmv (parts per million by volume) at the beginning of the industrial revolution (ca. 1850) to about 370 ppmv today.

Table 1. Global C flux budget.

Carbon Flows	Pg C
Annual atmospheric increase of CO_2	3.4
Sources	
Fossil Fuels	6.4
Land use change	1.1
Tropical deforestation	1.6
Sinks	
Terrestrial in temperate regions	2.0
Oceans	2.0
"Missing"	1.7
Potential sinks in croplands alone (50-100y[a]) 40-80	2.0

[a]IPCC, 1996

There is a strong consensus among atmospheric scientists that continued increase in the concentration of atmospheric CO_2 and other greenhouse gases such as methane (CH_4) and nitrous oxide (N_2O) will enhance the earth's natural greenhouse effect and lead to global warming (Intergovernmental Panel on Climate Change, IPCC, 1996). Some scientists argue from the fact that 1997 was the warmest and 1998 the second warmest years on record that the global climate change 'footprint' is already detectable.

CO_2, the greenhouse gas of primary concern with regard to climate change, is also essential to photosynthesis. Elevated CO_2 concentration [CO_2] stimulates photosynthesis and growth in plants with C-3 metabolism (legumes, small grains, most trees) and reduces transpiration (water use) in both C-3 and C-4 plants (tropical grasses such as maize, sorghum, sugar cane). Together these phenomena are termed the "CO_2-fertilization effect."

Table 1 gives current estimates of global sources and sinks for C. Fossil fuel combustion, land use change and tropical deforestation are adding ~ 9.1 Pg C y^{-1} (1 Pg is equal to 1 billion tonnes or 10^{15}g) to the atmosphere. About 3.4 Pg C y^{-1} remains in the atmosphere. Regrowth of forests in the temperate regions and the oceans each appear to be absorbing ~ 2.0 Pg C y^{-1}, leaving about 1.7 Pg C y^{-1} unaccounted for. Most of this "missing carbon" is probably going into the terrestrial biosphere primarily in the Northern Hemisphere. The CO_2-fertilization effect is, probably, also contributing to the increased capture of C in terrestrial ecosystems.

In its Second Assessment Report the Intergovernmental Panel on Climate Change (IPCC, 1996) estimated that it may be possible over the course of the next 50 to 100 years to sequester 40 and 80 Pg of C in cropland soils (Cole et al., 1996; Paustian et al., 1998; Rosenberg et al., 1998). Reference to Table 1 shows that if this is so, agricultural soils alone could capture enough C to offset any further increase in the atmospheric inventory for a period lasting between 12 and 24 years. These calculations are still crude and cannot be taken as certain, but they do suggest a potential to offset significant amounts of CO_2 emissions by sequestering C in the soils of lands currently in agricultural production. Of course, there is additional C sequestration potential in the soils of managed forests and grassland (which we do not address here). And, as is discussed below, there is a very large potential for C storage in the soils of degraded and desertified lands. However, a caution needs to be raised here: unless alternatives to fossil fuels are found, the energy demands created by growing populations and rising standards of living could greatly increase CO_2 emissions over the next century and the capacity of agricultural soils to sequester carbon could be exhausted to little long-term effect.

The decade of the 1990s marked the beginnings of a political recognition of the threats that greenhouse gas emissions--at increasing or even continuing rates--may pose to stability of the global climate. In response to this threat, the United Nations adopted a Framework Convention on Climate Change (UNFCCC) in Rio De Janeiro in 1992 (United Nations, 1992). The convention aims at the "stabilization of greenhouse gas concentrations in the atmosphere at a level that would prevent dangerous anthropogenic interference with the climate system." In December of 1997 the Parties to the UNFCCC met in Kyoto, Japan, and drafted a Protocol to place binding limits on greenhouse gas emissions and to begin the process of stabilizing their atmospheric concentrations (United Nations, 1997). The Protocol recognizes that its objectives can be met either by *decreasing the rate at which greenhouse gases are emitted to the atmosphere* or by *increasing the rate at which they are removed from it.* It was well recognized in the Kyoto negotiations that photosynthesis, by fixing C in standing and below ground portions of trees and other plants, provides a powerful means of removing CO_2 from the atmosphere and sequestering it in the biosphere. The Kyoto Protocol establishes the concept of credits for C sinks (Article 3.3) but allows credits for only a limited list of activities including afforestation and reforestation (Article 3.4). The Protocol does not allow credits for sequestration of C in soils except, perhaps (indeed, this is not yet clear), for carbon accumulating in the soils of afforested and reforested land. Although the capacity for doing so clearly exists, sequestration in agricultural soils is not now permitted to produce C sequestration credits under the Kyoto Protocol. This mitigation option was set aside in the Kyoto negotiations ostensibly because of the perceived difficulty and cost of verifying that C is actually being sequestered and maintained in soils. However, the soil carbon sequestration option is specifically mentioned in Article 3.4 for possible inclusion at a later time.

Another way of looking at the potential role of soil C sequestration is shown in Figure 1, produced with the integrated assessment model MiniCAM 98.3 (Edmonds et al, 1996a,b; Rosenberg et al., eds. 1999). The top line in the figure represents the anticipated increase in carbon emissions to the atmosphere from the year 2000 to the end of the 21st century under a so-called "business-as-usual" scenario of IPCC (1990). It also shows a more desirable emissions trajectory that allows atmospheric [CO_2] to rise from its current level and stabilize at a maximum of 550 ppmv by 2035. Annual C emissions are allowed to increase at first but then are lowered steadily to reach a level in 2100 between 6 – 7 Pg C y^{-1}. For the upper emissions line to be brought down to the desired level will require great changes from our current energy systems. The caption of Figure 1 identifies some of the technologies that will create such change in the 21st century. Increased efficiency in the uses of fossil fuels, development of non-carbon emitting fuels, improvements in power generation, a greater role for biomass, solar, wind, and nuclear energy and other technological advances will ultimately be needed to mitigate climate change. Figure 1 shows that soil C sequestration can play a very strategic role but cannot, in and of itself, solve the problem. Soil C sequestration alone could make up the difference between expected emissions and the desired trajectory in the first 3-4 decades of the 21st century, buying time for development of the new technological advances identified above. The calculations shown in Figure 1 are based on the assumption that from 2000 to 2100 agricultural soils sequester C at global annual rates ranging from 0.4 to 0.8 Pg y^{-1}, with rates twice as great in the initial years and half as great in the later years. It is further assumed that the full potential of soil C sequestration is realized without any additional net cost to the economy—not unreasonable in view of the known benefits of organic matter in soils. In addition, by allowing time for new technologies to be developed and for existing facilities to live out their design lifetimes, the costs of an avoided tonne of carbon emissions over the next century can be cut approximately in half.

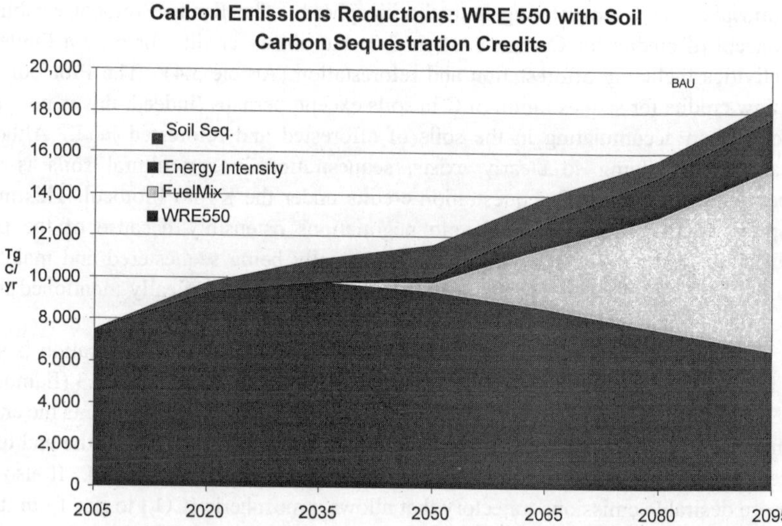

Fig. 1. Global Carbon Emissions Reductions: WRE 550 (Wigley et al., 1996, 550 ppmv atmospheric CO_2 concentration). This figure shows a hypothetical path to carbon emissions reductions from MiniCAM's business as usual (BAU) emissions pathway to the WRE 550 concentration pathway, under a scenario in which credit for soil carbon sequestration is allowed. Soil sequestration of carbon alone achieves the necessary net carbon emissions reduction in the early part of the century. From the middle of the century on, further emissions reductions must come from changes in the energy system (such as fuel switching and the reduction of total energy consumption).

How realistic are the estimates of potential soil C sequestration on which the economic modeling is based? The IPCC estimates for cropland assume the restitution of up to two thirds of the soil C released since the mid-19th century by the conversion of grasslands, wetlands and forests to agriculture. The experimental record confirms that C *can* be returned to soils in such quantities. Some examples: carbon has been accumulating at rates exceeding 1 Mg ha^{-1} y^{-1} in former U.S. crop lands planted to perennial grasses under the Conservation Reserve Program (CRP) (Gebhart et al, 1994). Soil C increases ranging from 1.3 to 2.5 Mg ha^{-1} y^{-1} have been estimated in experiments on formerly cultivated land planted to switchgrass (*Panicum virgatum*), a biomass crop (preliminary data, Oak Ridge National Laboratory). Further, there have been a substantial number of experiments over the last two or three decades with minimum tillage and no-till management of farm fields demonstrating that such practices lead to increases in soil C content (Lal et al., 1998a; Nyborg et al., 1995; Janzen et al., 1998).

Despite these indications that needed quantities of C can be sequestered in agricultural soils there are still important questions to be answered. Among them four appear to be critical: (1) Can methods be developed to increase still further the quantities of C that accumulate in soils and, perhaps more importantly, the length of time during which the C resides in soils? (2) Can opportunities for soil C sequestration be extended beyond the currently farmed lands to the vast areas of degraded and desertified lands worldwide. (3) can we develop quick, inexpensive and reliable methods to monitor and verify that carbon is actually being sequestered and maintained in soils? and (4) what are the policy and economic problems associated with implementation of soil carbon sequestration programs worldwide?

A workshop to explore these questions was organized by the Pacific Northwest National Laboratory, the Oak Ridge National Laboratory and the Council for Agricultural Science and Technology and was held in December of 1998 in St. Michaels, MD. The workshop was attended by nearly 100 Canadian and U.S. scientists, practitioners and policy-makers representing agricultural commodity groups and industries, Congress, government agencies, national laboratories, universities and the World Bank. Support for the workshop was provided by the Environmental Protection Agency, the U.S. Department of Agriculture, the Department of Energy, the Monsanto Company and NASA. White papers addressing the four key questions were prepared for presentation and discussion at the workshop. The papers, revised to take account of critiques and discussion and the recommendations engendered at the workshop are reported in Rosenberg et al., eds. *Carbon Sequestration in Soils: Science, Monitoring and Beyond. Proceedings of the St. Michaels workshop* (Battelle Press, Columbus, OH, 1999). Key findings of the workshop are given here.

New Science

The potential for carbon sequestration in all managed soils is large and progress can be made using proven crop, range and forest management practices. But this potential might be made even greater if ways can be found to restore more than the two thirds of the carbon that has been lost from conversion to agriculture and perhaps even to exceed original carbon contents in some soils and regions. This would involve a search for ways to effect greater, more rapid and longer-lasting sequestration. Promising lines of research are evolving that could lead to an improved understanding of soil C dynamics and the subsequent development of superior C sequestration methods. These studies aim to: improve understanding of the mechanisms of C stabilization and turnover in soil aggregates; improve description of the various carbon pools and transfer among them to better model the dynamics of soil organic matter; improve understanding of landscape effects on C sequestration and how it might be controlled through precision farming; apply genetic engineering to enhance plant productivity and favor C sequestration; and better understand the environmental effects of soil C sequestration (e.g., erosion, nutrient leaching, emissions of other greenhouse gases).

The Soil Carbon Sequestration/Desertification Linkage

It is estimated that there are some 2 billion hectares of desertified and degraded lands worldwide, 75% of them in the tropics, with degradation most severe in the dry tropics. The potential for carbon sequestration on these lands is probably even greater than on currently farmed lands. Improvements in rangeland management, dryland farming and irrigation can add carbon to soils in these regions and provide the impetus for changes in land management practices that will begin the essential process of stabilizing the soil against further erosion and degradation with concomitant improvements in fertility and productivity. Erosion control, agricultural intensification, forest establishment in dry regions, and biomass cultivation appear to offer the greatest potential for increased sequestration on degraded lands. Soil carbon sequestration offers a special opportunity to simultaneously address objectives of two United Nations Conventions--the Framework Convention on Climate Change and the Convention to Combat Desertification.

Monitoring and Verification

There is opposition to using soil carbon sequestration in the Kyoto Protocol calculations. One cause of the opposition is the perception that it will be difficult, if not impossible, to verify claims that carbon is actually being sequestered in the soils of fields around the world that may eventually number in the millions. It *is* currently possible to monitor changes in soil carbon content, but current methods are time-consuming and expensive and are not sensitive enough to distinguish year-to-year changes. If there are to be international agreements allowing soil sequestration to figure into a nation's carbon balance, agreed-upon means of verification will be required. Improved methods for monitoring changes in soil organic carbon might involve spatial integration based on process modeling and geographical information systems, application of high-resolution remote sensing, and continuous direct measurements of CO_2 exchange between the atmosphere and terrestrial ecosystems. There may very well be a market for new instruments that can serve as 'carbon-probes'. These verification and monitoring methods will have to be developed or tailored to operate at different scales (e.g., the field, the region). Verification of changes in soil C in individual fields will rely on laboratory analyses of soil samples or, perhaps a few years from now, on carbon probes. Estimates of soil C changes at the regional scale will be made with the aid of simulation models. High resolution remote sensing and GIS will be used to extrapolate C sequestration data from field observations and modeling results and aggregate them to still broader regions and to track trends in C sequestration with time.

Implementation Issues and Environmental Consequences

The prospect opened by the IPCC findings and the Kyoto Protocol that carbon may become a tradable commodity has not gone unnoticed in the agricultural and forestry communities. Beneficial land-management practices might be encouraged if credit toward national emissions targets could be gained by increasing the stores of carbon on agricultural lands. However, uncertainty about the costs, benefits and risks of new technologies to increase carbon sequestration could impede their adoption. Financial incentives might be used to encourage adoption of such practices as conservation tillage. Government payments, tax credits, and/or emissions trading within the private sector are also mechanisms that could be employed to overcome farmer reluctance. Despite uncertainty of many kinds, the process is beginning. Some utilities and other emitters of greenhouse gases, anticipating a future regime in which reductions in CO_2 emissions become mandatory, are already searching for cost-effective ways to offset or otherwise meet the limits imposed. Transactions are already being made. In October of 1999, the Trans Alta Corporation, a member of the Greenhouse Emissions Management Corporation (GEMCo, an association of energy utilities in western Canada) announced an agreement to purchase up to 2.8 million tonnes of carbon emission reduction credits (CERCs) from farms in the United States. The IGF insurance company will solicit the CERCs from eligible farmers or landowners, initially from Iowa and ultimately nationwide. We do not yet fully understand the social, economic and environmental implications of incentives that lead to a widespread adoption of soil carbon sequestration programs. Most foreseeable outcomes appear benign—for example an increased commitment of land to reduced tillage practices. Another likely outcome would be increased effort aimed at restoration of degraded lands and for retirement of agricultural lands into permanent grass or forest cover. Continuation and/or expansion of Conservation Reserve programs might also be encouraged and lead to improved management of residues in agricultural harvests. All of these actions have the potential of reducing soil erosion and its negative consequences for water quality and sedimentation. In addition, since increases in soil organic matter content increase water-holding capacity, irrigation requirements could be reduced. Conversion of agricultural lands to grasslands or forests would expand to provide wildlife habitat. Reduced soil disturbance and, possibly, diminished use of fertilizer could alter the volume and chemical content of runoff from agricultural lands. This would in turn reduce water pollution and improve water quality and the general ecology of streams, rivers, lakes and aquifers in these regions for use by non-agricultural water consumers.

But negative effects are also possible. Programs designed to move agricultural lands into forestry could negatively affect the traditional forest sector, leading to either deforestation of traditional parcels or reduced levels of management and lessened C sequestration. Such actions might offset much of the benefit of sequestering C in agricultural soils. Expanded use of agricultural lands for C sequestration might compete with the use of agricultural lands for traditional food and fiber production. The result might well be decreased production, increased consumer prices for crops, meat and fiber and decreased export earnings from agriculture. Reduction in intensity of tillage often leaves more plant material on the soil surface. Conservation tillage has been found to require additional use of pesticides to control weeds, diseases and insects. Increased use of pesticides may have detrimental effects on ecological systems and water quality. On the other hand, conversion of croplands to grasslands tends to decrease emissions of the strong greenhouse gas N_2O although it also tends to increase the decomposition of CH_4, another strong greenhouse gas.

Even in the case of such an apparently benign activity as soil carbon sequestration there is no 'free lunch'. The production, transport and application chemical fertilizers, manures and pesticides and the pumping and delivery of irrigation water needed to increase plant growth and encourage C sequestration all require expenditures of energy and, hence, the release of CO_2 from fossil fuels. It is necessary to determine to what extent the energy costs of the practices used to increase C sequestration actually reduce the net carbon-balance benefits. Of course, it is unlikely that soils will ever be managed for the primary purpose of C sequestration. Rather the fertilizers, manures, chemicals and irrigation water will continue to be used primarily for the production of food, fiber and increasingly in the new century for the production of biomass as a substitute for fossil fuel. C sequestration will demand little extra in the way of these inputs.

SUMMARY

Organic matter is an important constituent of soils, contributing greatly to plant productivity and ecosystem stability. Soil organic matter is also an important repository of carbon and a major component of the global carbon cycle and balance. In nature, soils act either as a source or a sink for atmospheric CO_2, depending on vegetation, weather, time of day and season of year. But land management is the most profound determinant of whether the net change in soil C is a gain or a loss. Land use changes, such as conversion of the temperate forests and prairies to agriculture, have contributed significantly since the industrial revolution began to the recorded increase in concentration of atmospheric CO_2. Today, deforestation in the tropics continues to add CO_2 to the atmosphere. Because of justified concern that the continued emissions of CO_2 and other greenhouse gases the atmosphere will lead to global warming, national policies and programs are emerging to slow, eliminate or offset these emissions. We know that agricultural practices that conserve soil and increase productivity also increase the content of C in soils thereby effectively removing CO_2 from the atmosphere. Integrated assessment of energy and economic options needed to stabilize atmospheric CO_2 during this century has shown that soil C sequestration can provide an important opportunity for mitigating the rise of atmospheric CO_2, especially if action is taken worldwide during the next three decades.

A stronger knowledge base is required before this can be accomplished. The St. Michaels workshop addressed the questions of: how best to improve the scientific basis for C sequestration in currently farmed lands and lands requiring protection and/or reclamation from desertification, how best to monitor natural and management-driven change in soil carbon content and how best to implement soil C sequestration programs. The nearly 100 scientists, practitioners and policy makers who attended the workshop emphasized the need for research leading to a more in-depth understanding of the mechanisms responsible for C stabilization and turnover in soil aggregates, of landscape effects on C sequestration, and of ways to combat desertification through C sequestration. High priority was given to research on the environmental impacts of soil C sequestration and on applications of genetic engineering to enhance plant productivity and increase C sequestration. The workshop also recognized the urgent need for fast, economic and reliable methods to verify and monitor soil C sequestration. A more thorough understanding of the social, economic and environmental implications of incentives that might lead to a widespread adoption of soil C sequestration programs was also deemed essential.

REFERENCES

Cole, C.V., C. Cerri, K. Minami, A. Mosier, N. Rosenberg, and D. Sauerbeck. 1996. Agricultural options for mitigation of greenhouse gas emissions. Chapter 23. *Climate Change 1995: Impacts, Adaptations and Mitigation of Climate Change* pp. 745-771. Report of IPCC Working Group II, Cambridge University Press, 880 pp.

Edmonds, J.A., M. Wise, H., Pitcher, R. Richels, T.M.L. Wigley, and C. MacCracken. 1996a. An integrated assessment of climate change and the accelerated introduction of advanced energy technologies: An application of MiniCAM 1.0. *Mitigation and Adaptation Strategies for Global Change*. 1:311-339.

Edmonds, J.A., M. Wise, R. Sands, R. Brown, and H. Kheshgi. 1996b. *Agriculture, Land-Use, and Commercial Biomass Energy: A Preliminary Integrated Analysis of the Potential Role of Biomass Energy for Reducing Future Greenhouse Related Emissions*. PNNL-11155. Pacific Northwest National Laboratory, Washington, DC.

Gebhart, D.L., H.B. Johnson, H.S. Mayeux, and H.W. Pauley. 1994. The CRP increases soil organic carbon. *Journal of Soil and Water Conservation*. 49:488-492.

IPCC. 1996. Climate Change 1995: *The Science of Climate Change*. Report of Working Group I. Cambridge University Press, New York. P. 4.

Janzen, H.H., C.A. Campbell, R.C. Izaurralde, B.H. Ellert, N. Juma, W.B. McGill, and R. P. Zentner. 1998. Management effects on soil C storage on the Canadian prairies. *Soil Till. Res*. 47:181-195.

Lal, R., L. Kimble, R. Follett, and B.A. Stewart, eds. 1998a. Management of carbon sequestration in soil. *Adv. Soil Sci*., CRC Press, Boca Raton, Florida.

Nyborg, M., M. Molina-Ayala, E.D. Solberg, R.C. Izaurralde, S. S. Malhi, and H. H. Janzen. 1998. Carbon storage in grassland soil and relation to application of fertilizer. Management of carbon sequestration in soil *Adv. Soil Sci*., CRC Press, Inc., Boca Raton, Florida. Pp. 421-432

Paustian, K., C.V. Cole, D. Sauerbeck, and N. Sampson. 1998. Mitigation by agriculture: An overview. *Climatic Change*. 40:135-162.

Rosenberg, N.J., C.V. Cole, and K. Paustian. 1998. Mitigation of greenhouse gas emissions by the agricultural sector: An introductory editorial. *Climatic Change*. 40:1-5.

Rosenberg, N.J., R.C. Izaurralde, and E.L. Malone, eds. 1999. Carbon Sequestration in Soils: Science, Monitoring and Beyond. Proceedings of the St. Michaels Workshop, December 1998. Battelle Press, Columbus, Ohio. 199 pp.

United Nations. 1992. United Nations Framework Convention on Climate Change. United Nations, New York.

United Nations. 1997. *Report of the Conference of the Parties on its Third Session.* Held at Kyoto from December 1 - 11, 1997. Kyoto Protocol, FCCC/CP/1997/7/Add.1, United Nations, New York.

Wigley, T.M.L., R. Richels and J.A. Edmonds. 1996. Economic and environmental choices in the stabilization of atmospheric CO_2 concentrations. *Nature*. 379:240-243.

12. CLIMATIC CHANGES — COSMIC OBJECTS, GLOBAL MONITORING OF PLANET, MATHEMATICS AND DEMOCRACY, SCIENCE AND JOURNALISM

DEMOGRAPHIC CHANGE AND WORLD FOOD DEMAND AND SUPPLY, SOME THOUGHTS ON SUB-SAHARAN AFRICA, INDIA AND EAST ASIA

TIM DYSON
London School of Economics, England

As in so many areas of life, demographic change is a major determinant of world food demand and supply. And while demographic shifts can affect food demand in several ways (e.g. through the process of urbanisation) the most crucial type of change is population *growth*. Thus analysts generally agree that demographic growth is the most important cause of world food demand growth; it is more important, for example, than rising incomes[1].

In 1950 the world's population was about 2.52 billion. Now it is around 6.05 billion. The 1998 UN medium variant (essentially 'best guess') population projections suggest that it will rise to about 7.50 billion by 2020, reaching nearly 9 billion around the middle of this century[2].

Recently it has become quite common to say that world population growth is no longer a problem. After all, birth rates are falling throughout most of the developing world - often faster than was anticipated - and therefore rates of population growth are falling too. Accordingly, projections of future demographic growth have been revised downwards. For example, the 1994 UN medium variant projections suggested a 2020 world population of 7.89 billion.

However, in this context three points seem particularly worth making. First, between now and the year 2030 the world population total will be rising at the rate of roughly an additional *billion* people every 14 years.

Second, virtually *all* of this growth will happen in the world's poorest regions—particularly South Asia, sub-Saharan Africa and East Asia.

And third, it is not just future demographic growth that matters, but *past* demographic growth too. To express this differently: it is past growth which poses us with current and future challenges of population *scale*. Even on an optimistic scenario, the world's population will probably remain well above 5 billion for most of the next 150 years[3]. During this lengthy period it will have to *sustain* itself within the global environment. Just because, in the second half of the twentieth century, humanity somehow managed to cope with demographic growth from 2.52 to 6.05 billion does *not* necessarily mean that it will manage to cope with the scale implications over the very long run. Anyhow, using the medium variant UN population projections, let's consider the food demand implications of future demographic growth during the next 20 years with particular reference to the world's poorest regions.

Between the years 2000 and 2020 the population of sub-Saharan Africa is projected to increase from 641 to 995 million–a rise of 55 percent. This projection includes some allowance, albeit speculative, for the effects of HIV/AIDS–which itself must be affecting food production in much of the region. In terms of average measures, for example of per capita cereal production or calorie intake, there has been little change in sub-Saharan Africa's dismal food position compared to the situation, say, 50 years ago. Indeed, average levels of per capita cereal output have fallen since the 1960s. Many factors have contributed to this poor food production performance (e.g. widespread political instability, and a long-standing neglect of the agricultural sector by governments), but past population growth rates of 3 percent per year have certainly made the task of raising per capita food output harder than it would otherwise have been. Another worrying feature is a long-run rise in cereal harvest volatility, largely reflecting an increasing frequency of major drought in the region[4].

Cereal yields in sub-Saharan Africa are extremely low - a little more than one ton per hectare. Indeed, partly because of the distinctiveness of its farming procedures and crops, this region's agriculture has been relatively neglected by the international research community. The so-called 'green revolution' was largely an Asian, rather than a sub-Saharan event.

Clearly, given the above population projection, during the next twenty years there must be roughly a 55 percent rise in sub-Saharan food output just to maintain current average levels of per capita consumption; and because of population growth even this may well be accompanied by a rise in the absolute *number* of undernourished people in the region. In addition, without a rise in crop yields, levels of poverty in the region will remain high[5] and what increase in total food production occurs will happen through an expansion of the cultivated area - often into fragile marginal and forested areas. So, while it is certainly *technically* feasible to raise yields at the required rate (or better), given sub-Saharan Africa's socio-economic, administrative and political circumstances, this will represent a very considerable challenge.

The UN projects that the population of South-central Asia will rise from 1.49 to 1.95 billion during the next 20 years. Taken together, India, Pakistan and Bangladesh account for 87 percent of the current total population. And, since I am currently researching it[6], I concentrate on India, which with more than a billion people alone accounts for 68 percent of this region's total.

India probably contains more poor and undernourished people than any other nation. Indeed, because of its huge population, a recent FAO report estimates that around 1995-97 India had 204 million undernourished people, compared to 180 million in sub-Saharan Africa[7]. If, as we project[8], the country's population rises by 32 percent (some 320 million) during the next twenty years, then to maintain current levels of per capita cereal production (which, in practice, means consumption) will entail output rises of a similar percentage. In India's case these rises *must* come mainly from raised yields, because most suitable land is already in cultivation.

The country faces many relevant problems, like growing water shortages in some areas, near-feudal local-level social structures in others, and highly inefficient agricultural subsidies (e.g. on electricity) which are politically very difficult to dismantle. That said, there are reasonable grounds to believe that India will be able to raise its food output somewhat *faster* than its population will grow. For one thing, at the aggregate level, the country's medium-term economic outlook looks comparatively good, partly because both the rate of population growth and the dependency ratio are now declining. Certainly there is the technical capacity to raise food crop yields; and new varieties and greater fertiliser use will be part of the answer too.

However, India also exemplifies many of the difficulties that are involved when discussing issues of population and food. A key issue is that hundreds of millions of people work on small farms, which because of demographic growth are getting even smaller. These farmers and labourers often cultivate coarse grains and rely solely upon the monsoon rains for their water. To provide these workers with livelihoods, crops must be grown which can be consumed and sold. Yet without various subsidies, there might be no ready market for these crops, or they might be grown more efficiently by larger commercial enterprises, or they might even be imported from other countries at lower prices than can be achieved locally. The alternative, i.e., of relying more upon market mechanisms and 'comparative advantage' to govern what crops are grown and where, would be extremely difficult politically and would involve additional pressures for rural out-migration.

An interesting feature which emerges from the Indian scene is that although average levels of per capita direct cereal consumption are very low by international standards, data from the National Sample Survey (NSS) organisation suggest that, except for the poorest 20 percent, all other income groups in the population have been eating less cereals, even though their incomes have been rising[9]. On the other hand, there have been significant increases in the per capita consumption, for example, of fruit, vegetables and milk. So even in a country where poverty and under-nourishment are widespread, there seems to have been an increasing *diversification* of food consumption patterns. Furthermore, a recent study indicates that, for methodological reasons, the NSS has probably been underestimating household consumption of pulses, vegetables, and meat by 46, 54, and 53 percent respectively[10]. Finally, between 1983 and 1993 the proportion of rural households reporting that they had 'two square meals a day' throughout the year rose from 81 to 94 percent, while the corresponding figures for urban areas were 93 and 98 percent[11].

These findings remind us of issues that often apply elsewhere in the world: first, that it is difficult to estimate things like levels of food consumption and nutritional status unambiguously; second, that even very poor people may choose to spend additional income on things other than food; and third, that the 'adequacy' of food consumption levels can be judged from different perspectives.

A plausible scenario for India's medium term future may be one in which the average diet continues to get more diverse, but at a low calorific level. If higher incomes materialise, as is to be hoped, then they may not be spent in ways that necessarily bring about commensurate improvements in diet and nutrition. Also, with increasing urbanisation in India, and indeed in all developing regions, people are leading less active lives. So especially in major towns *over-nutrition* is becoming an increasing problem.

The last developing region that will get particular consideration here is East Asia. About 86 percent of this region's 1.48 billion people live in China (the rest reside mostly in the Koreas and Japan). Despite China's comparative success in reducing its birth rate, the UN projections still suggest that the country's population will increase by another 177 million during the next 20 years, and that it will not be until around the year 2040 that China's then rather old population will start to decrease in size. Of course, during recent decades China has performed rather well in terms of its aggregate rate of economic growth. And, in some contrast to India, the country has managed to raise its average level of calorie intake significantly, while at the same time increasing greatly the diversity of the average diet.

That said, China's pattern of development has been uneven, not least between different regions. Thus the eastern coastal states have generally done rather better than other areas of the country. Partly as a consequence, there have been huge migration flows, both eastwards, and generally towards the towns. And one result of this is that in some rural areas there has been a marked feminization of the remaining farming population (a process which has counterparts, for example, in parts of East Africa).

So, despite its relatively favourable economic performance and its improved diet, with its massive population and increased inequalities accompanying its economic growth, the FAO estimates that there were still approximately 164 million Chinese who were undernourished in 1995-97.

In few countries in the world is the rough approximation between population and food so deeply imbedded into the psyche. Indeed, the two Chinese characters which together correspond to the word 'population' consist of a person and an open mouth. The issue of meeting the food needs of the population probably informed the decision to reduce the birth rate from the early 1970s. Of course, China is also a country with a limited land base, where future food production increases must come from increased yields. The feeding of an additional 177 million people in the next twenty years certainly represents a challenge, but without fertility decline that population growth figure would have been *very* much more.

Lastly in this quick review, brief mention should be made of Latin America (including the Caribbean) and the Middle East (including North Africa). Between 2000 and 2020 the populations of these regions are projected to rise by around 146 and 148 million respectively (i.e. 28 and 41 percent). Both regions, but particularly Latin America, contain significant numbers undernourished - according to FAO about 53 and 33 million respectively around 1995-97. In comparative terms Latin America has a relatively favourable agricultural resource-base and, of course, it is a major exporter of many different types of food. Latin America does not appear to be a region where meeting the basic food needs of the people is greatly constrained by the resource base. However, the Middle East may be somewhat different, especially apropos its scarce water resources. Currently the Middle East depends upon cereal imports to meet a very significant proportion of its total cereal requirements. Indeed, these cereal imports can be seen as a form of 'virtual water'. And considerable demographic growth will probably make this region even more dependent upon cereal imports during the next few decades.

CONCLUSIONS

Of necessity, this discussion has omitted a lot: in particular, food and agricultural production in North America and Europe, and related issues such as the evolution of future international trading arrangements, the expansion of the European Union, and the reform of the Common Agricultural Policy. Most certainly, all these and other issues will impact upon the food situation of the global poor.

In summary, however, there is little doubt that demographic growth has been, and still is, the single most important factor behind the growth of world food demand. And, while it is difficult to standardise for all the relevant factors, a good argument can be made that those world regions which have experienced rapid demographic growth have found it harder to meet the basic food requirements of their people. Thus the fact that sub-Saharan Africa's population doubled in size between 1975 and the year 2000 probably made it harder to improve the region's level of food consumption. And the fact that the population of East Asia increased by only 35 percent during the same time period probably helps to account for both its better food production and its better economic performance.

Also, in some parts of the world—not just sub-Saharan Africa—population growth is contributing to environmental damage (e.g. in hill areas) as poor people expand their cultivated area in an effort to eke out a bare living from the soil. And in other parts of the world, where most land is already in cultivation, population growth contributes to land fragmentation and out-migration to the burgeoning towns.

However, most analysts agree that—with the possible exception of sub-Saharan Africa—the prospects for the growth of food supply to match the growth of future demand are more upbeat than downbeat, at least over the medium run[12]. For average levels of food intake to be significantly improved in twenty years time will involve drawing on many things, such as improvements in farm support and education, more intense farm management procedures, greater use of inputs, better seeds, and the development of new techniques. By the year 2020 the world average cereal yield should be approaching 4 metric tons per hectare[13], a figure which should help obviate the need to expand the global cultivated area. With an increasing world population, the task of raising yields is vital to help conserve much of the natural environment because, essentially, there is a direct relationship between achieving higher yields and sparing land for nature[14]. However, and of course, this is not to say that all the institutional, political and other factors which together combine to keep millions of people poor and undernourished will be resolved by 2020. On the contrary. And whether there will be fewer hungry people alive in a world of 7.5 billion is hard to judge—because most demographic growth is happening among the poor.

Finally, I return to the issue of demographic *scale* and its operation over the *much* longer run. Recall that there are likely to be at least 5 billion people around on the planet for at least the next 150 years. It may be feasible to look, say, two decades into the future and envisage how the world's food needs can be met, albeit very imperfectly. But whether humanity can cope over the much longer run with the *indirect* consequences that stem partly from there being 5 billion (or more) people, is less certain. Here one has in mind, in particular, the challenge which will be posed to the *absorptive* capacity of the global environment. Moreover, it is as well to remember that in many ways modern farming and food production make their own significant contribution to this challenge, for example, through their use of large quantities of energy and synthetic nitrogen fertilisers.

ACKNOWLEDGMENT

The work on India reported here was helped by a research grant from the Wellcome Trust.

REFERENCES

1. See, for example, Alexandratos, N. (ed.) 1995. *World Agriculture: Towards 2010*, John Wiley, Chichester; Dyson, T. 1999. 'World food trends and prospects to 2025', *Proc. Natl. Acad. Sci. USA*, Vol. 96, pp. 5929-5936; and, Mitchell, D.O., Ingco, M.D., and Duncan, R.C. 1997. *The World Food Outlook*, Cambridge University Press, Cambridge.
2. United Nations 1999. *World Population Prospects, The 1998 Revision*, United Nations, New York.
3. United Nations 1999. *Long-range World Population Projections: Based on the 1998 Revision*, United Nations, New York.
4. See Dyson, T. 1996 *Population and Food: Global Trends and Future Prospects*, Routledge, London; also, Naylor, R., Falcon, W. and Zavaleta, E. 1997. 'Variability and growth in grain yields, 1950-94', *Population and Development Review* 23, no.1:41-58.
5. Lipton, M. 1999. Reviving the Stalled Momentum of Global Poverty Reduction: What Role for Genetically Modified Plants? Crawford Memorial Lecture.
6. Dyson, T. and Hanchate, A. forthcoming. 'The future of India: population and food', *Economic and Political Weekly*, Mumbai.
7. Food and Agricultural Organisation. 2000. *The State of Food Insecurity in the World 1999*. http://www.fao.org/FOCUS/E/SOFI
8. See Dyson, T. and Hanchate, A. forthcoming. 'The future of India: population and food', *Economic and Political Weekly*, Mumbai.
9. Joshi, P.D. 1998. *Changing Pattern of Consumption Expenditure in India and Some Selected States*, Ministry of Planning and Programme Implementation, New Delhi.

10. National Sample Survey, 2000. *Choice of Reference Period for Consumption Data*, Report No. 447, Ministry of Planning and Programme Implementation, New Delhi.
11. Bansil, P.C. 1999. *Demand For Foodgrains by 2020*, Observer Research Foundation, New Delhi.
12. See the references in endnote 1, Dyson in endnote 4, and Rosegrant, M., Agcaoili-Sombilla, M., and Perez, N. 1995. *Global Food Projections to 2020: Implications for Investment*, International Food Policy Research Institute, Washington, D.C.
13. Evans, L.T. 'Greater crop production', in Waterlow, J.C., Armstrong, D.G., Fowden, L. and Riley, R. (eds.) 1998. *Feeding a World Population of More than Eight Billion People*, Oxford University Press, Oxford.
14. Waggoner, P.E. 1998. 'Food, feed and land', in Crocker, D.A., and Linden, T. (eds.) *Ethics of Consumption: The Good Life, Justice, and Global Stewardship*, Rowman and Littlefield Publishers, Lanham, Maryland.

THE STATUS OF CLIMATE MODELS AND CLIMATE CHANGE SIMULATIONS

WARREN M. WASHINGTON
National Center for Atmospheric Research, Boulder, Colorado USA

Climate models are made up of several major components of the climate system. The usual components included are the atmosphere, ocean, sea ice, land/vegetation, and hydrology. The ecological and detailed chemistry aspects are not usually components of climate change model simulations. The atmospheric component is the most developed part of the modeled climate system and it has a long history from the early days of numerical weather prediction models. This component makes use of the basic laws of fluid dynamics and it takes into account the rotation of the earth and the fact that the atmosphere is shallow compared to the radius of the earth. Also, the atmosphere is assumed to be in hydrostatic balance which is a good approximation for large-scale motions that are characteristic of climate models. The physical processes that are included are the solar and infrared radiation, the precipitation processes in the form of rain and snow, cloud prediction, convection, transfers of momentum, water vapor, and sensible heat between the atmosphere and the earth's surface. The ocean component includes the same basic laws of fluid dynamics except the ocean is considered to be an incompressible fluid. The sea ice model components usually include the dynamics of ice motion and thermal dynamics of sea ice growth and melting. The latter is quite detailed to take into account the different types and thickness distribution of sea ice. It should be noted that sea ice acts much like a viscous plastic material in that it can be compressed as well as it can open up in the form of leads which can act to transfer large amounts of heat and moisture to the atmosphere. The land aspects of new generation climate models must take into account the different types of surfaces ranging from desert sand, grassland, forest, wetlands, swamp, lakes, and mixture of these types that can co-exist in a single atmospheric grid area. The ecological aspects of the models are still in a developing stage. In this component the transformation of plant species can take long periods of time, usually longer than the climate change predictions. Also, mankind has played a major role in changing the earth surface. The chemistry and biochemistry models are becoming an interactive component of new generation climate models but in a somewhat limited manner. At present, the important role of sulfate aerosol chemistry is included in some climate change models. Carbon and other chemical cycle models are not usually made an interactive component of climate models.

The computational design of climate models is increasingly becoming a very important consideration in the field. There is a transition from mostly vectors types of supercomputers to parallel types of supercomputers. The latter are based upon clusters or nodes with many processors. Two examples of such designs are the vector computer version based NCAR Community Climate Model described at www.cgd.ucar.edu/ccm, and the United States Department of Energy supported Parallel Climate Model (PCM) based upon parallel computing (see www.cgd.ucar.edu/pcm). The basic idea in both model paradigms is that a coupler ties the components together and transfers fluxes of energy, momentum, and water between the components. A coupler also allows for coupling of components that have different resolutions.

State-of-the-art climate models can simulate such regional features as the monsoons, El Niño, La Niña, Arctic Oscillation, and the North Atlantic Oscillation. These features are a very important part of the natural variability of the climate system. It should be added that these features were rarely represented in earlier generations of climate models. Now they are simulated with approximately the correct amplitude and frequency.

Simulations of climate change usually start from an equilibrium control state of the 1870s or so, in which the climate shows little observed change. The greenhouse gases, ozone, and sulfate aerosols were set to values close to the observed concentrations for that period. Historical simulations are begun from the 1870s control simulation in which the greenhouse gases, ozone, and sulfate aerosols concentrations are increased in the simulations that ends in the 1990s. From the 1990s into the future out to year 2100, the greenhouse gases, ozone, and sulfate aerosols are specified, based upon certain assumptions about how mankind will change the concentrations. Two possible scenarios that are typically used are the "Business as Usual", where the same trends of the 1990s are extrapolated until the year 2100 or to assume some type of stabilization of greenhouse gases and sulfate aerosols. In both cases, there are a variety of assumptions that can be made. Obviously, the climate change impact is a strong function of which forcing is used. One of the important aspects of climate model prediction is to estimate how much variability there is in the climate for, say, the decades near 2050. In order to get an estimate of this, climate simulations are usually performed with an ensemble of simulations of the order of 4 to 10. Figure 1 shows the global mean surface temperature anomaly from a control 1870 simulation. The bottom curve is the simulation with constant 1870 concentrations of greenhouse gases, sulfate aerosols, and ozone. The thin line curve shows no trend: however, there is year to year variability of which some is caused by regional features such as the El Niño phenomena, and the heavy dark line is smoothed by a low-pass filter. The other lines that start to slope upward are a function of time at about year 1960. They show the observed temperature change, the ensemble mean historical simulation of the "Business as Usual" and stabilization simulations. In one of the simulations, we include the effects of the solar variability. Also shown is the range of the predictions in between extreme members of the ensemble, which give some idea of the range of uncertainty using exactly the same climate models with different initial conditions from the 1870 control simulation. It should be noted that the range of uncertainty between different climate models is much larger than the range of prediction with a single model.

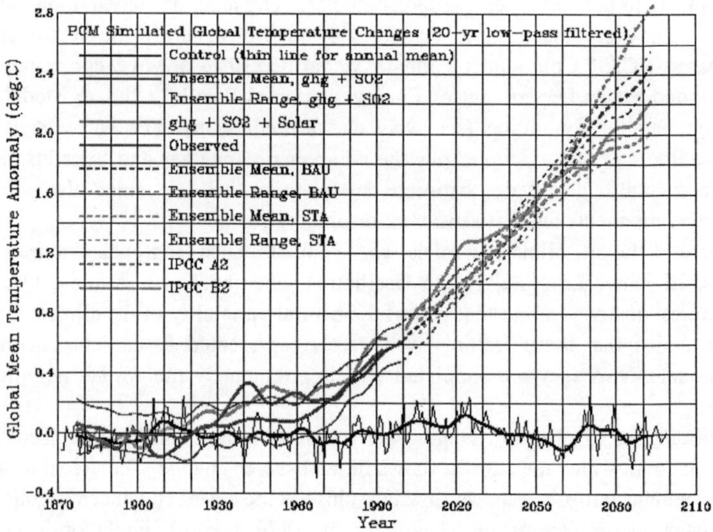

Fig. 1. Global Mean Temperature Anomaly (deg. C).

The larger range between different models gives some idea of the overall uncertainty of climate forecasts. It is suspected the reasons for the divergence of climate model forecasts is caused by our lack of knowledge about how many important processes work in the climate system. For example, essentially all climate models treat clouds relatively poorly. However, increased research emphasis and improved observations (both in situ and from satellite) will improve the simulations. Researchers are getting a better idea of how clouds function and how to model them. Before leaving this aspect of the report, it is worth noting that the geographical change in surface temperature shows the largest warming takes place in the higher latitudes, particularly in the wintertime. This warming pattern is consistent with the recent observations of regional warming. It should be further noted that there are regions that show little warming and some show cooling, such as the eastern United States. This can be explained in part by the effects of sulfate aerosols in highly industrialized regions where the direct effect is to reflect a higher percentage of solar radiation away from the earth's surface. The highly uncertain indirect solar aerosol cloud physics effect, which is the enhancing cloud reflection, would act to enhance the direct effect. By including the sulfate aerosols, climate models are capable of showing a similar pattern of warming and cooling that has been observed over the last 40 years. One forcing that is often not included in climate model simulations is the effect of volcanic eruptions. This effect is thought to be mostly on a shorter time scale and the effects will be almost impossible to predict for the future. Clearly, it will be a major player on year to year climate prediction. The question of attribution and detection of climate change is becoming better understood with each five-year assessment by the Intergovermental Panel on Climate Change (IPCC). It is expected that the next assessment will provide even more evidence of climate change that is consistent with model predictions.

Finally, there are new opportunities for a broader community of scientists to become involved in the more detailed interactions of atmosphere, ocean, sea ice, land/vegetation, and hydrological aspects of climate modeling. We can expect that the climate models will be major users of large parallel supercomputer systems that will have thousands of processors. In the past, climate modeling communities have been mostly relatively small teams of researchers. This paradigm is changing to include a wider spectrum of researchers who can interact through the Internet. They are no longer concentrated in one center. Improvements and innovations in climate change modeling research will flow much quicker in the 21st century. Ever improving climate models have been and continue to be a tool that is improving our understanding of the present climate system. They are the only scientific tools capable of being a window into the future about how mankind is changing the climate system.

FROM PUZTAI TO PERFECTION: A NECESSARY DREAM

ROBERT WALGATE
Chairman, World Federation of Scientists' Permanent Monitoring Panel on Science and Journalism, Open Solutions, Northwood, Middlesex, U.K.

[Introductory note: the Arpad Puztai "affair" concerned the claim of this Hungarian scientist, working in Edinburgh, that genetically introduced lectins were particularly harmful to human health. Ultimately it turned out that he had no data to substantiate his claim. But much of the media in Britain and abroad began to campaign for him, and his Director, Phillip James, who had originally stood by his researcher, was crucified on TV and in the press for "suppression" of his work.]

Science is a series of improving approximations; as each stage follows the last, we realize our previous mistakes.

Technology is even more approximate; engineers learn by their built mistakes.

Medical hygiene developed from error; as we learned from Paul Brown, iatrogenic medicine is the medicine of mistakes.

Making mistakes is how we learn.

And then we come to journalism!

Many Italian plants scientists in this very meeting are indignant at the errors of certain journalists just yesterday. There is not only a loss of trust in science; Italian scientists, it seems, and many others present, have lost all trust in science journalism.

But now I want to remind you of something you may have forgotten.

There is a journalistic ideal. A knight in armour, seeking truth against all adversity. Exposing corruption. Revealing heinous crime.

A few brave souls rise to it.

Some have died for it. Right here in Sicily, journalists have died attempting to reveal who is secretly involved in the web of the mafia.

Hundreds, right now, are being imprisoned and tortured for it.

Let me quote Amnesty International: "Governments have a human rights responsibility to secure freedom of speech and stop the harassment, torture and killing of journalists". That's Amnesty International on World Press Freedom Day in May this year.

According to Amnesty: "Governments around the world are continuing to control and suppress information by violating the human rights of the individuals whose job it is to report it."

"By exposing human rights abuses, journalists often become the victims of the kind of intimidation and harassment they have been reporting."

Now with your views of science journalism you may think it laughable. But these are the men and women, this is the ideal, this is the angry angel, that also calls some science journalists forward. A life-changing belief in truth.

The contrast between dying for human rights and misrepresenting a scientist for a good headline may seem too great for you to stomach.

But *that* is why our new PMP on Science and Journalism exists. So nobly and may I say daringly created by Professor Zichichi at the end of this same conference last year.

I say this was a truly millennial decision.

Because we desperately need a *new culture* of science journalism.

One which really takes the ideal of true journalism, the fearless search for truth, balance, and clear presentation to the public; a science journalism which as one of my members, Wolfgang Goede, puts it, *empowers the people*. And of course we must do that with facts, and not with misrepresentation, or the empowerment will be illusory.

And many of we science journalists know it.

At the 2nd World Conference of Science Journalists held in Budapest, Hungary, in 1999, 146 science writers and broadcasters from 29 countries called on all journalists of science, "including the natural and social sciences and humanities, and including our colleagues in the closely related fields of health and environment reporting, to recognize our increasing responsibilities to the people of the world to report accurately, clearly, fully, independently and with honesty and integrity".

I should explain that our PMP is formed from a nucleus of the journalists at Budapest, with some important extensions in geography and skills, so our starting positions are closely defined by that Budapest Declaration.

Let me quote just two more of the eight Budapest recommendations.

We addressed "editors, publishers, broadcasting organizations and other gatekeepers worldwide", and called on them "to recognize not only the wide public interest but also the increasing democratic and social importance inherent in science journalism, and to provide more support, space, programme time, staff and training" for science journalists.

Because you must understand that science journalism takes place in a context. Not only an editorial context, but also a business context. As you well know, the reports you see in the press, hear on the radio and see on the television are created by companies, often vast global enterprises, who are making money and trading political influence—because media and politics are closely linked, as the antics of Silvio Berlusconi in Italy make very plain. As for money-making, the Australian media moghul Rupert Murdoch is infamous for his remark "you can never *underestimate* the taste of the public".

Do you catch the meaning of that? This man who now owns so much publishing and broadcasting capacity worldwide, including that once august publication the London *Times* newspaper, is saying the *lower* you sink in your news, the more money you will make.

This atmosphere creates a competitive culture, a struggle for audiences, in which *every* publisher and broadcaster feels forced to engage. Even the highly repectable British Broadcasting Corporation, the BBC, is not immune; as my colleague Deborah Cohen tells me, BBC national radio now has no single, identifiable science programme.

So listening to the BBC in Britain the public can no longer make a regular predictable date with *Science Now,* the original radio science flagship, but must consciously search for the still good science content, under miscellaneous titles and at usually late times, in the media listings each week. BBC TV too is "dumbing down", with the best current affairs programmes on the rocks and the next in line turning to adversarial drama rather than in-depth analysis.

What can be done about this is, I fear, very little, except to shake a puny fist. We are confronting, in essence, the total world advertising budgets, which pay for space and time and want audience figures *now*. I don't know the numbers, but we must be confronting 100s if not 1000s of billions of dollars annually.

We addressed "editors, publishers, broadcasting organizations and other gatekeepers worldwide", and called on them "to recognize not only the wide public interest but also the increasing democratic and social importance inherent in science journalism, and to provide more support, space, programme time, staff and training" for science journalists.

Now whatever did we mean by that?

We meant that science journalists must earn, within their publications and media, the right to the increasing number of science and technology stories with a political content.

Because if they ever had that right – because largely they acted as your cheerleaders, innocent, like many scientists themselves, of political and economic motivations and contexts – they've lost it over the coverage of genetically manipulated crops, or "GM foods", or to use what I think was *Time*'s unforgettable but misleading coinage, "Frankenfoods".

In the UK, where the press was first most vociferous and angry, the trouble began with a letter by 22 scientists to the *Guardian* on 12 February 1999 on the Arpad Puztai affair (incidentally they supported Puztai, a reminder that scientists too can make mistakes). The topic was then front page news in the UK for a week, from 13-20 February.

In the Summer, the House of Lords Select Committee on Science and Technology commissioned research on the coverage of national tabloid and broadsheet newspapers on this subject between 8 January and 8 June 1999.

The research showed that a number of national newspapers – all the tabloids studies and several broadsheets – chose to campaign on the issue. According to your point of view at the time – even according to your *scientific* instincts over Puztai's incomplete and unpublished results – this was a bold decision, fully in concert with the ideals of journalism, or a naked, unprincipled chase for circulation. Of course at first it was a little of both. But it soon became mostly the latter.

What I want to draw your attention to is that the House of Lords' report found that "articles by non-scientific correspondents... were prominent... In particular, during the two-day period when the story broke... no news articles on GM foods were written by science and technology journalists and 45% were written by political journalists".

In other words, the science journalists had lost the ball.

Phillip James, the director of the institute where Puztai worked, and from which James suspended him, has been here this week. He has never written about what happened during those few months, or the six months earlier when the news of Puztai's results had first broken. But this morning he told me two outstanding things: that over the whole period, of all the journalists who approached him, only one out of ten were science journalists; and, damningly of all the journalists, *none* had read the many detailed press releases on the subject that his institute had produced. In the media, everyone was feeding off everyone else's story. The facts no longer seemed to matter. It had become a media feeding frenzy.

Because we, the science journalists had lost that ball, an issue that truly mattered to the public—a technical question in public health—a very raw and personal politics came to dominate the story.

Science—and the public interest—were both lost.

Of course it was a conflict among scientists too. Puztai probably believed in his data, and no doubt thought that further studies would confirm it.

But who was analysing this complex issue, with its scientific uncertainties and personal and potential commercial interests for the public good? A campaigning circulation-driven media in which the technically most competent writers had been sidelined.

It was the Puztai scandal—a scandal of the media that started right from the moment of his first revelations in mid-1998—that led me first to Budapest, and to help draft the global declaration on science journalism that I've described; and later to accept with great pleasure Professor Zichichi's invitation to create a Permanent Monitoring Panel on Science and Journalism.

What can we do to create the new culture - and frankly *new power* - of science journalism that we need?

Because let me be clear—it is essential *to our whole scientific civilization* that we scientists and science journalists, through Erice or by some other means do find that power.

Don't we stand here in Sicily, on this fortified, religious hill-top of Erice, on the ruins of successive great civilizations? Their very stones are in the wall around us, Punic on the bottom, tourists with their mobile phones on top. So why can't our civilization vanish? Could the 21st century not be a century of environmental ruin and cultural disaster, where science has been sidelined as much as science journalists, and obscurantism rules a smoking globe?

Because aside from the great confrontations of war and diplomacy and trade, what is truly changing our future? What makes this last 25 years truly, substantially different from the previous one? And what will make the next 25 years very different again? Clearly it is science, through its application in technology. We are in a different world from Galileo because of the science, and resulting technology, that was set in train by his clarification of the heavens. And we will be placed in a different world again by the Galileos of today.

But no-one is analysing and clarifying this exponentially accelerating process for the public, and especially the developments of technology, which involve questions of *choice* and *purpose* and *benefit to whom* and *risk to whom else*, which are in their uncertainties and choices as much political as scientific matters?

Where is democracy, where is understanding and choice in these issues, where it really matters? In practice it lies in boardrooms and stockmarkets and more occasionally, and I suspect with only a little more understanding, in the hands of Presidents and governments and international institutions.

Should we be satisfied with that? To leave it to the experts? I believe not, not because the experts may be wrong, though they may, but because they always work in this or that political context that may not reflect the political interests of the people at large. The political system of the world is a hierachical technocracy, not democracy; if you are happy with that, well and good; if not, and you believe in wider systems of choice for the public and the global good, we need the kind of honest, thorough, independent science journalism I am advocating.

And this is to say nothing of human values, which must enter the debate in force as we are gaining more and more ability to act on and change our own bodies, minds and progeny. Does a human being exist at the fertilisation of the egg? Can you throw away a 16-cell embryo? Or 32? Or when? Or not at all? Somewhere a line must be drawn, the choice is a human or even religious value, and your choice may be different from mine. So how to decide?

I'd like you to be clear that I'm not talking about a science journalism which will merely reflect and clarify the choices and advances of the scientists and the technologists. I am not speaking about one which will speak science and technology more clearly. Much more of this *is* needed, certainly, but above all I am speaking in the name of democracy; in the name of reclaiming the science-political news; in the name of reclaiming a story like that of Puztai and James where we would analyse science, technology, uncertainty, and politics, and - quite objectively and without political bias - lay out the threads of a complex story, which involves understanding and reporting the *process* and uncertainties of science as much as its product. And not least in the name giving human values their proper place within science.

Well, in the spirit of Erice, as you do on so many great issues, we shall aim here in our PMP to do exactly this for our own profession and our people.

But we shall focus above all on a product. We don't just want to *talk*, but in some way to *affect* the quality of science and technology reporting.

So we will be proposing to the Ettore Majorana School that they establish a School of Science Journalism here, one that will be as much a forum between scientists and journalists as a school, and that will closely involve developing country journalists where the Planetary Emergencies will hit and are hitting hardest and deepest; and we are requesting funds to launch a WWW journal, which will *critique* particular cases of yellow or poor science journalism, report scientific and technological issues and the planetary emergencies as we think they should be reported, so giving an example to the world on crucial stories, and be a concrete and regular output of the school, so our teaching becomes not academic but practical.

This journal should *also*, I want to stress very strongly, be a clarifying, widely available output of these extraordinary International Seminars on Planetary Emergencies, a meeting about which the world learns pitifully little.

And when it does learn, as we saw throughout the Italian media yesterday, in their reporting of the attempts to develop a HepB vaccine in plants, it learns without attending any of the sessions and learns *wrong*.

In science journalism, we've had enough of doing it wrong.

By contrast, what we *intend to do in our PMP is to do it—and train others to do it—right.*

MATHEMATICS OF INDIAN DEMOCRACY

PROF. K.C. SIVARAMAKRISHNAN
Centre for Policy Research, New Delhi, India

India is the largest democracy. Its electorate based on adult franchise for all persons above the age of 18 years is now more than 620 million. Elections to the Lok Sabha or the House of the People have been held 13 times. Elections to the State Assemblies have also been held on numerous occasions. This presentation describes the salient features of the system. Information is contained in these notes as well as the charts attached.

1. The large size of the country and the enormous size of the electorate make the organization of elections in India a vast and complex exercise. (Fig. 1)
2. All persons above the age of 18 years can vote. Because of the size, polling cannot take place simultaneously in all parts of the country, and has to be done in stages. A large number of polling personnel and security forces has to be mobilized. In terms of scale, the Indian general elections are the largest electoral event in the world (Fig. 2–'Largest Poll')
3. The Parliament consists of the Lok Sabha of the House of the People with 543 elected and two nominated members. The Rajya Sabha or the Council of States has 245 members. In addition, there are 25 states and 2 union territories, which have Legislative Assemblies. Some states have a bicameral organization with both the upper and lower house. (Fig. 3–'Edifices of Democracy')
4. The electorate has been rising steadily from about 173 million in 1952 to 620 million in 1999. The voting age was lowered from 21 to 18 which caused an increase of about 120 million from the 1984 to 1989 electorates (Fig. 4–'Who Votes?').
5. There is no restriction on the number of persons who can contest for a seat. In recent years, this number has been coming down. (Fig. 5–'How many in the Fray?' and Fig. 6–'Seats and Contestants')
6. The number of women elected to the Lok Sabha as well as the State Assemblies continues to be limited. However, the success ratio of women candidates is higher than men. (Fig. 7–'The Gender Bias' and Fig. 8–'Gender Advantage')
7. The results of the elections are declared on the basis of "First Past the Post." Whichever candidate has the largest number of votes becomes the winner. But usually the winning candidate does not have the majority of the votes polled and

certainly not the majority of the electorate in the constituency. (Fig. 9–'First Past the Post!' and Fig. 10–'What did the winners get?)

8. Vote share does not necessarily correspond to the share of seats. Parties, which have a better distribution of presence across the country, are able to secure more seats in the parliament. (Fig. 11–'Votes, Yes: But Seats?' and Fig. 12–'Vote share of parties')

9. The actual process of voting is by stamping the ballot paper with sign. Given the large number of candidates and parties, candidates are identified by symbols rather than names since a large proportion of the electorate is still illiterate. (Fig. 13–'Spree of Symbols')

10. From a situation of dominance by national parties, the position in recent years has shifted. Regional parties are increasing their influence. As a result, electoral verdicts, though clear in the states, become fractured at the national level, making coalitions necessary. (Fig. 14–'Coalition Games')

11. Electoral reforms have been perennial issue. A shift from the rule of "First Past the Post" has been suggested from time to time. Alternatives, which have been discussed, are a 'second run' of the top two candidates from the first run so that the winner secures at least 50%. A 'list system,' whereby, depending on the vote share, major political parties are allowed to select members from a pre-published list is another suggestion. Proportional representation through transferable vote has also been proposed. The Law Commission has recommended the adoption of the list system but this would mean increasing the size of the Lok Sabha by 25%.

12. As a unitary system of government with some federal characteristics, the balance of political power between the different states is a sensitive issue. As in most systems based on territorial constituencies, the Indian Constitution also provided for delimitation and adjustment of the boundaries every 10 years after each census. In 1976, the Constitution was amended, freezing delimitation until after the census of 2001. This was because some of the states in the south which were actively pursuing population planning measures feared that as a result of their reduced population, they would lose seats in the Lok Sabha which would be then picked up by some populous northern states. A freeze on allocation of the Lok Sabha seats between different states was therefore introduced. However, the freeze has brought about serious disparities between different constituencies across the country as well as within the states. (Fig. 15–'Freeze on Delimitation')

13. Since constituencies for the Lok Sabha comprise several State Assembly constituencies, which are called segments, the freeze applies both to Parliament and Assembly constituencies. Interstate disparity is a serious problem also at this level. Present consensus favors continuation of the freeze so far as interstate allocation is concerned, but favors delimitation of the same number of seats within each state.

14. The Electoral system has now been further enlarged by the 73rd and 74th Amendments to the Constitution, which come into effect in 1993. These amendments give a constitutional status to several thousand rural and urban local

bodies called Panchayats and Nagarpalikas. The number of elected representatives has increased phenomenally. The total of MPs in both houses of the Parliament and all the State Assemblies in the country has been about 5,000. But now the number of elected representatives to the rural and urban local bodies is more than 3 million. (Fig. 16–'Widening Base of Representation) The scale of representation is similar to the French system of Communes. But organic links between the different levels of elected bodies are not adequate.

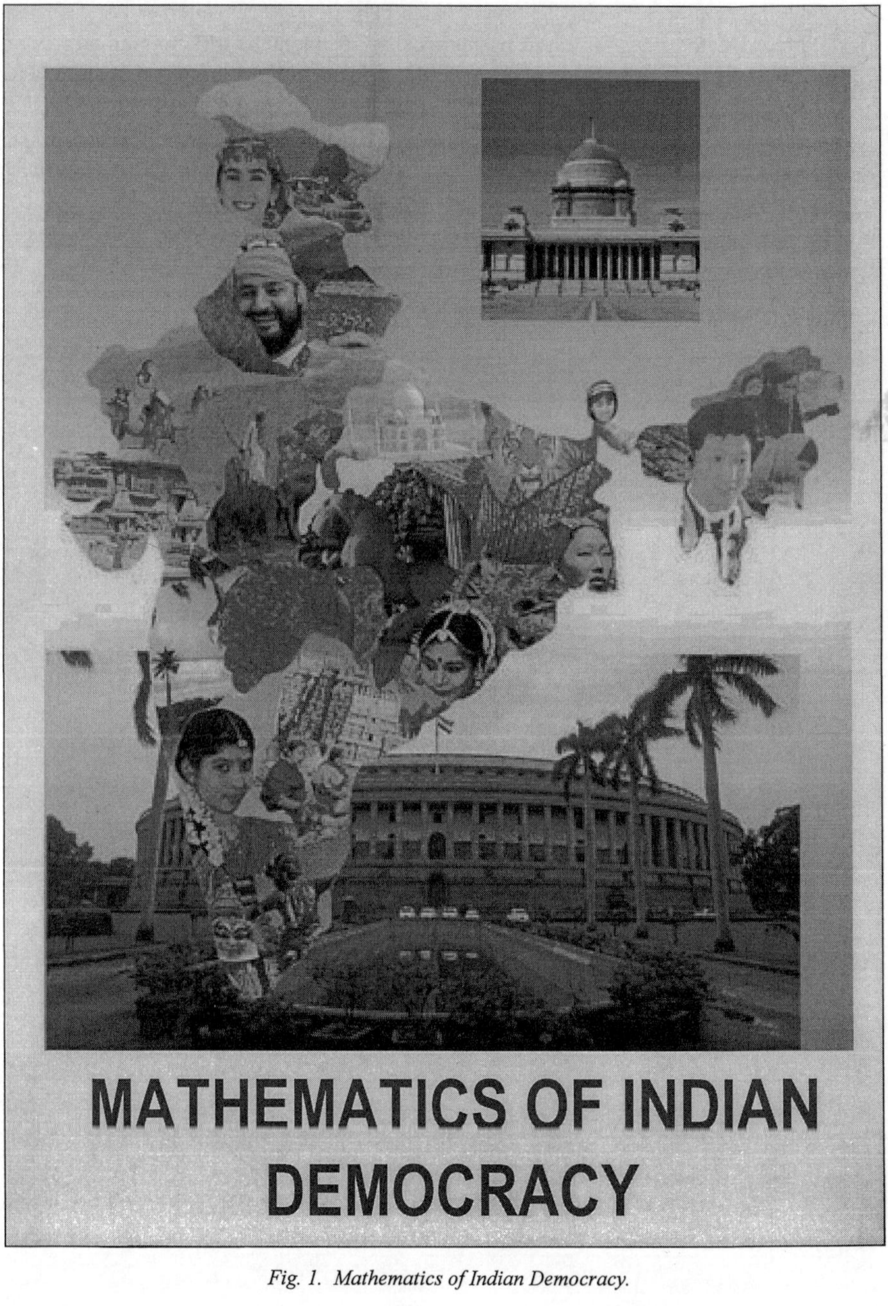

Fig. 1. Mathematics of Indian Democracy.

LARGEST POLL IN THE WORLD
(1999 General Elections)

Electorate	Men	323,826,138
	Women	295,733,806
	Total	619,559,944

Turnout		371,669,282
	Percentage	60%
	Rejected	1.91%

Polling Stations	774, 607
Polling Staff & Police	4,000,000+

Cost	50 paise per voter in 1971
	135 rupees now ($3 approx)

Fig. 2. Largest Poll in the World.

EDIFICES OF DEMOCRACY

Lok Sabha (House of the People) (489 in 1952)	545
Rajya Sabha (Council of States)	250

Vidhan Saudha, Assembly of Karnataka

27 State Assemblies	4061
5 State Councils	447
	5303

Election Commission of India organises the Elections

Fig. 3. Edifices of Democracy.

WHO VOTES ?

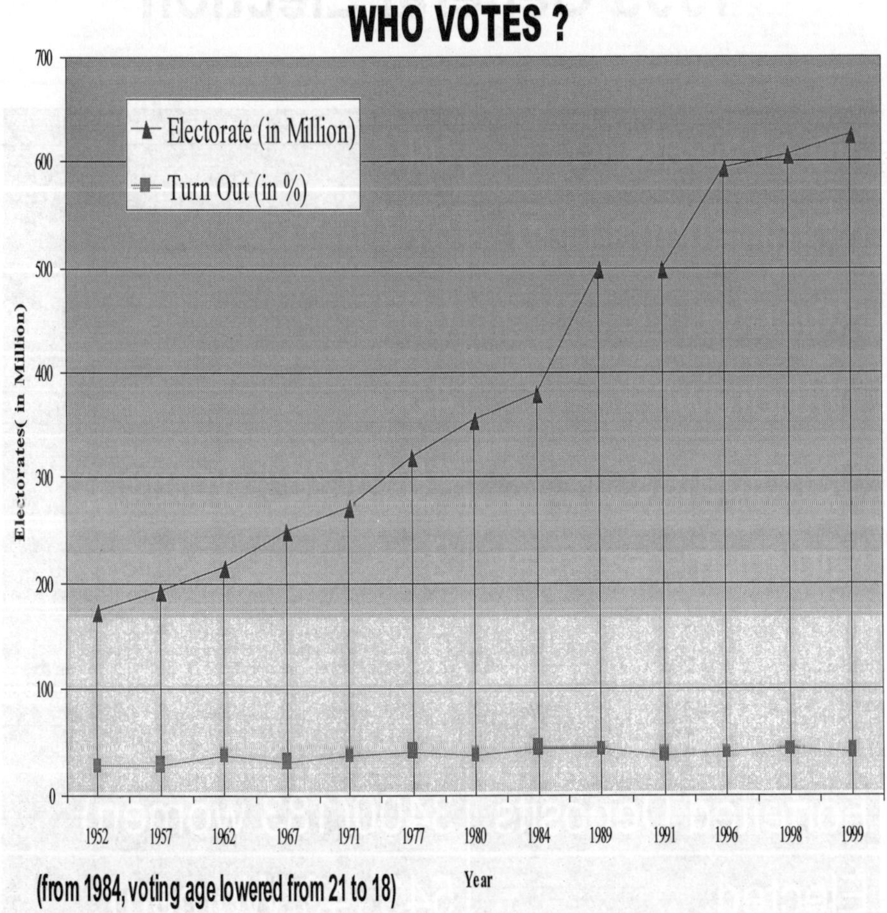

Fig.4. Who Votes?

HOW MANY IN THE FRAY ?
1999 General Election

No of Seats	543
Nominations filed	5771
Rejected	688
Withdrawn	435
Candidates	4648(284 women)
Minimum per seat	2
Maximum	32
Average	8.56
Forfeited Deposits	3400(183 women)
Elected	543 (49 Women)

Fig. 5. How Many in the Fray?

SEATS AND CONTESTANTS IN LOK SABHA ELECTIONS

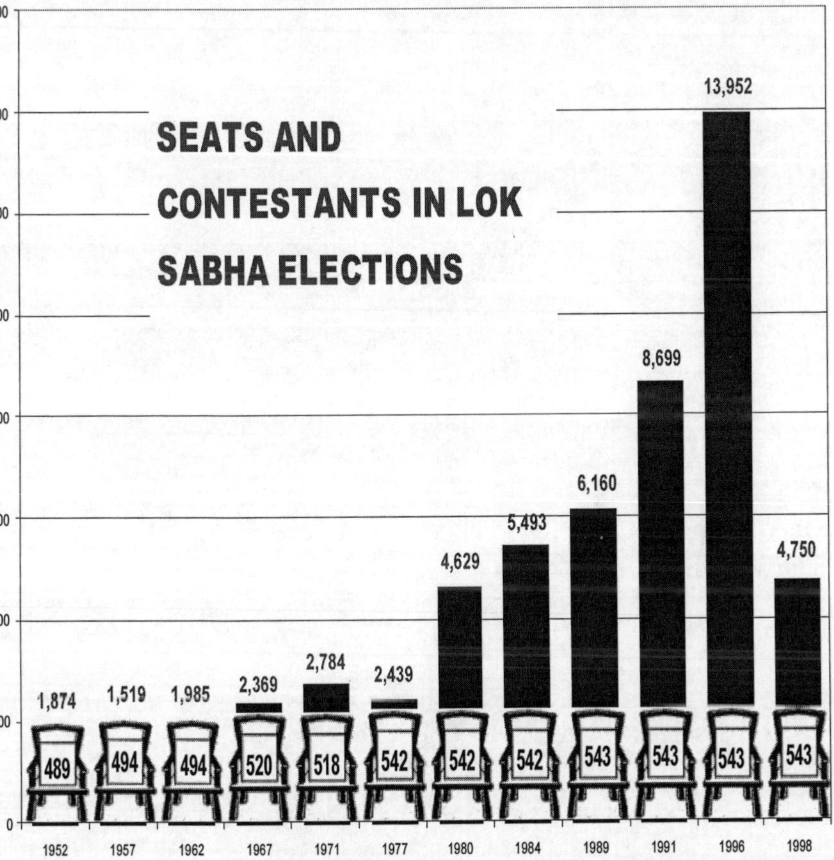

Fig. 6. Seats and Contestants in Lok Sabha Elections.

THE GENDER BIAS

Number of Women MPs still small

General Election	Year	No of Seats	No. of Women Members	% of Total
First	1952	499	22	4.4
Third	1962	503	34	6.8
Fifth	1971	521	22	4.2
Seventh	1980	544	28	5.1
Eight	1984	544	44	8.1
Ninth	1989	517	27	5.2
Tenth	1991	544	39	7.2
Eleventh	1996	544	40	7.4
Twelfth	1998	543	43	7.9
Thirteenth	1999	543	49	9.0

Fig. 7. The Gender Bias.

THE GENDER ADVANTAGE

NUMBER OF WOMEN NOT HIGH BUT SUCCESS RATE BETTER THAN MEN

Year	MEN			WOMEN		
	Contested	Elected	%	Contested	Elected	%
1957	1,473	467	31.7	45	27	60.0
1962	1,915	459	24.0	70	35	50.0
1967	2,302	490	21.3	67	30	44.8
1971	2,698	499	18.5	86	21	24.4
1977	2,369	523	22.1	70	19	27.1
1980	4,478	524	11.7	142	28	19.7
1984	5,406	500	9.2	164	42	25.6
1989	5,962	502	8.5	198	27	13.6
1991	8,374	492	5.9	325	39	12.0
1996	13,353	504	3.8	599	39	6.5
1998	4,476	500	11.2	274	43	15.7
1999	4,364	494	11.3	284	49	17.2

Fig. 8. The Gender Advantage.

Fig. 9. First Past the Post!

What did the winners get ?

How many got 50% + votes polled ?

91	96	98	99
200	146	177	203

At the top	70% of votes polled but 43% of the electorate
At the bottom	27% of votes polled but 14% of the electorate

BOTH WENT PAST THE POST !

MINORITY PREVAILS OVER MAJORITY !!

Fig. 10. What Did the Winners Get?

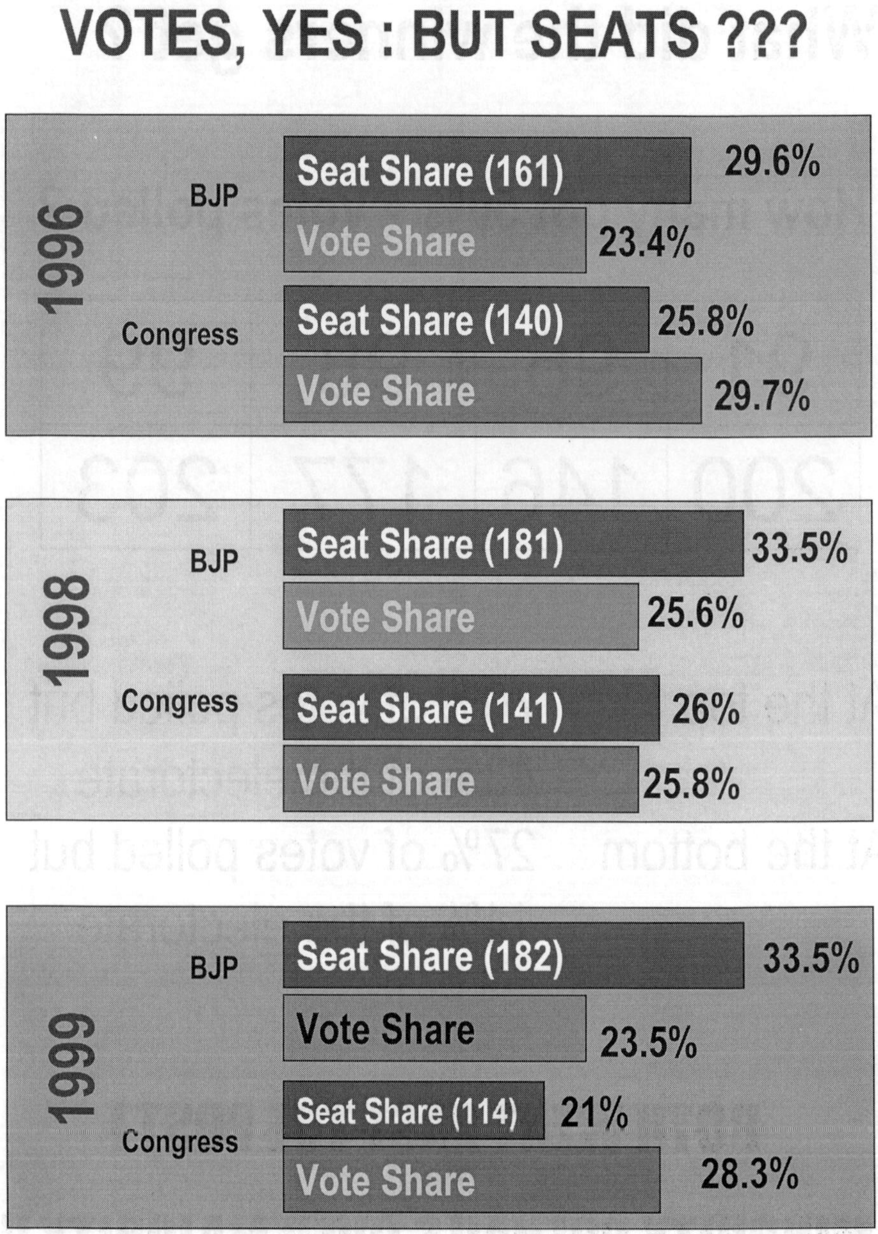

Fig. 11. Votes, Yes: But Seats???

VOTE SHARE OF NATIONAL AND STATE PARTIES

National Parties	9
State Parties	40
Registered Parties	122

Vote Share %

Year	National Parties	State Parties
1991	81	15
1996	69	25
1998	69	30
1999	67	27

Fig. 12. Vote Share of National and State Parties.

SPREE OF SYMBOLS

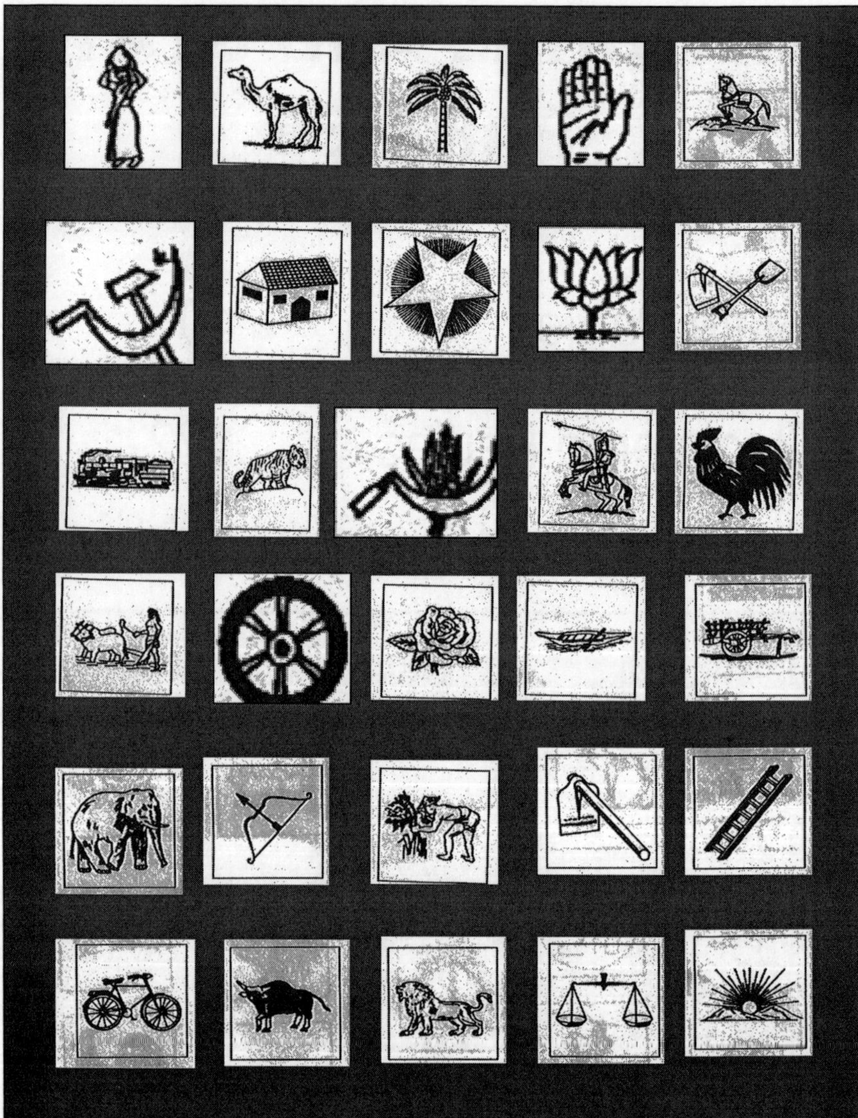

Fig. 13. Spree of Symbols.

COALITION GAMES

As played during 1996-1999

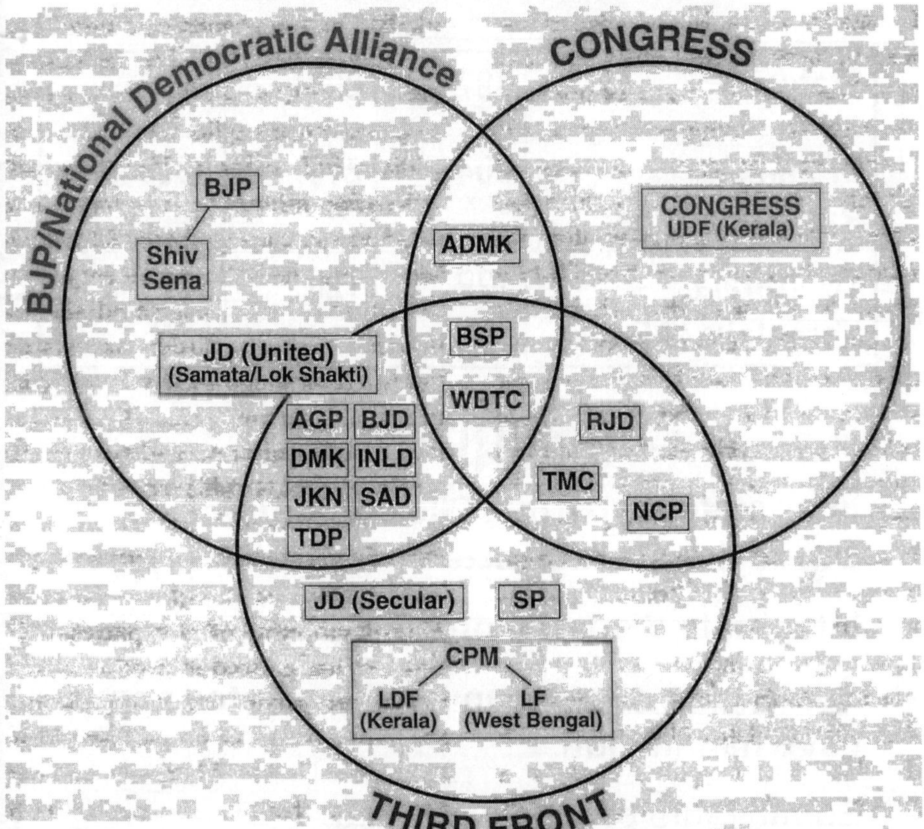

Source : Oldenburg

Fig. 14. Coalition Games.

FREEZE ON DELIMITATION

Why ?
fears of north – south divide

States	1971	2001	2016
UP, MP, Rajasthan,Bihar	204	218	233
TN, Kerala, Karnataka, AP	129	120	108

Consequences

- **Inter state disparity : 205 constituencies above the average size of 1.13 million votes**
- **Intra state disparity prominent**
- **Some have more than 2 million votes some have half million**
- **Equal value of votes affected**
- **Present consensus : Freeze Inter state allocation : delimit within state**

Fig. 15. Freeze on Delimitation.

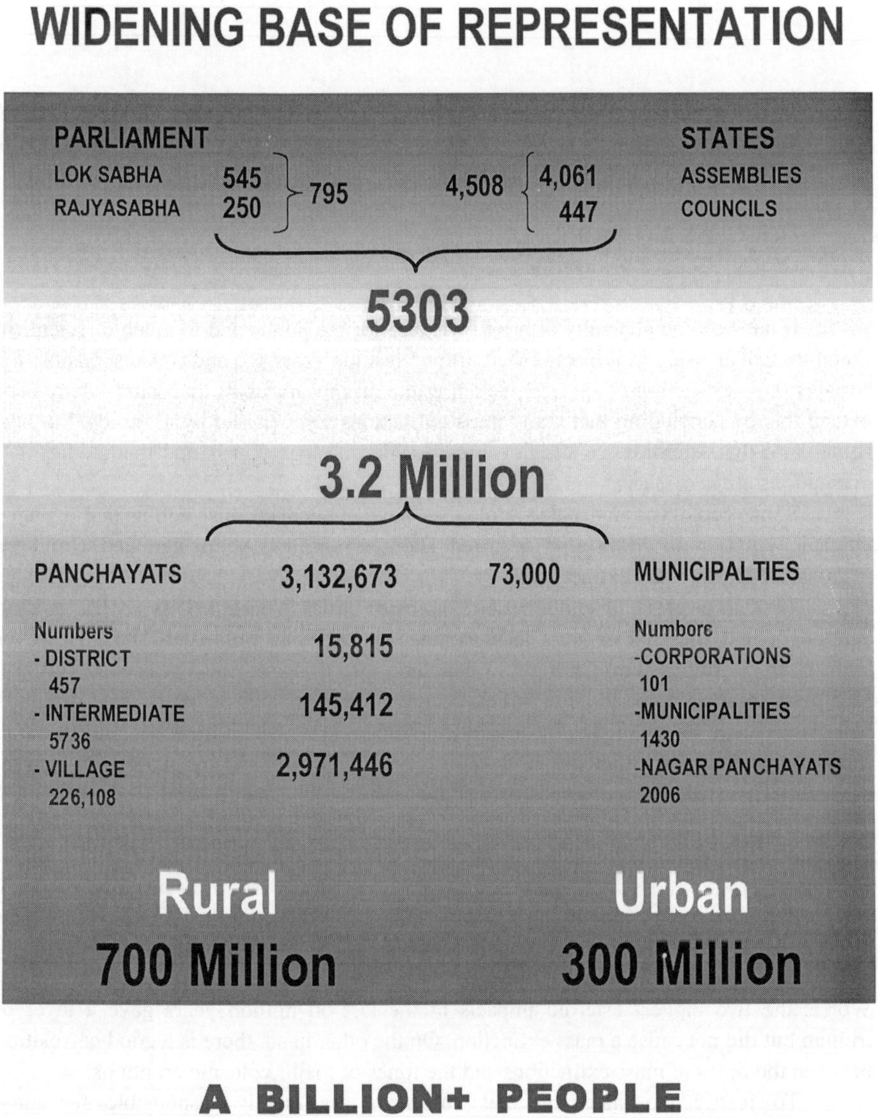

Fig. 16. Widening Base of Representation.

VOLCANOES NOT ASTEROIDS, CAUSED MASS EXTINCTIONS KILLING DINOSAURS ETC.; EXPLANATION FOR EARTH'S MAGNETIC FIELD REVERSALS

DOUGLAS R.O. MORRISON
CERN, Geneva, Switzerland

It has become generally accepted wisdom for the public and in much of scientific literature that an asteroid impacted the Earth 65 million years ago and was responsible for a major mass extinction of species, including the disappearance of dinosaurs. Many then extend this by concluding that many mass extinctions were caused by asteroids. Further, some warn that asteroids are a major threat to life on planet Earth, and that studies and precautions are necessary.

Some geologists and palaeontologists have challenged this, noting that a major volcanic eruption in India also occurred about 65 million years ago and could be responsible for the mass extinction.

That an asteroid of about 10 km diameter struck Yucatan 65.0 +/- 0.1 million years ago, and caused extensive local damage, is taken as well-established. However there is an enormous jump in logic to conclude that it caused the Cretaceous/Tertiary, K/T mass extinction world-wide which occured about 66 million years ago. Attempts have been made to prove this by theoretical calculations, which however require large assumptions. Here the experimental data on the extinctions of species and on the amount of material ejected, is considered. It is shown that the Yucatan asteroid was much too small and only corresponded in size to the explosion of the Toba volcano some 73,000 year ago which also caused major local damage but did not cause a mass extinction of species.

A second major jump in logic, is to assume that many other mass extinctions were also caused by asteroid impacts. Again considering the experimental data, there is no reasonable correlation between the time of large asteroid impacts and mass extinctions. Worse, the two biggest asteroid impacts in the last 60 million years gave a layer of iridium but did not cause a mass extinction. On the other hand, there is a good correlation between the times of mass extinctions and the times of major volcanic eruptions.

To justify the statement that volcanoes are mainly responsible for mass extinctions, some theoretical/experimental base is needed. The main problem of the geophysics of the Earth is: how does the heat produced by radioactivity escape from the core to the surface? There are two main mechanisms. The crust of the Earth is very rigid, and when two tectonic plates collide, one is forced down and is capable of moving through the mantle to reach the core, and the cooler rocks cool as they descend. Secondly, it has been shown that there is a liquid region called D", between the core and the mantle. Occasionally this liquid region can give a plume of hot material which can mount to the near surface giving a huge reservoir of hot material. Some 10% of this material breaks through the crust, and gives massive flood basalts of several million cubic kilometres in volume. It is these volcanic eruptions which can cause mass extinctions, being particularly deadly if they contain appreciable amounts of sulphur which make the surface anoxic (having no oxygen).

When volcanoes erupt, apart from lava, they emit many gases of which the two most important are carbon dioxide, CO_2, and sulphur dioxide, SO_2. The CO_2 is a greenhouse gas and raises the temperature of the world. The SO_2, on the contrary, re-emits much of the Sun's energy back into space resulting in a cooling of the Earth.

If the volcanic eruption occurs under deep water e.g. 5,000 metres deep in the Pacific, the sulphur is absorbed but the carbon dioxide escapes so that Earth becomes warmer. The largest well-known eruption was that of Ontong-Java under the Pacific ocean 120 million years ago, which emitted so much CO_2, that the world was exceedingly hot for 40 million years. On the other hand when an eruption occurs on land, both CO_2 and SO_2 are emitted, and if there is a moderate amount of sulphur, then a cooling is produced, e.g. the very small volcanic eruption of Pinatubo in 1992 gave off sufficient sulphur to cool the world by 0.5 degrees.

The Earth's magnetic field is found to reverse direction at frequent intervals, but so far no mathematical formula has explained the time series, and no satisfactory explanation has been offered. Here it is suggested that the D" liquid layer between the core and the mantle is in a chaotic state because of different rotations of the liquid core below and the very viscous mantle above, and because of erratic subduction crust entering the D" layer and plumes leaving it. Occasionally a very large plume will leave the D" layer, and this will temporarily stabilise the D" layer so that no magnetic field reversals will occur for a substantial time. Evidence for this comes from the Long Cretaceous Normal where the magnetic field did not reverse between about 125 million years ago until 80 million years ago, whereas usually magnetic field reversals occur every few million years. Now to create the Ontong Java eruption 120 million years ago, the plume had to leave the D" layer some 5 million years earlier, that is, about 125 million years ago, which is just the start of the Long Cretaceous Normal.

To avoid misunderstandings, this paper does *not* say that asteroids do not hit the earth. It says that in the last few hundred million years, none of the asteroid impacts have been big enough to cause major mass extinctions. Asteroids do hit the earth and can cause extensive local damage, so it is reasonable to study them in the hope that some action may be possible.

13. PERMANENT MONITORING PANEL REPORTS

REPORT OF THE ENERGY PERMANENT MONITORING PANEL

KAI M.B. SIEGBAHN
Institute of Physics, University of Uppsala, Uppsala, Sweden

General reviews were given by David Bodansky, Joseph Chahoud, and Douglas Morrison. Special reviews were given on China (Huo Yuping), Eastern and Western Europe (Douglas Morrison), India (Y.P Iyengar), Russia (Andrei Gagarinski), and Ukraine (Valery Kukhar).

All agreed that World energy needs were increasing: slowly in Industrialised countries and quickly in most Developing countries who wish to escape poverty and reach a standard equivalent to that of Western Europe (3 kW available, use of 5 tons of oil equivalent per person - half of the current U.S. values). Fossil fuels dominate (85% of total primary energy). The agreement at Kyoto was to reduce the emission of greenhouse gases (carbon dioxide and methane etc.) by industrialised countries. However the most urgent problem is health—particulate matter and acid rain is killing about a million people per year. In China, 71 major cities have acid rain and crop yields may be decreasing.

Energy production is increasing again in Russia. They are finding a shortage of natural gas and are considering increasing electricity generation from nuclear power by 50 to 100%.

China is heavily dependent on coal (75% now) which is causing severe pollution and health problems as does biomas burning. Oil and gas are, and will be, in short supply. It is considering all renewable energy sources and will try to develop them, but there appears little hope that they could replace coal. China is designing safer nuclear power stations as it would like to build about a hundred to reduce pollution.

India also is short of fossil fuel and has been designing a variety of safer reactors, especially with passive safety, and have a programme to build some more reactors.

In the Ukraine the remaining Chernobyl reactor will be shut down. Fossil fuels are in short supply or expensive. At present, 48% of the electricity is provided by reactors and research on safer reactors should allow the nuclear option to be retained.

The fusion situation was described by Jef Onega. Progress is being made with new experiments on the Joint European Torus, JET, and elsewhere, so that scientific break-even should be obtained soon. Despite USA withdrawal, Canada, the European Union, Japan, and Russia are proceeding with the Next Step, ITER, and sites are being proposed. However, despite the great reduction in radiation hazards compared to fission reactors, a major contribution from fusion should not be expected much before the end of this century.

Morrison described the Millenium Clean Energy Congress where all possible renewable fuels were presented and discussed. Some two billion people were said not to be connected to a power line. They mainly employ biomass and use little energy. If they were to use the same energy as Western Europeans, they would need to burn the equivalent of four tons of wood per person per year and there would be a problem of land availability. Hydro-electricity is the dominant renewable fuel and provides 19% of the world's electricity. It can and should be expanded, but will only retain its share as world consumption rises. Wind and solar (plus photovoltaic) power provide 0.1% of the world's electricity, and with subsidies, are expanding rapidly. They are particularly useful where there are no power lines. However none of the many renewables appear capable of becoming a major energy source replacing coal, without an unexpected technological breakthrough.

Nuclear fission power is very controversial because of (1) nuclear proliferation fears, (2) risk of radioactive accidents, (3) disposal of nuclear waste. In affluent countries, there is an active opposition. In Western Europe and North America, no nuclear reactors have been constructed recently or are planned—this will result in a serious fall in electricity production after 2010 in North America and after 2025 in Western Europe, but realistic proposals for replacements have not been made for this loss. In major non-affluent countries, there are proposals for a second wave of nuclear power with improved safety.

Bruce Stram noted that over the last 20 years, energy research funding in the USA had declined by 50%.

After due discussion, the Permanent Monitoring Panel for Energy accordingly makes the following conclusions and recommendations:

1. For developing nations to escape poverty and raise living standards to the level of Western Europe today, the world total energy production will greatly increase, perhaps by a factor of six by 2100.
2. Much of this increase can come only from increased fossil fuel production, mainly coal, as oil and gas supplies become exhausted. Coal is the worst polluting fuel as it causes major health hazards from particulate matter, acid rain etc. as well as increasing greenhouse gases.
3. Energy efficiency has and will help appreciably, and will occur for financial and other reasons, but it will not be sufficient to reduce in a major manner, the rise in world energy production.
4. Everyone would like renewable energy sources to become major sources of energy, but none appear to be capable.
5. Nuclear fission is judged to be capable of replacing fossil fuels and the disadvantages listed above are considered less serious than those from burning coal. The main problem may be opposition from part of the public. Thus it is

recommended that unless and until renewable sources can replace fossil fuels, nuclear power be expanded.

6. It is recommended that research be continued for (1) renewable sources in the hope of a breakthrough, (2) safety of reactors, (3) fusion.
7. It is recommended that discussions take place with journalists with the aim of learning how to communicate and discuss these opinions and recommendations with the public.

LINKING THE CONVENTIONS: SOIL CARBON SEQUESTRATION AND DESERTIFICATION CONTROL. A REPORT FROM THE DESERTIFICATION PERMANENT MONITORING PANEL WORKSHOP

DOUGLAS L. JOHNSON
Clark University, Graduate School of Geography, Worcester, Massachusetts, USA

The workshop was held on 24 August 2000 and addressed three issues: (1) recent developments in the emergence of soil carbon sequestration as a vehicle for implementing the Kyoto protocol on climate change as well as a mechanism for combating desertification within the world's semi-arid zones; (2) plans for pilot project development and research needs in order to implement pilot projects and to outline criteria by which to measure the success of such projects; and (3) identification of unresolved issues in the science, monitoring, and implementation of soil carbon sequestration.

RECENT SOIL CARBON SEQUESTRATION AND DESERTIFICATION CONTROL DEVELOPMENTS

The first morning session featured two presentations, each of which engendered spirited discussion, on recent global scale developments on the soil carbon sequestration front that might contribute to desertification control. Larry Tieszen of the USGS, EROS Data Center outlined three recent initiatives that showed particular promise and discussed both benefits and uncertainties in establishing carbon trading and sequestration. The World Bank, with resources contributed by Japan and Scandinavian countries, has established a prototype carbon fund with $150 million in available funds. The Pew Foundation and Environment Center is interested in supporting Clean Development Mechanism projects encouraged by the Kyoto Protocol to the United Nations Framework Convention on Climate Change (UNFCCC). In Kazakhstan, a number of institutions are supporting a rangeland rehabilitation project with important anti-desertification and soil carbon sequestration implications.

Ultimately, Dr. Tieszen maintained, projects that combat land degradation in dry areas stand to benefit from economic resources transferred from Annex 1 (industrialized) countries to developing countries if these Annex 2 states can generate Certified Emission Reduction credits (CERs). CER credits allow developing countries to participate in climate mitigation efforts and at the same time to raise funds for land management and development projects. The key to the program is offsetting carbon emissions in industrial areas with increased accumulation of carbon in soils elsewhere. These gains have to be verified from a baseline that establishes existing soil carbon quantities. Participants in the program enter into a contract to maintain and to increase the amount of carbon in the soil. These gains have to be real, they must be capable of verification, and they must occur in areas where land ownership is legally defined and clear. The biggest problem is who owns the carbon that a project sequesters - the farmer or the state? A second area of uncertainty is "leakage," the loss of soil carbon to the atmosphere due to land management activities undertaken because of the change in land use to sequester carbon in the soil and other causes (weather; season; vegetation cover; offsetting carbon losses in other places) and how to account for such losses. Also at issue is whether CER contracts can be entered into if land ownership is unclear (for example communal rather than private), and if certifiable collateral is lacking with which to cover the cost of paying back the buyers of CERs if the seller is unable to deliver carbon in the amounts specified.

Dr. Robert Ford, a natural resources advisor in the Global Bureau, Office of Agriculture and Food Security of U.S. AID reported on discussions underway that aimed to link the two conventions of paramount interest to the workshop: the Convention to Combat Desertification (CCD) and the UNFCCC. An expert workshop on "Carbon Sequestration, Sustainable Agriculture and Poverty Alleviation," sponsored by the World Meteorological Organization, U.S. AID's Office of Agriculture and Food Security, the United Nations Food and Agricultural Organization (FAO), and the International Fund for Agricultural Development (IFAD) is scheduled to take place in Geneva from 30 August to 1 September 2000. The Geneva workshop will explore a number of the themes initiated at Erice, particularly issues dealing with measurement and verification of soil carbon sequestration, and many of the Erice workshop participants will contribute to the Geneva seminar as well. The Geneva workshop, in turn, is a prelude to the continuing UNFCCC negotiations, which will resume in Lyons, France later in September.

In a brief, provocative presentation on the major concerns that U.S. AID had in development projects, Dr. Ford emphasized a deep interest in the restoration of degraded lands. Concerned that the rich often benefit differentially from development projects, he felt that this was less likely to be the case if the development focus was on the most degraded areas. On steep hillsides, for example, a shift in cultivation techniques from plowing to no tillage, when combined with perennial vegetation and polyculture, offered good prospects for long-term success. As a principle, blending modern agricultural technology with older, traditional methods and crops was producing good results in many places. In a more cautionary mood, Dr. Ford urged consideration of the role played by methane and other gases in the carbon sequestration process, questioned what might happen when soils become saturated with carbon, and expressed concerns about the reliability and verifiability of carbon accounting procedures. At the very least, it was likely that accounting and verification processes would have to be tailored to the conditions of the country selling CERs rather than simply employing the practices used in North America.

PILOT PROJECT IMPLEMENTATION: PLANS AND RESEARCH NEEDS

After a short break, the second session was addressed by Paul Bartel and Lennart Olsson. Mr. Bartel, an environmental monitoring advisor in USAID's Africa Bureau, Office of Sustainable Development, tackled a number of complex issues of environment, development, and demographic change that would impinge significantly on anti-desertification, soil carbon sequestrating pilot projects. He pointed out that issues of access to communal land and who has rights to what shares of the products of communal resources was a critical issue in project development. While communal tenure often is viewed by outsiders as a liability, in practice common property management systems can be a major asset because they permit the involvement of a large number of small holders and producers in the inevitably larger scale projects needed to make CERs a viable economic commodity. Although verification of soil carbon sequestration was a key issue, he noted that U.S. AID had more than fifteen years of experience in the development of effective methods for organizing communities both in support of development initiatives and for project evaluation. These principles could be directed both toward project development and implementation initiatives and toward the very real issues of soil carbon verification. For Mr. Bartels, the ultimate issue was income. It was critical that the maximum amount of income generated by CERs stick at the local level, a goal that was favored by the gradual development of local government from a coercive to a facilitative role. Moreover, while the individual income augmentation from an anti-desertification project might be small (e.g. $10/farmer), the sum was not insignificant; it could equal the equivalent of ten days of work for an African agricultural laborer. Were such sums to be pooled at a community level, they might also represent very significant contributions to communal projects (e. g. school; village well). Development projects designed to combat desertification and to store carbon could contribute in important ways to U.S. AID goals of increased global biodiversity, elevated income levels, and more robust civil societies.

Research needs were the topic of Dr. Olsson's presentation. He asserted that more knowledge was needed in two areas: (1) monitoring and verification of soil carbon fluxes; and (2) more accurate system analysis of the interaction between people, place, and environmental change at both local and regional scales. The monitoring and verification issue could be tackled in three ways. The first way was by means of baseline studies that determined before project implementation how much carbon was now in the soil, how much could potentially be added to the soil, and what were the best ways to put carbon into the soil and keep it there. The second approach was to use remotely sensed data to increase the geographic and real-time resolution of biophysical models (e.g. Century; EPIC) in order to reduce errors in the prediction of future environmental changes. A third need was to improve our understanding of the basic science involved in below ground storage of carbon. At least three years of detailed measurement of ground-atmosphere exchanges would be needed in order to feel comfortable about basic processes in semi-arid grasslands in Sahelian Africa, for example. This could be accomplished by establishing a number of flux towers (4-6) in districts that are likely candidates for anti-desertification projects. Knowledge of these basic fluxes would also help design projects in ways that would help to protect farmers from sudden and large economic losses due to environmental variability.

A better systemic understanding of the interaction of crops, livestock, fallow land, local market forces, off-farm income, and government commodity pricing policies, among other variables, was needed for realistic future project development. Both at the household and the community level our understanding of the forces driving change would be enhanced by better interactive models. From a management standpoint, the "dark horse" in Africa's Sahel is the role played by fire in semi-arid ecosystems and the impact that fire has on carbon storage. Mr. Bartel pointed out that there are models that have solved many of the complex problems considered during group discussion. Mr. Sheffner noted that the remote sensing images acquired to verify land use/land management changes accompanying soil carbon sequestration over large areas could be used for other land monitoring tasks, and such use would make acquisition of the imagery far more economical. The need for a simple "carbon probe" to assess soil carbon quickly and easily, rather than to have to collect soil samples for transport back to a distant laboratory for analysis, was expressed by Dr. Rosenberg.

IMPLEMENTATION CRITERIA AND MEASUREMENT OF PROJECT SUCCESS

After breaking into two discussion groups in order to consider implementation criteria and measures of pilot project success, the group reconvened in plenary session to discuss the results. Critical dissection of the subgroup presentations resulted in an improved understanding of how carbon sequestration/anti-desertification projects might be developed and evaluated.

Implementation

The implementation subgroup addressed its task by generating answers to the following questions:

1. *Who pays for monitoring and assessment of a "clean development" carbon sequestration project?*

The seller bares all the costs associated with monitoring and assessment. These costs should be included in the basic carbon price negotiated at the time the contract is signed.

2. *Who benefits from such a project?*

Primary beneficiaries must be at the local level. The idea is to pick project locations where experience already exists on which to build, and issues of how project benefits are to be shared between local participants and other actors at a regional and national scale have already been settled. In principle, government involvement then would be limited to setting standards and enforcing contracts.

3. *What is the appropriate scale for project activity?*
Project size would be determined by the scale needed to generate an assumed minimum contract size of 100,000 tons of sequestered carbon. This implies operating at a community, sub-provincial level that would aggregate a considerable number of agricultural villages. An area that is 40/60 km on a side (ca. 5,000 farmers) was assumed to be the appropriate scale in an agricultural context, with a correspondingly larger area and smaller population probably characterizing a pastoral project. Within this area, actual project activities are likely to be quite disaggregated spatially, a risk-avoidance activity that should benefit the larger project community in case rainfall in a given year was distributed erratically across the project area.

4. *Do self-help models exist on which to base carbon sequestration projects?*
The group is confident that numerous successful development models exist upon which carbon sequestration/anti-desertification pilot projects can be based. Two such successful indigenous models are the Africare activities in Senegal and the Nam movement in Burkina Faso. Others exist in other regions of the world that can be drawn on with confidence. A demonstration project that modeled a carbon sequestration process, from an initiating carbon sequestration contract through a brokered community demonstration project, would be very useful in demonstrating the practicality of the concept.

5. *What role might incentives play in initiating a project?*
There was general agreement that some combination of the following factors had to feature in an proposed project: (1) it had to be easy to conceive and carry out; (2) it must pay good money to the participants; and (3) it must reduce the risk that farmers and herders face in both coping with their environment and in investing time, labor, land, and capital in the carbon sequestration project.

6. *How might the project control for the impact of environmental variability?*
Both the buyer and the seller of carbon sequestration contracts would have to be protected from risk. For the farmer, this might involve an insurance program guaranteed by contingency funds established from the original project contract.

<u>Success</u>
Several measures of success were discussed in each of several categories.

1. *Farmers* would display evidence of success if:
    a. a project returned net positive benefits to participants at different income levels;
    b. income gains occurred ;
    c. soil quality improved and erosion losses were reduced;
    d. labor savings took place, making investments in economic diversification possible; and
    e. risk from climatic variability was reduced.

2. *Carbon* is sequestered equal to or greater than the amount contracted for originally. How to account for "leakage" remains a vexed issue.
3. The *cost/benefit ratio*, on balance, for the project was positive
4. *Questions*, amenable to future study and resolution, could be isolated that posed potential stumbling blocks to wide scale implementation
5. The project exhibits *potential to be adopted spontaneously* elsewhere in the region and indications of independent brokerage and trading of carbon emerges
6. Some combination of the following *social and community indicators* appear:

   a. employment improves;
   b. investment in infrastructure occurs;
   c. fewer people out migrate for employment;
   d. the role of women in the community improves;
   e. school attendance increases;
   f. morbidity and mortality levels decrease; and
   g. nutritional levels improve.

CONCLUSION

Participants noted that the Desertification Permanent Monitoring Panel's effort to develop anti-desertification initiatives, which had begun at Erice four years ago, was moving off in new and potentially fruitful directions. In these initiatives, many members of the Desertification PMP were playing a leading role. In Dakar, from 25-27 September 2000, an international workshop would take place to explore the potential role that soil carbon sequestration might play in anti-desertification efforts. In Geneva, August 20-September 1, 2000, a multifaceted examination of soil carbon sequestration verification issues would be addressed by scientists drawn from a broad spectum of institutions and agencies. And in early October, a meeting of African energy ministers in Tucson, Arizona would include a session that focused on the role that soil carbon sequestration might play in helping to meet the energy needs of their countries. Because the UNFCCC encouraged cooperation among and the transfer of development funds between more industrialized and less industrialized countries, there was considerable hope that the resources with which to arrest and reverse land degradation in dry places would become increasingly available. It was this expectation of future progress that encouraged the Desertification PMP to continue to address the unresolved research and verification issues that still attended the soil carbon sequestration issue.

WORLD FEDERATION OF SCIENTISTS PERMANENT MONITORING PANEL ON POLLUTION

RICHARD C. RAGAINI
Department of Environmental Protection, University of California, Livermore, CA, USA

The continuing environmental pollution of earth and the degradation of its natural resources constitutes one of the most significant planetary emergencies today. This emergency is so overwhelming and encompassing, it requires the greatest possible international East-West and North-South co-operation to implement effective ongoing remedies.

It is useful to itemize the environmental issues addressed by this PMP, since several PMPs are dealing with various environmental issues. The Pollution PMP is addressing the following environmental emergencies:

- degradation of surface water and ground water quality
- degradation of marine and freshwater ecosystems
- degradation of urban air quality in large (mega) cities
- impact of air pollution on ecosystems

Other environmental emergencies, including global pollution, ozone depletion and the greenhouse effect, are being addressed by other PMPs.

PRIORITIES IN DEALING WITH THE EMERGENCY

The PMP on Pollution monitors the following priority issues:

- clean-up of existing surface and ground-water supplies from industrial and municipal waste-water pollution, agricultural run-off, deforestation, and military operations
- reduction of existing air pollution and resultant health and ecosystem impacts from long-range transport of pollutants and trans-boundary pollution
- prevention and/or minimization of future air and water pollution
- training scientists & engineers from developing countries to identify, monitor and clean-up pollution

Furthermore, the PMP will provide an informal channel for experts to exchange views and make recommendations regarding environmental pollution.

MEMBERS OF PERMANENT MONITORING PANEL

The following scientists listed below were appointed by the chairman as permanent members, because of their interdisciplinary expertise in environmental matters:

Chairman Dr. Richard C. Ragaini, Lawrence Livermore National Laboratory, USA
Dr. Lorne G. Everett, University of California at Santa Barbara, USA
Prof. Majid Hassanizadeh, Delft University of Technology, Netherlands
Prof. Sergio Martellucci, University of Rome, Italy
Dr. Gennady I. Palshin, ICSC-World Laboratory, Ukraine
Prof. Paolo Ricci, University of San Francisco, USA
Prof. Vittorio Ragaini, University of Milan, Italy
Academician Albert Tavkhelidze, National Academy of Sciences of Georgia

ASSOCIATE PANEL MEMBERS

The following scientists listed below were appointed by the chairman as associate panel members:

Prof. Victor Baryakhtar, National Academy of Sciences of Ukraine
Prof. Robert Clark, University of Arizona, USA
Prof. Joerg Drewes, Arizona State University, USA
Dr. Vladimir Mirianashvilli, Institute of Geophysics, Georgia
Mr. David Rice, Lawrence Livermore National Lab, USA
Prof. Soroosh Sorooshian, University of Arizona, USA
Prof. Igor Zetsker, National Academy of Sciences of Russia
Ms. Kay Thompson, Department of Energy, USA

SUBGROUPS

The following subgroups listed below were organized by the chairman in order to more effectively monitor regional impacts in developing countries:

- Black Sea Pollution: A. Tavkhelidze, G. Palshin, R. Ragaini
- Groundwater Vulnerability to Water Reuse: J. Drewes, R. Ragaini
- Water and Air Impacts of Automotive Emissions in Megacities: S. Martellucci
- Water Pollution in Vietnam: P. Ricci, D. Rice and R. Ragaini

SUMMARY OF 1999-2000 ACTIVITIES

October 1999 (Geneva)

Professor Joerg Drewes, Arizona State University, and Dr. Richard C. Ragaini, LLNL, presented a World Lab proposal for funding, "Long-Term effects on Groundwater From the Practice of Irrigation With Sewage Effluent." This proposal was not funded.

December 1999

The World Federation of Scientists and the DOE signed a Memorandum of Agreement (MOA) for cooperation and coordination on projects of mutual interest. We hope that coordination of Black Sea activities will be the first implementation of this MOA.

December 1999 (Vietnam)

Professor Paolo Ricci, University of San Francisco, Dr. David Rice, LLNL and Dr. Ragaini, LLNL, made a fact-finding trip to Vietnam in response to an unsolicited water pollution project proposal submitted by the Vietnam Atomic Energy Commission (VAEC). As a result, a new project proposal has been submitted by the VAEC, and is being evaluated.

August 2000 (Erice)

Black Sea Activities

1. Professor Albert Tavkhelidze, Georgian Academy of Sciences, Dr. Gennady Palshin, World Lab - Kiev Office and Dr. Ragaini, LLNL, organized a Black Sea Workshop/PMP Meeting on Environmental Impacts of Oil Pollution. The Workshop resolution is shown below. Briefly, it calls for:

    - 6 WL scholarships to attend 2001 DOE Workshop on Water plume modeling
    - On-going WL training courses for Black Sea scientists
    - Black Sea hazards assessments by Kiev and Tbilisi World Lab offices
    - WFS-DOE cooperation on Black Sea web-site.

2. Three Workshop speakers, Dr. Valery Mikhailov, Institute of Sea Ecology, Dr. Ilkay Salihoglu, Middle East Technical University and Ms. Kay Thompson, DOE, presented talks on the Black Sea at the International Seminar:

    - Prof. Valery Mikhailov, Ukrainian Scientific Center of Sea Ecology, Ukraine *"Management of the Azov-Black Sea Ecosystem"*
    - Ms. Kay Thompson, U.S. Department of Energy, USA *"Forming Black Sea Coalitions to Support Sustainable Economic Development"*
    - Dr. Ilkay Salihoglu, Middle East Technical University ,Turkey *"Sub-Oxic Zone in the Black Sea"*

MTBE Activities

Professors Lorne Everett and Arturo Keller, both from University of California at Santa Barbara, presented talks on MTBE (methyl tertiary butyl ether) at the International Seminar.

- Prof. Lorne Everett, University of California at Santa Barbara, USA *"MTBE - America's Mega-City Mistake"*
- Prof. Arturo Keller, University of California at Santa Barbara, USA *"MTBE - International Use, Toxicology and Policy"*

<u>August 1-6, 2001 (Erice)</u>

We propose a meeting of a small planning group on automobile emissions in mega-cities during Professor Martellucci's 31st Course International School on Quantum Electronics, "Global Laser Automotive Applications."

<u>August 19, 2001 International Seminar (Erice)</u>

1. We propose to hold a Pollution PMP meeting.
2. We propose to hold a meeting of the WFS Black Sea Commission.

WORKSHOP REPORT

The "Black Sea Workshop on Environmental Impacts of Oil Pollution" was held in Erice, Italy on August 18-19, 2000. The purpose of this Workshop was to examine the environmental situation of the Black Sea as a result of oil pollution. The goal was to focus on what is known, what is not known, what activities are currently ongoing, and what activities are appropriate for the World Laboratory.

The Black Sea is of global interest on several levels. First, it has the worst environmental degradation of all of the world's oceans. Second, it is bordered by six countries, Russia, Turkey, Ukraine, Georgia, Bulgaria, and Romania, most of which have poor economies. Although international agreements, strategic plans, and national environmental programs are in place, these severe economic problems have slowed environmental monitoring, remediation and restoration significantly. Third, the Black Sea and the surrounding countries are vital for the distribution of the Caspian Sea basin oil and gas supplies to western countries.

The environmental crisis is a direct effect of both natural and anthropogenic causes, which have forced dramatic changes in the Black Sea's ecosystem and resources. Due to its unique geographical characteristics, the Black Sea is extremely vulnerable to anthropogenic pollution. It is very deep (more than 2 km), anoxic below 200 m, and the only outlet is through the narrow, shallow (100 m deep) Bosphorous Strait. The Black Sea is the largest anoxic water basin in the world. It's surface area is five times smaller than its catchment basin. The primary causes for the environmental deterioration include the enormous nutrient and pollutant load from three major rivers, the Danube, Dniester and Dnieper; and from industrial and municipal wastewater pollution sources along the coast. Fishery biodiversity and yields have declined dramatically in recent years. Water-borne cholera and hepatitis epidemics have broken out periodically in the northern coastal areas. Tourism has severely decreased. Economic losses from pollution exceed $500 million per year, as estimated by the World Bank.

In addition, there are oil developments, which will further degrade the Black Sea environment. A new pipeline, which terminates in Supsa, Georgia, was completed in 1999. Oil is flowing, and tankers are transporting oil through the Bosphorous.. A new Russian oil pipeline, which will terminate in Novorossiysk, Russia is under construction. Consequently, oil tanker traffic in the Black Sea will increase significantly.

The deterioration of the environmental situation in the Black Sea region is important to western countries, because it can affect the multi-national cooperation among the Black Sea and Caspian Sea countries, and impact the international community with strategic energy interests in the region. Many steps need to be taken to deal with the continuing environmental deterioration. These include control of pollution sources, including oil spill prevention and cleanup, catchment basin pollution control, coastal zone pollution control; and remediation of existing pollution. These steps will depend on the cooperation of the Black Sea countries, and their ability to overcome the massive economic, political, and ethnic problems facing the region.

As a result of the information presented in the Workshop, the participants drew up a resolution, which is shown below.

World Federation of Scientists
Planetary Emergency Panel (PMP) on Pollution
Workshop on Environmental Impacts of Oil Pollution in the Black Sea
August 18-19, 2000

Resolution

Recognizing that the Black Sea is the most environmentally-impacted water body in Europe*, the World Federation of Scientists (WFS) Planetary Emergency Panel (PMP) on Pollution has been concerned with the ecological impacts of the environmental pollution in the Black Sea for the past three years. In a February 1999 WFS workshop on this topic, it was concluded that the major environmental pollution impacts in the Black Sea are due to coastal discharges of untreated sewage, introduction of nitrates and phosphates from the Danube River and industrial discharges into the rivers and the Sea. Plans to mitigate these impacts have been addressed in the Black Sea Strategic Action Plan formulated by the riparian countries, and await funding implementation. It was concluded that the potential for oil pollution poses a significant environmental threat as construction of new oil pipelines from Central Asia increases the oil transport through the Black Sea by a factor of 2-3 in the future. This will increase the risk of oil discharges in the Black Sea from accidents, spills, deballasting, etc. Therefore, we recognize the important need to document the existing oil pollution, in order to prepare a baseline against which to evaluate any future changes.

This Workshop on Environmental Impacts of Oil Pollution in the Black Sea brought together scientists from the Black Sea countries to discuss the current state of knowledge of oil pollution, and produced the following action plan for the PMP:

1. Request World Laboratory support for 6 additional Black Sea scientists for training at a U.S. Department of Energy (US DOE) workshop on water plume modeling in Tbilisi, Georgia, in January 2001 through the World Laboratory Georgian Branch. The U.S. DOE is already providing support for 6 Black Sea scientists to be trained by the U.S. National Oceanic and Atmospheric Administration (NOAA). This joint activity will proceed under the Memorandum of Agreement between the U.S. Department of Energy and the World Federation of Scientists signed in 1999.
2. Request World Laboratory support to conduct periodic training courses in Erice for Black Sea scientists on environmental pollution issues of high priority. The first such training will address the adoption of standards for emissions and discharges of environmental contaminants into the Black Sea by all the riparian countries.
3. Request support for the World Laboratory Georgian and Ukrainian Branches to conduct high-priority hazard assessments for oil pollution in the Black Sea region using a Geographical Information System.
4. Collaborate with the U.S. Department of Energy to develop the new Black Sea Environmental Information Center web site (http://pims.ed.ornl.gov/blacksea) to facilitate communication and the exchange of information and scientific data among the countries of the region and the rest of the world.

These planned activities are intended to support the Black Sea littoral states to mitigate the environmental impacts of future increases in oil transport in the Black Sea and increase the benefit to its ecology.

* Dobris Assessment, European Environment Agency, 1995.

PROGRESS REPORT ON THE WORLD FEDERATION OF SCIENTISTS ACTIVITY IN LITHUANIA

Z. RUDZIKAS
ITPA, A.Gostauto 12, Vilnius, 2600, Lithuania. e-mail: tmkc.plls@wllb.lt

The Lithuanian members of the World Federation of Scientists (WFS) paid much attention to contribute to uphold the WFS principles, to promote international collaboration between scientists and researchers from various countries, to encourage free exchange of information without secrecy and without frontiers, to implement complex interdisciplinary studies of scientific and technological problems of vital importance, to fight against fifteen Planetary Emergencies. Professors Leonardas Kairiûkstis, Juras Pozela and Zenonas Rudzikas also participated in the activities of the relevant Permanent Monitoring Panels of the WFS.

We have used various means of transfering information (radio, TV, popular journals and magazines, newspapers, public lectures etc.) to disseminate the main ideas of the WFS as well as of the International Centre for Scientific Culture World Laboratory among Lithuanian society, scientists included.

We have completed the translation of Erice and Farnesina Statements as well as of Lausanne Declaration into the Lithuanian language (copies attached). They will be published in some Lithuanian journals or newspapers and will also be disseminated in paper and electronic versions.

The second direction of our activity was to look for young scientists who would be interested in research linked to one of the 15 Planetary Emergencies as set out by the WFS. In other words, we were looking for good candidates for the National Scholarship Programme (NSP) established by the WFS. The announcement of this Programme was published in the main newspaper for the Lithuanian scientists "Mokslo Lietuva" ("Scientific Lithuania", copy attached) and was also made known in a number of other ways.

We are very pleased to learn that the first batch of candidates, consisting of three fellows, have already been granted a one-year scholarship by the World Laboratory to undertake research corresponding to one or another Planetary Emergency. Let us characterize them shortly.

1. Ph.D. student *Sarûnas Antanaitis* (Lithuanian Institute of Agriculture). "Migration and Balance of Chemical Elements in the Cropping Systems in Different Levels of Chemization". Supervisor Prof. *Alfonsas Svedas*.

The task of his present research is to study the relationship between soil agrochemical properties and contamination with heavy metals, concentration of plant nutrients and heavy metals in groundwater and drainage run-off getting into open water sheds, migration of chemical elements, variation of yield level, plant production quality and cropping systems, differing in chemization level, plant nutrient and heavy metal balance.

The research is going to be continued in the third millennium. The results obtained will be used for the development of balanced cropping systems which could secure economically rational yield and the utilization of environmental resources, optimal quality of the products and minimal risk of polluting the environment with undesirable substances. The research is linked mainly to the third Planetary Emergency "Food".

2. Ph.D. student *Asta Kanapickaitė* (Lithuanian Forest Research Institute). "The Response of Scots pine (*Pinus sylvestris L.*) Different Populations to Atmospheric Pollution and Climatic Change in Lithuania". Supervisor Prof. *Leonardas Kairiûkstis*.

The objectives of the study are as follows:

- to carry out the research on Scots pine (*Pinus sylvestris L.*) exposed to physical and chemical environmental changes aiming at finding the indices or biomarkers reflecting these changes;
- to gain the knowledge about the effect of environmental changes on diversity within species of Scots pine (*Pinus sylvestris L.*) in Lithuania;
- to assess the effect of these changes on resistance to negative environment aiming at identification of the allowable limits of environmental changes.

As the result of the above-mentioned studies the knowledge about the response of Scots pine populations to environmental changes will be obtained. The limits of their resistance and possible indices or biomarkers will be identified as well. The results collected will be published as a scientific article. The study corresponds to the fifth ("Pollution") and seventh ("Climatic Changes") Planetary Emergencies.

3. Dr. *Egidijus Rimkus* (Dept. of Hydrology and Climatology of Vilnius University). "Lithuanian Climate Modeling". Supervisor Prof. *Kæstutis Kilkus.*

The main tasks of his investigations are as follows:

- to evaluate the possibilities of simulation of present climate changes in Lithuania using a number of the most popular climate models and to present new climatic change scenarios for Lithuania in the next century;

- to compare the rate of the observed and forecasted warming in different scale regions (Northern Hemisphere→Europe→Baltic Sea region- Lithuania);
- to suggest new prognostic scenarios of change of some climatological and agroclimatological indices in the next century on the basis of the best climate models.

This work will include not only local, but regional and global aspects as well. It deals with the seventh Planetary Emergency "Climatic Changes".

The second group of candidates for the National Scholarship Programme is under way.

In order to practically start the National Scholarship Programme, we had to apply (through the Ministry of Science and Education) to the Government of the Republic of Lithuania with request to include the WFS and ICSC-WL in the list of the international non-governmental organisations, whose scholarships are exempt from taxes. We hope very much for a positive answer (the sixteenth PE "Bureaucracy").

We are also taking measures to increase the number of the members of the World Federation of Scientists in Lithuania.

EXTENDING THE ACTIVITIES OF THE WORLD FEDERATION OF SCIENTISTS IN UKRAINE

DR. GENNADY I. PALSHIN
ICSC World Laboratory Branch Ukraine, Kiev, Ukraine

The Ukraine's research potential today is represented by 120 institutions of the National Academy of Science, 100 institutions of the Academy of Agrarian Science, 150 universities and other higher education institutions, and 600 research organizations (of Industry, Medicine etc.) nationwide. The overall staff employed (both scientific and technical) total 180,000 people, while in 1991 this figure ran to 300,000.

Over the last 2 or 3 years the group of WFS has made a significant effort to promote the Federation's activities in the Ukraine.

The objective of that effort was to demonstrate the Global Emergencies as they manifest themselves in the Ukrainian context, as well as to engage the leading Ukrainian scientists in their study.

The WFS office in Ukraine has gone through a series of meetings and discussions with officials of the National Academy of Science and Ministry of Education and Science as well as regional conferences held in the 25 provinces of Ukraine, where 2000 delegates took part.

With regard to the Planetary Emergencies, in the Energy sector, our task was to analyze the state of the power production industry in Ukraine, and to work out the concept of its further development in Ukraine until the year 2050.

Professors Baryakhtar, Kukhar, Shydlovsky and other leading researchers took part in the elaboration of the above-mentioned concept.

Results of this work have been discussed in scientific publications and conferences at both national and international level.

CHERNOBYL

The scientists, who are in the WFS in Ukraine, participated in the development of the Shelter Implementation Plan (SIP). Currently, they are involved in its accomplishment. The objective of the said plan is to build up a new "Shelter" containment structure preventing accidental release of radioactive matter.

Study of the pollution-related emergencies has been focused mainly on the Black Sea pollution problems.

In cooperation with the Ukrainian Research Institute of Sea Ecology lead by Professor Mikhailov, a number of project proposals, namely, dealing with pollution monitoring, development of water quality standards and assessment of the oil pollution risks were elaborated with and submitted for consideration by the WFS Pollution PMP (chaired by Dr. Richard Ragaini).

FOODS-ASSOCIATED EMERGENCIES

Although Ukraine is well known to be rich in black soil, the agricultural production here remains inefficient. Restructuring of the agricultural sphere under way in Ukraine has taken a rather bureaucratic turn. The collective farm system has been merely proclaimed abolished, and private farms have been announced to replace them. Practically nothing has been done to support that newly emerging private farm structure.

We have designed the advanced information system for the agricultural sector, which comprises data banks, computer-aided atlases, computer programs for farm budget calculation and farm management improvement etc. The above system will serve as a basis for the development of the nation-wide consulting service for private farmers, known in western countries as the Extension.

KANGAROO MOTHER CARE METHOD

Ukraine is in pressing need of this medical care method today. The number of prematurely born low weight children has increased dramatically over last 10 years—from 8 to 20%. Taking into account the lack of incubators and other expensive equipment, the KMC method has proved its efficiency in Ukrainian-like conditions.

Professionals from Ukraine were trained at the KMC WL Center in Colombia. Today, the said method is being adapted successfully at the Institute of Pediatrics in Kiev. Also, 6 affiliated centers are scheduled for establishment in other Ukrainian regions.

DEMILITARIZATION

Two international workshops on the environmentally safe demilitarization of conventional munitions and missile weapons were held as a part of our activities aimed at involvement of Ukrainian scientists in the solution of Planetary Emergencies problems.

Many WFS scientists participated in the above-mentioned projects. I would like to thank especially Prof. Schubert, Professor Martelucci, Dr. Ragaini, Professor Ortalli, Dr. Goebel and Dr. Manoli for their important contribution to these works.

PERMANENT MONITORING PANEL REPORT: LIMITS OF DEVELOPMENT/ SUSTAINABILITY

HILTMAR SCHUBERT
Franhofer Institut für Chemishe Technologie, Pfinztal, Germany

The PMP was dealing with two items:

- Megacities-Water as a Limit of Development
- Sustainability of Rural Regions, Focus Africa

MEGACITIES-WATER AS A LIMIT OF DEVELOPMENT

The workshop *"Megacities–Water as a Limit of Development"* took place on August 18 and 19, 2000. Scientific Coordinator was Prof. Dr. Geraldo G. Serra, University of Sao Paulo.

After an introduction by Prof. Dr. Hiltmar Schubert, who mentioned sense and aims of the workshop and the results of the last meeting in 1999 of the working group, he gave a short report of the World Conference on Urban Future "URBAN 21" in Berlin, July 4-6, 2000 opened by the United Nation Organization Secretary-General Kofi Annan.

The conference was organized by the "Global Initiative for a Sustainable Development of Urban Policies" (Brazil, Germany, Singapore and South Africa).

In Table 1 megacities which may have a population of over 10 Million in the year 2015, are mentioned.

After this introduction the working group was informed by three invited experts about general problems of water:

1. William J. Cosgrove, World Water Council
2. Prof. Soroosh Sorooshian, Arizona University
3. Prof. Paolo Ricci, Lawrence Livermore National Laboratory.

After discussions of these general presentations speakers of the following megacities reported the situation and problems concerning water issues:

Buenos Aires, Mexico City, Sao Paulo, New Delhi, Cairo, Texas Triangle.

After the discussion of the water situation of these megacities the working group came to the following conclusion:

- Water is an important item for the development of megacities, but will not be a limit, provided a qualified water management, modern technology and appropriate financial means are available.
- All megacities have different situations.
- Another workshop should be organized on water problems.

FUTURE ACTIVITIES OF THE WORKING GROUP

- Creating ideas and solutions with high efficiency and low cost for: Water Supply and Waste-Water Treatment.
- Waste Management and Public Transport as the next most severe problems.

Knowing that more than 50% of the global population are living in urban areas, the motivation for our activities should be to create ideas and innovations to turn the pessimistic prediction:

"The Dark Future of Megacities" into
"Megacities: The Sunny-Side of Living for our next Generations".

SUSTAINABILITY OF RURAL REGIONS, FOCUS AFRICA

The workshop *"Sustainability of Rural Regions, Focus Africa"* took place on August 19, 2000. The Scientific Coordinator was Prof. Margaret Petersen, Arizona University.

After a short introduction by Dr. Hiltmar Schubert on the workshop "Regional Forum Africa" at the occasion of the "Global Conference on Urban Future, URBAN 21" in Berlin, July 6, 2000, this workshop was opened by the South-African Minister Mrs. Mahanvele, and she referred to the past conference in Pretoria.

Professor Petersen reported in a short overview about the "HIV Situation in Africa" and the constraints of the population poverty caused by Aids.

Content of the workshop was discussions and reports about the "Impact of Economic- and Food situation in Sub-Saharian Africa" by HIV/Aids (C.A. Reynolds), the "Rural-Urban Migration" in Uganda (Frances Kuka) and "Rural Development" in Senegal (Colonel Diop). The last contribution dealt with the impact on "African Economic Development of Orphans by Aids" (Margaret Farah).

CONCLUSION

- The HIV/Aids problem in the Sub-Saharian Africa superimpose all efforts to improve the economic situation in those countries.
- The migration of the rural population in urban regions is the key factor for the sustainability of rural regions.

- Development of small-size enterprises, better supply of electricity, water and public traffic in rural regions.
- Draft for a proposal for a pilot research project to investigate the efficiency of training social workers for Aids-orphans to become household heads in Uganda.

FUTURE ACTIVITIES OF THE WORKING GROUP

- Mutual effects of urban and rural regions. African cities are the motor for rural regions.

Urban Future 21

Table 1: Mega-cities 1995 and 2015

Urban Agglomeration	**Population (thousands)**		**Annual Growth Rate %**	
	1995	**2015**	**1985–1995**	**2005-2015**
Africa				
Lagos	10,287	24,437	5.68	3.61
Cairo	9,656	14,494	2.28	1.97
Asia				
Tokyo	26,836	28,701	1.40	0.10
Bombay	15,093	27,373	4.22	2.55
Shanghai	15,082	23,382	1.96	1.85
Jakarta	11,500	21,170	4.35	2.34
Karachi	9,863	20,616	4.43	3.42
Beijing	12,362	19,423	2.33	1.89
Dacca	7,832	18,964	5.74	3.81
Calcutta	11,673	17,621	1.67	2.33
Delhi	9,882	17,553	3.80	2.58
Tianjin	10,687	16,998	2.73	1.91
Metro Manila	9,280	14,711	2.98	1.75
Seoul	11,641	13,139	1.98	0.32
Istanbul	9,316	12,345	3.68	1.45
Lahore	5,085	10,767	3.84	3.55
Hyderabad	5,343	10,663	5.17	2.83
Osaka	10,601	10,601	0.24	-
Bangkok	6,566	10,557	2.19	2.51
Teheran	6,830	10,211	1.62	2.30
South America				
Sao Paulo	16,417	20,783	2.01	0.88
Mexico City	15,643	18,786	0.8	0.83
Buenos Aires	10,990	12,376	0.68	0.50
Rio de Janeiro	9,888	11,554	0.77	0.84
Lima	7,452	10,562	3.30	1.32
North America				
New York	16,329	17,636	0.31	0.39
Los Angeles	12,410	14,274	1.72	0.46

Source: UNCHS (1996b), pp. 451-456.

NUCLEAR POWER PLANTS IN THE NEXT CENTURY

JURAS POZELA
ICSC-WORLDLAB, Semiconductor Physics Institute, Vilnius, Lithuania

INTRODUCTION

A great increase of Nuclear Energy (NE) production in the World energy consumption has been predicted for the 21st century. The increase of the NE part in the electricity production was marked in the last decade of the 20th century. However, at the end of the 20th century, great opposition to developing Nuclear Power Plants (NPP) arised. This opposition is for two main reasons: (1) the worry about radiation escaping from a nuclear power station, as happened with Chernobyl; (2) the discovery of great reserves of natural gas and oil which guarantee the high level of energy consumption in Europe without using NE to the end of the 21st century.

As a result of this opposition, the construction of new NPPs has stopped and increased use of fossil energy is proposed in many countries. Moreover, in Lithuania, the first unit of the largest NPP in Europe, Ignalina, is closing now.

The dangerous consequences of this trend for human health and ecological stability is considered in this report. The consequences for health of the radioactive fallout due to the Chernobyl accident are compared with the more dangerous consequences of air pollution due to the burning of fossil fuel for energy production. The consequences of expanding electricity production by building new fossil fuel power plants in Europe are considered.

HEALTH EFFECTS OF THE CHERNOBYL ACCIDENT

The Chernobyl accident has brought to the front line, the health consequences of nuclear accidents. The health effects of radiation have been carefully studied and determined by the United Nations Scientific Committee on the Effects of Atomic Radiation (UNSCEAR) and by the World Health Organization (WHO). The resulting data is in common use in all cases concerning the effects of radio active radiation.

The health effects of the Chernobyl accident have been studied and determined by the International Advisory Committee, which in 1991 published its findings in a report: The International Chernobyl Project. In 1995, WHO published the report. Health consequences of the Chernobyl accident, describing results of several thorough studies, some of which are continuing. The Nuclear Energy Agency (NEA) of OECD has published a report: *Chernobyl Ten Years On Radiological and Health Impact*. In the paper below, the data used on the health effects of the Chernobyl accident are taken from the report[1].

The only clearly detected health consequence caused by radiation of the Chernobyl accident is the childhood thyroid cancer. In the worst contaminated areas, 15-35 cases occur per year per one million children.

Hereditary effects have not been detected among the survivors of atomic bombs in Japan. The only detected late health effect has been an increase of cancer deaths. Atomic bombs did not produce radioactive iodine and hence have had no impact on thyroid cancer.

Many diseases other than cancer have been claimed to be caused by radiation from the Chernobyl accident. However, according to present knowledge, these diseases are psychological effects not related to radiation.

The radiation due to the accident will cause extra cancer deaths corresponding to the doses people have received, but the number of deaths is too small to be detected, because normal fluctuations in cancer death rates are bigger. The amount of deaths can only be estimated using the known radiation doses.

According to UNSCEAR and WHO, the probability of cancer death due to radiation is directly related to doses between 0-1000 mSv (mSv, millisievert is a measure for radiation exposure). If an individual gets a radiation dose of 200 mSv there is a probability of one percent that he will die of cancer in the following 50-70 years. For other doses, the probability is related to the dose rate.

According to the International Chernobyl Project, individual life-time radiation doses in the contaminated areas in Ukraine, Belarus and Russia due to the accident vary from low values to several hundred mSv. On the other hand, life-time radiation doses for every individual due to natural radiation vary from about 200 mSv to a few hundred mSv.

Hence, natural radiation doses are of the same order of magnitude as doses from the accident. Outside the contaminated areas of Ukraine, Belarus and Russia, natural radiation is much stronger that the radiation due to the accident. For example, in Finland, the life-time dose from natural radiation is in average 260 mSv. The dose from the Chernobyl accident will be 1-2 mSv in spite of the relatively heavy fallout.

The NEA report in 1995 estimates that during the first four years after the accident people living in the strict control zones have received a dose varying between 5-250 mSv. 1,800 persons have received a dose exceeding 150 mSv. The average dose received by all people (270,000) living in strict control zones was 36 mSv.

The life-time radiation dose can be expected to be at its maximum 50 percent higher than the dose during the first four years. Hence the average life-time dose of 270,000 people still living in the strict control zones will be less than 60 mSv. The maximum dose is around 400 mSv.

The dose of 400 mSv causes a probability of two percent and the dose of 60 mSv a probability of 0.3 percent of dying of cancer in the future.

The 60 mSv average radiation dose in the strict control zones of Ukraine, Belarus and Russia causes, according to the UNSCEAR data, a probability of 0.3 percent of dying from cancer in the future. Hence among 270,000 people in the strict control zones, about 800 extra deaths can be expected for 50 years due to the Chernobyl fallout. At the same time, about 2,500 cancer deaths are caused by natural radiation and the number of cancer deaths due to other reasons is about 50,000. In European countries, cancer deaths due to other reasons are 20–25 percent (see Table 1).

Table 1. Yearly death rates.

All death	100	percent
Deaths of cancer in Europe	12-25	percent

<u>Additions to yearly death rates due to different environmental reasons</u>
(in the case of radiation, the addition is caused by additional cancer, in the case of fine particles pollution by additional cancer and cardiovascular diseases)

Fine particles pollution from burning fossil fuels	3-9	percent
Natural background radiation in Europe	2	percent
Natural background radiation in Finland	3	percent
Chernobyl fallout, strict control zones	0.5	percent
Chernobyl fallout, Finland	0.03	percent

Numbers are based on the data from WHO, UNSCEAR, NEA and on the report by Dockery et al., The New England J. of Medicine, Dec. 9 1993. See also WHO European office report: Update and revision of the air quality guidelines for Europe, 1995.

In countries outside the former Soviet Union the increase of the death rate due to the Chernobyl accident in negligible. For example, in Finland, 1-2 mSv dose from the Chernobyl accident increases the death rate of about 0.03 percent. In comparison, the increase of the death rate due to natural background radiation is 3 percent. It is unlikely that surveillance of the general population will reveal any significant increase of leukemia or other cancers.

One can say that health consequences caused by radiation of the great Chernobyl catastrophe are not detected outside of a small local area near the disturbed unit of the Chernobyl NPP.

HEALTH EFFECTS OF FINE PARTICLES POLLUTION CAUSED BY BURNING OF FOSSIL FUEL IN POWER STATIONS

The growth of electricity production by building new fossil fuel power plants increases air pollution. In addition to carbon dioxide emissions and acid rain pollution, fine particles pollution has increased.

The burning of fossil fuel causes many health problems: some of the air pollutants cause local problems, but very small particles, micron size, can be carried thousands of kilometers.

Lately it has been found in studies in the USA, that fine particles pollution caused by burning of fossil fuel in power stations, buses and cars increases the death rate between 3-10 percent. This indicates that fine particles pollution is about ten times more dangerous than the radioactive fallout in the strict control zones in Ukraine, Belorus and Russia (see Table 1).

It may be expected that cleaner sources of energy, for example NE must be introduced in developed and developing countries to save lives.

Nuclear power has a great advantage for health in that it does not cause air pollution.

INCREASE OF ENERGY PRODUCTION BY POWER STATIONS BASED ON BURNING OF FOSSIL FUEL DESTROYS THE ECOLOGICAL STABILITY IN THE WORLD

At present 75% of the world's energy is consumed by the industrial countries and only 25% by the developing countries.

For developing countries to develop, they need more energy. The living standards can be expressed in terms of energy use per capita. The energy consumption per capita in oil equivalent in the USA, West Europe, Lithuania and developing countries is shown in Table 2.

Table 2. Per capita energy consumption in oil equivalent (1996).

Country	toe/cap	Ratio WE/Country
USA	9.16	0.45
WE	4.12	1
Lithuania (1991)	*2.54*	*1.6*
Lithuania (1996)	*1.26*	*3.3*
China	0.71	5.8
India	0.26	15.8
Bangladesh	0.075	50.9

It may be seen that if developing countries were to obtain a standard of living and an energy consumption as that of Western European countries, then they would have to increase their average energy consumption per capita by about 20 times[2].

It is assumed that the total energy consumption per capita will be almost constant over the next 100 years for the affluent states.

The developing countries use the fossil fuel for energy production because discovered reserves of natural gas, oil and coal are large and the building of new fossil fuel power plants is cheaper compared with NPPs.

However, this method of energy production is admissible to these countries until they achieve the present-day level of the used fossil fuel for the energy production per capita in industrial countries.

The energy production by burning fossil fuel in industrial countries causes many types of ecological pollution (air pollution, fine particles, acid rains, green house effect). The achieved pollution level can be considered as a threshold one for ecological stability in these countries. This threshold determines the admissible level to use burning fossil fuel per capita for developing countries. The exceeding of this level by developing or developed countries signifies a planetary ecological catastrophe.

This threshold level of using fossil fuel in industrial countries has to be decreased to prevent an ecological catastrophe in the world.

In order to prevent an ecological catastrophe in the world, the developed countries in Kyoto resolved to restrict and reduce the burning of fossil fuel in 2010 to the 1990 level.

This demand was fulfilled in Europe mainly by increasing the nuclear energy part in the total energy production (Tables 3 and 4).

Because of the limits for using fossil energy for electricity production, developing countries will have to develop, nevertheless, NPPs. There is no other alternative for increasing the energy production in these countries.

Seeking to prevent a planetary ecological catastrophe, the 21st century will have to be the Nuclear Energy Century.

Table 3. The increase of NE consumption in 1998, compared with 1990 in European countries.

Country	1998/1990 (%)
France	+ 23.8
Germany	+ 6.1
United Kingdom	+ 52.0
Sweden	+ 8.6
Bulgaria	+ 31.6
Czech Republic	+ 6.1

Table 4. The electricity production by NPP in 1998.

Country	%
Lithuania	81
France	78
Belgium	60
Ukraine	47
Sweden	46
Switzerland	40
Germany	36
Japan	35
United Kingdom	27
USA	20
Russia	13

CLOSING OF THE IGNALINA NPP IS THE WAY TO ECOLOGICAL INSTABILITY IN THE CENTRE OF EUROPE

In Lithuania in 1997, the energy consumption per capita (1.3 toe/cap) was three times less than in Western Europe (4.2 toe/cap). In 1990, the energy consumption per capita was 2.4 toe/cap (1.7 times less than in Western Europe). The Ignalina NPP produces more than 80% of the electricity in Lithuania.

The closing of the Ignalina NPP means an increase of the electricity production by burning of fossil fuel in power stations. This method contradicts the Kyoto declaration for European countries and increases the danger for the Lithuanian people's health because of the increase of fine particles pollution. The increase of energy production by burning of fossil fuel is the way to ecological destruction in the centre of Europe.

The closing of the Ignalina NPP does not increase the safety of the Plant because all active nuclear elements will be left in the Plant territory.

The closing of the Ignalina NPP means the loss of the specialists for NPP exploitation and one hundred percent unemployment in Visaginas town where the NPP is found.

In the not distant future, when Lithuania attains the European economic level, the total electricity production will have to increase twice compared to 1990. It has to be attained without increasing the burning of fossil fuel 1990 level. It can be done only by increasing the energy production by NPPs.

This will require rebuilding the closed NPP and the town. This is a very expensive choice for a small country.

CONCLUSIONS

1. The health consequences caused by the burning of fossil fuel in power plants exceeds that of the Chernobyl accident. NPPs are not so dangerous for health compared to the power plants using the burning of fossil fuel.

2. The increase in electricity production by building new fossil fuel power plants in Europe opens the way to an ecological catastrophe in the world.
3. The closing of the NPP in the European state of Lithuania is ecologically dangerous for Europe.
4. There are no alternatives for developing NPPs in all developed and developing countries of the World in the 21st century.

REFERENCES

1. Mr. Tiuri (Finland). Draft report for opinion on the health effect of the Chernobyl accident. Council of Europe, Strasbourg, 4 October 1996.
2. Douglas R.O. Morrison. Draft report for the World energy and climate in the next century. Erice, 1999.

HIV/MOTHER TO CHILD TRANSMISSION

Joint Report of AIDS/Infectious Diseases PMP and Mother and Child PMP

INTRODUCTION

Mother to child transmission of HIV is an urgent planetary emergency and a world wide human tragedy. At least 90% of all HIV infections in children are a result of mother to child transmission. It has been estimated that 4.5 million infants have been infected since the beginning of the epidemic, 3 million already died and that 600,000 new infections occur annually. We know that 90% of these infections are currently occurring in sub-Saharan Africa and that the HIV epidemic is reaching Asia and India with the largest populations in the world. HIV/AIDS has already substantially reversed the gains in infant and childhood mortality in many countries in sub-Saharan Africa. In resource-poor countries, without access to antiretroviral drugs, seroprevalence rates of HIV infection in pregnant women are documented to be as high as 40%, and transmission rates from mothers to their children range from 25-40%. In these countries almost all infected babies die in the first several years of life, the infected parents die, and the uninfected children are orphans.

In populations in the developed world with access to highly active antiretroviral drugs, mother to child transmission has been reduced to a rate of approximately 1%, and it is theoretically possible to eliminate it. We can effectively diminish MCT~HIV in the resource-poor countries.

As scientists, we are deeply concerned that we share the responsibility of promulgating the results of the research and advocate for accessibility of effective antiretroviral agents to decrease mother to child transmission of HIV. At the present *time, the most practical effective and safe antiretroviral intervention is nevirapine one dose to the mother at the time of delivery and one dose to the newborn.* It is becoming *accessible and it is affordable to the* resource poor nations with the greatest burden of infected people. There are other antiretroviral drugs that have been proven to reduce transmission with more complex regimens, which have not been widely accessible.

The intervention must be administered within the existing MCH care system. This includes personnel such as traditional birth attendants, district health workers, nurse midwives or physicians. This intervention may occur in the home, clinic, hospital or other similar locations.

MATERNAL CONSENT FOR THE INTERVENTION IS ESSENTIAL

The intervention should be associated with provision of appropriate information and education of the mother, When possible, care for the mother such as multivitamins, iron, iodine, treatment of STDs, prophylaxis and treatment of TB and of other opportunistic infections can be simultaneously provided. The availability of serological rapid tests should be encouraged for the diagnosis of HIV infection.

In high HIV seroprevalence areas, the drug intervention should be proposed to all seropositive pregnant women, to those who refuse testing, and possibly, to those who lack access to testing. In low seroprevalence areas, the intervention should be proposed to seropositive women.

Newborns should have their care continued in the usual maternal child health care setting, which includes immunization according to national program. Mothers should have their care continued with specific attention to the consequences of being HIV positive. This may include delay of the next pregnancy, prophylaxis and treatment of TB as well as opportunistic infections. Access to appropriate Anti Retroviral Therapy, should be encouraged where adequate infrastructure is available.

Primary prevention strategies must be continued and reinforced for men and women. There needs to be a specific focus on adolescents, males and females. Primary prevention has to be done with the collaboration of the community village, faith-based institutions, and local NGOs. The use of female and male condoms is still the best way not to be infected with HIV.

BREASTFEEDING: A CONTROVERSIAL ISSUE

The primary concern is safe and effective nutrition for all HIV infected infants. Breastfeeding is essential up to age 3-4 months for infant survival in many countries. Mother-infant bonding and infant development are enhanced by the early suckling and breastfeeding of the infant. It has been proven that there is a risk of acquisition of HIV from breastfeeding. The risk of acquisition is cumulative with time and there is an observational study suggesting that mixed feeding enhances the acquisition of HIV. The risk of mortality associated with alternative feeding practices decreases markedly after 6 months of age.

The local cultural traditions must be respected but availability of a safe alternative feeding should be assessed by the care provider. If a safe alternative is accessible, it should be recommended. This substitute could be prepared in the country of residence, must be microbiologically safe and could be administered by a cup and spoon.

If alternative feeding is not available, every effort should be made to obtain exclusive breastfeeding by discouraging all kinds of complementary feedings.

Early accelerated weaning at an age of 6 months may be encouraged as decreasing the period of exposure to breastfeeding may diminish the risk of acquisition of virus.

In some locations there are some established human breast milk banks using pasteurized breast milk. This requires appropriate collection, pasteurization, and safe transport and storage.

RESEARCH PRIORITIES

We must continue to increase our knowledge about how to diminish the transmission of HIV from breast milk. This includes further investigation of the impact of duration of exclusive breastfeeding upon the transmission of HIV. We need to further understand the pathogenesis of transmission of HIV by breast milk. Early weaning or administration of antiretroviral agents to the baby during the period of breastfeeding should be evaluated as potential means of decreasing HIV transmission. Understanding of cultural beliefs and traditions is essential for effective implementation of alternative feeding practices. Furthermore, feasibility studies of implementing human milk banks, and pasteurization of mother milk at home, using water bath, should be implemented in different geographical areas and cultural contexts.

There is still an urgent need to develop further strategies of antiretroviral and other interventions to decrease mother to child transmission. There is also a need to assess the efficacy and safety longitudinally in mothers and infants exposed to the interventions.

With regard to future intervention strategies such as therapeutic vaccine, basic research has to be encouraged to evaluate potential consequences of disrupting physiological immune balance in compartments (for example placenta and breast milk) on both HIV transmission and gestation.

CONCLUSION

We must help to extend the availability of effective interventions. Implementation of nevirapine and maintenance of the established beneficial effects of breastfeeding will help to address this planetary emergency.

14. MEGACITIES WORKSHOP — WATER AS A LIMIT TO DEVELOPMENT

WATER USE, ABUSE AND WASTE: LIMITS TO SUSTAINABLE DEVELOPMENT IN THE METROPOLITAN AREA OF MEXICO CITY

ALBERTO GONZÁLEZ POZO
Universidad Autónoma Metropolitana –Azcapotzalco, Mexico
VICTOR CASTAÑEDA
Urban Planner, Expert in Water Supply and Sewage Systems, Mexico

A MEGACITY WITH STRONG ANCIENT TIES TO WATER

In the beginning of the third millenium, the Metropolitan Area of Mexico City faces increasing problems of water availability and disposal proportional to its huge size. It is not only a technological or financial question of supply related to the increasing demand, but rather a balance that has to be sought between needs, resources, technology and social participation concerning sustainable development. A brief resume of its historic evolution is needed in order to understand some present aspects that relate to past experiences.

The Valley of Mexico has been the stage of important cultures and human settlements since 1,500 BC. Between 100 BC and 600 AD, a big city, Teotihuacan, inhabited by 100,000 to 200,000 people flourished covering 22 sq. Km. This precolumbian metropolis was next to the shore of the ancient Lake of Mexico, located in the middle of the basin. There is archaeological evidence of the water infrastructure systems used there, based upon cisterns and pools that re-collected water from springs, and rainwater from roofs, patios and squares[1], as well as superficial drains and ditches to evacuate to the lake rainwater excess. Teotihuacan declined and was abandoned ca. 800 AD for unknown reasons. (Was the scarcity of drinking water one of them?) Later on, in 1325, the Aztecs founded their own capital, México-Tenochtitlan, as an artificial island in the middle of the lake. It was smaller than Teotihuacan but nevertheless had some 100,000 inhabitants living in "chinampas", or floating gardens The water supply was provided by acqueducts coming from springs near the southern and western lakeshores. Very soon, the Aztecs became very good hydraulic engineers, because the fluctuations in the water level of the lake often threathened their city. They had to develop a complicated system of channels, wooden and earthen dikes, bridges and other devices in order to control the water of the lake. Nevertheless, the city was flooded in 1449 and 1498[2].

When the Spaniards destroyed Tenochtitlan in 1521, they decided to build a new settlement over their ruins: Mexico City. Its size was much smaller than that of the Aztec capital, and its settlement pattern was not based upon floating gardens but on dry ground.

They kept the same water sources used by the Aztecs, improved the acqueducts and consolidated the new city. In few decades Mexico City faced flooding episodes again (1555, 1580 and 1604), and in 1629, after a big one, it was almost abandoned. But the search for an artificial drain for the whole valley, at the northern hills, had already started in 1607 and continued until 1788 when a draining gorge, 12 kilometers long, was successfully finished[3]. At the end of this process, the original big lake of Mexico Valley was fragmented in five minor lakes (Zumpango, San Cristóbal, Texcoco, Xochimilco and Chalco).

Mexico became an independent nation in 1821, and during most of the XIX Century its capital didn´t grow, nor its hydraulic infrastructure. But between 1890 and 1910 important works were succesfully built: a new sewer tunnel next to the former gorge, 10 kilometers long; the major sewer ditch for Mexico City, 47 kilometers long; and a new water supply system, capturing water of springs in the southeastern part of the valley and delivering it to the city through a piped acqueduct 40 kilometers long. Two of the five lakes of the previous century almost disappeared, and the other three shrunk dramatically. At the end of this period, piped systems for water distribution and sewage were gradually built inside the urban area inhabited by ca.400,000 people.

The next two decades of civil war, from 1910 to 1930, started to attract new inhabitants coming from the countryside to Mexico City. The incipient industrialization process that begun in the thirties, triggered a population growth rate that rose approximately from 3% to 4% yearly. As a consequence, Mexico City grew from less than half a million inhabitants in 1910 to 1 million in 1930 and 1.5 millions in 1940. In the next four decades, the urban growth accelerated, turning the spread of urban areas outside the Federal District into a growing number of Municipalities of the neighbouring Federal State of Mexico[4]. The population of the Metropolitan Area of the Valley of Mexico grew continuously to 3.1 million in 1950, 5.3 million in 1960, 9.2 million in 1970, and 14.4 million in 1980[5]. In the last two decades, the rate of growth in the Federal District has been rather small, leaving the Metropolitan Municipalities the most important part of the urbanization process. Thus, the size of the Mexican Megacity increased to 15.5 million people in 1990 and reached 18 million this year.

As a result, its hydraulic infrastructure has become a huge maze of different systems and subsystems built during this century that allow the supply of water and the sewage of waste-and rainwater. The most important were: the starting point of tapping water from the underground through numerous public wells in 1940; a new acqueduct bringing water from springs and wells at Lerma, outside the Valley of Mexico, between 1947 and 1951; a second drainage tunnel in 1950; and the so-called Deep Drainage System (Sistema de Drenaje Profundo, described in the next pages) that condemned the small lakes still surviving to extinction from 1960 to 1975.

But, as stated in the 24th Session of the International Seminars on Planetary Emergencies held in Erice in August 1999, the size of the Mexican Megacity is greater if we consider not only the Metropolitan Zone of the Valley of Mexico (MZVM) but that of the Megalopolis of Central Mexico as well, a region of ca. 30,000 sq. Km. (8% urbanized), involving the Federal District and 189 municipalities in 5 Federal States around it[6]. The size and problems of the corresponding water supply and sewage infrastructures are proportional to this huge conglomeration.

GEOGRAPHIC AND ENVIRONMENTAL CONTRADICTIONS

Part of the problems to solve arise from unusual environmental features that must be underlined:

- Mexico (like Peru, Ecuador and Bolivia) is a country of highlanders. 70% of its 97,000,000 inhabitants (that is, 63,000,000 people) live in settlements located at 500 meters above sea level or higher, but most of the water resources available are precisely located in the lower, coastal zones.
- This basic contradiction is greater in the Valley of Mexico, home for 18,000,000 people. It is a closed basin of ca. 9,500 square kilometers at 2,240 meters above the sea level. Although there shouldn't be any problems with the provision of water in this area with a mean rainfall of 700 mm/year and a temperate subhumid climate (Cw Koeppen), there are many indeed. One third of the sources of drinking water are situated at distances between 50 and 175 kilometers from the urban area, at 1,00 to 1,400 meters below the floor of the Valley. Thus, important extra investments and energy are required to bring the liquid upwards to the Megacity.
- The huge urbanized areas needed for a population of 18 million people in the Valley of Mexico cover ca 1,400 sq. kilometers. These areas, formerly used for agricultural purposes, played an important role in the process of rainwater percolation to the underground water table. Now, mostly covered by buildings and impervious pavements, they contribute to worsen the shortage in the replenishment of the aquiferous layer.
- The tendency to drain the lakes within the valley is another contradiction, since part of the rainwater that flows to the sewage as well as the total used water with almost no treatment, contribute to deplete water resources from the metropolitan region and causes water pollution in the Moctezuma-Panuco River that flows towards the Gulf of Mexico.

SUPPLYING, TREATING AND DISTRIBUTING WATER TO A THIRSTY AND WASTEFUL MEGACITY

Freshwater resources for the Mexican megacity represent a total flow of 67.5 cubic meters per second, 35% from external resources harnessed in six different neighbouring basins (Lerma, Amacuzac, Cutzamala, Valle Oriental, Queretaro and Tlaxcala) within a radius of 127 kilometers around Mexico City and 65% provided by internal resources mostly consisting of underground water coming from springs and deep wells within the Valley of Mexico. But this pattern raises serious questions about sustainability of both external and internal resources. For instance, the daily tapping of groundwater in the Valley of Mexico causes the soft underground to shrink, resulting in a slow but continuous sinking of the floor of the Valley at rates of 10 to 40 centimeters yearly, with a mean of 1.50 meters per decade. This phenomenon affects 40% of the urbanized area that was formerly part of the lake, and endangers the stability of many buildings subject to differential sinking, as well as other negative effects upon the sewage system discussed below.

The total volume of drinking water is unequally distributed: 60% to the Federal District (with 8 million inhabitants) and only 40% to 28 municipalities of the neighbouring Federal State of Mexico (10 million inhabitants). The water is used for the following purposes:

Table 1. Metropolitan Area of Mexico City: Water Uses in The Federal District and Metropolitan Municipalities.

Federal District			***Metropolitan Municipalities***			***Total***	
Uses	*Cubic meters per second*	*%*	***Uses***	*Cubic meters per second*	*%*	*Cubic meters per second*	*%*
Domestic	23.72	67	Domestic	18.88	80.0	42.60	72.2
Industrial	6.02	17	Industrial	3.54	15.0	9.56	16.2
Services & commerce	5.66	16	Services & commerce	1.18	5.0	6.84	11.6
Total	35.40	100	Total	23.60	100.0	59.0	100.0

Source: Castañeda, Victor, "Gestión integral de los recursos hidráulicos" in Eibenschutz Hartmann, Roberto (Coord.), *Bases para la Planeación del Desarrollo Urbano en la Ciudad de México. Tomo II. Estructura de la ciudad y su región,* Universidad Autónoma Metropolitana – Xochimilco and Miguel Angel Porrua, Grupo Editorial, Mexico, 1997. p. 80. Based on official data from the Federal District Government.

The detailed statistics and data available for the Federal District show two important issues: first, the average amount of water for domestic purposes only reaches 279.9 liters per capita per day, (high for a developing country, but below the official recognition of 290 liters per inhabitant daily); and second, the estimates of losses caused by leaks in the water distribution infrastructure going from 10% and 30% of that volume, give a real amount from 173.5 and 223.1 liters per capita per day, respectively.

Table 2. Consumption of Water in Liters Per Capita Per Day in Selected World Cities.

City / Year	*Water consumption: liters per capita per day*
Panama, Panama 1991	778.0
La Habana, Cuba 1991	567.4
Moscow, Russian Federation, 1989	461.0
Kiev, Ukraine, 1991	440.0
Budapest, Hungary, 1990	436.5
Helsinki, Finland, 1990	321.0
Bangkok, Thailand, 1989	309.1
Warszawa, Poland, 1990	263.9
Wien, Austria, 1990	236.5
Tegucigalpa, Honduras, 1988	153.7
Lima, Peru, 1991	147.7
Jakarta, Indonesia, 1988	32.5

Source: UNITED NATIONS CENTRE FOR HUMAN SETTLEMENTS (HABITAT), *Compendium of Human Settlements Statistics 1995,* United Nations, New York, 1995. pp. 472-475

Table 3. Average Water Provision Per Capita Per Day and Estimated Losses in the Federal District, Mexico (1994).

Criteria	*Average provision: liters per capita per day*	*-30% of losses in the flow, by leaks: liters per capita per day*	*-10% of losses in the flow, by leaks: liters per capita per day*
All types of uses, including industrial	369.9	258.9	332.9
Domestic, commercial and services	307.0	214.9	276.3
Only domestic uses	247.9	173.5	223.1

Source: Castañeda, Victor, *op. cit. p. 82.*

The impact of such losses in the total flow is estimated between 3.5 and 10 cubic meters per second. The average is similar to the present industrial consumption (6.02 cubic meters per second), a volume capable of satisfying the domestic demand of 1,350,000 persons per day.

The infrastructure of water storage, treatment and distribution in the Federal District consists of ca. 500 kilometers of pipes conducting water to 279 storage tanks, 227 pumping stations, 16 full treatment plants, 360 chlorine treatment devices and more than 10,700 kilometers of pipes in the primary and secondary distribution networks.

The distribution of piped drinking water to private homes covers 71.5% of the dwellings in the Federal District. Another 24.8%, living in irregular and squatter settlements, get water in private or public standpipes outside their homes and the remaining 3.7% get it delivered by water trucks. The proportion of drinking water inside homes is lower in the Municipalities of the rest of the Metropolitan Area. The differences are geographic and partly related to levels of income, as shown in the following table showing water demand, suply and shortage in both the Federal District and the principal Metropolitan Municipalities:

Table 4. Supply and Demand of Drinking Water in the Metropolitan Area of Mexico City (1989).

Regions	*Demand (cubic meters per second)*	*Supply (cubic meters per second)*	*Shortage (cubic meters per second)*
Federal District			
North	8.15	7.14	1.01
West	4.80	4.78	0.02
Center	12.60	12.59	0.00
East	7.15	6.05	1.10
South	6.25	6.25	0.00
Subtotal	38.85	36.81	2.14
Metropolitan municipalities			
NZT	11.70	8.30	3.40
Cuautitlán	1.60	1.60	0.00
Coacalco	1.40	1.30	0.10
Ecatepec	7.10	4.50	2.60
Nezahualcoyotl	5.00	3.50	1.50
Chalco	0.80	0.30	0.50
Subtotal	27.60	19.50	8.10
Total	66.55	56.31	10.24

Source: Castañeda, Victor, *op. cit.* p. 77

A final commentary about distortions in the supply and demand of drinking water in the Metropolitan Zone of the Valley of Mexico links the great amount of leaks in the infrastructure (as said, between 10% and 30% of the volume available) with bad consumption habits by all types of users: domestic, industrial and services, thus giving a wasteful pattern of water consumption for the whole Megacity. Formal programs to rationalize the use of water start only from 1985 onwards, with partial successes like the substitution of old W.C's (with tanks of 20 liters each) by new ones (with tanks of only 6 liters each), as well as other measures to spare drinking water. But there is still a great job to do in this direction.

RAINWATER AND WASTEWATER DISPOSAL: THE WORST MIXTURE

The Metropolitan System of Sewage and Flood Control comprises 9,500 kilometers of primary sewage network (pipes of major diameters) and 1,260 kilometers of secondary network (pipes of 60 centimeters diameter or less), 79 pumping stations with a capacity of 630 cubic meters per second, several regulatory dams, 112 kilometers of sewage open ditches, 54 kilometers of piped rivers, and the "Deep Drainage Subsystem", consisting of 137 kilometers of tunnels (of 5 and 6.5 meters diameter) that run between 30 and 220 meters underground.

86% of homes in the Federal District are connected to the Sewage System; 8% are served by septic tanks and the remaining 6% has no sanitation at all. The situation in the Metropolitan Municipalities is even worse: 50% to 80% of homes integrated to the Metropolitan Zone several decades ago are connected to the Sewage System, but the percentage can still be lower for those incorporated more recently[7].

The volume managed by all these systems gives a capacity of 57 cubic meters per second (75% of wastewater of urban origin and 25% of rainwater excess drained in urban areas). Only 8.8% of that volume is treated for urban re-use, 5.9% drains to the Lake of Texcoco, and the rest, more than 85% goes out of the Valley where it is partially treated and used downstream for agricultural purposes.

The future of all these systems is threatened by the slow but relentless sinking of the valley floor caused by the high extraction of groundwater for drinking purposes described above. If the process continues, a point will be reached when the slope of the whole Sewage System will disappear or, even worse, invert itself. The flooding of 1607 would be nothing compared to such situation[8].

Centuries of technological solutions to discharge both waste water and excess of rainwater outside the Valley of Mexico are now under serious criticism. If large amounts of water are needed for the Megacity, why let them escape outside...? It is easy to ask this question, but the answer implies a careful evaluation of alternatives to solve a huge problem.

Nevertheless, Federal and Local Authorities have taken the first steps to restore the balance. Since 1965, there exists a "Plan Texcoco", partially realized in order to restore part of the ancient Lake of Texcoco (one of the five lakes of the valley a century ago). Once finished, the restored lake would receive, stabilize, treat and use wastewater and rainwater for agricultural and recreational purposes as well.

POLITICAL, ADMINISTRATIVE AND FINANCIAL ASPECTS OF WATER SUPPLY AND DISPOSAL.

The Metropolitan Area of Mexico City is politically divided among the Federal District and some 30 Municipalities in the neighbouring State of Mexico. There are task groups and programs of coordination between the two federal authorities, which include integrated plans for the water supply and sewage systems in the whole Metropolitan

Area. But a Metropolitan Authority is still missing to regard the problem as a whole. Most statistics and basic data (including costs and financing) are partial.

The daily cost of the water supply system is between 0.1581 and 0.1951 U.S. Dollars each cubic meter per second, totalizing between 197,047 and 230,771 U.S. Dollars. Considering the figures and estimates of losses shown in Table 3, the cost of water consumption per capita in the Federal District is as follows:

Table 5. Water Consumption Cost Per Capita in the Federal District (Includes all types of uses).

Unitary cost Dollars/m3	*Average provision 369.9 liters/capita/day*		*Consid. 30% of losses 285.9 liters/capita/day*		*Consid. 10% of losses 332.9 liters/capita/day*	
	Daily cost (Dollars)	*Monthly cost (Dls)*	*Daily cost (Dollars)*	*Monthly cost (Dls)*	*Daily cost (Dollars)*	*Monthly cost (Dls)*
0.1581	0.0244	0.7322	0.0171	0.5129	0.0220	0.6596
0.1871	0.0274	0.8225	0.0192	0.5758	0.0247	0.7403
0.1951	0.0286	0.8580	0.0200	0.6000	0.0257	0.7725

Source: Castañeda, *op. cit.*, p. 97

The cost of enlarging the flow with external sources will be higher. Just to harness additional 15 m^3/sec has a cost between 9,616 and 12,813 U.S. Dls daily.

The tariffs of drinking water are so low that they don't cover the operation costs, not only hindering the enlarging and modernization of the system, but the actions needed for the maintenance and rehabilitation of the infrastructure as well. The tariffs in the Federal District follow several criteria: for measured volumes or fixed prices distinguishing between domestic and non-domestic uses. But the whole system is subsidized, and the participation of private companies in the storage, treatment and distribution of drinking water is low.

BETWEEN GLOOMY AND OPTIMISTIC PROSPECTIVES

The future of water management in the Metropolitan Area of the Valley of Mexico depends on crucial political, economic, social and technological aspects to be solved. But above all, a comprehensive strategy is missing, in a vision that considers sustainable *urban development* and *urban technologies*. Urban development implies that timely choices must be made among competing urban technologies to achieve optimal welfare while paying attention to economic criteria (i.e., with explicit recognition of the relative scarcity of means and resources to be employed). Urban technologies have mainly physical, chemical and ecological character, while urban development, though always technologically rooted, deals with sociopolitical and humanistic issues. Thus the pursuit of welfare in urban settlements displays a dual -biophysical and cultural- nature[9]. Updating current practices by massive public investment and sticking to traditional hydraulic techonology with conventional design criteria, materials, devices, components and systems will not do. Neither budgets nor technological improvements are enough to meet the needs and demands. Current technologies are inappropriate, if not senseless, to guarantee Mexico City's viability as a megalopolitan settlement in the next century. The situation has been driven to a scenario of unsolved demands and growing vulnerability.

Some strategies that are now under consideration deserve the following commentaries:

- Stronger links between water resources planning and urban planning are needed, to avoid further urbanization on zones that allow the percolation of rainwater to the underground water sources. Strategic areas for building pools, dikes, dams, treatment plants and other hydrological controls must be secured in the land use plans.
- Increase, on a temporary basis, the capacity of water sources outside the valley and decrease the pumping of water from underground sources, to mitigate the sinking of the valley floor.
- In the last two years, a considerable effort has been made to fix hundreds of leaks in the water distribution system. Much more actions are needed in order to reduce the leakage to a minimum.
- Develop strong educational and participation programs related to the rational use of water for domestic- services- and industrial purposes and redesign the structure of tariffs under the principle of charging proportionally higher prices for profitable uses of water in industry, services and dwellings of high income sectors, and low prices for low-income domestic uses.
- A second stage of the plan to restore the Lake of Texcoco with more ambitious goals could be implemented. It covers an area of 10,000 Hectares, considering 8,760 Hectares for flooding with an average depth of 3.60 meters. The investment cost represents ca. 112 million U.S. Dls including a treatment plant of 2.50 m3/sec capacity, and the operation cost of the plant is of ca. 9 million U.S. Dls[10]. Other proposals are more radical: one of them tries to restore the former five lakes[11].

It took millions of years, since the valley was closed to the south in the Quaternary, to form the peculiar hydraulic system of the Valley of Mexico with a big lake in the middle of the closed basin. Only three centuries, from the sixteenth onwards, were neccesary to partly drain the big lake forming a system of five lakes. The twentieth century let a Megacity of 18 million people grow, built powerful and costly infrastructure that brought water from sources far away, as well as hundreds of wells that extracted groundwater at increasing volumes, and big, deep underground tunnels that expelled water outside the valley and almost dried the lacustrine environment.

It is essential to find in the next decades a sustainable pattern of urban development and urban technology that may restore a balance that Mexico City lost since long ago, trying to overcome its problems of excessive growth.

NOTES

1. The water in the northern part of the Lake, where Teotihuacan was located, was salty, unusable for drinking purposes.
2. Memoria de las obras del Sistema de Drenaje Profundo del Distrito Federal, Departamento del Distrito Federal, México, 1975. Vol. II, p. 81.
3. *Ibid*. pp. 81-128.
4. The Federal District is the official name of the territory where Mexico City is located. It is the seat of the Mexican Federal Government. The State of Mexico is just one of the 32 States that constitute the whole country of Mexico.
5. Negrete, Maria Eugenia and Salazar, Héctor: "Dinámica de Crecimiento de la ciudad de México (1900-1980)", in *Atlas de la Ciudad de México*, Departamento del Distrito Federal – El Colegio de México, México, 1987. pp 125-128.
6. González Pozo, Alberto, Alcántara Onofre, Saúl and Alonso Navarrete, Armando: "The Metropolitan Zone of the Valley of Mexico: Evolution, Contradictions and Perspectives", paper presented at the 24 th. Session of the International Seminars on Planetary Emergencies, Limites of Devlopment Monitoring Panel: Megacities and Sustainability, Erice, August 18-19, 1999.
7. Ruvalcaba, Rosa María and Schteingart, Martha: "Estructura urbana y diferenciación socioespacial en la zona metropolitana de la ciudad de México (1970-1980)" in *Atlas de la Ciudad de México*, Departamento del Distrito Federal – El Colegio de México, México, 1987, pp. 112-113.
8. A warning signal could be the flooding in the Summer of this year, that affected hundreds of homes in Chalco, a southeastern urban suburb inhabited by low-income families.
9. Zoreda Lozano, Juan J. and Castañeda, Victor: "Critical issues in urban technologies for sustainable development: The case of water infrastructure in Mexico City", in *Policy Studies Review*, A Publication of the Policy Studies Organization, The University of Tenessee Energy, Environment & Resources Center, Summer/Autumm 1998, Vol. 15, No. 2/3, p.159.
10. Quadri de la Torre, Gabriel: "Significado y desafíos institucionales de la recuperación del Lago de Texcoco" in Teodoro González de León et al. *La ciudad y sus lagos*, Clío, México, 1998. pp. 61-85.
11. Kalach, Alberto: "Vuelta a la ciudad lacustre" in *La ciudad y sus lagos*, Clío, México, 1998. pp. 43-59.

GLOBAL WATER QUALITY, SUPPLY AND DEMAND: IMPLICATIONS FOR MEGACITIES

PAOLO F. RICCI
University of San Francisco, San Francisco, California, USA
RICHARD C. RAGAINI
Lawrence Livermore National Laboratory, Livermore, California, USA
ROBERT GOLDSTEIN, WILLIAM SMITH*
Electric Power Research Institute, Palo Alto, California, USA

INTRODUCTION

Water is the most immediate and critical limiting factor to both human and environmental well-being. The reason is that water supply is, essentially, fixed. Therefore, although the *real* price of supplied water is currently low, changes in price can induce changes in water uses, and, therefore, have potentially large impacts through the multiplier effect. The protection of water bodies, through stewardship, acts as a binding constraint that can result in conflicts between the affected jurisdictions (e.g., some of the states of the United States sharing a common river through inter-basin transfers).

The following Table 1 lists a range of issues concerning the supply and demand of water and direct effects.

There are also transitional issues, such as tendency to urban sprawl, the decline of agricultural land supply and many sub-issues. Fundamentally, water issues are sectoral, involve water transfers and affect both the inputs and the outputs of the national and international economies. Perhaps the most important issue that affects projections—because they are long-term—and thus the development of the impact of the issues on water availability is the potential (or actual) structural changes in the sectors of the national economies.

An example of the *structural* change in the U.S. economy that has affected the relationships between supply, demand and quality of water to this date occurred in the period 1980 to 1985. From 1950 to 1980, the patterns of water use had shown a steady *increase* that was reversed by changing to a *decrease* from 1980 to 1985. Since 1985, the water uses have remained steady (Solley, 2000). Over the period 1950 to 1995, the U.S. population has increased at a constant rate from 150 million to 250 million.

* This work is developed as a component of a research grant from the EPRI, Palo Alto, CA. The usual disclaimers apply.

Table 1. Water Supply And Demand Issues.

Issue	**Direct Effect on**	**Comment**
Water conservation	Demand	Supply of water↑ ⇒cost↓
Water supply infrastructure	Supply	Cost
Water re-use	Supply	Acceptability ↓ ⇒ cost↑
Water losses	Supply	Cost↑ ⇒ infrastructure ⇒ cost↑
Sanitation	Demand	Infrastructure ⇒ cost↑
Institutional	Supply & demand	Costs↑ ⇒ trade↓ ⇒ revenues↓
Legal	Supply & demand	Costs ↑ ⇒ property rights ⇒ environmental equity ⇒ costs↑
Food (fisheries, livestock, crops…)	Supply	Availability↓ ⇒ cost of products↑ ⇒ international trade↓ ⇒ national security
Political	Supply	Costs↑ ⇒ cost of products↑ ⇒ international trade↓ ⇒ national security
Demographic	Supply & demand	Infrastructure ⇒ cost↑
Climate change		Draughts, floods, shifts in thermal energy balance ⇒Infrastructure ⇒ cost↑ ⇒ agricultural yields shift ⇒ national security ⇒ trade↓
Agriculture	Demand	Cost↑ ⇒ cost of products↑ ⇒ international trade↓ ⇒ national security
Megacities and megalopolies	Supply & demand	Demographic ⇒ infrastructure ⇒ cost↑ sanitation ⇒ water re-use↑ ⇒ costs↓
Technology	Supply & demand	Availability↑ ⇒ cost of products↓
Protection of vital ecosystems	Supply	Availability↓ ⇒ cost of products↑
Environment/recreation	Supply	Availability↑
Energy production and reliability	Supply & demand	Costs↑ ⇒ energy availability↓ ⇒ reliability↓

The structural changes that affected water supply and demand were principally due to federal laws controlling water pollution, technological changes in processes that use water as an input (including cooling towers and a movement away from once-through cooling) and increased recyling of water. The agricultural sector improved the delivery of water for irrigation and this sector lessened its reliance on ground water because of the increased costs of pumped water. At the same time, the agricultural patterns shifted from the West of the United States to the East, and there was a concomitant decline in farmers' economy. The number of irrigated acres peaked in 1980 and has been steady from 1985 to 1995 (at about 58 million acres). In the West, the irrigated acreage was 49 millions in 1980, declining to about 45 millions; while the East has steadily increased from about 2 million acres in 1950 to about 12 million in 1995 (Solley, 2000). Point source pollution includes mine drainage, industrial discharges, sewage overflows, pollution from feedlots, underground storage tanks leaks and spills. Non-point sources of pollution include agricultural, silviculture, construction, mining and urban run-off, pollution from septic systems and from landfills. Erosion can create a number of problems, such as decreasing water storage, affect tourism, increase dredging and irrigation costs, as well as hindering natural water filtration by hardening the crust of soils.

The supply for domestic water is decreasing, notwithstanding the increase in population, because of conservation, the detection and correction of leaks and increased price of water supplied. Domestic water uses have been approximately constant from 1980 to 1995.

The principal areas of the water situation that the on-going study paper addresses are:

- National and sectoral water demands
- National and sectoral water supplies
- Water quality and its impacts on water supply and demand at the levels adopted for supply and demand analysis.

Studying these three aspects jointly is important to define the issues and benefits that will arise from water supply, demand and quality. Demand and supply of water are linked through the costs of supply and the willingness to pay for water by the sectors of the economy; the quality of water ultimately determines the quantity of water available for private and public uses. The ideal policy scenario is to develop sustainable management practices and use management tools that, when implemented, can protect water quality and yet provide a plentiful supply of water at a reasonable social cost.

ASPECTS OF THE U.S. WATER QUALITY & SUPPLY SITUATION

It is useful to draw issues and solutions concerning global water quality and supply by studying the U.S. situation. The U.S. water situation is most clearly understood at the national, such as the Water Resources Regions or States, and sectoral levels such as water of demand by energy production, agriculture and so on.

The United States is characterized by reliance on both surface water (78%) and ground water (28%), the principal users are irrigation and livestock (41%) and thermal power generation (39%) (Solley, 2000). The prime movers for the changes were federal legislation (the Clean Water Act, 1972; the National Energy Policy Act, 1992) as well as conservation planning. The structural changes in the *aggregate* water projections were the result of the transitions from supply-side management of water to its demand-side management and increased users' awareness of the importance and costs of water

(Solley, 2000). The water situation and the issues that surround it are best understood by examining water demand, supply and quality at the sectoral level. The five broad sectors used for national estimates are:

- Irrigation
- Thermoelectric power generation
- Industrial & Commercial
- Domestic & Public
- Livestock

These five categories summarize the major uses of water in the United States. Demand projections have been obtained from the Department of Agriculture; those estimates have been checked for discrepancies in the original sources from which they were developed.

Generally, the most consistent and reliable water demand projections are limited to these five sectors and are available nationally and at the level of the USG Water Resource Regions (WWRs). The unit of analysis has traditionally been the USGS in twenty water resources regions (WRRs) which are large coastal and other contiguous areas of the country. We will rely on the most recent data because of changes in the structure of the economy in past, increased efficiency and changes in the way USGS data are combined. The water resources regions (from WRR 1 to WRR 20) are also used by Guldin (1989) to develop water budgets. For water quality, however, the data are given at the state or other major jurisdictions, by the U.S. Environmental Protection Agency.

The sectoral water uses are characterized by many complex and inter-dependent relationships that affect the probable equilibria points between water supply and demand, over the period 2000 to 2020, and even more so to 2050. The fundamental unknown in those long-term predictions is the potential for changes in the structure of the economy and in the sectors. It follows that developing the potential impact of the issues affecting water supply and demand has to account for those potential structural changes.

While population increases and shifts are a major reason for changing water demand and altering supply, style of life and aging affects the water situation perhaps more so. For instance, eating habits of the population can greatly influence water use in other sectors of the economy. Increasing consumption of meat would increase livestock water demand and agricultural water demand (a significant portion of agriculture produces feed for livestock.) The prediction of caloric intake, in the U.S., is to slightly increase from 3600 kcal/person in 1995, to approximately 3750 kcal/person in 2025. Combined with population increases, this change in dietary habits would lead to an increase in agricultural and livestock water demand. The cereal's yield is also projected to increase from 5 tons/ha in 1995, to 6 tons/ha in 2025. Increases in production efficiency within the agricultural sector of the economy can mean greater demand on water supply. Increases in agricultural water use leads to increased contaminated runoff.

The demographic effects of an aging population can affect the supply of water in a number of ways including the potential for retirees to move out of communities where the supply and quality of water are poor. There is relatively little transitional cost involved in those moves and therefore the speed of the move can be almost instantaneous – at the scale of 20-year forecasts. In terms of research on the infrastructure, the quality of the roads and their numbers, traffic volumes and congestion are the largest contributors to development than the networks that supply ad treat water. A critical unknown is the effect of droughts on supply, the dislocation that such droughts can cause and the local and regional resilience to drought effects.

UNIT OF ANALYSIS

We find that water quality problems are local and regional. Although there are global phenomena that affect water availability, the solutions are most likely to be geared to water basins because the effect of global changes is neither uniform nor has the same intensity and severity. There are 2149 basin watersheds in the United States, the smallest being 700 sq. mi. Multiple water uses, different and heterogeneous landscapes and multiple users characterize these basins with often conflicting interests. Watersheds are targeted for funding by the U.S. federal government, as exemplified by watershed assistance, watershed restoration action strategies, watershed pollution prevention, and watershed assistance grants. Depending on the availability of data, our work will use WRRs as the units of analysis for water demand and supply as well as state-specific information. For water quality we will use the states of the United States and such jurisdictions as Puerto Rico and Washington D.C. because the data provided by these jurisdictions are relatively consistent, are reviewed by the U.S. EPA and can be matched to the USGS Water Resources Regions.

A brief description of the coastal and non-coastal zones of the United States is (President Clinton's Clean Water Initiative (1994), Table 3) follows in Table 2.

Table 2. Descriptive Statistics of the Coastal and Non-coastal Zones of the U.S.

	Total U. S.	**Coastal zones**	**Non-coastal zones**	**Non-coastal zones as a % of total**
Counties (1990)	3,130	678	2,452	78.3
Population (1990)	246,750,237	127,351,147	119,399,090	48.4
Land area, (1990) mi^2	3,521,131	635,166	2,885,965	82.0
Surface water, (1990) mi^2	241,151	122,402	118,749	49.2
Farms (1987)	2,081,085	405,844	1,675,241	80.5
Farm land, (1987) acres	958,775,957	98,677,897	860,098,060	89.7

The urban population of the United States is 185.7 million, the rural population is 61.4 million, there are 93.9 million households in the U.S. with 2.63 persons per household and there are 23.2 million rural households. Approximately 60 million households are served by urban/storm water systems and about 10.5 million households have non-urban storm water (President Clinton's Clean Water Initiative, 1994, p. D-8). It has been estimated that each 1 billion of environmental investment creates 13,000 direct and 20,000 indirect jobs. The pollution control industry employs approximately 460,000 people (EPA, 1993), the national expenditures under the Clean Water Act are approximately 50 billion dollars (annualized at 7% per year, in $1992) and about 5 billion for drinking water (1992 $, at 7%) (EPA, Environmental Investments: The Cost of a Clean Environment, 1990)

PHYSICAL AND ECONOMIC SCARCITY OF WATER

The scarcity of water can be physical as well as economic: the fundamental difference for this Work is that economic scarcity depends on the state of the technology and the costs associated with it. Physical scarcity is something that is outside our immediate control and thus cannot be realistically changed. The fundamental questions are: when will there be gaps or disequilibria between the supply and demand for water and what are the determinants of the shift for the sectors of the economy considered in this Work? The predictions that we have developed in this work suggest that most impacts will be essentially gradual: no critical discontinuity is expected in the period 2000 to 2020. However, the gradual pattern will probably not be maintained if the time horizon of the predictions is moved to 2050 because of local and regional impacts of climate change.

Our analyses are based on the data compiled by the USGS and other agencies. The USGS data is used because it provided the most consistent national compilation of water use data since 1960. In addition, all other major water studies reviewed for this report relied on the circulars provided by the USGS; while predictions have varied, use of common data allows for higher accuracy of analysis when comparing reports. (Water Resources Council, 1978; Solley et al., 1983, 1988, 1998; Brown, 2000). The USGS has defined the following classes of water use (Brown, 2000); where "with." stands for withdrawals (consumptive + return flows/recharge): livestock, domestic and public, industrial and commercial, thermoelectric and irrigation. The trends and the critical uncertainties are shown in Table 3.

Table 3. Trends and Criticial Uncertainties.

Water Demand by User	Uncertainties	Trends
Livestock	Taste, income, diseases,	Insufficient data for 1990, 1995: possible higher increases in water uses.
Domestic and public	Housing stock, modernization, conservation, epidemics.	Historical rates of change: 1.5%, 0.9%, 0.8%, 0.3%. Cannot establish a reliable number.
Industrial and commercial	Efficiency, consumer confidence, employment rates, taxes.	Stable: 5% per year.
Thermoelectric power	Efficiency of production, appliances and telecommunication, federal and state policy.	Historical rates of total energy uses: 6%, 3%, 1.1%, 0.4%. Used 0.6% to 0.14%. Use 1.3% to 0.6%.
Agricultural irrigation	Urban sprawl, energy costs, technological changes, taste, income, transportation, climate change, federal and state policy.	East: 0%. West: 0.08$ to 0.04%.

The summary of the aggregate projections made by Brown (1999, 2000) are shown in Table 4.

Table 4. Summary of Aggregate Projections.

1995 Baseline Water Use (BGD is billion gallons per day; gpd is gallons per capita per day)	Source of Water	Year 2040 projections, as intervals, if available, Middle pop. Forecasts by Census	Key Predictors of Water Use.
Livestock (~ 5.5 BGD); (21 gpd; constant '60 to '95)	Self-supplied	[5.5 BGD (1995) to 7.7 BGD (2040)]; 21 gpd.	Population; (withdrawals/capita)
Domestic and public (≈32 BGD); 121 gpd	Self-and public-supplied	[32 BGD (1995) to 45 BGD (2040)]; 121 gpd.	Same
Industrial and commercial (≈36 BGD);	Self- and public-supply	[37 BGD (1995) to 39 BGD (2040)]; [7.4 gpd to 3.9 gpd]	Population; per capita income; withdrawals/$ of income
Mining	Self-supplied	NA	None stated
Thermoelectric (≈ 130 BGD)	Self-supplied	[132 BGD (1995) to 143 BGD (2040)]; [504 gpd to 389 gpd]	Population; KWh/capita; freshwater KWh /Total KWh;
Hydroelectric power	Self-supplied	NA	None given
Agricultural Irrigation (≈ 130 BGD)	Self-supplied	[134 BGD (1995) to 130 BGD (2040)]; [514 gpd to 354 gpd].	Acres irrigated; withdrawals/acre.

Sources: various Tables and discussion in Brown (1999, 2000).

Overall, the *consumptive* use in 1995 was about 100 BGD, approximately 29% of the total water withdrawals (Brown, 2000). Broadly, from 1900 to 1990, population increased by about 1.2% per year while water withdrawals increased at a faster rate, approximately 2.4% per year; in terms of gallons per capita per day (gpd), Americans withdrew, in 1900, 430 gpd but 1350 gpd in 1990. In this time period, relative to the total, withdrawals from public supply was approximately constant (at 12%), self-supply by industry decreased from 25% to 6%, irrigation declined from 50% to 40% but thermoelectric demand increased from 12% to 40% (Brown, 1999, 2000). The demographic projections use the Bureau of the Census data (1992) and the Bureau of Economic Analysis (1992) as developed to the state and county levels by T. C. Brown (1999, 2000). The income data are developed from data developed by the Bureau of Economic Analysis (1992) and aggregated to the WRR by Brown (1999, 2000).

PREDICTIONS OF FRESH WATER SURPLUSES AND DEFICITS, BY WRR

The following discussions summarize the water surplus and deficits (measured in billions of gallons per day, BGD) and total fresh water withdrawals – based on the population forecasts by the U.S. Bureau of the Census. Guldin (1989) excludes Alaska and Hawaii. His estimates of the population is higher than Brown (a difference of approximately 10 million people in 2000) and the total water withdrawals differ by approximately 30 BGD in 2000. However, the two authors' numbers are very close to one another, relative to the best known projections by Water Resources Council (1968), Wollman and Bonem (1971), the Senate Select Committee (1961), the National Water Commission (1973). The population and water withdrawals projections by Wollman and Bonem (1971), the Water Resources Council (1968), the Senate Select Committee (1961) and the National Water Commission (1973) range from approximately 310 to 330 million people (in year 2000), while water withdrawals were projected to range from 550 BGD to 1000 BGD, also in 2000. The 2000 water withdrawals projection for 2000 made by the National Water Commission (1978) is the lowest of all projections (about 310 BGD), even though the population is slightly higher than Brown's. We concluded that for the purpose of developing research products for EPRI, the numbers provided by Guldin (1989) and Brown (1999, 2000), and their projections, are consistent with what is currently understood about the influence of water supply and demand. Guldin (1989) and Brown (1999, 2000) numbers are shown next in Table 5.

Table 5.

	Surpluses and Deficits (-) in Billions Gallons per Day, BGD, from Alternative Demand Projections in 2020, by WRR, Given Average Rainfall Conditions and (Global Climate Change) Effects (Guldin, 1989). Percent Change in Total Withdrawals, Given Low, Middle and High Census Population Forecasts from 1995 to 2040, (Brown, (1999)					
Water Resource Region (WRR)	**20% Lower Demand, BGD**	**Low Pop.(% change)**	**Normal Demand, BGD**	**Middle Pop. (% change)**	**20% Higher Demand, BGD**	**High Pop. (% change)**
New England	7.65 (3.15)	-11	7.54 (2.88)	15	7.43 (2.61)	42
Mid-Atlantic	24.32 (19.37)	-17	23.71 (18.73)	7	23.1 (18.09)	32
South-Atlantic Gulf	17.86 (-3.76)	2	16.37 (-5.33)	26	14.88 (-6.90)	52
Great Lakes	9.64 (3.82)	-22	9.23 (2.92)	2	8.83 (2.02)	28
Ohio	14.92 (7.07)	-20	14.25 (6.18)	3	13.58 (5.30)	28
Tennessee	4.33 (-0.04)	-10	4.21 (-0.17)	19	4.09 (-0.30)	50
Upper Mississippi	9.64 (5.21)	-18	9.15 (4.61)	6	8.66 4.01)	31
Lower Mississippi	69.9 (-35.18)	13	60.45 (-43.85)	27	51 (-52.51)	42
Souris-Red-Rainy	3.5 (3.09)	12	3.47 (3.05)	29	3.44 (3.01)	47
Missouri	14.79 (3.57)	-6	10.85 (0.19)	3	6.9 (-3.20)	12
Arkansas-White-Red	8.87 (-2.66)	-16	6.42 (-4.81)	-5	3.98 (-6.96)	7
Texas-Gulf	7.17 (-3.59)	-12	5.44 (-5.33)	6	3.7 (-7.06)	25
Rio Grande	-0.2 (-1.87)	-28	-0.75 (-2.47)	-25	-1.3 (-3.06)	-22
Upper Colorado	-0.52 (-5.70)	28	-1.11 (6.35)	30	-1.69 (-7.00)	32
Lower Colorado	-7.82 (-13.23)	-1	-9.62 (-15.04)	5	-11.42 (-16.85)	12
Great Basin	0.89 (-1.18)	4	-0.06 (-2.13)	9	-1.01 (-3.08)	15
Pacific Northwest	65.22 (34.08)	-9	62.42 (30.77)	0	59.63 (27.47)	9
California	34.55 (18.53)	-4	28.57 (12.94)	3	22.59 (7.36)	9
Total Contig. U.S.	223.88 (74.37)	-8	197.82 (48.31)	7	171.75 (22.25)	24

Developed From Table 23, Guldin (1989) and Table 7, Brown (1999). Table 23 in Guldin (1989) provides additional information to 2040.

WATER SUPPLY AND DEMAND ISSUES

The most obvious factors affecting water budgets are short- and long-term climatic variations and changes as well as precipitation and human activities linked in complex ways. This section contains a discussion of the factors that create issues with the numbers developed from Guldin (1989) for freshwater as groundwater and as surface water. The following lists the specific issues that affect water supply availability and quality:

- Pollution
- Droughts

- Floods
- Excess water
- Trans-boundary
- Pollution control
- Water treatment and other infrastructure improvements
- Investment
- Ecosystems at risk
- Changes in the hydrological cycle
- Rehabilitation of degraded areas
- Food management
- Supplemental irrigation
- Strengthening of: economic, legal, institutional arrangements at local, regional, state, national and international levels
- Capacity building
- Climate change
- Biotechnology applied to crops and livestock
- Low consumption crops
- Population growth in urban areas
- Sanitation
- Irrigation
- Demand management
- Water use conflicts
- Energy supply and cost
- Policy

The following functions will affect the supply and demand of water:

- Integration between appropriate jurisdictions
- Financial viability
- International effects
- Trade effects
- Sectoral shifts

Watershed Management. This is a key issue because watershed management maintains or improves the quality and quantity of water flows. Watersheds are critical to sustain development and ecological well-being. In the United States, 28% of the watersheds are classified as Class I (regimen attainment), 50% are in Class II (special emphasis) and 22% are in Class III (investment emphasis). Class III watersheds require technological investments to attain the goals of resources management (Guldin 1989, at 11, 12). Those investments should be directed towards environmental, economic and social goals such as rehabilitation including reforestation, land use planning, farm conservation, stabilization of channels and streams, as well improving the local economy. Class I watersheds have attained a dynamically stable equilibrium that is consistent with average precipitation and drainage, as well as productivity. Class II watersheds do not require capital investment to reach the equilibrium of Class I watersheds, but are particularly sensitive to cumulative changes in certain activities (such as events that have little impact at the acre level) and are not resilient to either cumulative impacts or sudden changes in exploitation.

Loss of Wetlands. The losses are due to conversions to urban and suburban uses as well as changes in the agricultural patterns at the regional level. Roughly, using the OTA numbers (1984) in Guldin (1989) the loss of wetlands to agriculture is about 12 million acres, due to urban development it is about 1 million acres while the gains are about 1 million acres for agriculture. Most of the losses of various types of wetlands (approximately 95% of the losses) are due to human activities with the rate estimated to be approximately 300,000 acres per year (1989), down from 550,000 acres per year in the period 1950 to 1970. In the U.S. about 5.2 million acres of wetlands have a good potential (and about 17 million have some potential) for conversion to productive purposes such as agriculture but the Food Security Act of 1985 contains language that can prevent some conversions from taking place by withdrawing farm support funds.

Irrigation: This is the largest use in terms of withdrawals and consumptive uses. Irrigation accounts for approximately half of the groundwater withdrawals. The critical issue is the sustainability of those withdrawals, given land use changes, local climactic changes. The net result from these is that water prices will increase too much, affecting development. The total water withdrawals range from 142,500 Mgpd in 2000 to 173,400 Mgpd in 2040, including relatively minor quantities as wastewater. Irrigation (gravity or pressure-fed) is deemed to increase at a lower rate from 2000 to 2040. The reasons are that pumping costs are increasing, as are energy costs, and aquifer yields are declining. The rate of return from agriculture is also decreasing: if these trends change then irrigation may become attractive again (Guldin, 1989). The pricing policies of the U.S. Bureau of Reclamation may change because they are not user-favorable and some subsidies cause disequilibria. In any case, increases in cost will force technological changes and innovation opening a window of opportunity for EPRI.

Instream uses: Supply of water and uses (such as navigation, hydropower generation and cooling, recreation and dilution) are affected by changes in flow regimens. Ecological activities and the very survival of some species can be threatened, depending on the length of the water shortages. The preferable remedy is watershed management, rather than capital investments.

Surface water: Most of the water for supply is stored in reservoirs (approximately 90%). Such storage is affected by diminishing marginal returns and the availability of water, when needed, may fail because of drought or other factors. The construction of reservoirs is increasingly constrained by local activities.

Groundwater: There are approximately 5,000 cubic miles (55,000 trillions gallons) of ground water in the coterminous U.S., with a recharge rate of about 1 trillion gpd (Guldin, 1989). The fresh water pumping rates in 1985 was about 83 billion gpd and the overall water supply for the U.S. is positive, but there are issues. One of them is that agricultural irrigation is the largest user of this water, namely about 56 billion gpd (or abut 24% of the total withdrawals) with the highest users in California, Texas, Nebraska, Arkansas, and Florida. (Guldin, 1989). The rates of withdrawals have increased from 1960 to 1980 because of irrigation in the east using central pivot systems, urbanization, energy production droughts, and increased inability to build reservoirs and for inter-basin transfers. The municipal water withdrawals are expected to increase from 20,000 Million Gallons per Day (Mgd) in 2,000 to 34,000 Mgd in 2040. Thermoelectric water withdrawals from groundwater will increase from 700 Mgd in the year 2000 to 676 Mgpd in 2040, and cooling withdrawals will decrease from 703 Mgd in 2000 to 676 Mgd in 2040. Irrigation will increase from 56,000 Mgd in 2000 to 64,000 Mgd in 2040 and livestock will change from 1,500 Mgd in 2000 to 1,800 in 2040.

Thermoelectric steam cooling: This is the second largest water use with the total (surface and groundwater) withdrawals estimated ranging from 157,000 Mgd in 2000 to 228,000 Mgd in 2040.

Water shortages. These are expected to occur by the year 2040 and principally affect the Lower and Upper Colorado River, the Rio Grande, the Great Basin, California and the Lower Mississippi River. In particular, irrigation is the predominant water use in the areas likely to experience shortages. A possible solution to the scarcity problem is through market instruments, an area in which EPRI has considerable experience. Specifically, the lower region of the Colorado River – even during average conditions – faces significant water deficits; in dry years the deficit is about 300% of the in-stream flow and the groundwater overdrafts are about 400%. The Rio Grande region is characterized by high water use. The Great Basin region will incur water deficits due to the growth in irrigation demand as will California.

Groundwater shortages are predicted for the High Plains of Texas, Oklahoma, Kansas, Nebraska, Wyoming, Colorado and N. Mexico. The Central Valley of California can also experience shortages. Similarly, the Southeastern and Atlantic Coastal Plains are expected to face shortages, as will the lowlands of Arizona. The adverse effects include land subsidence, salt-water intrusions, changes to the local flow patterns, damage to property and so on (Guldin, 1989). The water quality aspects of the work that we develop in this work are summarized below in terms of the amount and percentage of designated uses met, in 1987, by type of water body:

THE STATUS OF THE UNITED STATES' WATER QUALITY

This discussion that follows deals with the quality of waters in the United States at the national and state levels, including however some principal jurisdictions that are not states of the Union. The differences in units of analysis, namely the WRRs for the water supply and demand, and the states for quality are not particularly significant because our on-going work is directed to long-term projections. Thus, the overall sense of the directions and magnitude of potential changes can be captured with the currently available information.

There is considerable difference between the reports that attempt to characterize the water quality of the United States. For our work, we have selected the data sets that the states provide to the U.S. EPA, under section 305(b) of the Clean Water Act. This data set is increasingly homogeneous with respect to the protocols that the states and other jurisdictions have adopted and include the input from a number of stakeholders, such as Indian Tribes. Nevertheless, this apparent homogeneity has some problems; these include differences in the summaries developed by each jurisdiction and provided to the U.S. EPA, changes in the monitoring networks, or stations, or both, differences in the methods for assessment. Nevertheless, the information developed by the U.S. EPA, on these reports, is the best available for the purpose of our work.

According to the National Water Quality Inventory (1998), the majority of the water bodies in the U.S. are adversely affected by "moderate to high levels of agricultural run-off." Furthermore, about 1/3 of the U.S. waters are characterized by fish advisories leading to no fish consumption; and about 1/5 of the country has high levels of wetland loss (p.4). The status of American waters in 1996 is summarized as follows (from sampling data, U.S. EPA (1998) in Table 6.

Table 6.

	Status and Percentage Meeting Overall Water Quality[1] in 1996			
Water Body	**Full support all uses (*Good*)**	**One or more uses are threatened (*Good*)**	**One or more uses are impaired (*Impaired*)**	**Not Attainable**
Rivers and Streams	56%	8%	36%	< 1%
Lakes, Ponds and Reservoirs	51%	10%	39%	< 1%
The Great Lakes	29%	1%	97%	< 1%
Estuaries	58%	4%	38%	< 1%
Ocean Shoreline Waters	79%	9%	13%	0%

[1]Developed from Figures 3, 6, 9, 12 and 15, U.S. EPA (1998a); *Good*, *Impaired* and *Not Attainable* are short descriptions used by the U.S. EPA to characterize the status in the columns' headings.

The wetlands of the Nation are affected by sediments and siltation, nutrients, filling and draining, pesticides, total suspended solids, chlorides and pollution from metals, habitat and water flow changes, as well as increased salinity. The principal causes of water pollution are agriculture, hydrological changes, urban run-off, construction, resources extraction and grazing. Shown in Table 7, the U.S. EPA 1998 has ranked (1 being the highest) the five principal causes of water pollution as (U.S. EPA (1998), Table 2, p. 9 and Table 4, p. 13; Figures 4, 7 and 13) and EPA (2000) Figures 4, 7, 10, 13 and 16:

Table 7.

Rank	**Rivers (Percent impaired)**	**Lakes (Percent impaired)**	**Estuaries (Percent impaired)**	**Ranking of Major Sources of Pollution**
1	Siltation (18%)	Nutrients (20%)	Nutrients (22%)	1. Agriculture
2	Nutrients (14%)	Metals (20%)	Bacteria (16%)	2. Point and non-point sources
3	Bacteria (12%)	Siltation (10%)	Priority toxic organic chemicals (15%)	3. Municipal point sources
4	Oxygen-depleting substances (10%)	Oxygen-depleting substances (8%)	Oxygen-depleting substances (12%)	4. Upstream sources
5	Pesticides (7%)	Noxious aquatic plants (6%)	Oil and grease (8%)	5. Agriculture

The principal sources of pollution affecting the Nation's water are shown in Table 8 (U.S. EPA, 1998, Tables 3, 4, pp 12, 13):

Table 8.

Sources	**Example**	**Body Principally Affected**
Industry	Pulp and paper, heavy industry, food processors, textiles	Estuaries
Municipal	Public sewage treatment overflows	Rivers, lakes
Combined sewer systems	Outflows	Rivers, lakes, estuaries
Storm water and urban run-off	Paved or other hard surfaces	Lakes, estuaries
Agriculture	Pastures, crop production, feedlots, animal operations	Rivers, lakes and estuaries
Silviculture	Forest management, logging, roads	Rivers, lakes
Construction	Land and road development	Rivers, lakes
Resources extraction	Mining, drilling oil, tailings run-off	Rivers
Land disposal	Leachate and discharges	Rivers
Hydrological modifications	Open channels, dredging, reservoir construction, flow regulation	Rivers, estuaries and lakes
Habitat modification	Riparian vegetation, stream-bank, wetlands drainage or filling	Rivers

Agriculture, industry, poor sewage treatment (such as septic tanks), leaking underground storage tanks and landfills affect the ground waters of the Unite States. A fundamental concern is the stewardship of natural resources; there is a unified watershed policy (DOI). The concern is quantified by noting that about half of the American watershed has from serious to moderate water quality problems. It follows that an important area of concern is the credibility of the water data used (e. g., what is the credibility of the data in such data bases as STORET, the National Geographical Data, the Watershed Boundary Data, the National Elevation Data, the Land Cover Data).

Much work has been committed to establish the nature of the water pollution problem in the United States (U.S. EPA 1998). The objectives of this Work are met by summary measures of water quality such as those described by indices of water quality. The U.S. EPA (1998) has reported at the state level and other jurisdiction, rather than at the Water Resource Region, using watersheds as their unit of analysis, the first use of such an index. This Section is based on that information. That information is mandated under Section 305(b) of the Clean Water Act. This index combines seven indicators of conditions of a watershed and eight indicators of the vulnerability of the watershed's rivers, lakes and estuaries. It is a linear aggregate of the fifteen indicators, shown below. The minimum set of indicators that was used to produce the IWI was a weighted combination of "[a]t least 4 of 7 condition indicators and 6 of 8 vulnerability indicators … " with the indicator determining the "rivers meeting all designated uses " was given a larger magnitude than any other indicator (U.S. EPA 1998, p. 55).

WATER QUALITY SUMMARY

Earlier results about the quality of U.S. water consist of data for 1988 (U.S. EPA, EPA 440-4-90-003 (1990), which provide the initial condition for understanding the quality of the waters of the United States. This data is summarized in Table 9 below:

Table 9. Percent of Individual Uses Supported, adapted from U.S. EPA, EPA 440-4-90-003 (1990) – Numbers are percentages.

	Rivers and Stream		**Lakes, Ponds & Reservoirs**		**Estuaries**		**Great Lakes Shorelines**		**Ocean Shorelines**	
Use supported	Good (%)	Fair (%)	Good (%)	Fair (%)	Good (%)	Fair (%)	Good (%)	Fair (%)	Good (%)	Fair (%)
All uses	70	20	74	17	72	23	8	18	NA	NA

U.S. Environmental Protection Agency (1990), Tables 1-1, 2-1, 3-1 Good is *fully supporting* and fair is *partially supporting*. All uses mean designated uses: fisheries, contact recreation and drinking water. NA is not available.

The 1988 assessment is based on about 520,000 assessed miles of rivers (48 jurisdictions), 16,000,000 assessed lake-acres (40 jurisdictions); and 26,700 square miles of assessed estuaries (23 jurisdictions).

In 1988, the major causes of river pollution were (rank ordered from highest percent of river-miles impacted): siltation, nutrients, pathogens such as bacteria, organic matter, metals, pesticides, suspended solids, salinity, flow alterations, habitat modification pH and thermal discharges. Table 10, shown next, depicts the national aggregate *percentages of impaired waters* (aggregated over major, moderate and minor impacts) (U.S. EPA, EPA 440-4-90-003, 1988):

Table 10. Summary of Percent of U. S. Impaired Waters 1988 by Cause of Pollution – All numbers are percentages.

Waters	Silt	Nutr'ts	Bact.	Organic Enrich't	Metals	Pestic.	SS	Salt/ OG*	Flow	Habit. Modif.	pH	Therm Flow Alter#
Rivers	42	27	19	15	11	10	6	6	6	6	5	4
Lakes	25	49	9	25	7	5	8	14	3	11	5	3#
Estuar.	7	50	48	29	10	1	NR	23*	NR	NR	<1	#

Developed from EPA 440-4-90-003 (1990), Tables A-1, 2-2, and 4-2. SS is suspended solids. # is Flow Alteration, there is no thermal effect reported for lakes. * is oil and grease (OG). Percentages are calculated as the sum of the quantities (in miles or acres) of pollution-specific impaired miles or acres divided by the total impaired miles or acres, for all jurisdictions. Lakes are affected by priority organics (8%). Estuaries are affected by priority organics (4%), unknown toxicants (5%) and other inorganics (<1%).

The 1988 water quality assessment of the Great Lakes (meaning Illinois, Indiana, New York and Ohio) is that priority organics are the major pollutants affecting 761 shoreline miles out of 819, the second is metals affecting 215 miles, and the third is nutrients, affecting 76 miles (U.S. EPA, EPA 440-4-90-003, 1988).

The ground water withdrawals in 1985 were approximately 76 BGD, the bulk of the withdrawals occurred in California, (10 BGD), followed by Arizona (4 BGD), Arkansas (4 BGD), Idaho (4BGD), Kansas (4 BGD and Nebraska (4 BGD), this water being supplied principally to agriculture. The national trend is an increase in ground water withdrawals, from 33 BGD in 1955 to a maximum 82 BGD in 1980, declining to 76 BGD in 1985. (EPA 440-4-90-003, 1988, pp 121, 122). In 1988, 9 states reported excellent ground water quality, and 17 reported good ground water, the remaining states giving no opinion (EPA 440-4-90-003, 1998, p. 122). Most states and other jurisdictions indicated that UST were the sources of pollution with the highest priority, followed by abandoned waste sites, agricultural activity and by septic tanks, municipal landfills were sixth in priority, followed by oil and gas brine pits. Mining waters, sewer leaks, cyanide heaps, construction and manufacturing were also of equal and highest priority.

The EPA has more recently developed the IWI indicator/index system with such stakeholders as the states, Indian tribes, and others. The index is based on indicators of individual beneficial uses of water: aquatic life and wildlife habitat support, fish and shellfish consumption, drinking water supply, recreational swimming, boating, agricultural irrigation, livestock consumption, ground water recharge, and cultural benefits as a function of the degree of well-being of the water bodies for those uses. The information described below was reported to the U.S. EPA by the states, Indian tribes and other jurisdictions mandated to do so under the CWA Section 305(b). The Environmental Protection Agency (1998) reports the overall status of the Nation's waters, in 1996, as shown in Table 11:

Table 11. 1996 Percent of Individual Uses Supported, adapted from U.S. EPA EPA841-R-97-001 (1998) – Numbers are percentages.

	Rivers and Stream		**Lakes, Ponds & Reservoirs**		**Estuaries**		**Great Lakes**		**Ocean Shorelines**	
Use supported	Good (%)	Fair (%)	Good (%)	Fair (%)	Good (%)	Fair (%)	Good (%)	Fair (%)	Good (%)	Fair (%)
Aquatic life	60	23	55	25	61	27	12	9	91	3
Fish consumption	84	14	60	32	75	22	2	34	91	5
Swimming	76	10	63	21	83	15	96	3	82	5
Secondary contact	78	16	62	23	76	22	96	4	93	5
Drinking water supply	79	19	81	7	NA	NA	98	<1	NA	NA
Shellfishing	NA	NA	NA	NA	69	16	NA	NA	84	6
Agriculture	93	3	84	10	NA	NA	89	11	NA	NA

US Environmental Protection Agency (1998), Tables 2-3, 3-3, 4-3. 4-8 and 12-4. Good is *fully supporting* and fair is *partially supporting*.

Table 12 immediately below has also been developed from water quality data provided by the U.S. Environmental Protection Agency (2000):

Table 12. 1998 Percent of Individual Uses Supported, adapted from U.S. EPA EPA841-S-00-001 (2000) – Numbers are percentages.

	Rivers and Stream		**Lakes, Ponds & Reservoirs**		**Estuaries**		**Great Lakes**		**Ocean Shorelines**	
Use supported	Good (%)	Fair (%)	Good (%)	Fair (%)	Good (%)	Fair (%)	Good (%)	Fair (%)	Good (%)	Fair (%)
Aquatic life	58	20	58	23	54	29	36	12	87	4
Fish consumption	87	5	54	35	63	34	4	29	84	10
Swimming	69	11	69	15	88	5	97	2	80	8
Secondary contact	76	14	78	10	81	15	99	1	93	5
Drinking water supply	87	6	82	9	NA	NA	98	0	NA	NA
Shellfishing	NA	NA	NA	NA	70	14	NA	NA	89	11
Agriculture	97	2	89	3	NA	NA	100	0	NA	NA

U.S. Environmental Protection Agency (2000), Figures 3, 6, 9 12, and 15. Good is *fully supporting* and fair is *partially supporting*.

The assessment of ground water quality, hence availability for supply, suggests studying the potential contamination from a number of sources, such as leaking underground tanks, septic systems and so on. This nation-wide effort is based on state-specific and other stakeholders' efforts. Because of the differences, namely: no reports, reports at the hydrological unit, reports at the state level only and so on, an "evaluation of ground water quality data is not possible." (U.S. EPA, EPA-816-R-98-011, 1998, p. 24). Nevertheless a qualitative understanding of the ground water situation for 1996 (Modified from U.S. EPA, EPA-816-R-98-011, 1998, Table 1, p. 24, 25), based on 162 aquifers and other hydrological units in twenty-nine states, is as follows in Table 13:

Table 13. Number and Source of Ground Water Contamination in the U. S. (1996).

Sources of Contamination	Number of sites	Sites listed and/or with confirmed releases	Sites with confirmed ground water contamination	Sited with completed cleanup
Leaking UST	100,921	40,363	17,827	19,379
UST sites (no releases found)	2,210	NA	NA	NA
Septic Systems	10,656	10,594	NA	NA
State Sites	7,017	5,751	2,614	3,166
Underground injections	5,006	1,077	911	204
CERCLIS	2,399	1,332	645	49
RCRA Corrective Action	2,114	283	289	52
MN Dept. Agriculture	600	164	50	NA
DOD and DOE	404	234	166	39
Miscellaneous	229	905	514	32
Non Point Sources	171	190	62	36
National Priority List	167	250	204	24
Landfills	149	78	74	0
Wastewater land application	116	NA	24	0

UST is Underground Storage Tanks; NA is Not Available; CERCLIS is the Comprehensive Environmental Response, Compensation and Liability Information System; RCRA is the Resource Conservation and Recovery Act; MN is Minnesota.

The ground water quality is determined from finished water from PWS wells (61%), untreated water from PWS wells (24%), ambient monitoring networks (52%), untreated water from private wells (36%), special studies (6%) and facility monitoring wells (EPA–816–R – 98 -- 011, 1998, p. 31, 32 and Figure 19, p. 32). The quality parameters include nitrate, VOC, SVOC, bacteria, pesticides ionizing radiation, a number of metals, inorganics, TDS, hardness, specific conductivity, alkalinity, nutrients and so on. Table 14 below provides a summary of the number of wells impacted and the rank-ordering of the ground water pollutants, at the national level, based on the number of reporting states:

Table 14.

MONITORING TYPE	Wells with number of MCL excesses and (number of states reporting excesses/total number of states reporting) rank-ordered by importance (I highest) for:					
	I. Nitrates	**II. VOCs**	**III. SVOCs**	**IV. Pesticides**	**V. Metals**	**VI. Bacteria**
Networks	267 (8/15)	30 (7/10)	5 (3/3)	5 (2/8)	195 (7/11)	10 (1/3)
PWS untreated water	85 (5/7)	77 (5/6)	10 (3/4)	2 (1/2)	100 (2/2)	1 (1/1)
Private untreated water	2,233 (9/10)	96 (2/3)	4 (1/3)	101 (4/5)	113 (1/1)	0 (0/1)
PWS finished water	230 (11/18)	152 (6/17)	18 (3/14)	0 (0/1)	175 (4/6)	404 (3/3)
Special studies	309 (2/2)	19 (1/1)	0 (0/0)	0 (1/1)	0 (0/0)	101 (1/1)

Tables 3 through to 8, EPA 816-R-98-011 (1998), pages 34 though to 39. The MCLS for individual constituents of VOCs, SVOCs, metals, pesticides and bacteria are omitted for brevity's sake. The MCL for nitrates is 10 mg/L of water.

COSTS

The study of water demand, supply and quality must include the costs associated with treatment. For public wastewater treatment, the nation-wide monetary needs have been estimated to be (U.S. EPA EPA841-R-97-008, p. 403):

Table 15.

Needs for Publicly Owned Wastewater Treatment Facilities and Other Concerns ($ 1996)	
Category of Need Under Title II Clean Water Act	**Total (in billions $ 1996)**
Secondary treatment	26.5
Advanced treatment	17.5
Infiltration/flow correction	3.3
Replacement/rehabilitation	7.0
New collector sewers	10.8
New interceptor sewers	10.8
Combined sewer overflows	44.7
Storm water	7.4
Non point sources from agriculture and silviculture	9.4
Ground water, estuaries, wetlands, urban runoff	2.1
Total	139.5

Adapted from Table 14.1; (EPA 841-R-97-008, 1998, p. 404).

Another aspect of costs is that associated with either remediation or prevention. For instance, in Massachusetts, gasoline contamination from underground storage tanks releasing about 2000 to 3000 gallons resulted in about 5 million dollars cost and it took approximately 10 years to complete the aquifer remediation. In New Jersey, the costs associated with establishing a new well-field to replace a system contaminated by a landfill was about $500,000 per well ($5,000,000 total). The UST federal program (LUST Trust Fund) has disbursed approximately $570,000,000 from 1986 to 1996. The states have raised approximately 1.3 billion dollars (1997) for these cleanups. The total number of sites is approximately 400,000, of which 162,000 were cleaned up and 115,000 are being cleaned (EPA - 816-R- 98-011, 1998, p. 66). The Sole Source Aquifer program has ranged from approximately $570,000,000 in 1992 to approximately $1.8 billion in 1996; this money is allocated as federal financial assistance to prevent significant public health risks (EPA –816-R-98-011, pp. 58, 59 (1998)). It also appears that the National Priory List Sites (CERCLA/Superfund sites) where the ground water was classified (453), 426 of these do not affect the ground water. However 622 NPL sites do report ground water contamination; overall 702 of 1121 NPL sites are associated with ground water contamination (EPA –816-R-98-011, 1998, p. 69).

Since 1972, when about 40% of the U.S. population was served by municipal wastewater treatment, the EPA has spent over 64 billion dollars in that form of treatment, U.S. EPA (EPA, EPA841-R-97-008 (1998)). The percentage increased to over 60% of the population, by 1992. The annual expenditure portfolio under the Clean Water Act (Table 19.1, p. 510 and Table 22, President Clinton's Clean Water Initiative (1994)) is shown in Table 16.

Some of the benefits from fishing include 19 billion dollars in wages and approximately 1.3 million jobs; about 50 million fishermen spent about 24 billion dollars on fishing-related activities. (Sport Fishing Institute, 1994). Commercial fishing contributes approximately 17 billion dollars to the U.S. economy, with shellfish contributing almost half of that value.

The U.S. EPA has estimated the range of benefits from controlling urban run-off, toxicants to human health from swimming and fishing, and CSO (President Clinton's Clean Water Initiative (1994)), as shown in Table 17.

Table 16.

Quantified Current and Planned Spending, in million dollars per year						
	Private	**Municipalities**	**Agriculture**	**State Water programs**	**Federal Agencies**	**Total**
Pre-1987 Act[1]	25,286	17,190	191	373 (560)	9,564	52,604 (52,791)
NPS Control and Watershed	NR	389 – 591	240 – 389	125 (150)	234	988 – 1,339 (1,013 – 1,364)
Storm Water Phase I	3.990 (16,235)	1,650 – 2,555 (1,785 – 2, 760 P. II)	NR	NR	NR	5,640 – 6,545 (18,020 – 18,995)
CSOs	NR	3,450 (14,140)	NR	NR	NR	3,450 (14,140)
Other Costs	943 – 1,073	88	NR	NR	NR	1,031 – 1,161
TOTAL	30,219 – 30,349 (46,454 – 46, 584)	22,767 – 23,874 (35,242 – 37,324)	431 – 580	498 (710)	9,798 (11, 181 – 14,279)	63,713 – 65,099 (94,018 – 99,577)

[1]Administrative cost to the EPA only, not to be used for projections. Abandoned mines are not included in Table 19.1, but the total spending in parentheses includes them; the costs in parentheses are *current and potential spending,* not *current and planned spending..*

Table 17.

Aggregate Benefits of Pollution Control for Urban Sources (CSOs, storm water and toxics) in millions of 1993 dollars.	
Freshwater recreational fishing and swimming	650 – 4,670
Marine recreational fishing	40 – 440
Marine non-consumptive recreation	30 – 300
Marine and freshwater commercial fishing	40 – 190
Withdrawals and diversions	30 – 80
Human health effects	40 – 320
Undiscounted total benefits	820 – 6,000
Total benefits at 7%, 15 years	560 – 4,100
Total benefits at 3%, 15 years	660 – 4,900

Modified from *President Clinton's Clean Water Initiative* (1994), Table ES-6.

This data excludes many non-monetized benefits such as marine boating, restoration of biodiversity, other human health effects and so on. On the basis of the discussions provided by the EPA, the numbers underestimate the annualized benefits. In 1986, for instance, the states spent between less than $5 per capita to more than $15 per capita in water quality and quantity related programs; the percentages of the states budgets dedicated to water programs ranged from less than 0.1% to more than 0.3% (EPA 440-4-90-003, Table 12-1, p. 179).

TYPES OF ADVERSE EVENTS AFFECTING WATER SUPPLY, DEMAND AND QUALITY AND STRATEGIES

There are different types of adverse events (accidental or premeditated) that can affect water quality and quantity. These include:

- *Transient events*, (such as spills or other types of releases), that can be contained rapidly.
- *Rare events*, such as the unsuspected arrival of a contaminated ground water plume that either cannot be contained rapidly or at all, or events that disappear rapidly but have significant or catastrophic outcomes.
- *Routine events*, such as the severing of a water main during construction, that are characterized by small overall cost, and that are possibly unavoidable.
- *Gradually-increasing events,* such as regional response to climate change or other long term conditions and that affect the supply of water.

In terms of predictions, water quality and quantity vary temporally and spatially. The unpredictability of transient events, the variability in events that are caused by relatively slow changes (such as global climate change) and the associated inactivity due to the uncertainties, all contribute to a complex situation. The vulnerability of water sources and resources, limitations to the yields and the potential for depletion require competent management that must include collaborative efforts that involve the private sector.

- Ensure viable and sustainable water flows for in-stream uses such as fishing, wildlife and tourism.
- Improve the quality of the watersheds by maintaining water quality and quantity, managing run-off, improving riparian areas and soil productivity.
- Use non-structural methods to avoid flood damage.
- Implement non-point sources abatement and control for activities such as silviculture and range management.
- Increase the protection and the stock of wetlands.

The obstacles to these strategies include some subsidies that distort the price structure and therefore limit competition and raise prices. However, removing such obstacles would require a fundamental change in current practices. Water managers should give high priority to activities and interventions that optimize off-stream uses.

Monitoring of water systems at the watershed level for channel conditions should be in place. The reduction of non-point pollution through TMDLs is increasing. Income, property and other taxes should be designed to increase the stock of wetlands and other non-glamorous resources.

REFERENCES

W.B. Solley, *Water-Use Estimates in the United States 1950-95, with projections to 2040*. USGS, Reston VA (2000).

H. Bower, *Report Highlights Global Water Shortage, Pollution* (Unpublished Manuscript, 4/2000.

T.C. Brown, Projections of U.S. Freshwater Withdrawals, *J. Wat. Res. Res.*, 36: 769 – 780 (2000).

Ibid., *Past and Future Freshwater Use in the United States*, U.S. Dept. of Agriculture, Forest Service tech rep. RMS-GTR-39 (1999).

W.B. Solley, E.B. Chase and W.B. Mann IV, *Estimated Uses of Water in the United States in 1980*, USGS Circular 1001, Wash. DC (1983).

W.B. Solley, C.F. Merck and R.R. Pierce, *Estimated Uses of Water in the United States in 1985*, USGS Circular 1004, Wash. DC (1988).

W.B. Solley, R.R. Pierce and H.A. Pearlman, *Estimated Uses of Water in the United States in 1990*, USGS Circular 1081, Wash. DC (1993).

W.B. Solley, R.R. Pierce and H.A. Pearlman, *Estimated Uses of Water in the United States in 1995*, USGS Circular 1200, Wash. DC (1998).

Water Resources Council, *The Nation's Water Resources*, U.S. Gen. Printing Office, Wash. DC (1978)

W. Viessman and C. DeMoncada, *State and National Water Use Trends to the year 2000*, U.S. 96th Congress, 2nd Session, Senate Comm. Pub. Works, Comm. Print 96-12.

R.W. Guldin, *An Analysis of the Water Situation in the United States: 1989 – 2040*,Ge. Tech Rep. RM 177, Rocky Mountain Forest and Range Experimental Station, Fort Collins, Co (1989).

Sport fishing Institute, *Economic Impact of Sport Fishing in the United States*, Wash. DC (1994).

Office of Technology Assessment (US Congress), *Wetlands: their use and regulation*, Rep. No. OTA-O-206, Office of Tech. Assessment, Wash., DC (1984).

U.S. Environmental Protection Agency, *The Quality of Our Nation's Water: 1996*, EPA841-S-97-001, Office of Water, Washington DC (1998a).

U.S. Environmental Protection Agency, *National Water Quality Inventory: 1996 report to Congress Groundwater Chapters* EPA-816-R-98-011, Office of Water, Washington DC (1998b).

U.S. Environmental Protection Agency, *Environmental Investments: the costs of a clean environment*, EPA-230-11-90-083, EPA Office of Policy, Planning and Evaluation, Nov. 1990.

U.S. Environmental Protection Agency, *The Quality of Our Nation's Waters: a summary of the National Water Quality Inventory: 1998 report to Congress*, EPA841-S-00-001, Office of Water, Washington DC.

U.S. Environmental Protection Agency, *National Water Quality Inventory: 1988 Report to Congress*, EPA440-4-90-003, Office of Water, Washington DC, (1990).

The California Water Plan Update, Bull. 160-98, Executive Summary (Dept. Water Res., Sacramento, Ca, Nov. 1998).

WATER RESOURCE MANAGEMENT IN THE TEXAS MEGACITY: A PRIMA FACIE CASE FOR COMPREHENSIVE RESOURCE MANAGEMENT

GEORGE O. ROGERS, CHRISTOPHER D. ELLIS
Department of Landscape Architecture and Urban Planning, Texas A&M University, College Station, Texas, USA

ABSTRACT

The Texas-MegaCity is shaped as a triangle formed by Dallas in the North, Houston in the Southeast and San Antonio in the Southwest. This area contains three metropolitan areas that are each larger than a million residents with more than 13.2 million residents in the entire region. The Texas-Triangle is greater than 25,000 square miles and contains three of the top five fastest growing metropolitan areas in the United States. The triangle impacts nine river basins and five aquifers. While rainfall is adequate to support the population most of the time, recent years of drought have resulted in shortages, and some areas are dealing with this on an ad hoc basis. In addition to problems of water *quantity*, the potential threats to water *quality* are located in the same general area as the key sources of water. As the population grows in the Texas Triangle there is a clear need to manage water resources in a comprehensive manner. Currently, water resources are managed by multiple jurisdictions associated with municipalities, county governments, watersheds and aquifers, and even the state agencies that often take opposing points of view. This has led to an hoc management of existing resources, characterized by some promising approaches to water management. This paper concludes that to meet the needs of a rapidly growing MegaCity like the Texas Triangle, comprehensive, regional management of water resources will be required.

INTRODUCTION

This paper examines the nature of water resources in the Urban Texas Triangle. The Urban Texas Triangle is an area comprised of the geographic region delineated by the metropolitan cities of Dallas/Fort Worth in the north, San Antonio/Austin in the southwest and Houston in the southeast. This area (Fig. 1) contains approximately 25,000 square miles and 13.2 million residents. This paper examines the geographic distribution of existing population, water resources and potential threats to the quality of water. While the co-location of the vital water resources with the population is understandable and expected, the co-location of potential threats to the quality of water is sufficient cause for early alarm.

THE URBAN TRIANGLE AS MEGACITY

Population in the Urban Triangle was 13.2 million in 1998, with 9.2 million residing in the four primary metropolitan areas of Dallas/Fort Worth, San Antonio, Austin and Houston. The concentration of population is also allocated along the transportation corridors provided by the Interstate highway system, specifically Interstates 10, 35 and 45, which provide the southern, western and eastern boundaries of the Triangle. Figure 2 shows that both the current population and rate of growth in the population and the rate of growth in the population are distributed in the Texas Urban Triangle.

Fig. 1. The Urban Texas Triangle

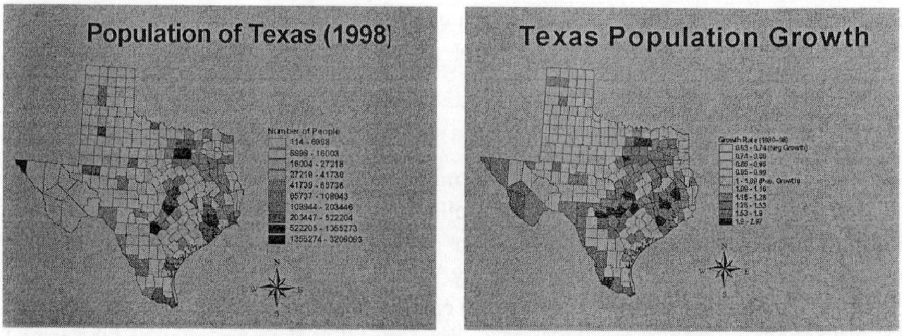

Fig. 2. The Urban Texas Triangle is home to nearly half the population of Texas and is one of the fastest growing areas in the United States (Texas State Data Center, 1998).

Examined over the last two decades, population in the Triangle has grown from 8.95 million to over 13.2 inhabitants. Three of the metropolitan areas in the Triangle are among the five fastest growing metropolitan areas in the United States: Dallas, Austin and Houston. Moreover, while the growth in population is relatively linear, averaging approximately 300,000 per year (2.2% growth annually), the proportion of population in the state of Texas residing in the Triangle has grown in a non-linear manner from 42 % living inside the triangle in 1990 to 45.7% in 2000 (Fig. 3). The 1990 annual median income per capita in the Texas Urban Triangle is $24,113, while the annual income per capita in the rest of Texas is only $20,952 (U.S. Census Bureau, 1990). Figure 4 shows the median income distribution across Texas.

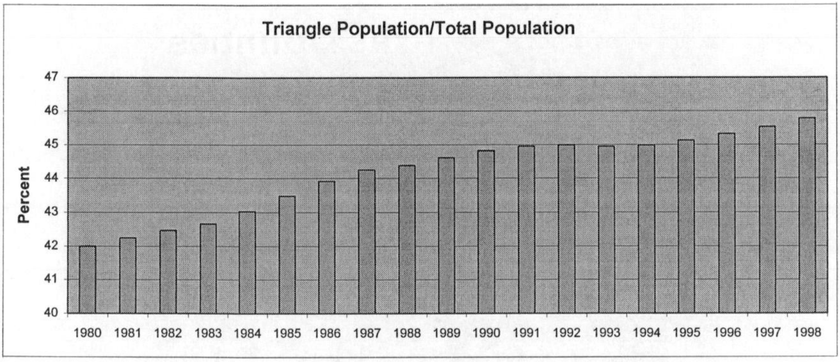

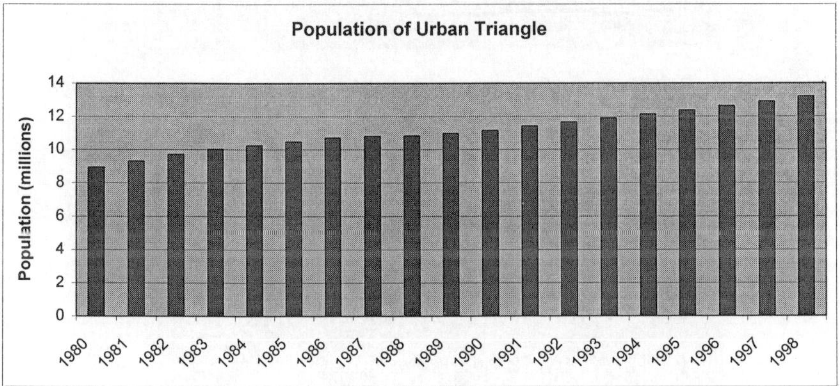

Fig. 3. Comparison of population growth in the Texas Urban Triangle and the State of Texas (Texas State Data Center, 1998).

While the concentration of the population in the Urban Triangle is high in the metropolitan areas, much of the approximately 25,000 square miles in the Urban Triangle remains rural in nature. More importantly, the density in the metropolitan areas and along the transportation corridors has intensified with an average increase of 115 people/square mile over 18 years. The remainder of the state increased by an average of only 16 people/square mile over the same time period. Moreover, some of the most agriculturally productive lands in Texas Triangle are located along the transportation corridors and the urban/population development in the Triangle overlaps these areas. Figure 5 presents the spatial distribution of land uses in the Triangle.

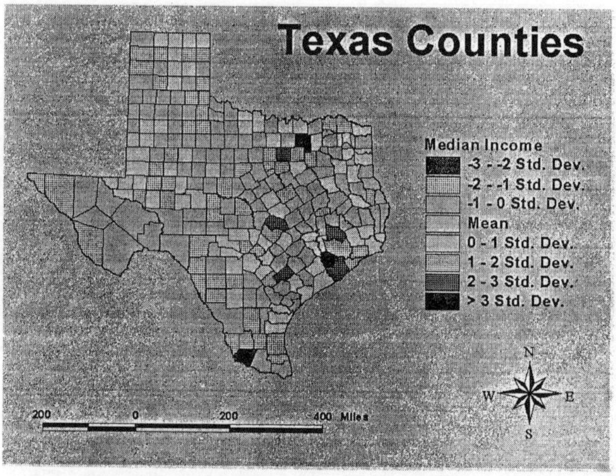

Fig. 4. Distribution of median household income across Texas in 1990 (U.S. Census Bureau).

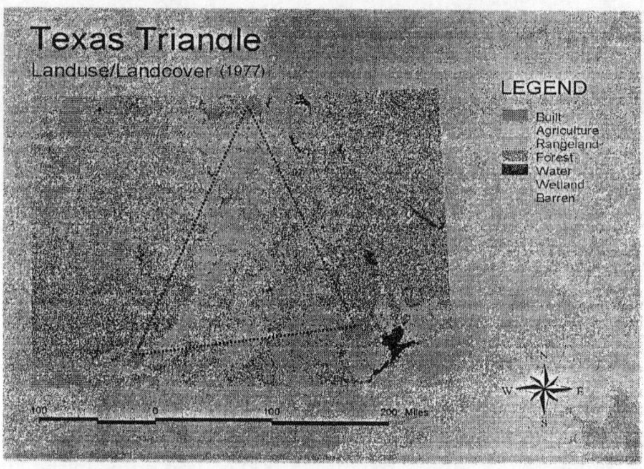

Fig. 5. Land Use/Land Cover in the Urban Texas Triangle (U.S. Geological Survey, 1977).

METHODS, MEASUREMENT, DATA, AND ANALYSIS

Data was acquired from numerous state and federal agencies such as the U.S. Census Bureau, U.S. Geological Survey, National Oceanic and Atmospheric Administration, and the Texas Natural Resources Information System. Socioeconomic data is aggregated at the county level while natural resource data was collected at scales appropriate for the phenomena being studied (generally 1:24,000 to 1:2,000,000). Whenever possible, the most recent releases of data are used in this analysis. However, some natural resources data are not collected on a regular basis and may be up to 25 years old or greater. Due to the diversity in methods of collecting the data used, they are not described in further detail here. Readers are urged to obtain available metadata records directly from the referenced agency.

WATER RESOURCES

Rainfall in the Triangle (Fig. 6) averages 39 inches per year, ranging from 7 inches in the western portion of the triangle to 49 inches in the eastern and coastal regions of the triangle. This means approximately 55 million acre-feet of rain falls on the Triangle per year. If 64% is used for irrigation (Water for Texas, 1997, p. 3-1), and another 9.9% is used for manufacturing, approximately 25% are available for direct support of the Triangle's population in other ways. Put differently, almost 14 million acre-feet of water are available to support the population. This means that if available water were allocated to support population at a rate of 325,000 acres-feet per capita, rainfall in the area would support approximately 40 million people inhabiting the Texas Triangle. This assumes single use of water but of course water can be used, cleaned, and re-used many times.

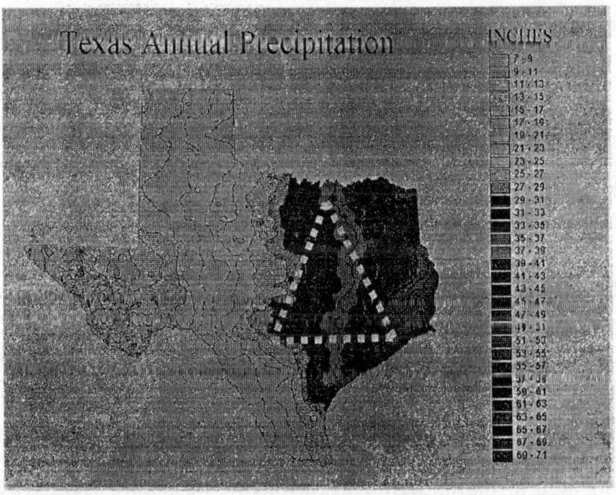

Fig. 6. Texas annual precipitation inches (Source: NOAA).

The rainfall is collected in six river basins covering the Urban Triangle. The river basins are presented in Figure 7. Three major river basins in the Triangle include the Trinity River Basin, running from Dallas/Fort Worth in the north to Houston in the southeast, the Brazos and Colorado River basin, running from New Mexico in the west to southwest of Houston. These three river basins account for 59% of the land area of the Texas Urban Triangle.

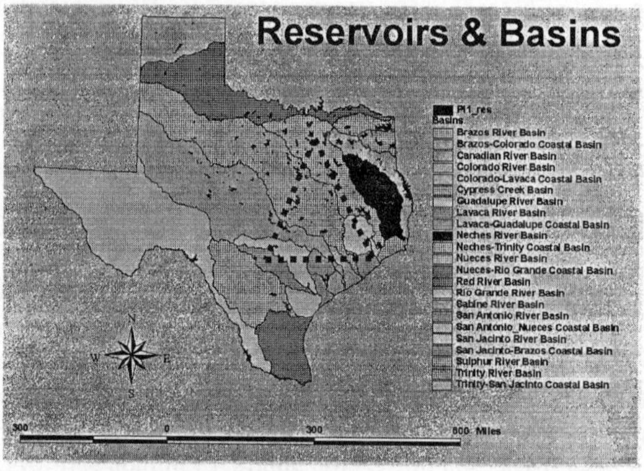

Fig. 7. River basins and reservoirs in Texas (Source: Texas Natural Resources Information System)

There are four major aquifers in the Urban Triangle and a fifth aquifer of some significance (Fig. 8). The northernmost aquifer is the Trinity Aquifer. The Trinity Aquifer runs from the Red River on the border with Oklahoma through Dallas southwestward to near San Antonio. The major use of the water contained in the Trinity Aquifer is for irrigation of agriculture. The Edwards Aquifer is the smallest, although not least important aquifer, in the Triangle. The Edwards Aquifer runs from the northeast of Austin, Texas to near the Rio Grande River (Mexican border). The water quality in the Edwards Aquifer is quite good, and is a major source of drinking water in Austin, Texas, in the northern part of the aquifer, and used primarily for agriculture in the southernmost regions of the aquifer. The Carrizo Aquifer runs from the northeastern border with Arkansas to the southwest through San Antonio to the Mexican border at the Rio Grande River. Carrizo Aquifer water is used primarily for municipal water supplies (31%) and agricultural irrigation (81%) (Water for Texas 1997, p. 3.218). The Gulf Coast Aquifer is the southernmost aquifer in the area. It is geographically located from the border with Louisiana in the northeast through the Houston area to the Mexico border in the southwest. The Ogallala Aquifer is important, not because it is located in the Triangle, but rather because it is located at the uppermost portions of the major river basins that flow through the Texas Urban Triangle. To the extent it is used in the Texas Panhandle and allowed to flow into these basins, it is a potential issue impacting the Triangle.

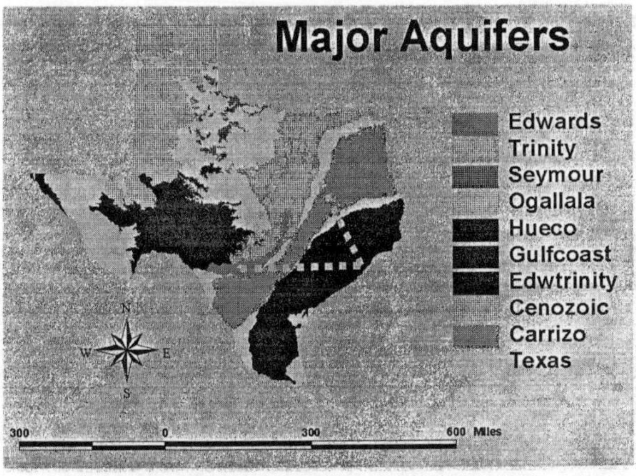

Fig. 8. Major aquifers of Texas (Source: Texas Natural Resource Conservation Commission).

THREATS TO WATER QUALITY

The potential threats to water quality considered are presented in Figure 9. There is no attempt to show how these potential threats are directly related to degradation of water quality in the paper. These data only show that some of the potential threats to water quality are located in the Urban Triangle. Moreover, we do show that the threats are concentrated in the areas characterized by the highest population and the fastest growing population areas of the Triangle. As would be expected, the number of sanitation land fills is directly related to population concentrations, with there being about one landfill site per seven square miles in the Triangle and about one landfill site per 1100 square miles in the other areas of the state. (Fig. 9a)

The concentrations of public water systems in the Triangle that exceed fecal coliform standards are generally located in those areas of the Triangle that have the highest concentrations of populations. (Fig. 9b) This data on fecal coliform provide the most direct indication of some already existing contamination of water supplies in the Texas Urban Triangle.

Figure 9c presents some preliminary evidence that toxic releases have occurred in the Urban Triangle. Moreover, these releases are co-located with the production of hazardous materials and the population concentration of hazardous materials and the population concentrations of the Triangle. Figure 9d offers some preliminary evidence that emissions into the air are also concentrated in the Texas Triangle. The Dallas metropolitan has now (in 2000) met federal standards for emissions for automobiles. The Houston area has yet to meet these standards and has recently surpassed Los Angeles as the smoggiest city in the United States. These releases have been concentrated in the Urban Triangle in the same fundamental areas that produce hazardous chemicals, releases toxins to surface water and sewer systems, and where the population is located. Adjusting the size of releases and emissions by land area does not significantly alter the pattern reflected in the geographic distribution of the data.

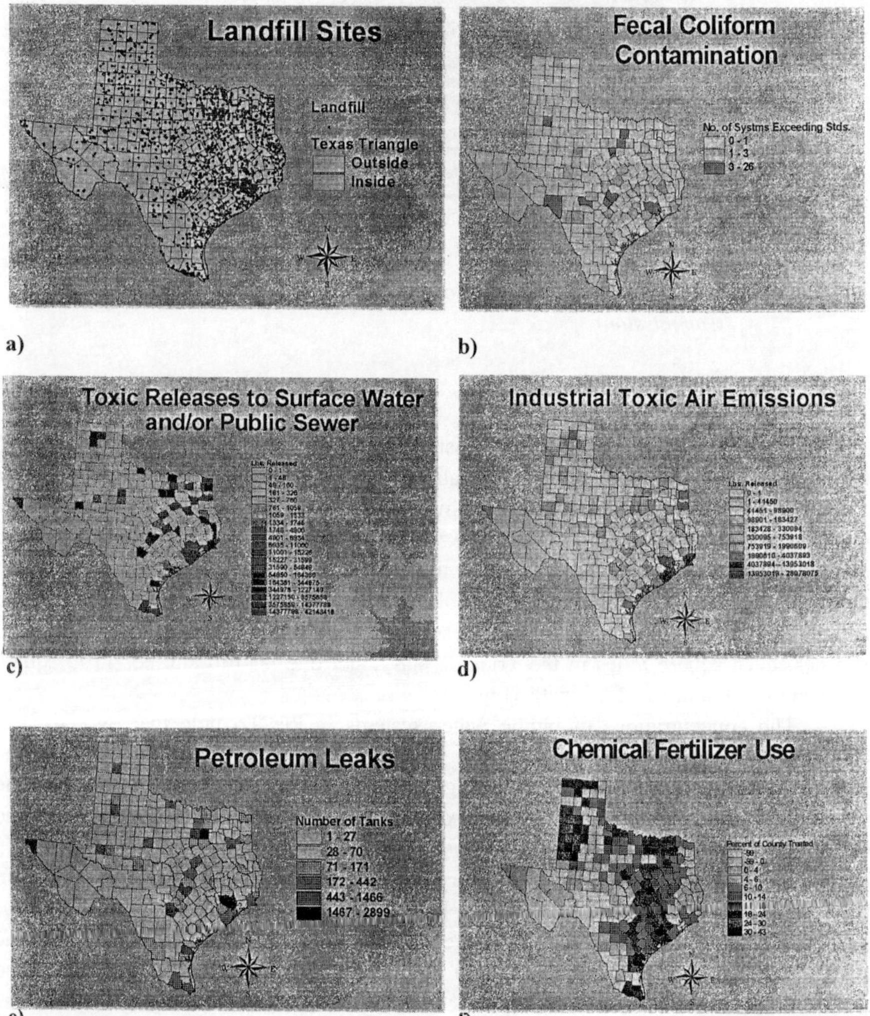

Fig. 9. Threats to water quality (Texas Center for Policy Studies, 2000).

Figure 9e shows the geographic distribution of the number of leaks in petroleum tanks. All counties have reported leaks occurring within their boundaries. The familiar pattern of the Texas Triangle once again emerges. As would be expected, the number of leaks is spatially distributed with the residential population.

The geographic distribution of chemical fertilizer use is presented in Figure 9f. This data indicates that the application of chemical fertilizer is not only concentrated in the Texas Urban Triangle, but also in the areas upstream to the Triangle in both the Brazos and Colorado River basins.

PROJECTIONS FOR THE FUTURE

This section of the paper presents projections of population in the Texas Urban Triangle and uses those to project the usage of water on the basis of population. Figure 10 shows the projected population numbers, collected at the county scale from the Texas State Data Center and amalgamated to represent population in the Texas Triangle up to the year 2050. These projections indicate that the population will grow from 13.2 million in 1998 to 27 million in 2050 in the Texas Triangle. The pattern of water use seems to depend on population (Fig. 11). Here, county level water use data was normalized by population and displayed to show the per capita water use spatial distribution.

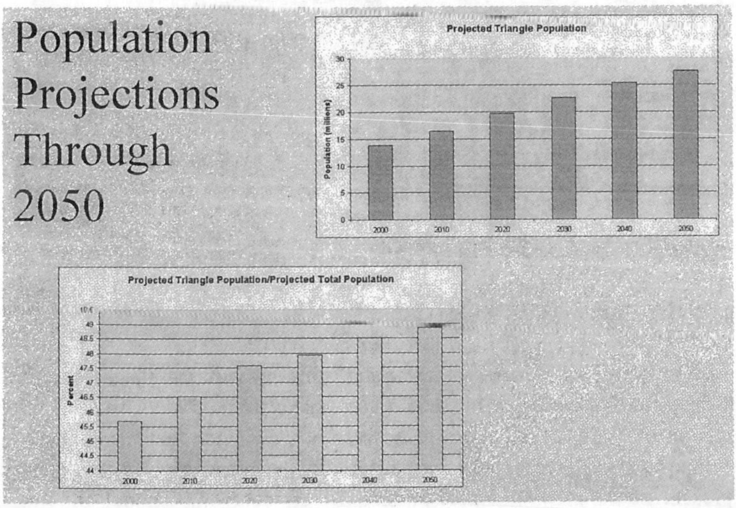

Fig. 10. Texas Population Projections through 2050 (Source: Texas State Data Center).

Water use projections, based on expected population, means that water use will expand from 5 million acre-feet to 10 million acre-feet by the middle of the next century.

The proportion of water use in the Urban Triangle is projected to expand from 21% to over a quarter, or 26% of the State of Texas' use of water. This indicates that anticipated water use might be met without the need for recycling water. It becomes clear that without the consideration of recycled water resources, the quantity of water is adequate for projected urban growth. Hence, water quantity and projected use alone does not seem to create a meaningful limit to development in the Texas Urban Triangle.

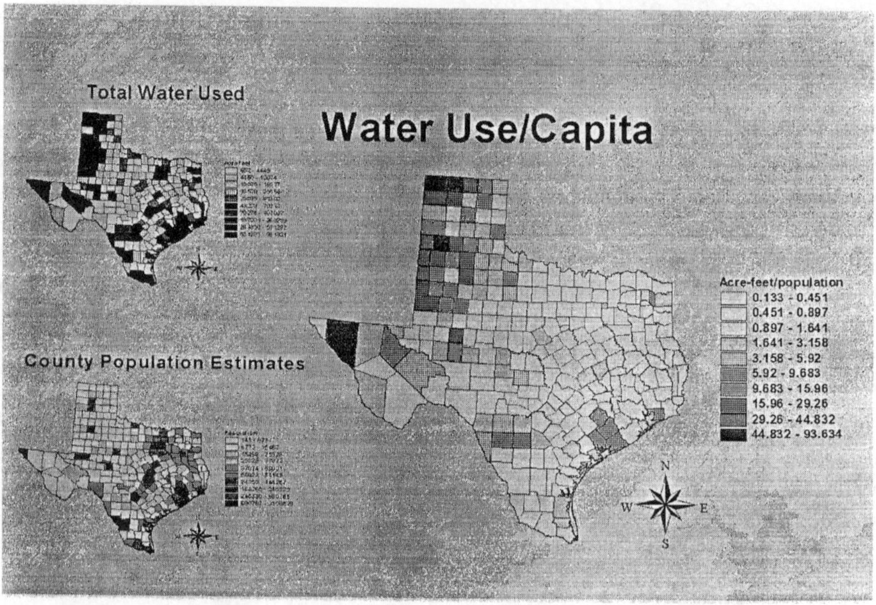

Fig. 11. Texas water use per capita (Source: Texas Environmental Almanac, 2000).

PROSPECTS FOR WATER RESOURCE MANAGEMENT

This section presents a series of "anecdotes" that provide examples of activities in the Urban Triangle that have had impacts on, or are related to, water issues. It begins with two examples of situations where water resource consideration have (and are) directly guided land-use and development in the Woodlands and Austin, Texas. Next the section will present some preliminary discussion of an economic model of water resource allocation being discussed by some serious entrepreneurs. Next two communities, San Antonio and Cuero, where flooding has been a repeated problem, will be discussed. And finally, a summary of one environmental clean-up effort in the Colorado River Basin will be summarized. These examples provide a caricature of the prospects for water resource management in the Triangle. These anecdotes are not intended to be a comprehensive or systemic analysis of the prospects for water resource management, but rather highlight the kinds of activities (related to water) underway in the area.

The Woodlands development was one of New Town communities with roots in the Johnson Administration's vision of better cities in the United States which gave rise to the New Community Act of 1970. The design of Woodlands was influenced by Ian McHarg's vision of limiting impact on water resources, and the environment. The intensity of the development was matched to the soil-type to enable a "zero-impact" design principle. Areas with soils that prevent water penetration (eg. clays) were reserved for the most intense development of shopping areas, multiple-family dwelling units and other concrete/asphalt dominated uses. Adjacent to these areas, water retention ponds collect excess water from heavy rains. These retention ponds provide a visual amenity for the area and prevent increases in down stream flooding associated with increased runoff. Areas with porous-soils (eg. sands) were reserved for parks and wildlife refuges. In areas where residential development was permitted, residents are required to preserve ground cover. Strictly enforced deed restrictions provide use guidelines to allow both water conservation and preservation. The proportion of each residential property covered in lawn is restricted to conserve water use and the allowable concrete and asphalt coverage is restricted to assure penetration of available water. These design features, together with the social/cultural emphasis on mixed use of land and heterogeneous community, have helped make the Woodlands one of the most successful communities developed under the New Town Act of 1970.

The City of Austin obtains the majority of its drinking water from the Edwards Aquifer. The Edwards Aquifer is also recharged in the Austin area via several creeks in the area (e.g. Barton, Onion) which collect surface water in the area and provide it to the aquifer through the karst topographic features in the aquifer recharge zone. The benefit of the water is clear, but the potential to collect environment pollutants along with the water in developed areas of the recharge zone is real. The City of Austin recognizes this problem and has developed a comprehensive plan (Fig. 12) that encourages development in areas that are outside the recharge zones. For example, a well known computer company recently was encouraged to locate their facility outside the recharge zone by giving tax incentives only if they would locate outside the recharge zone. The Barton Creek/Edwards Aquifer Authority was created in 1993 by the Texas Legislature to "to manage, preserve and protect the Edwards Aquifer". The authority is working toward that goal by a) encouraging growth management, b) the use of xeriscaping that promotes water conservation via low maintenance, c) protecting the recharge from pollution, and d) assuring recharge by protecting and restoring the flow of clean water into the aquifer.

Recent droughts in Texas have highlighted potential problems with water quantity. Figure 13a shows the geographic distribution of the water deficits compared to historical average rainfall. Figure 13b shows the impact these shortages have had on selected reservoirs. This situation has given rise to recent discussion among entrepreneurs to develop private enterprise water projects that would purchase water rights in the Texas Panhandle and build a water delivery system, presumably, a pipeline or canal, to transport that water to the San Antonio area, a distance of approximately 300 miles. The economic losses in agriculture alone, due to the drought, total a staggering $600 million. This situation recently led to a national disaster declaration. Hence, the current drought has created a potential for entrepreneurial activity associated with water resources. While the effectiveness of business organizations at meeting public needs is not questioned here, the potential vulnerability created by water resources being owned and managed by private enterprise is of concern. Without effective regulatory authority, complete privatization of water resources would create the opportunity for "public blackmail" associated with both the quantity and quality of this vital resource.

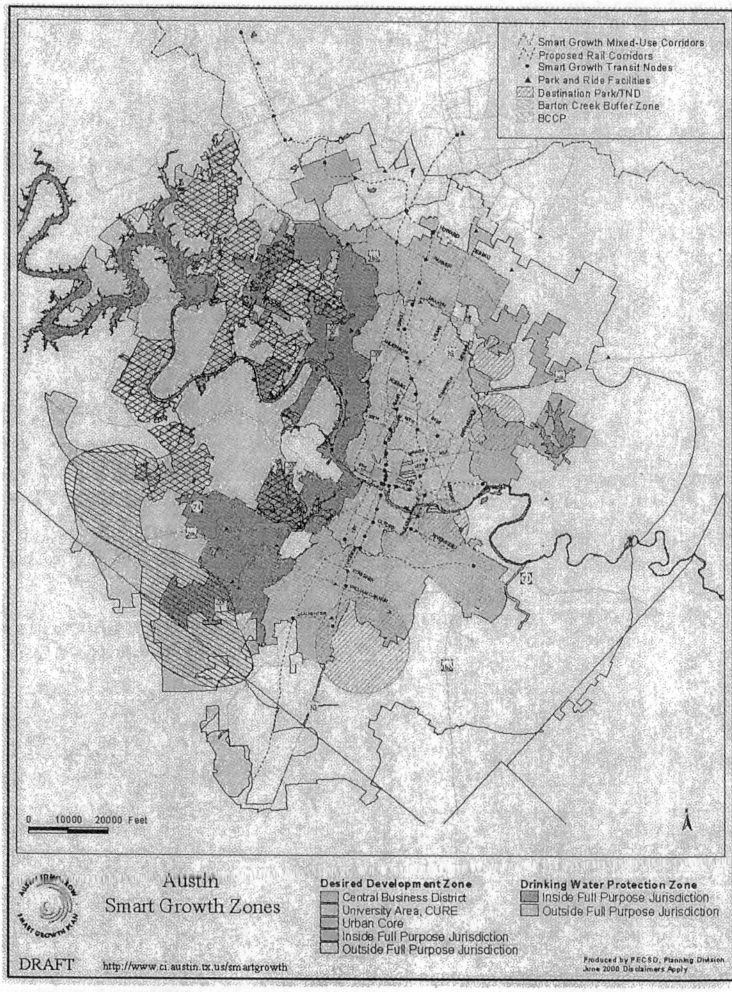

Fig. 12. City of Austin Smart Growth Zones (Source: City of Austin).

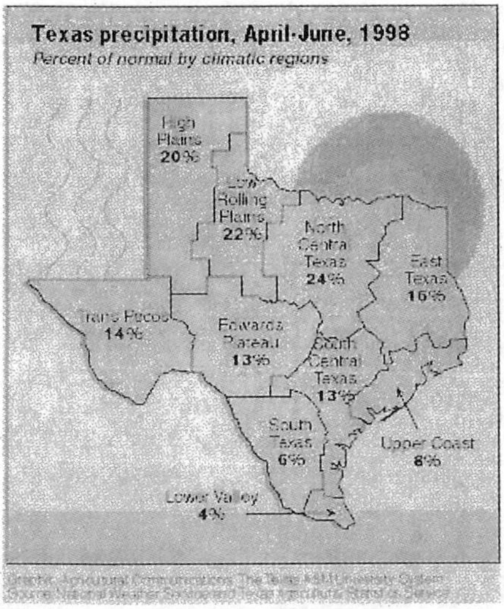

Fig. 13a. Drought conditions are depleting water supplies and affecting water quality (Source: National Weather Service and Texas Agricultural Statistics Service).

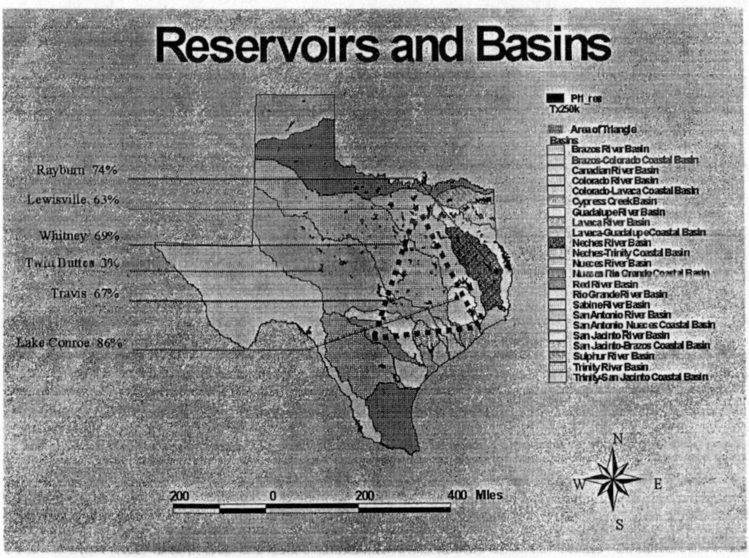

Fig. 13b. Drought causes significant shortages in drinking water supplies (Houston Chronicle, 2000).

In the City of San Antonio (and other places) an innovative approach to flood control involves intercepting flood waters upstream from the city and channeling the water into a siphon, where it passes beneath the city and exits below threatened areas (Fig. 14). This "flood-tube" has been used not only to capture flood waters and channel them effectively through the city, but also as a storage device for use during long dry periods that may impact river use and its esthetic value. The San Antonio flood tube was designed by the United States Army Corp of Engineers. Construction began in 1987 and was completed in 1997. Since its completion, it has been used to flush the river running through the city by pumping water out of the tube and into the river upstream, and replacing it by pumping it out of the river and back into the tube downstream. Water in aquifers is no longer used for this purpose in San Antonio. The water stored in San Antonio's flood tube is also used in a limited way for an irrigation source for public parks and other public areas.

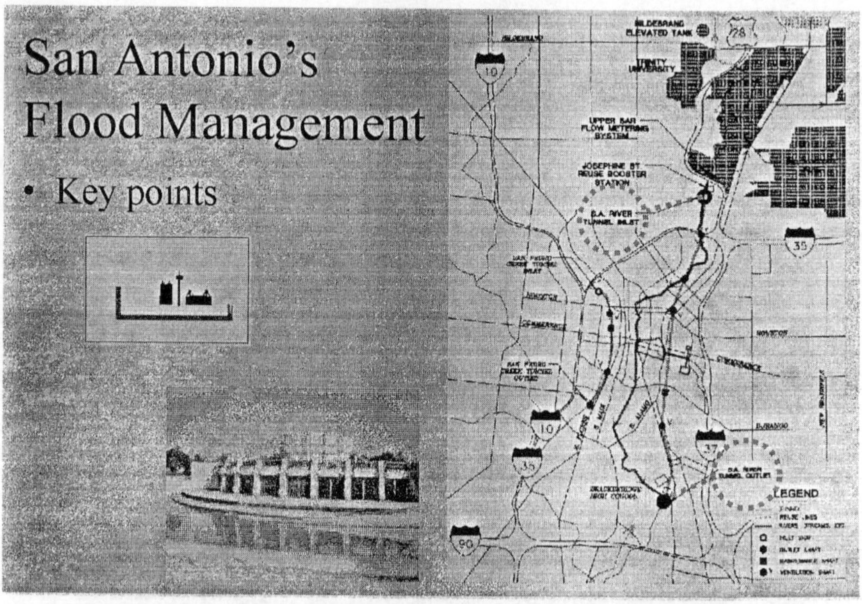

Fig. 14. San Antonio's Flood Tunnel is also used to maintain water quality.

Cuero, Texas is a small town of about 7000 residents located near the southern edge of the Texas Triangle in DeWitt County. The per capita annual income in Cuero is approximately $9500 compared to the per capita annual income of $21,800 for the State of Texas as a whole. Cuero first recorded a flood in 1913, then again in 1936, 1952, 1972, 1987, 1991 and 1993. It is interesting to note that the average time between floods from the first five events is a flood about 18.5 years, while the time between the last four events averages about 3.7 years. These data may reflect an increasing propensity to flooding associated with growing urbanism in the southeast corner of the Texas Urban Triangle. The most recent flooding event occurred in the third week of October 1998. The flood was a result of both an extreme rainfall event and a runoff event from downstream areas. The flood crested at about 50 feet above flood-stage, which was approximately 30 feet greater than National Weather Bureau predictions. This was an unusual event to be sure, but it illustrates the kind of extreme events that can occur in the Texas Urban Triangle. The kind of events to which the growth in the Triangle may be contributing.

The Lower Colorado River Authority (LCRA) was formed by the Texas Legislature in 1934 as a conservation and reclamation district. Its mission is to improve the quality of life in the Central Texas served by the Lower Colorado River. In 1988, a group of citizens became concerned about the water quality in the basin, and formed the Clear Clean Colorado Foundation. Initially the Todd foundation provided support for students in Austin to monitor the quality of water along Walnut Creek at three locations. Five new monitoring teams were added in 1989, and over the next two years support was added by a local school district, and a clothing company. In 1991, the effort received a four-year grant from the National Science Foundation to help support the program. During this period the LCRA took over operations of the River Watch Foundation. In 1993 the U.S. Environmental Protection Agency supported an effort to develop quality control/quality assurance procedures for chemical parameters. River Watch currently supports monitoring groups at about 70 locations involving about 20 local school districts in the LCRA drainage area. The program involves students from local schools monitoring the quality of water in tributary streams. The program has enjoyed tremendous success in involving local citizens in the process of monitoring and improving water quality in the basin.

REGULATORY RESOURCES

There are two distinct genre of socio-political organizations that control and impact both the quality and quantity of water resources in the Texas Triangle. These resources may be thought of as primary and secondary. Primary water resource management organizations have as their primary function the protection and delivery of the water supply; they are in most instances charged in the most direct way, with assuring adequate supply of quality water resources. Secondary organizational structures are responsible for water supply and quality, but only as one of many other responsibilities. The primary organizations are Water Districts, Basin Authorities and Aquifer Authorities. Secondary structures generally fit into the categories of Municipal Governments, County Governments, Regional councils of government and their planning organizations, and state and federal agencies.

There are numerous Water Districts and Special Utility Districts involved in the Texas Triangle. These authorities are primarily involved in end-user supply of water. Most are run by a board of users elected by users in their districts. There are numerous river basin authorities associated with the Texas Triangle. The six river basins plus the

Colorado, Brazos, Trinity and San Jacinto are split into upper and lower basin authorities. In addition, the Lavaca, Guadalupe, and San Antonio basins combine in the southern part of the Triangle to form another basin and attendant authority. These basin authorities are authorized by the State of Texas and managed by authorities appointed by the governor. There are four aquifer authorities acting in the Texas Triangle. These authorities operate under the auspice of the state legislature with boards appointed by the governor. These aquifer authorities often take direct action in their load areas (e.g. Edwards Aquifer Authority) to protect water quality and assure water supplies. This often involves coordinating with local municipal and county government organizations and the public at large.

There are 64 county governments and multiple municipal governments in each county of the Texas Urban Triangle. Municipal governments are of an elected mayor and council structure and many have an administration hired by the elected structure to handle municipal business. Municipal governments most frequently operate municipal wastewater treatment plants for the vast majority of residents in the area. County governments are typically structured as a county judge and several county commissioners that are elected by the residents. The county government is usually responsible for governmental services outside the municipalities. The county governments are also most likely to be the regulatory authority in charge of residential wastewater treatment outside municipal areas via county public health organizations. Many of the municipal and county governments operate utility companies that distribute water, electricity and natural gas and other public services.

There are various state and federal agencies that have at least some jurisdiction over water resources. State agencies involved include the Texas Water Development Board, the Texas Natural Resource Conservation Commission and the Texas Wildlife and Fisheries Department. Conflicts can arise between the agencies with conflicting purposes. The Texas Water Development Board is primarily charged with developing water resources while the Texas Natural Resource Conservation Commission is charged with environmental compliance. Federal agencies can include the Environmental Protection Agency (EPA), the Army Corp of Engineers, the Bureau of Land Management (BLM) and the Federal Emergency Management Agency (FEMA). The EPA and the Corp and BLM often come into conflict over the development of new reservoirs while FEMA is often called upon to administer flood insurance mitigation and protection programs.

CONCLUSIONS

To be clear, this paper has not attempted to conduct a comprehensive analysis aimed at assessing the quantity and quality of water resources in the Texas Urban Triangle. This paper has attempted to examine, in the broadest of terms, the nature of the geographic distribution of population, water resources and potential threats to the integrity of those resources in the Texas Urban Triangle. This paper also presents some limited anecdotes of water activities in the area. These examples are intended to exemplify the range of activities in these areas to protect and assure the quantity and quality of water in the area and the range of issues in the area. This paper also examines the regulatory structures involved in management of water resources in the Texas Urban Triangle. This paper is not intended to present definitive analysis of water resources and their management in the Triangle. Rather, it examines the existing evidence, to the extent required, to address the question of: does water limit development, and what are the critical issues relating to water in the urban development in the Triangle?

The primary conclusion of this limited qualitative analysis is that quantity of water resources alone is not sufficient to limit growth of urban development in the Texas Triangle. That is to say that water scarcity has not stopped or slowed urban development in any appreciable way, and the anticipated available water anymore than we currently use water on a per capita basis. And this conclusion assumes a conservative "single use" water strategy, which is already being violated. This being the case, what is likely to inhibit urban development in the future? And why has water been such an important topic of discussion in recent months in the Texas Urban Triangle?

Before attempting to address these questions directly, the discussion of evidence regarding potential threats to water quality is in order. The co-location of potential threats to water quality with both population and the concentrations of urban population alone gives rise to sufficient alarm, to require a more detailed analysis than possible here; and the development or refinement of management structures to address the quantity and quality of water in the Triangle in a comprehensive examination. Secondly, the concentration of population and potential threats to water contamination materialize. Thirdly, potential threats to water resource quality located upstream of the Triangle indicates both impact of external areas on the Triangle and impact of the Triangle on the external areas need to be considered. Finally, the implications of the water as a carrier for human and industrial waste for the use-and-re-use of water needs to be considered in a more comprehensive manner throughout the Triangle.

This paper presented a small selection of anecdotes to highlight some of the activities regarding water resources in the Texas Urban Triangle. The long-term development strategies of zero-impact on water runoff in the Woodlands provides one small example of successful development without significant impact on water resources. The comprehensive planning in Austin with respect to the intricacies of the Edwards Aquifer recharge area provides both short-term and long-range benefits to the local area. The clean up of the Colorado River provides an example of public participation and extensive participation of stakeholders in developing the program to help clean up the basin for all users. The flood tube in San Antonio provides an example of an innovative technological solution to a flooding problem that provides solutions to additional problems related to water in the local area.

The multiple jurisdictions involved in the regulation of water in the Urban Triangle create a fractured if not disjointed management structure for the water resources in the region. The multiplicity of regulatory organizations can be found to be overlapping and confusing, and are easily seen to lead to a potential for jurisdictional disputes. For example, some are charged with the distribution of water for use, while the aim of others is clearly conservation. Missions can be found to conflict, making it difficult to effectively manage the available water resources, especially when missions, objectives and strategies are not coordinated between agencies and levels of regulatory authority. Achieving a comprehensive management strategy that encompasses the full-cycle use-and-re-use of fundamental resources is critical to assuring availability of key resources necessary to support life in the emerging megacity

ACKNOWLEDGMENTS

We would like to express our sincere appreciation to the World Federation of Scientists for their support in hosting the International Seminar on Planetary Emergencies in Erice, Italy where this paper was presented. We would also like to thank Professor Jesus Hinojosa and Jennifer Evans for their help in initiating research on the Texas Urban Triangle, and Ric Jensen for sharing data and references from the Texas Water Resources Institute.

REFERENCES

Texas Center for Policy Studies. 2000. Texas Environmental Almanac 2nd Edition compiled by M. Sanger and C. Reed. University of Texas Press, Austin.

U.S. Census Bureau, United States Department of Commerce (http://www.census.gov/)

National Oceanic and Atmospheric Administration. Texas Annual Precipitation (Available at: http://www.noaa.gov/)

AgNews, 1998. Texas A&M University Agricultural Program. (http://agnews.tamu.edu/)

Abram, L. June 6, 2000. "Texas' major lakes at lowest level in two decades." Houston Chronicle

City of Austin. 1998. City of Austin Smart Growth Zones, (http://www.ci.austin.tx.us/smartgrowth/)

Texas Natural Resource Conservation Commission, (http://www.tnrcc.texas.gov/)

U.S. Geological Survey. 1977. Land Use/Land Cover Maps of Texas (Available at: http://www.tnris.state.tx.us/index.htm)

U.S. Census Bureau. 1990. County Level Median Income. (Available at: http://www.census.gov/)

Texas State Data Center. 2000. Texas Population Estimates and Projections Program (Available at: http://txsdc.tamu.edu/)

15. WORKSHOP ON ENVIRONMENTAL IMPACTS OF OIL POLLUTION IN THE BLACK SEA

ENVIRONMENTAL IMPACTS OF OIL POLLUTION IN THE BLACK SEA. SUMMARY OF THE POLLUTION PERMANENT MONITORING PANEL WORKSHOP

RICHARD C. RAGAINI
Lawrence Livermore National Laboratory, Livermore, California, USA

The workshop was held on 18-19 August 2000 and addressed five issues:

1. a review of what is known about Black Sea oil pollution;
2. identification of unresolved issues in the Black Sea;
3. current and projected actions in the Black Sea;
4. possible WFS/World Lab actions;
5. final resolution.

OIL POLLUTION ISSUES IN THE BLACK SEA

On the first morning, introductory remarks were made by the three organizers:

Dr. Gennady Palshin, ICSC-World Laboratory, Ukraine Branch;
Prof. Albert Tavkhelidze, National Academy of Science of Georgia;
Dr. Richard C. Ragaini, Lawrence Livermore National Laboratory, USA.

During the initial discussion, Dr. Ilkay Salihoglu, Middle East Technical University, Turkey pointed out that there was a NATO-sponsored oceanographic cruise scheduled for the Black Sea September 22 - October 15, 2000, but it did not include measurements of pollution, radioactivity or petroleum. The focus was on nutrients and heavy metals. Also, Turkey is launching a new coastal research program. He said oil pollution is an increasing crisis that requires more attention. Other environmental problems are already getting priority attention.

Prof. Valery Mikhailov, Ukrainian Scientific Center of Sea Ecology, Ukraine, spoke on *"Problems of Oil Pollution in the Azov-Black Sea Basin."* He pointed out that the Azov Sea is not contaminated with hydrogen sulfide, as is the Black Sea, because the Azov Sea is shallow and less than 30-40m deep. The Danube river dumps 54,000 tons of oil into the black Sea each year, while the contribution from all other sources is 210,000 tons per year. The sediment in Sevastopol harbor contains 12g/kg oil hydrocarbons, and the primary cause is that ships do not have oil treatment facilities. This is three times the regulatory requirement. Due to these high levels of hydrocarbons and other stresses, the fisheries have declined dramatically in recent years. Another marine problem, which is not widely known, is the danger of unexploded mines left there from World War II. The largest problems in the Black Sea are sewage and industrial discharges, which are being addressed by the BSEP (Black Sea Environmental Program based in Istanbul). While oil pollution is a minor problem today, the oil transport in the Black Sea is expected, in the near term, to increase by a factor of three. Similar problems face the Caspian Sea where monk seals have been dying off, due to sewage and industrial discharges.

Dr. Lado Mirianashvilli , Georgian Academy of Sciences, Georgia spoke on "*Application of Geoinformation Systems for Operative Responding to Oil Spill Accidents.*" He emphasized that their main worries in Georgia are leaks from the oil pipeline from Baku, Azerbaijan caused by seismic activity. Georgia has been funded by the World Lab for seismic monitoring.

Dr. Ilkay Salihoglu, Middle East Technical University, Turkey spoke on "*Effect of the Black Sea Oil Pollution on the Turkish Strait System and the North Aegean.*" He discussed the history of the formation of the Black Sea 12,000 years ago, and today there is a two-flow system with fresh water on the top flowing from the Black Sea into the Mediterranean, and saline water on the bottom flowing from the Mediterranean into the Black Sea. This density gradient governs the behavior of the Black Sea, and prevents mixing, thus maintaining the permanent anoxia condition below 200 m. Twelve million people live in the Bosphorous Strait, which is very vulnerable to tanker accidents similar to the Nacia tanker accident in 1994. So Turkey is very concerned about the increase in oil tanker traffic in the Black Sea and in the Bosphorous Strait.

Ms. Kay Thompson, U.S. Department of Energy, USA spoke on "*Establishing a Data Base of Black Sea Pollution.*" She pointed out that a regional oil spill contingency plan has been formulated, and presented to the ministers of the six Black Sea countries for approval. She described the Black Sea web site maintained by the U.S. Department of Energy, which will contain historical environmental data from the six Black Sea countries. Ukraine has already started loading its historical data. The web site will also contain scientific papers, and Romania has agreed to contribute their papers. She proposed that the WFS should function as a peer reviewer to ensure the quality of the data. She also mentioned that the U.S. Navy is interested in sponsoring a scientific workshop on the Black Sea, which looks at the issues concerning submerged munitions.

Ms. Melissa Lapsa, Oak Ridge National Laboratory, USA, spoke on "*Communication Across the Black Sea via Internet Technology,*" and she described in detail the U.S. Department of Energy Black Sea web site. It was pointed out that the Istanbul Center has already run calibration programs with the IAEA Laboratory in Monaco on heavy metals, radioactivity, and hydrocarbons.

Dr. Ender Okandan, Middle East Technical University,Turkey, spoke on "*Detection and Assessment of Oil Pollution in the Black Sea and Bosphorous Strait.*" She pointed out that 2900 tankers passed through the Bosphorous strait in 1995 carrying 50M tons of oil, and this total increased to 82M tons of oil in 1999. In 1995 the total number of ships increased to 47,000, while in 1999 this had increased to 48,000 ships. In 2000 the Caspian oil fields are producing 1.3M barrels/day, and this is expected to increase to 3.6M barrels/day in 2015. She emphasized that the natural and anthropogenic sources of oil must be quantified, and the fate and transport of these oil leaks must be studied.

Mr. Dumitru Dorogan, Romanian Marine Research Institute, Romania, spoke on *"Oil Pollution Risk Assessment in the Black Sea and Romanian Coastal Waters."* He claimed that the threat to the Black Sea from land-based sources of pollution is potentially greater than in any other sea. One of the largest threats is waste water discharges, which are estimated at 571M cubic meters per year, of which 144M cubic meters are untreated. In 1990 the U.S. Academy of Sciences estimated that accidents dumped 121,000 tons of oil per year into the Sea, while operational discharges dumped 411,000 tons, compared to 2.13M tons in 1973. According to IMO, normal operations account for 70% of the pollution from ships with 21% from accidents. With the growth of the oil industry in the Caspian Sea area, the oil traffic in the Black Sea is expected to more than double in the near future. The capacity of the existing pipeliness from Baku, Azerbaijan to Supsa, Georgia and Novorossiysk, Russia is estimated at 550M barrels/yr of oil increasing to 7B barrels/yr in 2005. A new pipeline under construction from Kazahstan to Novorossiysk will carry about 3.5B barrels/yr by 2005. The accidents and risk to Istanbul from this increase in oil is expected to increase by 5-6 times.

UNRESOLVED ENVIRONMENTAL NEEDS IN THE BLACK SEA

The 1997 Black Sea Action Plan listed the most important unresolved needs:

- significant reduction in ship discharges
- construction of oil treatment facilities in major Black Sea ports
- creation of a harmonized system of enforcement and fines for unsafe shipping
- implementation of National Contingency Plans
- implementation of a Regional Contingency Plan
- creation of a harmonized regional risk assessment system
- institute measures to avoid further introduction of alien marine species.

There is now a regional plan, which has been presented to the ministers of the six littoral countries for approval. The six littoral countries are not funding any significant oil pollution work now, although there are some activities being done with outside funding.

CURRENT AND PROPOSED ACTIONS IN THE BLACK SEA

1. NATO/IAEA are conducting an oceanographic cruise September 2000.
2. EU TACIS and PHARE programs are continuing:
   a. support for the RAC in Odessa
   b. 1998 workshop in Odessa.
3. Proposals from the ERAC in Varna, Bulgaria:
   a. establish a Black Sea Oil Identification System
   b. management of pollution caused by shipping and industrial activities.
4. U.S. proposals:
   a. proposal submitted to DOE on environmental monitoring
   b. bio-diversity proposal submitted to the Turner Foundation

   c. possible workshop to be sponsored by the U.S. Navy.
5. Ukraine proposals:
   a. regional monitoring program proposal
   b. environmental quality objectives supported by TACIS.
6. Romanian proposals:
   a. address marine beach erosion.
7. Georgian proposals:
   a. call for independent EIA of pipelines for seismic risks, including the Baku-Ceyhan pipeline and the Novorossiysk - Samsung pipeline.

POSSIBLE WFS/WORLD LAB ACTIONS

1. Call for establishment of a regional Black Sea authority to negotiate regional legal and regulatory infrastructures.
2. Sponsor six students to attend January 2001 DOE/NOAA Tbilisi training workshop on NOAA GNOME water pollution software model.
3. Sponsor workshop on adoption of regional regulatory standards for environmental enforcement

FINAL RESOLUTION

Recognizing that the Black Sea is the most environmentally-impacted water body in Europe*, the World Federation of Scientists (WFS) Planetary Emergency Panel (PMP) on Pollution has been concerned with the ecological impacts of the environmental pollution in the Black Sea for the past three years. In a February 1999 WFS workshop on this topic, it was concluded that the major environmental pollution impacts in the Black Sea are due to coastal discharges of untreated sewage, introduction of nitrates and phosphates form the Danube River and industrial discharges into the rivers and the Sea. Plans to mitigate these impacts have been addressed in the Black Sea Strategic Action Plan formulated by the riparian countries, and await funding implementation. It was concluded that the potential for oil pollution poses a significant environmental threat as construction of new oil pipelines from Central Asia increases the oil transport through the Black Sea by a factor of 2-3 in the future. This will increase the risk of oil discharges in the Black Sea from accidents, spills, deballasting, etc. Therefore, we recognize the important need to document the existing oil pollution, in order to prepare a baseline against which to evaluate any future changes.

This Workshop on Environmental Impacts of Oil Pollution in the Black Sea brought together scientists from the Black Sea countries to discuss the current state of knowledge of oil pollution, and produced the following action plan for the PMP:

1. Request World Laboratory support for 6 additional Black Sea scientists for training at a U.S. Department of Energy (U.S. DOE) workshop on water plume modeling in Tbilisi, Georgia, in January 2001 through the World Laboratory Georgian Branch. The U.S. DOE is already providing support for 6 Black Sea scientists to be trained by

the U.S. National Oceanic and Atmospheric Administration (NOAA). This joint activity will proceed under the Memorandum of Agreement between the U.S. Department of Energy and the World Federation of Scientists signed in 1999.

2. Request World Laboratory support to conduct periodic training courses in Erice for Black Sea scientists on environmental pollution issues of high priority. The first such training will address the adoption of standards for emissions and discharges of environmental contaminants into the Black Sea by all the riparian countries.
3. Request support for the World Laboratory Georgian and Ukrainian Branches to conduct high-priority hazard assessments for oil pollution in the Black Sea region using a Geographical Information System.
4. Collaborate with the U.S. Department of Energy to develop the new Black Sea Environmental Information Center web site (http://pims.ed.ornl.gov/blacksea) to facilitate communication and the exchange of information and scientific data among the countries of the region and the rest of the world.

These planned activities are intended to support the Black Sea littoral states to mitigate the environmental impacts of future increases in oil transport in the Black Sea and increase the benefit to its ecology.

* Dobris Assessment, European Environment Agency, 1995.

Signed:

Kay Thompson, USA	Melissa Lapsa, USA
Richard Ragaini, USA	Sergio Martellucci, Italy
Albert Tavkhelidze, Georgia	Dumitru Dorogan, Romania
Lado Mirianashvilli, Georgia	Ender Okandan, Turkey
Valery Mikhailov, Ukraine	Joseph Chahoud, Italy
Ilkay Salihoglu, Turkey	Vittorio Ragaini, Italy
Zenonas Rudzikas, Lithuania	Leonardas Kairiukstis, Lithuania

APPLICATION OF GEOINFORMATION SYSTEMS FOR OPERATIVE RESPONDING TO OIL SPILL ACCIDENTS

LADO MIRIANASHVILI, ALBERT TAVKHELIDZE
Georgian Academy of Sciences, Tbilisi, Georgia

This paper deals with quantification of the oil spill threat, evaluation of existing technologies and methods, and identification of the most appropriate approaches to oil spills both in the Black Sea basin and along pipelines. The study and resulting recommendations will improve response capabilities for the Black Sea waters, coast and land protection.

Activities to be accomplished in case of oil spill can be subdivided into several stages and substages:

1. Preparatory period: a) collecting data for data bases; b) application of the Geoinformation Systems (GIS) technology for the development of informative data bases with the purpose of identification of the most sensitive areas, evaluation of risk factors associated with environmental, biological, human-use and economic resources, and elaboration of a scenario for planning response operations; c) identification of the most appropriate methods for localization of slicks: i) processing space images in infrared/ultraviolet frequency ranges, as well as the ones obtained during radar monitoring; d) selection of cleanup methods: mechanical, chemical (detergents), biological; e) development of mathematical methods for modelling of slick spreading.

2. Emergency activities: a) localization of slick in open sea by means of space images or other methods; b) calculation of possible directions of slick spreading; c) selection of the most appropriate methods for clean-up operations.

3. Post-emergency period: a) activities for complete elimination of oil spill consequences; b) evaluation of effectiveness of response operations.

PREPARATORY PERIOD

The best approach to dealing with oil spills is to be well prepared for emergency. Preliminary work minimizes the negative effect of oil spill on the environment. GIS technology is one of the most comprehensive and modern methods that can be applied to managing oil spills. Expert Systems (in ArcInfo, ArcView) can be developed on GIS basis to improve planning activities, to increase efficiency of both oil spill emergency response and post-emergency operations. Expert System model of media must include all necessary parameters to give the most exhaustive information and assist in receiving the correct decision. The parameters are as follows: terrane model, networks of rivers and highways, information on underground water, locations of resort zones and public beaches; data on water temperature, salinity, direction of water currents and winds (meteorological parameters); threatened, endangered, and rare species. Most of the biological information includes species identification, temporal presence by month, and breeding characteristics. The above information is essential since many cleanup methods are habitat-specific.

Another parameter of GIS will be the sea bottom relief especially along the shoreline. The primary importance of shoreline slope during oil spills is its effect on wave reflection and breaking. Steep segments are usually subject to abrupt wave run-up and breaking, which enhances natural cleanup of the shoreline. Flat areas, on the other hand, promote dissipation of wave energy further offshore, which allows for longer residence time of oil within the zone. GIS information system will also comprise data on offshore rocks, concrete barriers and other protective features, substrate types, e.g. sediments which will be divided by grain size. The latter feature gives important information since sediments have potential for penetration and burial of the oil and thus the potential for prolonged exposure of important infaunal organisms which may be susceptible to oil-spill effects. Penetration and burial in sediments increases the persistence of oil, leads to potential long-term biological impacts, and makes cleanup much more difficult and intrusive. Penetration and burial are very different. In case of sediments the depth of penetration is controlled by the grain size of the substrate, as well as the range of grain sizes. Deepest penetration is expected for coarse sediments (gravel) that are most uniform in grain size. On gravel beaches, oil penetration up to one meter can occur under heavy oil accumulations. If the sediments are poorly sorted, such as on mixed sand and gravel beaches, penetration is usually less than 50 centimeters. The most rapid burial usually occurs on coarse-grained sand beaches, because they have the highest mobility under normal wave conditions. During storms, oil in gravel beaches can be buried by the building of gravel berms or bars.

Substrate type also affects the trafficability. Fine-grained sand beaches are typically compacted and hard, and they are the most likely substrate type to be trafficable. Use of equipment on gravel beaches tends to cause significant disruption.

Wave-energy flux is an important consideration in determining the potential of oil-spill impacts on coastal habitats. It is the primary determinant of the degree of exposure, also referred to as the hydrodynamic energy level, at the coastline. Wave-energy flux is basically a function of the average wave height, measured over at least one year. Where waves are typically large (e.g., heights >1 meter occur frequently), the impact of oil spills on the exposed habitats is reduced because: 1) offshore-directed currents generated by waves reflecting off hard surfaces push the oil away from the shore; 2) wave-generated currents mix and rework coastal sediments, which are typically coarse-grained in these settings, rapidly removing stranded oil; and 3) organisms adapted to living in such a setting are accustomed to short-term perturbations in the environment.

Shorelines can be subdivided according to energies of waves observed at the site: high-energy, low-energy and medium-energy shorelines. The shore which all year round is regularly exposed to large waves can be attributed to high-energy type. Low-energy shorelines will be those usually sheltered from wave. To the medium-energy shorelines will be attributed the ones which have seasonal patterns in storm frequency and size of waves. Persistence of oil depends on the type of shoreline: High energy means rapid natural removal, usually days to weeks. Low energy means slow natural removal, usually years. Medium energy means that stranded oil will be removed when the next high-energy event occurs, usually days or months after the spill. These kinds of features are used to identify those shorelines which have the potential for longer than usual oil persistence.

Infrastructure and environment model will be created to monitor pipelines. The following data will be fed in GIS: thickness of pipe walls; internal and external protecting coating of pipes; information about cathodic protection; depth of laying; soil composition; depth of underground waters; seismicity of adjacent regions; places of former leakages; remoteness from inhabited areas, cultivated land, springs, rivers, lakes, swamps, archaeological and cultural monuments, touristic routes, roads, etc. The latter is essential for choosing shortest routes by rescue teams in case of emergencies. Based on both observed and calculated data one can identify stretches recommended for reinforcement or replacement. Data base on pipelines will assist in minimizing adverse effects on the environment during possible accidents.

EMERGENCY

For localization of oil-spill a Side-Looking Airborne Radar (SLAR) and infrared/ultraviolet (IR/UV) images can be used. In a few cases a microwave radiometer (MWR) is used for quantitative thickness/volume determination of "thicker" layers (> 0.1 mm), and laserfluorosensor for quantitative thickness/volume determination of "thin" layers (<0.01 mm) and for characterization of oil.

The capability of the SLAR to detect oil on the sea surface has been known since the mid-seventies and so has the capability of the IR/UV scanner. The capabilities of the nitrogen and excimer lasers in relation to oil have been known since the late seventies.

Three commercial brands of MWR are used for routine surveillance. They are all manufactured and used in Europe. The oldest MWR is a single frequency system which is not considered reliable, moreover accuracy of its performance is not known. Another dual-frequency radiometer is claimed to have accuracy of 78% within the first hour after discharge of oil. So far, the three-frequency microwave radiometer is only used by one surveillance group. A 50% accuracy of performance is guaranteed.

Digital methods for processing of space images will assist in evaluation of the polluted area sizes and its spreading with time. Images with resolution of 10 to 20 meters are quite sufficient. ERDAS IMAGENE will be used to process the images. The same software will be used to tie images with geographic coordinates. A method for assignment of the area of interest will be applied. The above methodology and technique is mastered at the Geoinformation Systems Laboratory, Georgia.

For several days the disaster region may be clouded and therefore distance monitoring methods inefficient or complicated. Mathematical modelling can be of help to calculate possible direction and rate of slick spreading for such cases. Another and probably the most significant advantage of modelling is that it gives possibility of long-term planning that is the most important feature for handling oil spill accidents. Therefore we can install barriers that can stop oil spreading as well as plan cleaning operations in advance, before the slick reaches a certain point. Data on underwater currents, predominant wind direction, salinity for each area will be retrieved from GIS and fed into the model. The corresponding software package can be developed in Georgia as well.

The data bases need to be periodically updated, e.g. every 3-5 years.

The fresh oil reaching the shore would adhere to the sand and rocks and evaporate to produce a varnish-like black coating. Therefore, it's necessary to clean up the disaster area. Containment and diversion techniques include: skimmer, pneumatic boom, horizontal air and water jets, plunging water jets, diversion paravanes, and floating paddle wheels. Most are limited to two knots but some can work at higher velocities at steep angles to the current.

Widely used are detergents. The application of detergents could affect the physical conditions of the oil but would require one bbl of detergent for 5 to 10 bbls of oil and the residual negative effects on the environment are difficult to predict.

Probably application of a biological method is the most environment friendly approach. The biological method is based on inoculating spilt oil with oil degrading microorganisms. Instead of forming the usual oil coating that makes the coastal areas oil destructive, the oil is changed by the biological activities to produce an oil emulsion which does not adhere tightly. Biological degradation continues until the oil environmental conditions are favorable. Though one should keep in mind that the biological method takes much more time for cleaning than the others.

Special attention must be granted to liaison with governmental structures responsible for emergencies and environment protection, as well as co-ordination of activities between different groups. Not only operativeness of forwarding information to the above bodies is important, but the feedback is essential as well. Contacts with corresponding governmental structures should start at the initial stage of developing Expert Systems. Co-operation and co-ordination will make things much easier and more effective during emergencies.

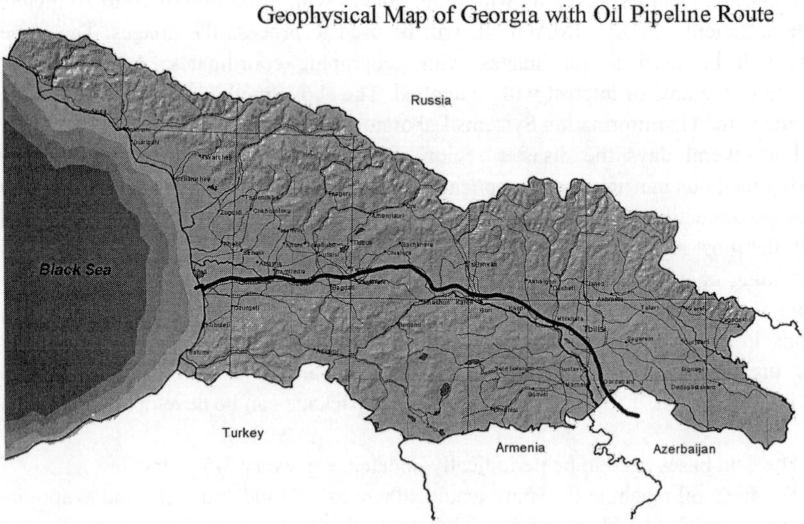

The black and white version of a colour map compiled by I.Kolesnikov et al.

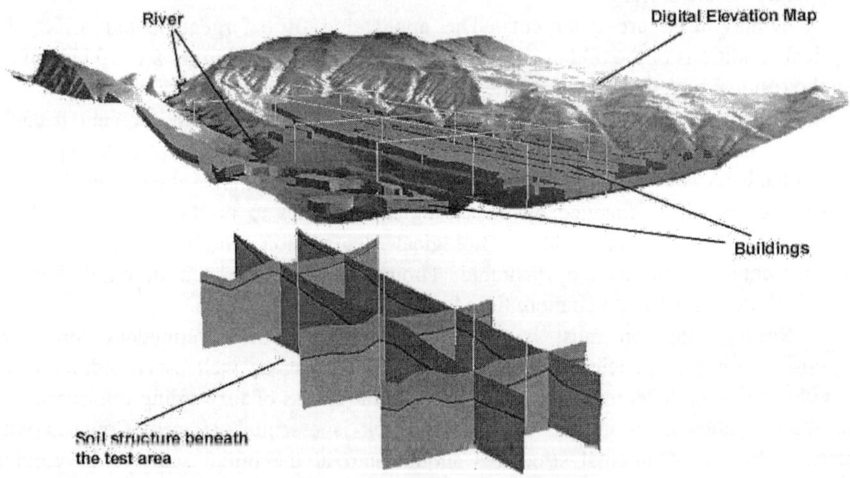

The black and white version of a map compiled by M. Elashvili.

Environmental Sensitivity Index Map

The Georgian Black Sea Coast

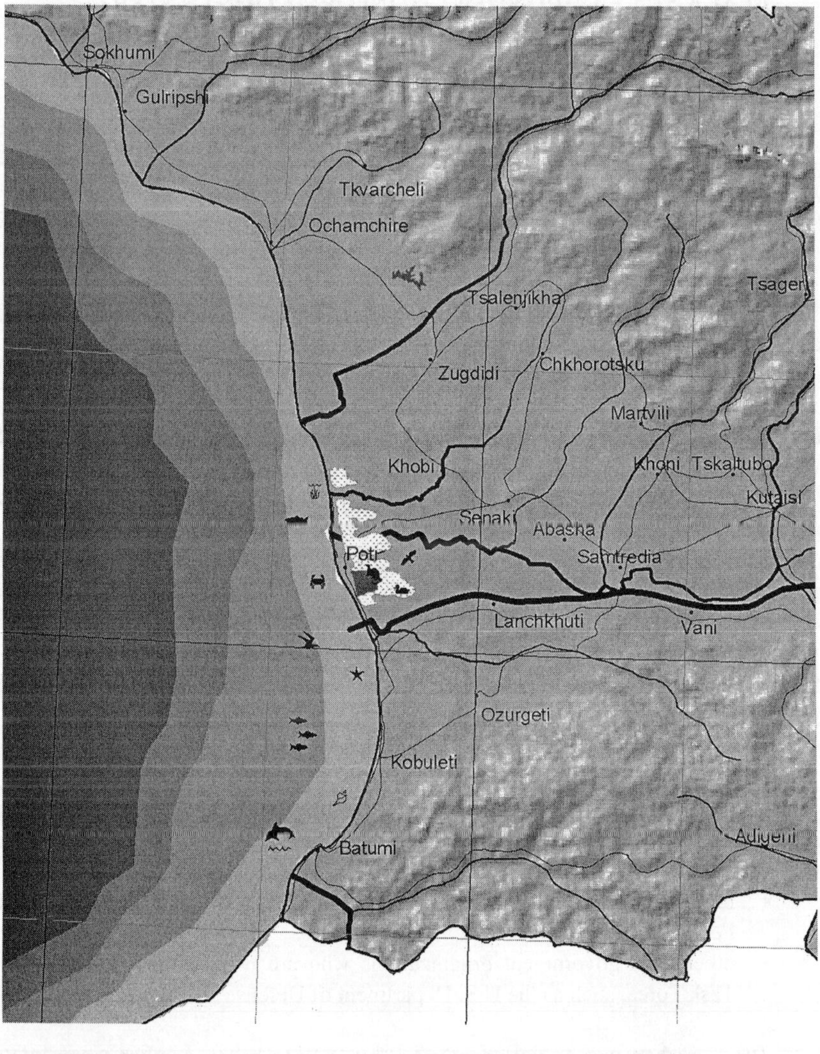

The black and white version of a colour map compiled by I.Kolesnikov et al.

BLACK SEA ENVIRONMENTAL INFORMATION CENTER

(http://pims.ed.ornl.gov/blacksea)

KAY THOMPSON
U.S. Department of Energy, 1000 Independence Avenue, S.W. Room 7G-050 Washington, D.C. 20585
MELISSA LAPSA
Oak Ridge National Laboratory, P.O. Box 2008, Building 4500N, MS-6189, Oak Ridge, TN 37831

The amount of oil transiting the Black Sea is expected to double over the next ten years, from the current one million barrels per day to more than two million. The U.S. Department of Energy's (DOE's) Office of International Affairs has undertaken a program in the Black Sea region called the "Black Sea Environmental Initiative." This initiative is headed by a DOE-led Interagency Task Force to address significant Black Sea environmental issues, including oil spill response and prevention. One objective of the "Black Sea Environmental Initiative," is to foster cooperative relationships, improved communications and strengthened environmental management tools for all the stakeholders in the region.

Working with delegates from Bulgaria, Georgia, Romania, Russia, Turkey, and Ukraine, DOE and Oak Ridge National Laboratory (ORNL) coordinated a workshop on a regional oil spill emergency response system for the Black Sea on September 14-17, 1999, in Odessa, Ukraine; the workshop was co-sponsored by DOE and the National Academy of Science, Ukraine.

The workshop included over 50 participants from:

- government and port authorities in the Black Sea countries (Bulgaria, Georgia, Romania, Russia, Turkey, and Ukraine)
- international oil companies
- international organizations such as the International Tanker Owners Pollution Federation, and
- other U.S. government organizations who are members of the Interagency Task Force, such as the U.S. Department of Defense.

The workshop was an important effort by DOE to bring together representatives from the Black Sea region, oil companies, and other organizations to begin a dialog on Black Sea environmental issues and to facilitate the creation of a regional capability to respond to oil spill threats on the Black Sea. The first regional follow-up meeting to the Odessa workshop was also a success. Held in Constanta, Romania, in July 2000, this workshop gave countries of the region an opportunity to meet and discuss progress made and current research initiatives presently underway.

The "Black Sea Environmental Information Center" web site (http://pims.ed.ornl.gov/blacksea) was unveiled at the Odessa workshop. Created by ORNL for DOE, the web site facilitates information flow and dialog between the countries of the region. The site, which is in a stage of rapid development, is dedicated to providing information and training on environmental issues and problems related to the Black Sea. The content of the web site and its functionality are defined by the participants in the "Black Sea Environmental Initiative" and participants in the regional Black Sea environmental workshops. The site contains an area where web site users can post and reply to questions related to the Black Sea environment and can register as a point of contact. A series of training links is provided to help prepare for environmental emergency response situations. Web site visitors are also able to review information provided by the Black Sea states on laws in the Black Sea countries. The site also hosts a chat feature where scheduled meetings can be conducted on-line.

The web site provides one-stop shopping for information on: oil spill clean-up, monitoring and other commercial technologies, scientists' requests for research partners, various countries' laws, regulations, and standards relating to the environmental condition of the Black Sea, publication of scientific papers, and on-line discussions of those issues.

Currently, development is proceeding on providing information collected by the Ukrainian Scientific Center of the Ecology of the Sea (UkrSCES). The information will include compiled data, maps, graphic files, and background information on UkrSCES. The data will consist of a catalog of oceanographic data on the Black Sea (including chemistry and pollution), geophysical data, meteorology, and aerology, for a period of 31 years. Data includes maps and graphics, statistical evaluation of the data, distribution of the chemical, biological, and geophysical elements; and pollution in various regions of the Black Sea.

This information will be stored in an SQL database and will be accessible from the web site. A portion of the data, maps, and graphic files are already accessible (from the "research" button on the web site home page). DOE and ORNL have partnered with the Department of Defense's Partnership for Peace Information Management System (PIMS) to provide the infrastructure necessary to support access to this web-based information to the Black Sea states. This infrastructure includes satellite uplinks and hardware necessary to support the "Black Sea Environmental Information Center" web site.

Six country-specific workshops were planned in each of the countries to facilitate progress on national laws and regulations to protect the Black Sea. The first one was held in Tbilisi, Georgia, in June 2000. DOE plans to work with each country to implement a workshop on legal and legislative issues that are critical to effective oil spill response systems and to identify legislative issues essential to regional cooperation on oil spill response. These workshops also cover international agreements. Recognizing that each of the six Black Sea countries has a unique legislative system, group of existing laws, and laws in preparation, DOE is planning a separate workshop in each country.

As participation, collaboration, and cooperation is essential for:

- policy makers and administrators in government agencies, including federal, state and municipal government organizations;
- non-governmental groups, and community organizers;
- manufacturing, commercial, industrial, agricultural, transportation, and residential sectors;
- financial institutions;
- citizens likely to be affected by the policies adopted; and
- schools who educate tomorrow's decision-makers,

the "Black Sea Environmental Information Center" web site is designed for many audience groups and in the future, will customize information retrieval results by audience category.

Statistics on the usage of the web site confirm positive results from the "Black Sea Environmental Information Center" web site. The web site has consistently attracted users from around the world (over 30 countries) each month and users from the Black Sea region are finding it a useful tool for communications and information.

Later this year the web site will be expanded to include information on all the existing petroleum pipelines and proposed additions to the transportation network surrounding the Black Sea.

We hope that the scientific community will use the "Black Sea Environmental Information Center" web site to share information, conduct on-line meetings, and strengthen their own network for collaboration. Discussions are already underway with the Romanian Institute for Marine Research and Development, in Constanta, Romania, to contribute data and research papers, with NOAA to conduct online training exercises, and with the UN to collaborate on a biodiversity area of the web site. We further hope to expand this web site to the Caspian and Azov Seas.

IMPORTANCE OF ASSESSMENT OF OIL POLLUTION ALONG BLACK SEA COAST AND BOSPHOROUS STRAIT-TURKEY

ENDER OKANDAN, FEVZI GÜMRAH, BIROL DEMIRAL
Petroleum Research Center, Middle East Technical University, 06531 Ankara-Turkey. Petroleum and Natural Gas Engineering Department, Middle East Technical University, 06531 Ankara-Turkey

ABSTRACT

The Black Sea ecosystem continues to be threatened by inputs of pollutants and has been the subject of several internationally funded research. Measurements in different regions of the Black Sea have shown that the main source of pollution is stemming from river discharges. However among these surveys oil pollution is not well documented.

Presently, man-caused oil pollution is inevitable due to tanker traffic, exhaust emissions from marine vehicles, accidental oil spills, discharge of oily ballast waters and pollution that results from drilling and production activities in oil and gas fields. There is a background of oil concentration that naturally occurs in the Black Sea waters due to natural oil seepages at the sea bottom.

In this paper an analysis of the tanker traffic across the Bosphorous is given. The risk of accident occurrence across the Bosphorous will inevitably increase with the increasing oil cargo from the Caspian region. The identification of the source and type of oil pollution will become crucial in such case, if the source of pollution is debated or if the fate and effect of hydrocarbon spills is to studied for the ecologic receptors.

The analysis results of a field study are discussed where the source of oil contamination in a dam lake was debated. The suspected crude oil and a diesel oil sample were tested by GC and GC/MS for their biomarker types and contents, which were then used to compare them with the biomarkers of the extracts of the contaminated waters. The results showed that the contamination was from the crude oil produced in the nearby oil field.

INTRODUCTION

The Black Sea, which is the largest land-locked sea, has a surface area of 461,000 km^2 and 547,000 km^3 of water body. The continental shelf of the northern part occupies 25% of the total area which is less than 200 m. The two deep basins, one in the eastern and one in the western part, exhibit depths reaching 2,212 m. at its deepest point.

The Black Sea is surrounded by 6 states, however the total catchment area includes 23 states of Western and Northern Europe. The link between the Black Sea and the Mediterranean is through the Bosphorous and Dardanelles Straits which are connected through the Marmara Sea. In the Black Sea, stratification of water column results from the salinity difference of waters from the Mediterranean (average 22 per thousand) and that of the Black Sea (17.5-19 per thousand). The thickness of the upper permanant layer is between 100-200 m and its characteristics prevent the penetration of oxygen to the bottom. The organic matter of the Black Sea has been sinking and decomposing in the deep waters, creating an anoxic environment below 200 m depth. The decompositon of organic matter using the oxygen of nitrates and sulphates results in the production of hydrogen sulphide which causes an environment best fit for anaerobic bacteria.

The surface water temperature of the sea ranges from 5-9°C in winter to 21-25°C in summer. The temperature of deep water stays constant at 9°C below the depth of 200 meters. The water currents occur in an anticlockwise manner along the shore, while two spiral flows occur that divide the basin into two with additional small and anti-clockwise eddies[1]. The water current system is obviously the main parameter for the transport of contaminants in Black Sea.

The fragile ecosystem of the Black Sea was the subject of several international scientific collaborations which produced ample data for the biological diversity of the region. The identification of several contaminants and their origins were documented. Among these, details of oil pollution are not available. The increase in oil tanker traffic with the future increase in oil production from the Caspian Region warrants a detailed discussion of the topic.

In terms of oil and gas reserves, Azerbaijan, Kazakhstan, Turkmenistan, and Uzbekistan are the four key countries in the Caspian Region. The oil production from Azerbaijan's Azeri-Chirag-Guneshli field and in Kazakhstan: Tengiz, Karachaganak and the new discovery in Kashagan, will reach the western markets through Black Sea if other transportation alternatives are not realized.

A controlled but steady growth in demand of around 3%/year will cause increasing volumes of Kazakh oil to be exported, building to about 2 million barrels/day (b/d) by 2015 according to the Wood Mckenzie report[2]. By 2020, Azerbaijan and Kazakhstan are expected to dominate regional production and contribute 1.1 million b/d and 2.5 million b/d respectively, totaling 3.6 million b/d which is three times the expected production of 1.3 million b/d in 2000.

This projection clearly shows that the total volume of oil and refinery products through the Black Sea transportation route will increase in the years to come. The accident risk and thus environmental risk will inevitably become a topic for the international community. The identification of oil contaminants on one point will be of importance as well as remediation measures. A monitoring program will become important to observe and predict how oil contamination will affect the biological environment in the upper layer of the Black Sea waters as well as its almost stagnant anoxic bottom layer.

TANKER TRAFFIC ACROSS THE BLACK SEA AND BOSPHOROUS

The Black Sea is connected to the world's oceans through the Bosphorous and Dardanelles Straits which are examples of the narrowest straits around the world open to commercial traffic. The Bosphorous has a length of 31 km and at its narrow point the width is 700 m. Dardanelles is 62 km long and 1.3 km at its narrowest point.

The report produced by GEF Black Sea Environmental programme in 1997[3] showed the tanker traffic during 1995. The data reflect the tanker movement before the start of transportation of Caspian oil to the west (Fig. 1).

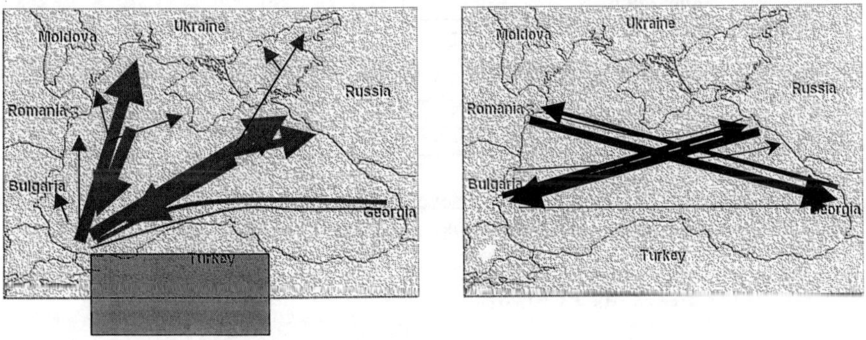

Fig. 1. Tanker traffic in Black Sea, 1995 [(1)].

Table 1 gives statistical data on tanker traffic through Bosphorous during 1995-1999[4,5].

The number of tankers carrying hazardous material (crude oil, refinery products, LPG, chemicals) increased by 22% from 1997 to 1999. During 1999, 81.5 million metric tons of hazardous material was carried through the Bosphorous[4,5]. By the year 2015, the amount of oil and products that will be carried through the straits, if other routes are not realized, will exceed the capacity of the straits for the safety of Istanbul city and the environment.

Table 1. Number of ships traveling across Bosphorous.

		YEARS				
		1995	1996	1997	1998	1999
Total number of ships		46 945	40 952	50 942	49 304	47 906
Nationality of ships	**Turkey**	16 368	18 879	19 937	18 615	16 219
	Malta	2 547	3 181	3 718	4 666	5 661
	Ukraine	5 693	5 729	5 714	5 304	5 453
	Russia	7 251	7 851	7 134	6 061	5 224
	Syria	2 200	1 993	2 188	2 203	2 321
	Others	12 895	3 319	12 251	12 455	13 028
Length of ships	**< 150 m**	-	33 716	44 455	42 629	40 716
	150m-199m	-	5 758	4 628	4 732	5 022
	200m-249m	-	1 047	1 312	1 456	1 624
	250m-299m	-	415	513	448	516
	> 300 m	-	16	34	39	28
Type of cargo	**Dry cargo**	-	-	24 302	24 931	26 429
	Hazardous material	-	-	4 303	5 142	5 504
	Others	-	-	22 337	19 231	15 973

Bosphorous has been the site of several accidents. Figure 2 shows the location of those accidents that occurred between 1982 and 1994.

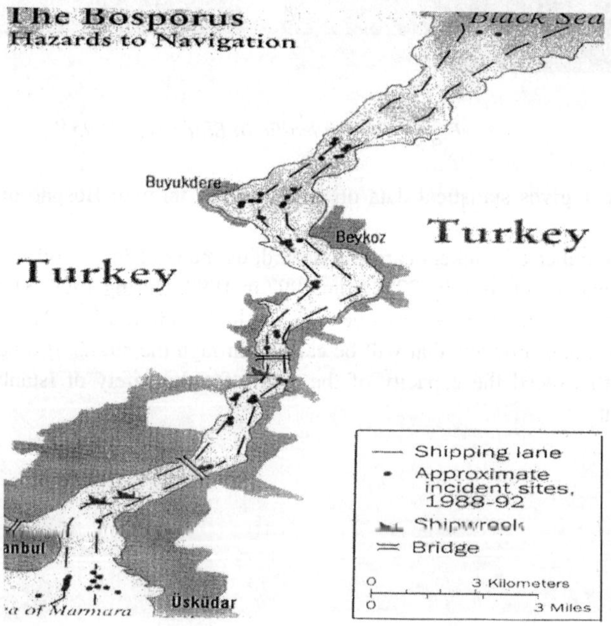

Fig. 2. Accident occurance in Bosphorous during 1988-1992.

The most recent accident was in the spring of 2000 when a Russian tanker carrying fuel broke, spilling 750 tons of oil along the shores of Ahirkapi (Florya), on the Marmara coast of Istanbul. The remnants of the fuel were not completely cleaned by the methods undertaken by the ship owner as reported on July 25, 2000.

About 30 beaches and deltas were identified along the Turkish Black Sea coastline which are affected by pollution in the sea. Presently there are more than 10 million people in Istanbul.

OTHER SOURCES OF OIL CONTAMINATION IN THE BLACK SEA

Apart from oil spillage from tanker accidents, the illegal ballast water disposal from tankers which is not assessed yet needs considerable attention. This will require international or bilateral agreements to be put into operation.

It has been reported that the PAH's in exhaust emmisions of marine vehicles can become siginificant if emission standards are not met. In old ships such standards are not expected to be present, however control and monitoring programmes must be launched by refusing ships not meeting these criteria to cruise in the Black Sea.

Another source is the oil contamination due to oil and gas drilling, and production activities which may increase in the years to come, as the prospects are promising in the southern part of the sea. There is already drilling and production activity in the nortwestern part of the sea. The natural oil seepages and their discharge rates must be studied to assess the natural oil concentration levels in the sea. An example of such an oil seepage occurs off the coast of Rize along the Black Sea coast of Turkey at a water depth of 1,102 meters[6].

FATE OF TOTAL PETROLEUM HYDROCARBONS IN THE ENVIRONMENT

Crude oil, the source material of nearly all petroleum products, has hydrogen and carbon as the main constituent while oxygen, sulphur, nitrogen and some metallic elements such as vanadium, nickel and copper exist in the complex chemical compounds of crude oil. In the refining process petroleum products are enriched with light hydrocarbons, leaving most of the organic compounds containing sulphur, nitrogen, oxygen and heavy metals in the residual material.

Changes in the product composition occur when petroleum products or crude oil are released into the environment. This process of change is called weathering. The main weathering processes are dissolution, evaporation and biodegradation. In case of oil spills on soil and water surfaces, photodegradation can become significant[7]. Each of the weathering processes affect the fate of hydrocarbons differently. Aromatics are more water soluble than aliphatics, and aliphatics tend to be more volatile. When a fuel mixture is released into the environment, the principal water contaminants are aromatics, while aliphatics will be the main air contaminants. Solubility and volatility of compounds generally decrease with an increase in molecular weight.

The rate of weathering of each chemical compound is retarded because fuels are complex mixtures. For example, solubility of pure benzene in water is 1,800 mg/l; for gasoline with 1% benzene concentration the equilibrium concentration in water will be 20 mg/l of benzene. This shows that the solubility and volatility of hydrocarbons decrease when they are in a mixture[7].

The third process of weathering, namely biodegradation is always active when petroleum hydrocarbons are released into the atmosphere. The bacteria and microorganisms present in oxic and anoxic environment tend to biodegrade the hydrocarbons. However degradation is more rapid under aerobic conditions.

For human beings the toxic effect of aliphatics (C_5-C_8) and that of aromatics, especially polyaromatics, (PAH) has been established. However the effect of hydrocarbon fractions on ecologic receptors is not well documented. For the Black Sea, the fate of hydrocarbon contaminants, degradation as well as transportation by atmospheric events within a two layer zone of the sea water body, need to be studied.

AN EXAMPLE FOR THE IDENTIFICATION OF THE SOURCE OF OIL CONTAMINATION IN WATER SAMPLES

The study was conducted in a dam lake where continuous crude oil spill was occuring from a near by oil field. The two sides of the controversy were debating if the source of contamination was from crude oil from the oil field or diesel oil which is used in establishments near the lake.

Fingerprint analysis comprised of GC and GC/MS was used to make the differentiation. Two crude oil samples from the oil field and diesel oil samples from a near by refinery were collected. The saturates of samples were separated by column chromatography and were first analyzed by gas chromatography (Fig. 3). The diesel oil, which is a refined product of crude oil, as expected, had no or very little C_1-C_9 fractions and heavier components were not present. On the other hand, crude oil had shown a different chromatogram that is characteristic of the crude oil. The pristane to phitane ratio of the crude oil was 1.52, 1.8 as measured on the two samples collected. For the diesel oil it was 0.76.

Two water samples from two different sites on the dam lake were collected in dark-colored bottles and sealed with high purity dichloromethan,. When received by the laboratory, the first bottle which did not have dichloromethane seal was used to determine oil concentration using fluoresence spectrometry. Samples had 0.06 and 0.04 ppm oil concentration. Figure 4 shows the results of different oil samples as they appear on the spectrograms. The spectrums however are not conclusive as to decide on the source of oil.

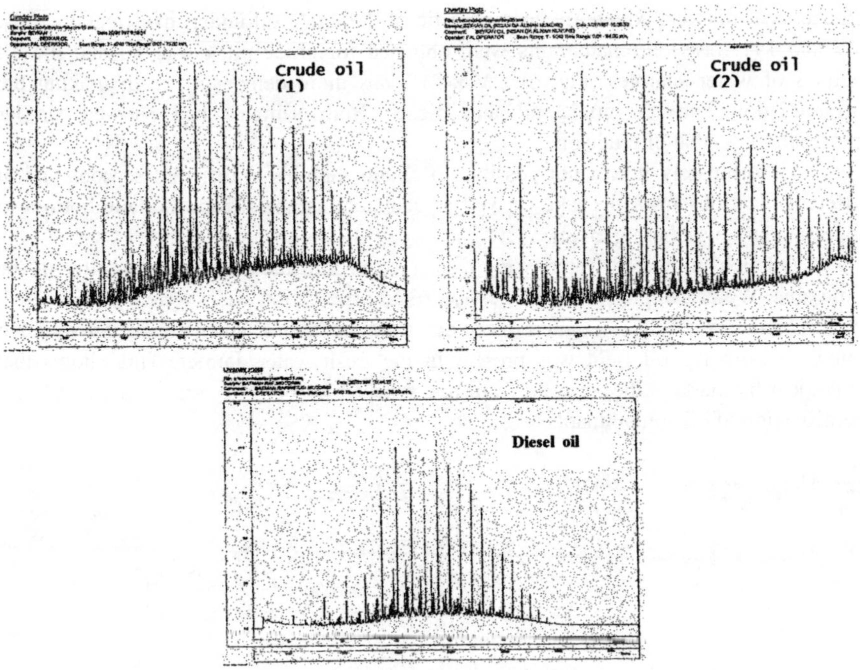

Fig. 3. Gas chromatography results for the crude oil and diesel oil.

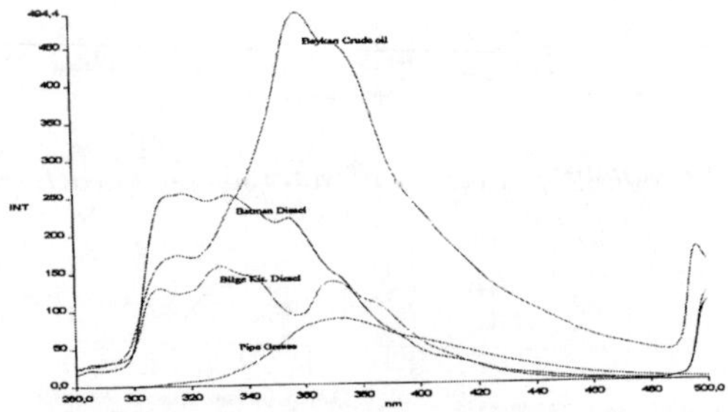

Fig. 4. Fluorescence spectrogram of different oils.

The water samples saved for fingerprint analysis were extracted by dichlorometane, and saturates were separated by column chromatography and put in special vials. The gas chromatography results are presented in Figure 5. Both samples showed no low carbon number fractions and it was not conclusive whether they are of crude oil or diesel oil origin. Then it was decided to do fingerprint analysis to identify geochemical markers which are characteristic of the crude oil under investigation. The 217 and 191 mass chromatograms were obtained for crude oil, diesel oil and the oil extracts of water samples (Fig. 6). M/z 217 was more demonstrative in making the differentiation between crude oil and diesel. Both samples had showed similar spectrograms to that of crude oil.

Unfortunately it was not possible to obtain samples at different seasons and at different locations from the lake as well as sediment samples from the lake bottom to understand the weathering effect.

Oil field produced waters were also injected into the fresh water aquifer in the same area. Water samples from the newly drilled observation wells were also collected. Figure 7 shows the fragmentogram of the extracted oil sample which clearly showed that crude oil from the oil field was present in the fresh water aquifer. This shows that geological biomarkers, because they are not attached by bacteria, are most suitable for identification of oil contaminants.

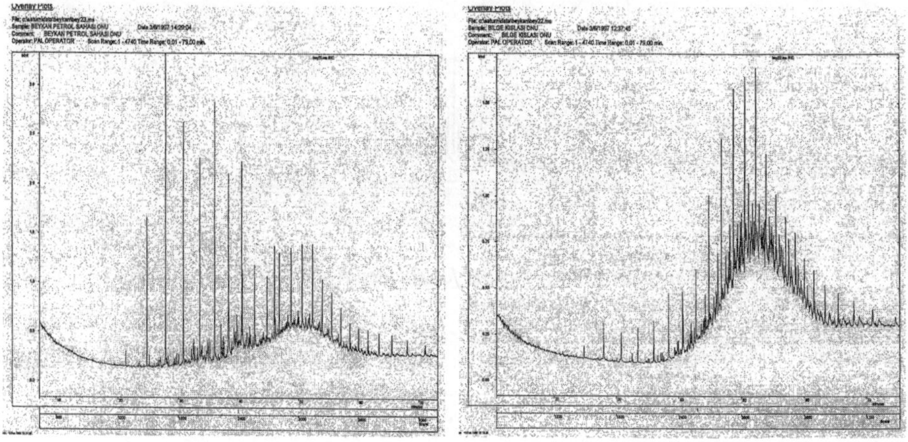

Fig. 5. Gas chromatography results of water sample extracts from the dam lake.

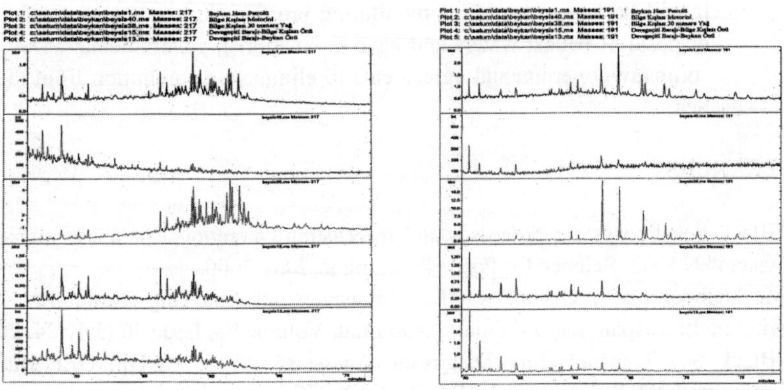

Fig. 6. Comparison of 217 and 191 m/z fragmentograms of oil contaminants and water sample extracts from the dam lake.

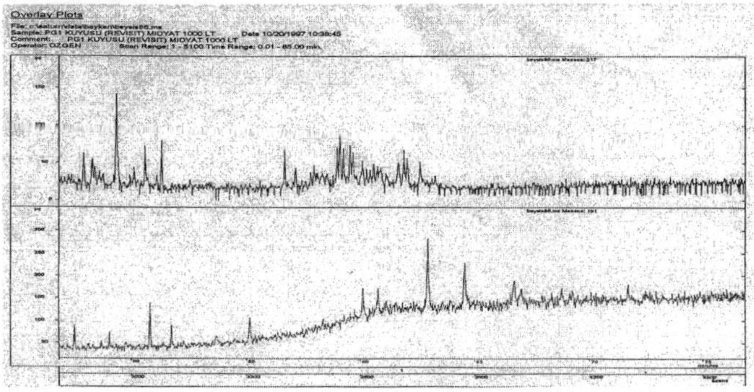

Fig. 7. 217 and 191 m/z fragmentogram of the oil extracted from the water sample taken from the observation well after 1 ton of swabbing.

CONCLUSIONS

- The Black Sea, which has a very delicate ecosystem due to its climatological and geological conditions as well as contaminant transport from different sources, will face severe oil pollution from the increasing tanker traffic accompanied by accident risks and from oil and gas production activities now and in the future.
- The fate of hydrocarbon pollution in the Black Sea environment, how different weathering processes in two layered water body will affect oil spills if and when they occur, must be studied with a regional monitoring program.
- An example is given in this paper that shows the identification procedure for the source of oil contamination in a dam lake.

- It will be timely to start a monitoring program for the tanker traffic, illegal disposal of ballast waters and have international, governmental as well as oil company-governmental agreements to eliminate oil pollution from the Black Sea.

REFERENCES

1. "Black Sea Ecosystem processes and forecasting/Operational database management system" NATO- Science for Peace Programme, May 2000.
2. McCutcheon, H. , Osbon, R. "Risk management, financing availability keys to winning in Caspian region" Oil&Gas Journal, Volume 98, Issue 30 (July 24, 2000)
3. "Black Sea Transboundary Diagnostic Analysis" 1997, UNDP, GEF Black Sea Environmental Programme, ISBN 92-1-126075-2.
4. "Istanbul ve Çanakkale Bogazlarndan Geçis Yapan Gemilerle ilgili Istatistiki Bilgi ve Degerlendirmeler, 1999", Deniz Trafik Düzen Baskanligi, Istanbul-Turkey.
5. Private communication, Savunma Arastirmalar Merkezi. ITÜ Vakfi-Istanbul.
6. TPAO Private communication.
7. "Composition of Petroleum Mixtures", Total Petroleum Hydrocarbon Criteria Working Group Series, Volume 2, 1998, Amherst Scientific Publishers, USA.

OIL POLLUTION RISK ASSESSMENT IN THE BLACK SEA AND THE ROMANIAN COASTAL WATERS

DUMITRU DOROGAN
The National Institute for Marine Research and Development "Gr.Antipa" Bd. Mamaia no.300, Constanta 8700, Romania

ABSTRACT

The paper outlines the increase of oil pollution risk in the Black Sea due to the increase of petroleum activities mainly from the Caspian area, the approach to risk and oil contamination level of the Romanian coastal waters. Accordingly, new research, management and co-operative programs for the Black Sea area are obviously necessary at the multinational level, the present paper trying to capture the attention of the scientific community.

INTRODUCTION

The Black Sea, with its surface area of 423.000 km^2, (one fifth the Mediterranean surface) 547.000 km^3 volume, an average depth of 1270 m and a maximum depth of 2212 m, is comparable with the Baltic Sea or North Sea being designated also as a "specially protected area" according to MARPOL 73/78 rules 10, pt.1.c.

About 25% of its area (144.000 km^2) is occupied by its northwestern continental shelf of Ukrainian, Romanian and Bulgarian territorial waters with a depth of less than 200m.

The Black Sea is the biggest continental basin in the world, almost a closed hydrographic system, the largest intercontinental sea, with tideless and brackish waters, with a shore line of 4358 km long; the Romanian coastline 245 km, the Russian 475 km, the Ukrainian 1628 km, the Turkish 1400 km, the Bulgarian 300 km, the Georgian 310 km (BSEP 1997[3]).

Being a huge draining basin, its total catchment area covers a vast surface of the European Continent and includes parts of 22 countries (6 coastal countries and 16 other Eastern and Central European states).

At least 182 million people live in the Black Sea basin (47% - 81 million, live in the Danube basin alone) occupying a 1.874.904 km^2 catchment area of all the rivers discharging into the sea, nearly five times larger than Black Sea's total surface area.

Bulgaria and Romania border the Black Sea to the west, Ukraine to the north, the Russian Federation and Georgia to the east and Turkey to the southern rim.

The smaller 139.000 km^2 and shallower (average depth of 8 m and 12 m maximum depth) Azov Sea is considered a part of the Black Sea connected by the Kerch Strait.

The Black Sea's only link to other seas is with the Mediterranean through the Bosphorus and Dardanelles—the Turkish straits—which have a functional depth of 35 and 65 m respectively. These straits permit a limited change of water 612 km^3/yr with salinity of 18 coming through the surface outflow from the Black Sea to the Aegean Sea, and 312 km^3/yr with salinity of 22,5 carried as an underflow from the Mediterranean and Aegean Seas (Topping'98[1], Unluata'90[18]). The average salinity of the open Black Sea is 17-18‰ at the surface and 22-24‰ at a depth of 2000 m (BSEP'97[3]). In the northwestern area, the Danube, Dniestr and Dniepr provide over 70% of all freshwater quantities discharged in the Black Sea, thus its salinity is quite low compared to other marginal seas.

As the biggest of the rivers (2850 km), with about 209 km^3/yr (6300m^3 /sec), fresh water flow from a catchement area of 12 countries (817.000 km^2) the Danube has a major influence and is the greatest potential threat to seawater quality and the health of marine life. It represents 3/4 of the northwestern rivers runoff and 2/3 of the total riverine input (370 km^3/yr) into the basin, and carries wastes amounting to more than the total discharge to the North Sea (NATO'99[10]).

The Dniester (10,2 km^3/yr), the Dniepr (51,2 km^3/yr) and the Don (21,9 km^3/yr) are the second group of fresh water inputs to the basin (BSEP'97[3]).

The total fresh water contribution of Turkish rivers, none of which are regulated, is only 35 km^3/yr.

However, annual sedimentary loads of Turkish rivers at 34.600.000 tons, exceed all other rivers discharging into the basin, except the Danube, which brings 51.700.000 tons of sediments (BSEP'97[3]). On the other hand, the annual organic matter load of Turkish rivers (275 t/yr.) is 1/3 of that of the Danube (913 t/yr.) and the discharges of all other rivers on the northern part of the Black Sea is 977 t/yr. (Izdar[7]).

During the last 30 years the damming of the Danube and other rivers has decreased the fresh water input to the Black Sea by up to one fifth and the decrease in sediment load has resulted in erosion along the Romanian coast. After the building of the II Portile de Fier dam on the Danube in 1985, the sediment flux decreased by about 20-30 millions tons/yr.

The wind, solar and wave energies in the Black Sea are insufficient to completely mix the lighter fresh surface water (brackish waters ~ 18‰) with the underlying denser sea water anaerobic H_2S-rich more saline waters ~ 22,5‰. As a result, the difference in density between these two water layers and the lack of penetration of oxygen from the surface to the bottom revealed the Black Sea as a "uniqum hydrobiologicum", with waters permanently anoxic below the depth of 180 m. 85-90% of its total water volume is anoxic - the largest anoxic water body on our planet with about 2,5 - 3 millions tons of H_2S is almost unchanged throughout the last 500 years (Degens [5]).

The threat to the Black Sea from land based sources of pollution is potentially greater than in any other sea on our planet (Mee'98[9]).

The total waste water discharges from total communities (10.385.000 people) in the Black Sea coastal area is estimated at 571 175 000 m^3/yr (BSEP'97[3]), with 55 m^3/yr/resident (Bulgaria–36.300.000 m^3/yr, Georgia–26.675.000 m^3/yr, Romania–22.000.000 m^3/yr, Russia–44.000.000 m^3/yr, Turkey–237.600.000 m^3/yr, Ukraine–204.600.000 m^3). 144.572.500 m^3/yr. untreated sewage is discharged into the Black Sea annually.

A complete assessment of the pollution of the Black Sea basin is not yet possible because of the lack of accurate data, although the chemistry of the Black Sea has been a subject of scientific investigation for about a century. The lack of common calibration standards and selectivity, and various methods used by countries for the measurement of petroleum hydrocarbons in seawater sediments and biota, implies considerable variability in the accuracy of data supplied by each country.

The consequences of pollution are however relevant: effects on tourism, loss of biodiversity, changes in the hydrological balance due to the constructions of dams on rivers, the collapse of fisheries, mass mortality of benthic organisms, penetration of new species in the Black Sea, eutrophication etc.

Today about 80% of the species in the Black Sea are immigrants from the Mediterranean (BSEP'97[3]). The Black Sea has a very small species list and appears to be particularly vulnerable to exotic species, but all human beings are in some sense immigrants.

OIL POLLUTION RISK

Risk = seriousness x probability.

The reduction of probability could be achieved by prevention measures and the reduction of seriousness has to be done through protective measures (technological and organizational).

Assessing the impacts of oil spills requires the collection of a variety of biological, chemical and socio-economic data.

Most of the oil pollution in the hydrosphere has no emergency origin and substantially differs from both extracted (transported) oil and trade oil materials. Specific molecular composition of hydrocarbons, significant content of non-hydrocarbon high-molecular tar substances, asphaltens, asphaltogenic acids and anhydrides and other differences are enhanced as a result of physical, chemical, biological and biogeochemical transformations in water, especially at the geochemical barriers.

Contents of oil hydrocarbons cannot be considered as an adequate characteristic of the level and mechanism of oil pollution influence on water basins and organisms.

The oil pollution is due to the rivers inputs, accidental tankers situations, debalasting, sewage, offshore oil platforms oil transport, etc.

The shipping industry was the first mode of transportation, the sea being seen as God's highway. Thus the environmental impact of this activity was acknowledged early.

In 1954 the international shipping community, through the International Maritime Organization (IMO), adopted the Marpol 73/78 Convention for the Prevention of Pollution from Ships. Between 1960 and 1970 the number and the size of vessels increased dramatically. More that 1,7 billion tons of crude oil are transported annually by ships.

Ships can pollute the environment as a result of normal operations and as a result of an accident. The historical experience is that major spills from oil exploration and production operations are far less common than those from oil tankers. Ship design has an important contribution to the operational pollution and their continuous improvement is trying to reduce this problem.

Pollution occurs more frequently from large crude carriers than from smaller ships and is more frequent from vessels discharging than from vessels loading.

In 1990 U.S. National Academy of Science estimated that tanker and non tanker accidents accounted for 121.000 tons of oil entering the sea per annum compared with 411.000 tons entering the sea from operational discharges.

These 411.000 tons are a real achievement compared with 2,13 million tons of oil entering the sea due to operational pollution recorded in 1973, being five times less and continues to decrease (Lyras[8]).

According to the IMO's Group of Experts on the Scientific Aspects of Marine Pollution (GESAMP), normal sea operations appear to be responsible for approximately 70% of the total pollution originated from ships compared with 21% due to accidents. According to statistics, collected by the International Tankers Owners Pollution Federation, between 1975-1995 about 800 accidents occurred involving the loss of more than 7 tons of oil. For major spills (over 700 t) the number of incidents per annum has dropped from 21 to 3 and for small spills (below 700 t) from 68 to 20. Over 80% of recorded oil spills are less than 1000 tons. Only 5% of recorded spills are greater that 10.000 tons (IPIECA[6], ITOPF[21]).

The Black Sea Transboundary Diagnostic Analysis (BSEP'97[2]) shows that oil discharge into the Black Sea, amount to more than 110.000 tons annually (Table 1). 52%, meaning 45.500 t/yr., are represented by industrial and domestic sources (due to inadequate treatment stations or oil loading discharge) and 48% (53 500 t/yr.) are carried by the Danube. Illegal oil dumping which could be very high is not included in the estimates.

However there is a considerable amount of uncertainty in the data due to the large temporal variability of land based point sources of pollution and the difficulty associated with the assessment of diffuse pollution sources.

The annual traffic of the oil and oil refined products trade in the Black Sea is estimated at 60 million tons (outbound traffic: 36,5 mill tons of crude oil, 8 mil. tons of refined products; inbound 5,8 mill tons of crude oil and 1 mill tons of refined products in 1995) and an average of 1000 oil tankers are passing annually through the Turkish straits (Table 2).

Table 1. Oil inputs to the Black Sea (unit: t/y).

Source of Pollution	Bulgaria	Georgia	Romania	Russian Federation	Turkey	Ukraine	Total
Domestic	5649.00		3144.1		7.30	2121.90	30016.30
Industrial	2.72	78.00	4052.50	52.78	752.86	10441.00	15379.86
Land-based				4200.00		5169.20	9369.20
Rivers	1000.00			165.70		1473.00	2638.70
Total	6651.72	78.00	7196.60	4418.48	760.16	38299.10	57404.06
Accidental oil spills(average for the last 10 years)							136.00
Danube River							53300.00
TOTAL in the Black Sea							110840.00

Table 2. Black Sea Oil Trading (Oil tankers > 10,000 DWT only) (Lloyd's analysis of Petroleum Exports - 1995).

Type of trade	Crude Oil (million of tons)	Number of tankers	Products (million of tons)	Number of tankers
Black Sea export	42.06	484	10.93	375
Internal Black Sea trade	11.81	166	1.34	53
Total Black Sea oil transportation	54.41	650	12.27	428

The average age of the tankers operating in the Black Sea is about 20 years. The average size of all tankers operating in the Black Sea in 1995 was approx.30.600 DWT and for crude oil export tankers the average size is aprox. 81.000 DWT. The biggest oil tankers permitted to sail through Bosphorus are limited at 150.000 TDW. In 1997, oil tankers accounted for 7000 out of 50.000 vessels in the Bosphorus strait. The Bosphorus has for years been under threat from oil tankers. With a length of 31 km, its narrowest part is less than 1 km wide (698 m), and 12 routes are used by civilian sea traffic, making it the world's most risky sea passage.

There were 174 accidents reported between 1980 - 1996 and 500 have taken place in the last 50 years. As result of 50 accidents, the city of Istanbul was endangered by huge fires which caused severe sea pollution and historical buildings damage in the coastal zone.

The Black Sea is under the threat of ship source pollution of which oil pollution is mainly responsible. The quantity of oil discharged from ships is small compared to the amount of oil which could be released in a single day during a tanker accident.

The present dominant volume of oil trade has its origin in the export of crude oil and refined products from the Russian Federation and the Ukraine via the Bosphorous to the world. Other important oil trade routes are the import of crude oil and products to Romania and Bulgaria as well as the internal trade between the littoral states of the Black Sea. Within the last years great attention has been given to the export of crude oil from the Coastal State bordering the Caspian Sea, in particular Azerbaidjan and Kazahstan.

Estimates of the Caspian basin oil reserves vary widely, ranging from 200 to 300 billion barrels of oil.

With the growth of the oil industry in the Caspian area and as a consequence of an expected general increase in the economic activities of the region, the traffic in crude oil and refined products will more than double in the near future. It will be a big challenge for the Black Sea as a "special area". In the actual geopolitical situation, the Black Sea region will become one of the most dynamic and long-term planned area in the global economic system. Due to its most favorable geographic and economic position, ports, recreational and industrial infrastructure will be developed accordingly, and some of the energy and industrial problems will be solved at an international scale.

From the oil producers statistics, Azerbaidjan has, and is estimated to export: in 1977 - 1,90 mill.tons, in 1998 - 3,75 mill.tons, in 2003 - 10 mill.tons, in 2010 - 31,75 mill.tons and Kazahstan: in 1998 - 7 mill.tons, in 2005 - 35 mill.tons, in 2010 - 35 mill.tons.

As consequence of some tense socio-economical and ecology economical conflicts, and in order to avoid some dictatorial policies or economical embargoes, new transport corridors have been developed with possible huge ecological incidence.

The actual capacity of the existing pipelines (Baku to Novorossiysk and Supsa) is estimated at 550 million barrels of oil, which will increase to 7 billion barrels by 2005. The third larger pipeline from Kazahstan to Novorossiysk, which is under construction, will carry about 3,5 billion barrels/year by 2005 (Ragaini[10]), Novorossiysk will become a more intensively used petroleum port. The rate of accidents and the risk threatening Istanbul will increase by five or six.

It's obvious that a large oil impact could change the whole biogeochemical system of the Black Sea leading to an ecological collapse and an unlimited crisis.

Presently, from the Ukrainian waters are discharged annually into the Black Sea and Azov Sea about 4000 tons of oil products. One should keep in mind that the Ukrainian shelf with a similar length off the Turkish shelf includes almost 90% of the tourist and reserve areas.

The present concentration of total petroleum hydrocarbons (TPH) in sediments, in coastal areas of the Black Sea, is quite high: from 2,1 to 310 mg/g in the Ukrainian coastal area, from 7 to 170 mg/g in Russia and 12 to 69 mg/g in Turkey (BSEP'98[1]). These concentrations are similar to those found in the western Mediterranean, but are an order of magnitude higher than those observed in abyssal plain sediments, and substantially higher compared with Antarctica and the Great Barrier Reef.

The actual characteristic of most sediments is dominated by chronic or degraded petrol, suggesting that the source is long-range wind transport.

Vertical profiles in water depths of 130 m showed that the highest polycyclic aromatic hydrocarbon (PAH) concentrations were found in the deepest samples (120 m) and these PAHs were enriched in pyrolytic components, which was attributable to sediment resuspension (BSEP'98, Bayona[1]). The evidence of "fresh oil" (n alcane concentrations) inputs is only in the vicinity of the Danube and the city of Sochi (BSEP '98 -Mee[1]).

The concentrations of polycyclic aromatic (PAH) found in the Danube river estuary are comparable to those found in other medium polluted rivers of the western Mediterranean, i.e. Ebro (Dachs[3]).

The most contaminated stations in term of "total" hydrocarbons (concentrations > 100µg/g) are associated with discharges from Odessa (635 µg/g), inputs from the Danube (638 µg/g) and the port of Sochi (368 µg/g) - (BSEP'98[1]).

Sevastopol Bay has an average concentration of 5 mg/l compared with the average open sea concentration of 0,1 mg/l.

Along the Bulgarian coastline frequent unidentified oil pollution has been observed. The research postulated the explanation that oil from the underground deposits is coming out through the cracks of the seabed (Radev'98[10]).

This is only a minor and a particular situation.

An economic evaluation of the sensitive areas affected by pollution shows that from tourism only, 20% of a recovered environment could bring in a return $550 mill annual income to the coastal countries.

ASSESSMENT OF OIL POLLUTION IN ROMANIAN WATERS

Fortunately in the Black Sea Romanian coastal waters, no large-scale oil spill incidents involving tankers have occurred. Approximately 100 oil spills due to minor incidents have been reported and penalized in the last 10 years, mostly resulting from inept handling of ship's machinery and equipment, particularly in ports (e.g. the reported pollution in 1995-10, in 1996-11, in 1997-14). No major accident caused by ships running aground, or in collision with other vessels has been reported.

A population of about 600.000 inhabitants affects the Romanian coastal zone, (8% of the population of the Danube basin) concentrated in the Constanta area (400.000 habitants).

The most sensitive areas are the tourist beaches from Constanta up to the Bulgarian border.

Ecological sensitive areas are the the Danube Delta Reserve, the 2 Mai Reserve and littoral lakes.

The Danube Delta, including the Razelm-Sinoe lake complex with an area of 442.000 ha is the largest coastal wet land complex in Europe to be declared a biosphere reserve protected by the RAMSAR Convention and the World Convention on Natural and Cultural Heritage.

There are over 80 species of fish in the Danube Delta, 275 species of birds, including 175 species of nesting birds (5% of the world population of the Dalmatian pelican).

Other sensitive areas are the industrial and artisan fishing zones all along the coast.

Pollution risk zones are harbor areas, high maritime traffic zones, offshore oilrig zones and used water discharge points along the Romanian coast.

As an example, PETRO-MIDIA Navodari, the Romanian large Petrochemical Industrial Complex, is discharging daily, using a combined mechanical-chemical and biologic treatment, about 1500-2000 m^3/hr waste waters with a quality score for oil residues of 3 mg/l in the northern area of Mamaia beach.

Oil figures prominent among Romanian imports and it is an economical activity highly dependent on shipping. The transit in 1997 was 3267 ships.

Particularly in Constanta and Midia harbors it is most likely that vessels could cause large-scale oil spill incidents, or the submarine pipe which links the offshore oil rigs with Midia Petrochemical Complex, could be damaged and 5334 m^3 of crude oil could be dispersed from 50 m depth in the mass water, with ecological consequences hard to imagine. But the risk is no different compared with other offshore exploration and exploitation areas in the world.

Presently, the submarine pipes are considered the safest method of energetic fluids transport.

The Romanian crude oil and gas from offshore oilrigs is transported through 85 km pipes with a flow rate of 75 m^3/h at a pressure of 2 bars, up to the Midia Petrochemical Complex.

In the last ten years, at the international level, the statistics of the oil pipe accidents shows that 31% are due to external forces (storms, earthquakes), 27% due to corrosion, 5% due to unproper welding, 31% due to the failure of controlling devices and 6% due to inept supervising activities.

Generally oil pollution risk assessment for submarine oil pipes has to take into account the following aspects:

- risk assessment due to a third party;
- risk assessment due to corrosion;
- risk assessment due to design;
- risk assessment due to inept supervising and operation;
- risk assessment through the environment impact.

A comprehensive risk assessment for a submarine oil pipe has to lie on particular statistical data (environmental parameters, naval traffic in area, quality of construction and operational technology, similar events on international level, etc.).

The identification of potential consequences related to pipeline spills involves trajectory analysis, baseline and sensitivity analysis and the quantification of environmental and infrastructure impact.

One of the serious impacts on pipeline can be associated with extreme earthquakes. The highest intensities are not observed in Romanian waters, but in the following regions: the Kaliakra Cape (Bulgaria), the Crimean peninsula, a large part of the southern and some parts of the eastern coastline.

Concluding, PETROMAR, the Romanian offshore oil company possess a safety subsea $12^{3/4"}$ oil pipeline, but a periodical intelligent goodwill inspection is recommended.

The treated water discharged into the sea from the Romanian oil production rigs has a content of 70-100 mg/l dispersed oil. Presently, due to a new treatment station the concentration is 15 mg/l in accordance with the Black Sea Convention limits (Piescu[16]), the existing technology being equal with those from the North Sea and the total discharged oil annually being only 5 tons compared with 10-12 tons in 1995.

No blow-out risk assessments have been performed by PETROMAR, but a blow-out during workover may however occur.

Low production per well combined with low and decreasing well pressures indicate that the discharge amounts from a single well blow-out might not be severe (in the order of hundreds of m^3).

The riser is protected by an isolating valve on the seabed that normally should prevent large spills, therefore it is small (less than 10 m^3).

Corrosion, ship anchoring and seabed construction works are potential risks, but probabilities of damage are considered low. According to present observations, the ropes, which are laid down into H_2S, zone break down six times quicker (or more often) than ropes which are not laid down into H_2S level (Shumilov[19]). Therefore, special investigations of corrosion processes are an actual necessity.

The low vapor pressure of the oil would prevent the pipeline from being emptied quickly in case of rupture. However riser and pipe rupture due to corrosion and anchoring damages are considered to represent a risk according to 1996 NOVATECH-Norway Environmental Assessment Study.

At present, no oil spills have been recorded due to Romanian offshore oil drilling and exploration. We are fortunate because there are no available resources yet to combat a pollution, there are no risk analyses, drift predictions, oil spill scenarios and no implemented intervention Plan in case of emergency with international agreements for intervention.

A monitoring program for the oil impact due to oil exploration equipment has to be developed according to the requirements of the Paris and Oslo Commission (OSPAR).

The evolution of the hydrocarbon pollution of the Romanian waters has been studied as a monitoring program by comparing the amplitude and the dynamics of the hydrocarbon pollution phenomenon with the pollution contribution originating from the main influential terrigenous factors in the area (Piescu[13]).

Total petroleum hydrocarbons in seawater are currently measured by spectroscopic techniques, mostly by spectrophotometric detection in UV, in correlation with the effects of modifying some specific sea water quality parameters such as oxidability and sulfur content of the samples taken from the 0 and 10 m depth line in the following coastal zones:

- Mamaia Bay, affected by waste waters from the Navodari-Midia Petrochemical Plant and Midia Port with a shipyard;
- Constanta area, affected by used urban waters with high oil content discharged through the North and South sewage treatment stations and by

Constanta shipyard, Constanta Port (operational discharge) and Tomis Marina Port;

- Southern area from Eforie to Mangalia affected also by used urban waters discharged through Eforie and Mangalia sewage treatment stations and by Mangalia shipyard;
- Danube mouth between Sulina and Gura-Buhaz.

TPH detected in 1986 in Romanian surface waters (0-10 m depth line) ranged from 0,15 to 5,1 mg/l. Medium values of TPH concentrations measured show a discontinued presence of oil as the pollutant, the concentration value being specific to each area: - 0,85 mg/l in 1986; 0,57 mg/l in 1991; 2,65 mg/l in 1992; 5,16 mg/l in 1993.

In 1995 the medium level of TPH in Constanta Harbor was 1,241 mg/l showing chronic oil pollution compared with other areas of Romanian water (0,582 mg/l at 10 m depth line of the Romanian shore).

In 1997 the medium level of TPH in the Romanian coastal waters (10 m depth line) diminished 4,6 times compared to 1995 level and 4,1 times to 1996 level.

In 1998 the TPH medium level was evaluated at 0,796% mg/l in front of the Constanta harbor and 1,344 mg/l in the harbor area (Piescu[17]).

However, the dynamic analysis of oil pollution between 1995-97 pointed out a decrease in the process, being 10,3 times lower for the Danube discharge zone into the sea.

In 1999 the TPH in the seawater of the Danube mouth was 36% more.

In the southern waters of the Romanian shore TPH was 2,5 times more.

TPH in sediments in the Danube mouth area is 4,16 times more that those in 1998 and 6,5 times more than in the samples taken from the southern waters of the Romanian offshore (Piescu[18]).

It has been noticed that the main pollution sources are operational discharge losses of ships and offshore activities, but the distribution of TPH revealed that the oil concentration level is also a consequence of discharged used water in the seawaters.

The maximum values of TPH (6 mg/l) were measured from the Eforie beach samples of illegal discharged residues in open sea.

In terms of volume, oil pollution is by far the greatest form of marine pollution caused by ships. Particularly oil pollution is essentially nontoxic. Crude oil (aromatic hydrocarbons) is more toxic than residue - more volatile and more toxic to plants. The toxic fractions, the lighter ones, will be rapidly lost from floating oil through evaporation and dissolution and the persistent fractions of crude and fuel oil do not appear to possess toxic properties, but can affect biota by adhering to them and disturbing, or cutting off respiratory exchange.

The ecological balance is restored or even enhanced within a few years in contrast to other forms of pollution, but this type of pollution attracts enormous media attention.

As an example, following the Torrey Canyon oil pollution (1968), it was noticed that some plants in area were extremely vigorous, so growth stimulation in some cases has to be investigated.

However at 0,1 ml/l oil and oil products, zooplankton died during the first 24 hours.

The survival of fish depends mainly on the way the oil is introduced. When oil products were emulsified in seawater, the damage is much greater than with oil films on the surface. It appears that the mechanical action of droplets is important. The survival of different species as young *Mugil saliens* in relatively high concentrations on the first day, showed that the sensitivity to oil pollution varies between species.

The crabs and *Balanus sp.* are active also in the oil film for several days. Larval form of benthic organisms and fishes were more resistant to oil pollution than plankton larvae, which are more resistant to oil pollution than developing eggs. Oil and oil products in the sea are highly toxic to developing fish eggs and cause their destruction in concentrations of 10^{-3} - 10^{-5} ml/l.

Plankton and benthic crustacea are also highly sensitive to oil pollution (Mironov'72[22]).

Ultimately, the marine environment assimilates spill oil through the long-term process of biodegradation.

SUMMARY AND OUTLOOK

Contamination of the marine environment represents a hazard for marine life and possibly to man as a consumer of seafood

The total discharged quantities of pollutants in the Black Sea are not known.

The Black Sea is not an open ocean which can clean itself, the natural recovery being a very long process.

Shipping is responsible for 12% of marine pollution compared with land based discharges 44%, atmospheric inputs 33% and dumping 10%.

A properly operated Vessel Traffic System will contribute to less vessel stranding and collisions. Modern navigation technology can provide superior assistance to pilots. The industry has to continue to revise and update codes of good practice, operating procedures and apply the latest technologies in ship's design.

As concern the reduction and regulation of oil operational discharges, the development of the Black Sea Action Plan (BS-SAP) clearly defined the terms and the policy. In support, the Black Sea technical document Transboundary Diagnostic Analysis (TDA - BSEP'97[2]) examines the major problems and clearly demonstrates that the Black Sea environment can still be restored and protected.

Thus in 1997 in section A.4. of the TDA are emphasized the following problems to be solved and the proposals to the deadline of implementation in 2000:

- significant reduction in the amount of discharge of harmful substances and in particular discharges of oil from ships;
- reception facilities for oil in all Black Sea major ports and for chemicals (by the end of 2002);

- environmentally sound and safe shipping activities in the Black Sea by both local and foreign flag vessels; a harmonized system of Port State Control and a harmonized system of enforcement, including fines;
- National Contingency Plans in place in all Black Sea coastal States;
- coordinated action by Black Sea coastal states in the event of a marine accident - the Regional Black Sea Contingency Plan;
- harmonized national classification and risk assessment system;
- conduct an in-depth study on measures to avoid any further introduction of exotic species in the Black Sea.

Some are in progress, but some are still in their stage of proposals.

The implementation of the BS-SAP is currently behind schedule due to delays in completing the national legislation. Therefore, Romania did not implement its National Contingency Plan, not having prepared the legal base for the management of oil spill prevention and a combating system. The recent ratification of the OPRC Convention in January 2000 will oblige Romania to have its National Contingency Plan (NCP) in force by the end of 2000.

However the draft of the NCP is already taken into consideration by the Regional Black Sea Contingency Plan which has to harmonize the following aspects: emergency equipment, content of reporting forms, oil spill data classification of the scales of spillage, evaluation methods of coastal sensitivity to hazards, a spill decision support system including models for oil movement forecasting and a Computerized Expert System for Risk Assessment of Emergencies and Contingency Planning in the Black Sea Region.

Regional and national oil spill response contingency plans are critically important. The implementation of the Regional Contingency Plan is essential before the huge Caspian oil starts moving through the Black Sea.

The need for solid background data on levels of oil in the Black Sea is imperative for the Black Sea countries to increase their level of oil pollution monitoring and to improve their ability to accurately measure the various forms of oil in each marine compartment. The existing knowledge is related to a limited number of stations and several hot spots. It is important to enlarge the sample grid for monitoring work to assess the extent of region-wide oil contamination in the Black Sea.

However, it is relevant that the annual input of oil to the Black Sea from the river Danube is similar in value to the total land-based discharge from all Black Sea countries.

Concentrations of aliphatic hydrocarbons in the seawater and sediments are comparable to the concentrations found in the Western Mediterranean (PAH concentrations are lower in the Black Sea).

Inputs of PAHs in the Black Sea should be determined at the individual compound level, because toxicity of oil greatly depends on the individual component rather than on the total oil.

Determining the influx and efflux data of petroleum hydrocarbons through the Turkish straits will allow for a sustainable management in order to preserve the Black Sea ecosystem and to provide resources in the Black Sea.

The Black Sea riparian countries have to implement a sustainable management for environmental protection including strategies, action plans, reporting and evaluation methods.

In order to implement such a system, the communication capabilities amongst the scientists of the Black Sea countries are being significantly upgraded through a future web site dedicated to decision makers and public stakeholder use.

Romanian annual reports on the environment protection have to become the main monitoring instruments for the alignment to international standards and requirements.

In spite of all international cooperation, the Black Sea is still not on the priority list of European countries, not being so important for politicians.

Other economic and political priorities in this region and inefficient cooperation between ministries involved, are significant obstacles to a sustainable water management in the Black Sea.

The sustainable development of the Black Sea will require continued and enhanced international cooperation in a variety of fields, but without adequate national machinery, no international system can hope to function. The Black Sea Strategic Action Plan represents one of the last important frameworks for the sustainable regional management. The international community will have to contribute effectively and in a coordinated manner. Two of most recent important cooperation aspects in progress supported by U.S. Department of Energy are communications and training.

With a concerted effort, future generations will enjoy the Black Sea.

CONCLUSIONS

30 years of research demonstrate that a major spill will not cause permanent environmental damage except in truly exceptional circumstances.

In oil spills, countermeasures can be completely effective only if all of the oil is recovered immediately after the spill. The technology to achieve this goal does not exist. Spill prevention will produce far greater returns than clean-up.

Oil pollution is essential nontoxic, but on large scale could be a disaster. Only polyaromatic hydrocarbon levels of concentrations and repeated exposure to them are of high concern since some components are carcinogenic or mutagenous (Blumer[22]).

General actions to be taken to avoid risk:

- prevention of spills - permanent review and improve of contingency planning;
- international regulations & conventions;
- improvement of navigational aids;
- harmonized systems, legislation standards, designs, equipment;
- training, common methodologies and frameworks, exercise programs;

- improved monitoring and compliance with shipping protocols and agreements;
- sustainable management integration (minimize environmental impacts, making environmentally sound technology available to all - Global Initiative);
- standardize management techniques;
- sustainable and equitable development policy (transfer to developing countries of the technology needed for safe storage, transportation, processing disposal of wastes);
- regional and international co-operation among regulatory, operational and environmental personnel and decision-makers;
- integrate environmental costs into market prices;
- policy and economic instruments (revenue from environment and energy taxes, establishing of an environmental fund, deterrent strategies;
- efficient and consistent global information system - compatibility and deadlines for future assessment and reporting;
- research and development integration and co-ordination, sectorial and on world scale;
- establishing regional centers for developing and communicating scientific information and advice on appropriate environmentally sound technologies in a collaborative international network;
- improving long-term scientific assessment;
- promoting technologies that minimize adverse environmental impacts;
- improved integrated risk assessments (solid data base; an efficient decision support);
- implementation of the Aarhus Convention (Access to Information Public Participation in Decision-Making and Access to Justice in Environmental Matters) and of the Amsterdam Treaty the right access to documents held by EU institutions;
- integration of Policy, Science and Law.
- promoting environmental awareness, educational programs, and a well-managed media, political and public information response.

REFERENCES

1. BSEP, (1998) - Black Sea Pollution Assessment, edited by Laurence B Mee and Graham Topping United Nations Publications, New York. Black Sea Environmental Series vol. 10, 1999. ISBN 92-1-129 506-8.
2. BSEP, (1997) - Black Sea Transboundary Diagnostic Analysis (ed.by L.D.Mee) United Nations Publications. New York ISBN 92 - 2-1126075-2.
3. BSEP, (1997) - Biological Diversity in the Black Sea - A study of Change and Decline (ed.by Zaitsev Yv. and Mamaev V.), United Nations Publications, New York. Black Sea Environmental Series.vol.3 ISBN 92-1-126042-6.

4. Dachs J., Bayona, J.M.Ch. and Albaiges J., (1997) - Spatial, vertical distribution and budget of polycyclic aromatic hydrocarbons in the western Mediterranean seawater.
5. Degens, Et. and Stoffers, P (1980) - Environmental events recorded in Quaternary sediments of the Black Sea.Jr.Geol.Soc.London p.137.
6. IPIECA REPORT SERIES (1991) - Vol.two - A guide to contingency planning for oil spills on water.
7. Izdar E, Konuk T, Ittekkok V, Kempe S, and Degens E.T. (1987) - Particle flux in the Black Sea Nature of the Organic Matter. In: Particle Flux in the Ocean SCOPE/UNEP Sonderband.
8. Lyras J (1998) - The Black Sea in Crisis - The Shipping World and Protection of the Sea, p. 189 - 192.World Scientific Publishing Co.,UK ISBN 981-02-3769-3
9. Mee L.D. (1998) - The Black Sea in Crisis: A Need for Concerted International Action. Ambio, 21, p.278 -286. World Scientific Publishing Co.,UK ISBN 981-02-3769-3.
10. NATO Science Series, (1999) - Environmental Degradation of the Black Sea: Challenges and Remedies, Kluwer, Academic Publishers, Dordrecht Environmental Security. Vol. 56, 1999.
11. Ragaini C.R., (1999) - Monitoring Black Sea Environmental Conditions. World Federation of Scientists Workshop, Erice, Italy.vol 3.
12. XXX - Romanian National Symposium Series, ACVADEPOL 3rd Ed 1995; V.Piescu - Oil Pollution Assessment of the Romanian Waters from Navodari to Vama-Veche.
13. XXX - Romanian National Symposium Series. ACVADEPOL 4th ed.1996. V.Piescu - Hydrocarbons content in Romanian coastal waters in 1995.
14. XXX - Romanian National Symposium Series ACVADEPOL 5th ed.1997. V.Piescu - Report on physic-chemical parameters of the offshore oil exploration in the Romanian Black Sea waters.
15. XXX - Romanian National Symposium Series, ACVADEPOL 6th ed.1998. V.Piescu - Hydrocarbons content analysis in the Romanian marine coastal zone.
16. XXX - Romanian National Symposium Series, ACVADEPOL 6 th ed.1998. M.L.Gresoiu - Hydrocarbons Pollution of the Black Sea - Romanian or European Responsibility?
17. XXX - Romanian National Symposium Series, ACVADEPOL 7th, ed.1999. V.Piescu - Constanta harbor's abiotic parameter quality stage.
18. XXX - Romanian National Symposium Series, ACVADEPOL 8thed.2000. V.Piescu - Pollution identifications in seawaters and sediments due to fluvial discharges and concentrations of the pollutants in marine organisms.
19. Unluata, V., Oguz T., Latif, M.A., and Ozsoy, (1990) - Physical Oceanography of the Straits. J.L.Prast (Ed) NATO/ASI Series, Kluwer Academic Publishers.
20. XXX - Varna 15-18 March 1998 Symposium - "Environmental Aspects of the Exploration, Production and Transportation of Oil and Gas in and through the Black Sea".

- Radev, M. (Sofia - Bulgaria) - Is Hydrocarbon Pollution Through the Seabed of the Southern Part of the Bulgarian Black Sea Sector a Result of Recent Petroleum Generation and Migration.
- Shumilov, V. (Kiev,Ukraina) - Corrosion Risk of oil and Gas Exploration Production and Transportation Equipment and Pipeline in H_2S- zone of the Black Sea.

21. ITOPF Handbook, 2000/2001 UK.
22. Marine Pollution and Sea Life (1972), editor Ruivo M., FAO. FISHING NEWS, London UK ISBN 0852380216.
    - Mironov O.G - Effect of Oil Pollution on Flora and Fauna of the Black Sea.
    - Blumer M. - Oil Contamination and the Living Resources of the Sea.
23. European Environment Agency (1999) - Environment in the European Union at the turn of the Century, Copenhagen. ISBN 92-828-6775-7.
24. United Nations (1992) - Earth Summit, Agenda 21, Rio de Janeiro.Brazil.

ENERGETIC CONSUMPTION OF DIFFERENT TECHNIQUES USED TO PURIFY WATER FROM 2-CHLOROPHENOL

VITTORIO RAGAINI, ELENA SELLI, CLAUDIA L. BIANCHI, CARLO PIROLA
University of Milano, Dept. of Physical Chemistry and Electrochemistry, Via Golgi 19 – 20133 Milano (Italy)

ABSTRACT

The degradation of organic pollutants in water is a topic of fundamental importance nowadays, especially when it is performed only using "clean" methods, avoiding the use of chemicals.

In the present work the degradation of 2-chlorophenol in water has been kinetically investigated using the following different techniques, employed either separately or simultaneously, always with the same experimental set up: light irradiation (315 - 400 nm), sonication, photocatalysis with different types of TiO_2, photocatalysis with sonication. An energetic comparison among these different techniques has been performed, focused on an industrial application of some of them.

INTRODUCTION

The degradation of organic pollutants of water using "clean" methods and reagents, *i.e.* ultraviolet irradiation (UV), ultrasounds (US), ozone, hydrogen peroxide, has been the subject of many recent papers and of some patents[1-3]. Mechanical, electrical, thermal, thermal-biological, chemical, biological, acoustic methods have been considered and a great number of both hydrophobic and hydrophilic organic pollutants have been shown to be degraded by ultrasounds[1,4-9]. Combined techniques or techniques alternative to US, *i.e.* UV, O_3, US/O_3, have been considered or tested by Pétrier *et al.* for phenols and municipal wastewater treatment[5,6,10]. A high energy plant using US (25 kHz, 3 kW) has been also proposed for the degradation of chloroaromatics, azo dyes and many other organic substances[11]. Combined US/O_3 techniques, leading to marked increases in reaction rate, have been kinetically investigated very recently[12-14].

The yield of UV treatment of organic pollutants in water is generally greatly increased in the presence of semiconductor particles, such as TiO_2 or ZnO, which are able to absorb the incident light and thus to act as photocatalysts through electron-hole separation[15-17]. On the other hand, the combination of US and TiO_2 has been reported to have a positive effect on the degradation rate of trichlorophenol[18]. The use of suspended photocatalysts could be considered a necessary inconvenient, as they must be removed from the liquid after the treatment.

The aim of the present paper is a comparison between the kinetic results of 2-chlorophenol degradation, chosen as model pollutant, by using several of the above-mentioned techniques, giving in addition an estimation of the energy consumption in each case.

The following techniques have been compared: US / UV / O_3 / US + O_3 / UV + O_3 / US + TiO_2 / UV + TiO_2 / US + UV + TiO_2 / UV + TiO_2 + O_3 / US + UV + TiO_2 + O_3 / US + UV + TiO_2 + Air-O_2 mixtures.

As our attention is mainly focused on the efficiency of the different techniques, on possible synergic effects and on the energetic aspects of degradation processes, no detailed analysis of reaction intermediates and products has been made.

EXPERIMENTAL

Materials

2-chlorophenol (2-CLP) was purchased from Aldrich (purity > 99%). Degussa P25 titanium dioxide (80% anatase, 20% rutile, surface area 50 m^2 g^{-1}, average particle size 30 nm, density 3.8 g cm^{-3}, according to the manufacturer; surface area 35 m^2 g^{-1}, according to our BET analysis) was generally employed as photocatalyst. Distilled water was used in the preparation of solutions and suspensions. Argon was a Sapio product and oxygen an Air Liquide product.

Apparatus

All degradation runs were carried out employing the experimental set up sketched in Figure 1, which allows one to investigate the effects of the different degradation techniques, employed either separately or simultaneously, without any modification in the geometry of the system.

The reactor was a cylindrical pyrex vessel, closed on top with a plastic cover, continuously stirred during degradation treatments by means of a magnetic stirrer. A stainless steel serpentine immersed in the treated solution or suspension, containing recirculating water passing through a thermostat, ensured temperature constancy at (30 ± 1)°C.

The ultrasound source was a W-385 Heat System-Ultrasonics apparatus, emitting at 20 kHz, with a maximum emission power of 20 W and a tip diameter of 12 mm. The power effectively absorbed by the apparatus was measured amperometrically; its emitting power was calibrated calorimetrically, by sonicating distilled water in a glass dewar vessel[19].

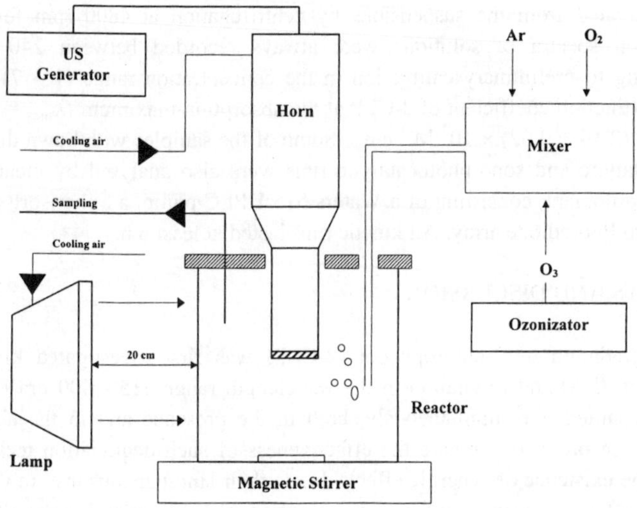

Fig. 1. Experimental set up employed for sono-photocatalytic degradation, also under bubbling of different gas mixtures.

The light source was a Jelosil, model HG 200, 250 W iron alogenide lamp, equipped with filters, emitting in the 315 – 400 nm wavelengths range. It was placed at 20 cm from the reactor wall and cooled during the runs through forced air circulation.

A couple of preliminarily calibrated silicone oil flowmeters allowed to control gas fluxes in the reactor and the composition of gas mixtures. Runs under normal atmospheric conditions were carried out without bubbling any gas in the sample, after having verified that identical kinetic results were obtained with and without air bubbling. When operating under different atmospheres, a gas flux of 4 L h^{-1} was generally employed, which was increased to 12 L h^{-1} in the case of O_2/O_3-Ar mixtures. Ozone was obtained from oxygen in a pilot plant (Ozono Elettronica, Italy), emitting a O_2/O_3 mixture containing 17 g m^{-3} of O_3.

Procedure

In runs employing either ultrasounds or light irradiation, or both, the pertinent sources were switched on at least 30 min before introducing solutions in the reactor. Irradiated solutions or suspensions contained a fixed initial 2-CLP concentration of 5×10^{-4} mol L^{-1}. Titanium dioxide was added to the solutions at the beginning of every run, directly in the reactor. Standard suspensions contained 0.1 g L^{-1} of TiO_2. Samples (3 mL) were then withdrawn at different reaction times, through a port of the reactor cover, and analysed spectrophotometrically in a Perkin - Elmer Lambda 16 apparatus. Prior to analysis, TiO_2

was separated from the suspensions by centrifugation at 4800 rpm for 60 min. The absorption spectra of solutions were always recorded between 240 and 330 nm. According to preliminary calibration in the concentration range $(1 - 7) \times 10^{-4}$ M, the molar extinction coefficient of 2-CLP at the absorption maximum (λ_{max} = 273.2 nm) was taken as $(2.14 \pm 0.02) \times 10^3$ M^{-1} cm^{-1}. Some of the samples withdrawn during sonolytic, photocatalytic and sono-photocatalytic runs were also analysed by means of a Waters HPLC equipment, consisting of a Waters 515 HPLC pump, a Spherisorb ODS 1 column and a 996 Photodiode array. All kinetic runs lasted at least 6 h.

RESULTS AND DISCUSSION

The degradation of 2-chlorophenol (2-CLP) was first investigated kinetically under sonication (US) and irradiation in the wavelength range 315 - 400 nm (US), employed either separately or simultaneously, both in the presence and in the absence of TiO_2 particles, in order to compare the effectiveness of such degradation techniques, and to assess the existence of synergic effects due to their simultaneous use. In the last series of runs the effect of the presence of ozone was systematically tested under all the previously investigated experimental conditions.

The absorption spectra successively recorded at different reaction time in the absence of ozone exhibit a progressive decrease of absorbance and an isosbestic point at 248 nm, without any significant modification in shape in the wavelength range 250 – 300 nm, apart from a shoulder appearing around 290 nm. Only one intermediate oxidation product was evidenced by HPLC analysis, most probably chlorohydroquinone[20], which has an absorption maximum at *ca.* 290 nm and does not interfere in the spectrophotometric analysis of 2-CLP, exhibiting negligible absorption at 273.2 nm.

<u>Sonolytic degradation</u>

Only a 10-20% degradation was obtained after 6 h when employing sonication alone under air atmosphere. The reaction rate was measured as a function of the solution volume, at fixed US emission frequency (20 kHz) and power (7.5 W).

Kinetic runs were carried out under standard sonication conditions (20 kHz, 7.5 W) also in the presence of 0.1 g L^{-1} of TiO_2 particles. No difference in 2-CLP degradation rate was remarked in any case, outside the experimental uncertainty, respect to the rate of sonolytic degradation in solution.

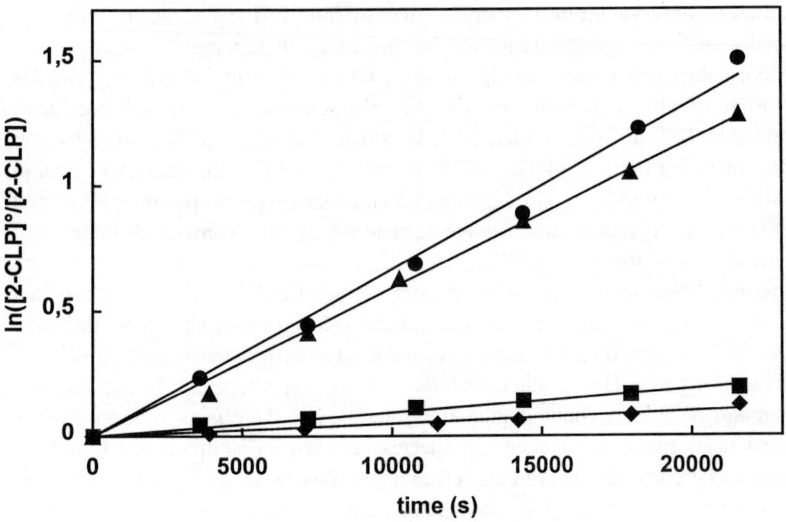

Fig. 2. First order kinetic plots of 2-CLP degradation under different techniques: (◆) UV irradiation only; (■) US irradiation only; (▲) photocatalysis in the presence of TiO_2 particles (UV + TiO_2); (●) sono-photocatalysis (US + UV + TiO_2). Reaction volume: 330 mL, air atmosphere.

Photoinduced and photocatalysed degradation

One order of magnitude higher reaction rates were obtained when operating in the presence of photocatalyst, leading to a 70-90% degradation of 2-CLP in 6 h. This result is not surprising, as titanium dioxide notoriously is a highly efficient photocatalyst, able to absorb light at wavelengths below 370 nm, while only a very minor fraction of the impinging radiation can be directly absorbed by 2-CLP under the adopted experimental conditions. Thus, a completely different photodegradation mechanism is at work in the presence of titanium dioxide: the main reaction path does not imply light absorption by 2-CLP and direct photolysis, but, instead, light is mainly absorbed by the semiconductor, leading to electron-hole separation and consequent oxidation of 2-CLP adsorbed on the semiconductor by photoproduced holes or by OH radicals produced at the semiconductor-water interface[15-17,21].

Simultaneous sonolysis and photocatalysis

First order rate constants obtained when degradation was carried out in air under both sonication at 20 kHz and 7.5 W and UV irradiation in the presence of TiO_2: a synergic effect of the two degradation techniques was observed only for reaction volumes greater than 300 mL. The observation of a synergic effect of sonolysis and photocatalysis, also operating under relatively low US frequency, has been recently reported[22].

Several concurrent effects should be taken into account in order to explain this behaviour. First of all, sonication modifies the characteristics of the semiconductor particles: indeed, BET analysis evidenced that, when 200 mL and 500 mL of water suspensions containing 0.1 g $L^{-1}$of TiO_2 (P25) were sonicated for 6 h, the surface area of the photocatalyst increased from 35 m^2 g^{-1} to 48.5 and 44.5 m^2 g^{-1}, respectively. It is well known that sonication decreases the average particle size, with a consequent increase of surface area and of catalytic activity.

Moreover, US irradiation provides an extra source of OH radicals, which, together with the holes photogenerated in the semiconductor valence band, should be the main cause of the oxidation of organic molecules in photocatalytic processes with TiO_2[15-17,21]. OH radicals also produce H_2O_2, which undergoes decomposition under UV irradiation, leading to an enhancement in photodegradation rates[23]. Also the effects, induced by US, of accelerated mass transport of chemical species between the solution phase and the photocatalyst surface and the continuous cleaning of this latter by acoustic cavitation might have some role in increasing the photocatalytic degradation rate.

Argon – oxygen atmosphere

In order to avoid the formation of nitrite and nitrate ions by sonication in the presence of air[5], the degradation reaction was investigated in the absence of nitrogen, under constant rate bubbling of Ar-O_2 mixtures of different composition. These conditions assured slightly higher reaction rates with respect to simple flowing of the same gas mixture in the reactor. The whole composition range of Ar-O_2 mixtures was explored, by operating under simultaneous photocatalysis and sonolysis conditions. First order rate constants values measured for O_2 contents above 50 vol.% were *ca.* twice the value measured under normal atmosphere.

A series of kinetic tests were also carried out either under air or under a gas mixture containing 80 vol.% of oxygen and 20 vol.% of argon, with a reaction volume of 330 mL. The results of such kinetic analysis are reported in Table 1.

Table 1. Rate constants of 2-chlorophenol sonolytic (US), photocatalytic (UV + TiO_2) and simultaneous sonolytic and photocatalytic (US + UV + TiO_2) degradation, measured under standard conditions in air (k_{air}) and under a 4 L h^{-1} flux of a 80% oxygen/20% argon mixture ($k_{O_2\text{-}Ar}$).

Run	$10^5 \times k_{air}$ (s^{-1})	$10^5 \times k_{O_2\text{-}Ar}$ (s^{-1})	$k_{O_2\text{-}Ar}/k_{air}$
US	0.61 ± 0.05	0.69 ± 0.05	1.13
UV + TiO_2	5.89 ± 0.08	13.4 ± 0.8	2.27
US + UV + TiO_2	4.92 ± 0.11	9.14 ± 0.13	1.86

Similar results were obtained with a reaction volume of 400 mL, i.e. in the presence of synergism between sonication and photocatalysis. Substitution of air with an

O_2-Ar mixture induces a marked increase in the rate of photocatalytic degradation (rate constants are practically doubled, *i.e.* $k_{O_2\text{-Ar}}/k_{air} \cong 2$), while a much smaller increase can be observed in sonolytic degradation. The ratio between rate constants measured under simultaneous sonolytic and photocatalytic degradation ($k_{O_2\text{-Ar}}/k_{air}$ = 1.86) reflects the contributions of the two techniques to the overall degradation of 2-CLP. An increase in the flux rate of the O_2-Ar mixture from 4 to 12 L h^{-1} did not induce any further increase in the reaction rate.

Effects of ozone addition

All kinetic runs were finally repeated in the presence of ozone, under bubbling of a mixture containing 80 vol.% of argon and 20 vol.% of O_2/O_3 (O_3 content in the O_2/O_3 mixture: 1.5 vol.%). This reaction condition is shortly denoted as O_3 in the text and in the figures. Different reaction intermediates were obtained in these cases, as evidenced by marked variations in the shape of the absorption spectra recorded at different reaction time. Thus 2-CLP concentration could not be monitored by spectrophotometric analysis at 273.2 nm, as in previous tests in the absence of ozone, and only an estimation (in excess) of its residual concentration could be given as a function of time, and other species in the reaction system absorbing part of the light at the monitoring wavelength.

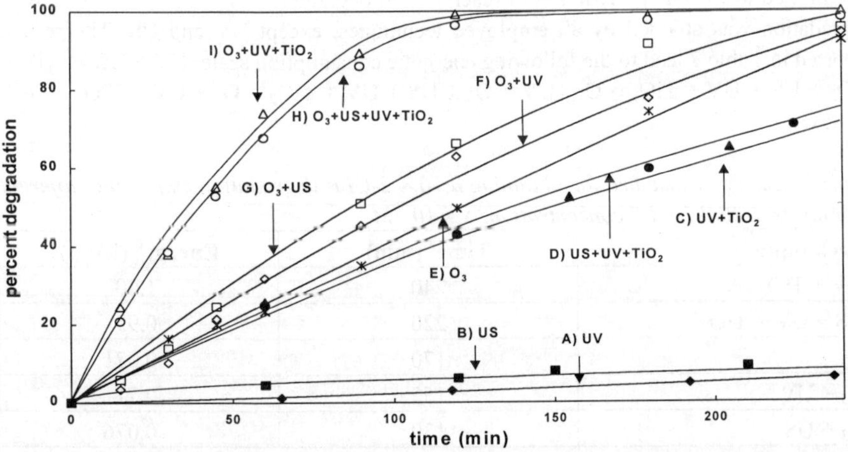

Fig. 3. Percent 2-CLP degradation as a function of time obtained in the absence (full symbols) and in the presence of ozone (open symbols) under irradiation (diamonds), sonolysis (squares), photocatalysis (triangles), sono-photocatalysis (circles) and under simple O_3 bubbling (in Ar(80%)-O_2/O_3(20%) mixture). Reaction volume: 400 mL.

Figure 3 illustrates the results of all kinetic runs in terms of percent degradation as a function of time. Under the adopted experimental conditions 2-CLP degradation under simple Ar-O_2/O_3 mixture bubbling (Fig. 3E), without US or UV irradiation, occurred at a higher rate respect to all other conditions discussed so far (Fig. 3A-D). Illumination (Fig. 3F) and, even more, sonication (Fig. 3G) in the presence of O_3 induced a further increase in the extent of degradation.

The highest extent of 2-CLP degradation, however, was measured under photocatalytic conditions in the presence of O_3 (Fig. 3I), while simultaneous sonication apparently caused a small reduction of rate (Fig. 3H). Therefore the following reactivity scale emerges from the results of 2-CLP degradation carried out employing different techniques or combined techniques (Fig. 3): O_3 + UV + TiO_2 ≥ O_3 + US + UV + TiO_2 > O_3 + US > O_3 + UV > O_3 > US + UV + TiO_2 > UV + TiO_2 >> US.

ENERGETIC CONSUMPTION

Although energetic consumption aspects must be obviously considered when dealing with water purification techniques, little information about this topic can actually be found in the literature.

The energetic consumption relative to the different experimental techniques tested in this work could be easily calculated from the data reported in Figure 3; this calculation was limited to the energy required to reach a 70% degradation of 2-CLP, as this percent degradation was attained by all employed techniques, except UV and US. The results reported in Table 2 lead to the following energetic consumption scale: UV > US >> UV + TiO_2 > US + UV + TiO_2 > O_3 +UV > O_3 + US + UV + TiO_2 > O_3 + UV + TiO_2 > O_3 + US ≥ O_3.

Table 2. Energy consumed for obtaining a 70% 2-CLP degradation employing different techniques. Initial 2-CLP concentration: 5×10^{-5} M.

Technique	Time[a] (min)	Energy[b] (kWh)
UV + TiO_2	240	1.00
US + UV + TiO_2	220	0.95
O_3	170	0.071
O_3 + UV	155	0.71
O_3 + US	130	0.076
O_3 + US + UV + TiO_2	60	0.28
O_3 + UV + TiO_2	55	0.24

[a] Evaluated from the degradation curves reported in Figure 7.
[b] Calculated on the basis of the following power consumption: US generator 10 W (effective energy supply), UV lamp 250 W, ozonizator 25.

As usual, O_3 means bubbling of a mixture Ar (80 %) – O_2/O_3 (20%); $O_3:O_2$ =

1.5:100. Although kinetic runs carried out under O_3 conditions imply the lowest absolute energy consumption, it is worth underlining that through simultaneous sonication (*i.e.* under O_3 + US conditions) a 70% 2-CLP degradation can be obtained in a 24% shorter time, while the amount of energy required is only 7% higher. Such higher energy consumption could be reduced by using US emitters with an energy efficiency greater than 75% (as that employed in the present study).

CONCLUSION

The analysis of the kinetic results obtained in this study by employing many different experimental techniques for the degradation of 2-CLP in water point to the conclusion that a synergetic effect is present coupling two or more different techniques one with the other. Especially O_3+UV+ TiO_2 with and without US lead to a full degradation of 2-CPL in less than two hours.

Notwithstanding the good degradation efficiency reached by all the treatments, the main problem of all the used techniques is the great energetic consumption: for the future this will be the problem to be solved for all the "clean" techniques which allow to purify water without adding chemical agents, but which need electricity to work.

REFERENCES

1. A. Tiehm, U. Neis (Eds.), Ultrasound in Environmental Engineering, TU Hamburg – Harburg, Reports on Sanitary Engineering, Vol. 25, 1999.
2. Proceedings of the 2nd Conference: Applications of Power Ultrasound in Physical and Chemical Processing, Toulouse (France), May 6-7, 1999, Chairperson A.M. Wilhelm, Progep, Toulouse (France).
3. Proceedings of the 7th Meeting of the European Society of Sonochemistry – May 14-18, 2000 – Biarritz – Guéthary (France), Chairperson H. Delmas, Progep, Toulouse (France).
4. N. Serpone, R. Terzian, H. Hidaka, E. Pelizzetti, J. Phys. Chem. 98 (1994) 2634.
5. C. Pétrier, M. Micolle, G. Merlin, J.-L. Luche, G. Reverdy, Environ. Sci. Technol. 26 (1992) 1639.
6. C. Pétrier, Y. Jiang, M.-F. Lamy, Environ. Sci. Technol. 32 (1998) 1316 and references therein.
7. A.J. Colussi, H.-M. Hung, M.R. Hoffmann, J. Phys. Chem. A 103 (1999) 2696.
8. H.-M. Hung, M.R. Hoffmann, J. Phys. Chem. A 103 (1999) 2734 and references therein.
9. H. Destaillats, H.-M. Hung, M.R. Hoffmann, Environ. Sci. Technol. 34 (2000) 311.
10. E. Naffrechoux, S. Chanoux, C. Pétrier, J. Suptil, in ref. 2, p.185.
11. O.V. Abramov, V.O. Abramov, A.E. Gekhman, H. Delmas, V.M. Kuznetsov, T.J. Mason, I.I. Moiseev, in ref. 3, p.85.
12. L.K. Weavers, F.H. Ling, M.R. Hoffmann, Environ. Sci. Technol. 32 (1998)

2727.
13. J.-W. Kang, M.R. Hoffmann, Environ. Sci. Technol. 32 (1998) 3194.
14. L.K. Weavers, N. Malmstadt, M.R. Hoffmann, Environ. Sci. Technol. 34 (2000) 1280.
15. N. Serpone, E. Pelizzetti, Photocatalysis Fundamentals and Applications, Wiley, New York, 1989.
16. J.M. Herrmann, C. Guillard, P. Pichat, Catal. Today 17 (1993) 7.
17. M.R. Hoffmann, S.T. Martin, W.Y. Choi, D.W. Bahnemann, Chem. Rev. 95 (1995) 69.
18. I.Z. Shirgaonkar, A.B. Pandit, Ultrason. Sonochem. 5 (1998) 53.
19. T.J. Mason, Practical Sonochemistry, Ellis Horwood, Chichester, UK, 1991, p.45.
20. J.-C. D'Oliveira, G. Al-Sayyed, P. Pichat, Environ. Sci. Technol. 24 (1990) 990.
21. C.S. Turchi, D.F. Ollis, J. Catal. 122 (1990) 178.
22. P. Théron, P. Pichat, C. Guillard, C. Pétrier, T. Chopin, Phys. Chem. Chem. Phys. 1 (1999) 4663.
23. N.H. Ince, Wat. Res. 33 (1999) 1080.
25. C. Guillard, J. Disdier, J.M. Herrmann, C. Lehaut, T. Chopin, S. Malato, J. Blanco, Catal. Today 54 (1999) 217.

16. TRANSGENIC PLANTS AS VACCINES: IMPACT ON DEVELOPING COUNTRIES WORKSHOP

TRANSGENIC VACCINES IN PLANTS–PROSPECTS FOR GLOBAL VACCINATION

GIOVANNI LEVI, PH.D.
Laboratory of Molecular Morphogenesis, Advanced Biotechnology Center-IST, Largo Rosanna Benzi n10, 16132 Genova, Italy

With more than 13 million deaths a year, infectious diseases are still the major life threat for children and young adults in the world. Diarrhoeal diseases alone claim nearly two million lives a year among children under five. Although effective and relatively cheap treatments for many infectious diseases do exist, their limited accessibility and improper use lead to the gradual erosion of their strength due to the development of anti-microbial resistance. As it has been pointed out by the WHO, we may only have a decade or two to make optimal use of the medicines presently available.

For a brief period, in the middle of this century, infectious diseases seemed vanquished, as antibiotics seemed to be able to eradicate them. Since then, antibiotics have been largely used in human and veterinary medicine. But antibiotics are now under assault, because natural selection has equipped many of the human and animal pathogens they were supposed to attack with resistance genes. Human diseases such as tuberculosis, gonorrhea and cholera are making an alarming comeback, and even foodborne zoonoses are again becoming a prominent problem for public health, also because of the large production losses they may cause. Antibiotics are partly responsible for another worrying phenomenon: whereas some microorganisms, which once were very important and prevalent agents of infectious diseases, have become less prevalent and have progressively lost importance, others previously little known (so-called emerging pathogens) are attracting attention as important causes of infectious disease. Some of these microorganisms were always noxious to humans and animals, but previously they were rare; others are totally new, such as HIV. In addition, some bacteria, which once were symbiotic, have acquired virulent features. Re-emerging and emerging infectious diseases are thus a cause of major public concern. Counterintuitively, lifting the selection pressure (that is, removing antibiotics from the environment) does not seem to be sufficient to solve the problem. Of course judicious use of antibiotics could help prevent the appearance of new antibiotic-resistant mutants, but it seems not to have any effect on the populations of antibiotic-resistant bacteria that already exist. It soon became evident that future strategy to control emerging and re-emerging infectious diseases should chiefly rely on vaccination. Vaccines are again becoming a priority in medical (human and veterinary) research. Yet, although vaccines exist for many diseases and more and more vaccines are in progress of development (children now face 15 vaccinations for 10 diseases with another 50 vaccinations for another 30 diseases in the works), high costs to produce and administer them and logistical barriers such as transportation and refrigeration limit access to them, particularly in the developing world, where other conditions contribute to the spread of disease and medical care often is limited or unavailable.

Keeping this in mind, it becomes of critical importance to rapidly develop new strategies of global vaccination. An ideal vaccine should be safe, cheap to produce, temperature stable and easy to deliver and administer to children in developing countries. The generation of plant-derived edible vaccines might provide an answer for the generation of at least some of these vaccines. Medicinal plants, that were the basis of most drugs used in the past, seemed to have been largely superseded by synthetic chemicals during the last fifty years. Some botanicals, such as digitalis, have remained in use, but they have been heavily outnumbered by formulations that, while they may be modeled after older herbal concoctions, differ from them in both their molecular form and their method of preparation. Herbal remedies are now sold as nutritional supplements , and people think of them as a sort of soft therapy, with minimal side effects, to be used to prevent diseases or to treat minor disturbances. Until a few years ago, the conventional wisdom was, indeed, that medical plants were a product characterizing a specific market niche, of minor importance. This might not be true in the near future. Plant engineering technology, developed in the early 80s to improve crop yields and disease resistance, is about to produce dramatic advances in drug production and delivery. The May 1998 issue of *Nature Medicine* presented, in two papers, the first clinical trials using genetically engineered plants to deliver immunotherapies in humans.

The first paper[1] described a successful trial of a potato-based vaccine intended to prevent infection of the bacterium Escherichia coli (E. coli), which can contaminate food or water supplies causing severe diarrhea. Potatoes were genetically engineered to produce a part of the toxin released by E. coli. The vaccine-containing potatoes were developed and grown by Charles Arntzen and Hugh S. Mason of the Boyce Thompson Institute for Plant Research in Ithaca. The trial was designed to determine the safety and efficacy of this vaccine, and it was approved by the U.S. Food and Drug Administration (FDA) in 1997. It was conducted at the University of Maryland School of Medicine's Center for Vaccine Development under the direction of Carol Tacket. The six-month, double-blind study, enrolled 14 healthy adults; 11 were chosen at random to receive the genetically engineered potatoes and three received pieces of ordinary potatoes. The investigators periodically collected blood and stool samples from the volunteers to evaluate the vaccine's ability to stimulate both systemic and intestinal immune responses. Ten of the 11 volunteers (91 percent) who ingested the transgenic potatoes had fourfold rises in serum antibodies at some point after immunization, and six of the 11 (55 percent) developed fourfold rises in intestinal antibodies. The potatoes were well tolerated and no one experienced serious adverse side effects.

The second article[2] described a trial of a plant-generated antibody designed to prevent the oral bacterial infection that contributes to dental cavities. Julian Ma and his colleagues at Guys Hospital, London, tested a monoclonal secretory antibody against Streptococcus mutans generated in an engineered tobacco plant, created by the Planet Biotechnology, Inc. The tobacco plants were engineered to assemble a chimeric human IgA/IgG secretory antibody. Approximately 1 kg of plant material was required to prepare sufficient antibody for one course of treatment. The team applied a solution of the antibody to a subject's teeth, previously sterilized, showing that the plant secretory antibody afforded specific protection in humans against oral streptococcal colonization for at least four months, while those not receiving antibodies showed signs of recolonization by day 21 and complete recolonization by day 88. The authors thus claimed that they have demonstrated that transgenic plants can be used to produce high affinity, monoclonal secretory antibodies that can prevent specific microbial colonization in humans. These findings could be extended to the immunotherapeutic prevention of other mucosal infections in humans and animals.

The editorial of *Nature Medicine*, dedicated to the two articles, concluded stating: "We are not likely to be growing vaccines in our vegetable patches for a few years. Nevertheless, these papers are significant advances in plant-based immunotherapeutics, still a relatively small field, and make this area of biomedical research a priority."

In November 1998, indeed, Mycogen Corporation, a major agricultural and biotechnology company, announced that it had entered into license agreements with Washington University for exclusive commercial rights to human and animal health applications of technology to genetically alter plants to produce and deliver edible vaccines.

In the same month Agritope, Inc., an agricultural biotechnology company specializing in the development of new fruit and vegetable varieties for sale to the fresh produce industry, was awarded a grant from the U.S. Department of Commerce, Advanced Technology Program (ATP). The ATP grant provided funding of approximately $1 million to support genetic engineering research to control the ripening process of fruit. Agritope identified and patented a single gene that can be inserted into plants and expressed to regulate the plant's ability to produce the ripening hormone ethylene. The focus of Agritope's ATP-funded research is to use the tools and techniques of plant genetic engineering to precisely regulate the ripening process in apples, peaches, pears and bananas. Interestingly enough, the banana portion of the program is combining the expertise of Agritope with the resources of the Boyce Thompson Institute for Plant Research, Inc. (BTI), the institute that is pursuing the program of edible vaccine production in the fruit. Vaccines in fruit obviously need fruit in which one can regulate the ripening process. Therefore research on edible plant vaccine leads to the mainstream of agricultural biotechnology.

The potentials and limits of the technology of vaccine production in transgenic plants have been evaluated in a recent meeting between European, American and Chinese scientists organised in Erice, Sicily by the World Federation of Scientists (WFS) in collaboration with the European Biotechnology Node for Interaction with China (EBNIC). In this meeting we had the participation of both Dr. Charles Arntzen and Dr. Julian Ma, the two most prominent authorities in the field at the moment.

Plants can be used to produce vaccines either in the form of subunit vaccines or recombinant pathogenic plant viruses for active immunisation or as antibodies for passive protection. Plant-based vaccine production would present several major advantages compared to present available technologies. First of all the cost of production could be reduced of up to three orders of magnitude. It has been estimated that for large production of tomato-based edible vaccines the cost could be as low as 0.0025 U.S.$ per dose. Furthermore the administration of plant-derived edible vaccines would not need the use of disposable needles nor of highly specialised medical personnel and could be well received by children when presented in fruits such as bananas. Most of the vaccines produced in plants would be also relatively stable to temperature making it possible to overcome the need for cold storage. Last, but not least, plant derived vaccines would be free from pathogens such as prion proteins which might be present in preparations of vaccines derived from animal products. The cheap cost of production would make it possible to conceive the use of therapeutic approaches based on the use of large amounts of antibodies for passive immuno-protection.

Transgenic plant products could be delivered in their native form (e.g. bananas, carrots etc.) or could be the starting material for the production of processed edible vaccines to be administered as capsules. In any case they are going to be considered as pharmaceutical products to be handled under the same regulations as any other active drug. Appropriate measures, such as use of seedless or male-sterile varieties and contained cultures, are taken to prevent mixing of these transgenic plants with the wild-type population. While regulations for the production of active compounds in plants are being developed in the USA, it appears urgent that similar actions are also taken in the EU.

Approval from the U.S. Food and Drug Agency (FDA) was obtained to conduct three human trials (two for prototype diarrhea prevention, and one against Hepatitis B infection). Similarly China has initiated the production of transgenic plants vaccines against cholera and other severe infectious disorders. In the EU, although several laboratories are involved in transgenic plant vaccine research, local and communitarian regulations, which do not yet discriminate clearly between transgenic plants as source of food and transgenic plants as therapeutical agents, have made it difficult to reach the level of production or even of clinical trials. Activities are now in progress with international agencies to introduce the concept of plant-based vaccine production to developing countries that desire less costly vaccines to prevent infectious diseases, and develop in-country capacity for vaccine manufacture.

Obviously the possible diffusion of this technology to developing countries will have to go hand in hand with the development of appropriate training and monitoring programs to assure that proper use of plant vaccines is made. Indeed depending on the nature and the dose of orally administered antigens systemic and local immune unresponsiveness can be induced instead of protective immune responses. A profound knowledge of the molecular and cellular mechanisms of mucosal immune response is needed to choose the appropriate therapeutical targets and strategies. Furthermore international organisms should be able to control questionable applications of this technology such as vaccination of populations without informed consent or even military application.

Although the first clinical trials on human volunteers have shown that edible plant vaccines are effective in inducing an immune response, an experiment demonstrating directly their capacity to protect from disease transmission is still to be done. Most of the technical difficulties concerning the generation of appropriate vectors to generate vaccines in bananas, tomato, carrots and even seaweed have already been solved and it is reasonable to predict that it will not take long before we will be able to test their efficacy in large scale immunisation tests opening the way for plans of global vaccination.

REFERENCES

1. Arntzen, C.J., 1998, "Pharmaceutical foodstuffs Noral immunization with transgenic plants." Nature Medicine (vaccine supplement); 4(5):502-03.

2. Ma, J.K-C, Hikmat, B.Y., Wycoff, K., Vine, N.D., Chargelegue, D., Yu, L., Hein, M.B., Lehner, T., 1998, "Characterization of a recombinant plant monoclonal secretory antibody and preventive immunotherapy in humans." Nature Medicine; 4(5):601-605.

TRANSGENIC PLANTS EXPRESSING HUMAN GLUTAMIC ACID DECARBOXYLASE (GAD65), A MAJOR AUTOANTIGEN IN TYPE 1 DIABETES MELLITUS

MARIO PEZZOTTI

1 Dip. Scientifico e Tecnologico, Università di Verona, Verona, Italy

ALBERTO FALORNI

Dip. di Medicina Interna e Scienze Endocrine e Metaboliche, Università di Perugia, Perugia, Italy

Transgenic plants are emerging as an important system for the expression of many recombinant proteins, especially those intended for therapeutic purpose.

Our interest is focused on production and characterization of transgenic plants expressing human GAD65, the major autoantigen in Type 1 diabetes mellitus (T1DM). Transgenic plants that express high levels of recombinant human GAD65 could be the source of food for oral administration of the autoantigen.

Little information is available on the pathogenesis of T1DM in man, due to obvious limitations: heterogeneous genetic background, different environmental factors, unavailability of pancreatic tissue for histology studies. The non-obese diabetic [NOD] mouse is an inbred rodent strain which develops spontaneous insulin-dependent diabetes. Diabetes seen in this animal model is remarkably similar to the human disease. Several lines of evidence suggest that the CD4 T-cell is the immunological effector. CD4 T-cells have been divided into two subsets, Th1 and Th2, based on the cytokine secretion patterns. Th1 and Th2 cells, mutually exclusive, mediate different immunological responses. It has been suggested that autoimmune diabetes develops in genetically prone individuals when Th1 and Th2-cell dependent effects are unbalanced, with a predominance of Th1-type cells in the site of insulitis.

Several immunoprevention strategies for secondary prevention have been tested in human T1DM. However, based on the fact that at least 90% of T1DM subjects have no affected relatives, primary prevention of T1DM in the general population is the ultimate goal. None of the therapies so far investigated combines all the features—efficacy, safety, specificity, low cost and applicability to the general population—required for primary prevention. Induction of oral tolerance would satisfy all these requirements. Since GAD65 is the major autoantigen associated with human T1DM, studies of oral tolerization with hGAD65 have a strong rationale. However, because of the high cost of producing large quantities of recombinant hGAD65, these studies have so far been unpractical. Autoantigen-expressing transgenic plants overcome this problem, since they would allow the recombinant protein to be produced on a large scale at a relatively low cost.

Recombinant autoantigens expressed in transgenic edible plant organs would prevent primary T1DM easily and simply, as they would be directly ingested with the diet in the absence of expensive purification procedures.

Our recent experiments on the expression of human GAD65 in tobacco and carrot [Porceddu et al. 1999] showed that the expression levels of the immunoreactive and correctly folded recombinant protein were similar to those reported for other in planta expressed human proteins. However, a major issue concerning the induction of oral tolerance is the dose of autoantigen to be fed. High doses could induce deletion or anergy of specific T-cell clones while low to intermediate doses could activate regulatory T cells in the gut, with subsequent active suppression. The dose of GAD65 needed to induce oral tolerance is still unknown. Results obtained in the NOD mouse have shown that repeated oral administration of at least 1 mg of insulin or GAD67 is necessary to reduce the incidence of autoimmune diabetes. We demonstrated that hGAD65 can be expressed in planta, but the expression levels observed were not adequate to plan studies of induction of oral tolerance in animals fed transgenic plants.

Human GAD65 is membrane-anchored by signals located in the NH_2-terminal region. The absence of these signals in the NH2-terminal region of GAD67 is responsible for its localization in the cytosol. Our recent study showed the targeting of hGAD65 to membranes of chloroplast tylacoids and mitochondria in transgenic tobacco.

Therefore we tried to improve the expression levels of hGAD65 by the targeting of hGAD65 to the plant cell cytosol. As cytosolic expression levels of some heterologous proteins resulted higher than those of membrane-bound proteins in transgenic plants, the expression of hGAD65 could be enhanced by targeting the protein to the cytosol through site-directed mutagenesis of critical residues responsible for membrane interaction. This was achieved by constructing a chimeric molecule generated by substitution of the first 31 amino acids of GAD65 with an homologous region of GAD67. The rationale for this strategy is indirectly provided by the evidence that GAD67, the cytosolic isoform of GAD, can be expressed in transgenic plants at 10-fold higher levels than those we recorded with GAD65. It must be noted that in human T1DM, all the B-cell and T cell GAD65 epitopes are located in the middle and COOH-terminal regions of the enzyme. Moreover, the NH_2-terminal region of GAD65 is apparently not seen by the immune system. Thus, deletion of the signals located in the region encompassing amino acids 1-31 of the enzyme should not preclude its use for induction of immunological oral tolerance.

The level of expression obtained in planta of such chimeric protein (GAD67/GAD65) was five fold higher than that observed with GAD65 but still too low to perform oral tolerance studies in NOD mice.

Autonomously replicating plant viruses can also express foreign genes in plants at very high levels. We have engineered potato virus X (PVX) where hGAD65 was under the control of a copy of coat protein promoter and infected *Nicotiana benthamiana* and potato plants. GAD65 recombinant protein was transiently expressed at level of 2.5% of total soluble plant proteins. These results pave the way to future studies on oral tolerance induced by feeding plant organs that express recombinant human GAD65.

REFERENCES

A. Porceddu, A. Falorni, N. Ferradini, A. Cosentino, F. Calcinaro, C. Faleri, M. Cresti, F. Lorenzetti, P. Brunetti and M. Pezzotti (1999) Transgenic plants expressing human glutamic acid decarboxylase (GAD65), a major autoantigen in insulin-dependent diabetes mellitus. Molecular Breeding 5: 553-560.

ACKNOWLEDGMENTS

This work was in part supported by a grant of Ministry of University and Scientific Research, Project "Ottimizzazione dell'espressione in planta del maggior autoantigene umano per studi di tolleranza orale nella prevenzione del diabete mellito autoimmune". The financial support of Telethon (grant E.0955) is also gratefully acknowledged.

GENETICALLY ENGINEERED THERAPEUTIC ANTIBODIES

ZELIG ESHHAR, PH.D.
Department of Immunology, The Weizmann Institute of Science, Rehovot, Israel

While the first monoclonal antibodies (mAb's) were described more than a quarter of a century ago, it took more than 20 years for the regulatory authorities to allow these powerful reagents to be used in the clinic. Although to date only nine mAbs have been approved by the FDA, there are tens of clinical trials in advanced phases that show potential therapeutic efficacy. The soaring stock prices of the Biotech companies that produce mAbs reflect the market and public appreciation of antibodies as powerful therapeutic means. The scientific community did not have to wait for this realization, and has long benefited from the tremendous advances mAbs brought to the fields of cell biology and diagnosis. In fact, it is largely due to basic research that the initial unsuccessful clinical attempts turned into the success story of today. Most of the antibodies approved for treatment of patients, and those that are undergoing clinical trials, are genetically engineered antibodies, which are derivatives of the classical, hybridoma-produced mAbs. The main problem that limited the use of murine mAbs for therapy lies in the immune response developed in patients against the foreign (mouse-derived) antibody. The production by patients of human anti-mouse antibodies (HAMA) following the administration of mouse mAbs is especially critical when repeated treatment is required. In acute cases, antibodies such as OKT3, which have been used to suppress allograft rejection, work quite successfully. Here it was important to use antibody of the right class to achieve the optimal effector function (e.g. cytotoxicity). The control of antibody sub-class and elimination of immunogenicity of non-human antigenic determinants have been achieved to a large extent using genetic engineering techniques to produce humanized, chimeric mouse-human and fully human antibodies.

The innovations that paved the road towards therapeutic antibodies include the development of means to clone the variable regions of the mAbs and to express them as functional antibodies in mammalian cells, and as recombinant antibody fragments in bacteria. A further development, which gave a quantum leap to the field, was the introduction of phage display antibody technology. This technique enabled not only the selection and generation of new antibodies from the natural and synthetic repertoire of the human variable gene loci, but also the reshaping of existing murine antibodies into human ones. The fact that antibodies are composed of defined domains enabled their exon shuffling and the grafting of murine variable (V) regions onto human constant (C) heavy and light chains with the desired effector activity to obtain chimeric antibodies.

Human IgG1was shown to be the preferred choice for recruitment of cytotoxic effector functions. Later on, the complementarity determining residues (CDR) of the V regions that make the actual contact with the antigen were grafted onto the framework regions of human V-region and expressed with human C regions to yield humanized antibodies. These chimeric and humanized antibodies were nevertheless found to be immunogenic in humans, and repeated administration often resulted in production of anti-idiotypic antibodies. Therefor, as an alternative, fully human antibodies have been generated, assuming (wrongly!) that these would not produce an anti-idiotypic response. In addition to phage antibodies selected from human V-region repertoire, two other techniques are being used today. The first technique reconstitutes SCID immunodeficient mice with lymphocytes taken from patients that overcame an otherwise fatal disease by the development of protective antibodies. These are then immortalized by a fusion with hetromyelomas that were selected to maintain the human chromosomes. This technology is however limited by the limited persistence of the human immune system in the SCID mice. Probably the best system developed for the generation of fully human mAbs consist of transgenic mice, whose own antibody loci were deleted and replaced by large portions of the human light and heavy gene loci. These mice can be immunized with any antigen and their B cells, as well as the resulting hybridomas, generate human antibodies with a wide spectrum of specificity and affinity. Such Xenomouse technology combines the ease of the generation of mAbs using the classical technology with the human repertoire, and has already yielded unique antibodies that are under phase II trials. I believe that this will be the preferred technology in the future. Nevertheless, the induction of anti-idiotype antibodies in patients was not fully eliminated by using such human antibodies. Apparently other solutions will be needed for cases when repeated administration is needed. Possible approaches include the induction of tolerance to the antibody before treatment, inclusion of immunosuppressive treatment, and use of a panel of different antibodies to the same target to avoid anti-idiotype production.

In parallel to the developments of the various generations of therapeutic antibodies, technologies for their mass production have been exploited. The best source of properly glycosylated antibodies is mammalian cells. Various expression systems have been developed using strong promoters and amplification techniques. The most commonly used producer cells include COS, CHO and mouse myeloma cell lines. The main disadvantage of antibody production by mammalian cells is the low concentration of antibodies secreted by the cultured cells. More recently, high concentrations and large amounts of high quality antibodies have been produced in the milk of transgenic goats and sheep. Bacteria can not assemble whole glycosylated antibodies, but can serve as a reliable source for antibody fragments. Yeast and insect cells, via baculovirus vectors, produce whole antibodies; however, the composition of the oligosaccharide side chains differ significantly from the human ones, and antibodies produced in yeast result in antibodies defective in their ability to fix complement. Since 1989, when antibodies were first expressed in tobacco leaves, plants were found to be an inexpensive option for producing large amounts of antibodies, including therapeutic IgA for the treatment of dental caries. Another advantage of plant produced antibodies ('plantibody') is the ability to store them for a long period of time (e.g. in potato tubers). Recently, transient expression systems have been developed in plants that allow feasibility studies that can yield results within a few weeks. This is a significant improvement over the stable expression system in which 5-10 months elapsed before the product could be analyzed.

The potential of therapeutic antibodies is very bright. In the very near future, the number of antibodies that will be approved for clinical trials will be more than double. New antibodies against novel antigens discovered by advanced technologies will enter into clinical trials. Efficient production technologies will significantly reduce the cost of the final product and we shall see more antibodies at the bedside to treat not only life threatening diseases, but hopefully also for use in preventive medicine.

For detailed reviews see the August issue of *Immunology Today*, marking the 25th anniversary of monoclonal antibodies.

PRODUCTION OF VACCINE IN PLANT EXPRESSION OF FMDV PEPTIDE VACCINE IN TOBACCO USING A PLANT VIRUS BASED VECTOR

LI-GANG WU, JI-HUA FAN, QING-QI ZHANG, HUI-HUI ZHU, ZHENG-KAI XU[1*]
National Laboratory of Plant Molecular Genetics, Shanghai Institute of Plant Physiology, The Chinese Academy of Sciences, Shanghai, 200032, China
ZHI-AI ZHOU
Institute of Veterinary Science, Shanghai Academy of Agricultural Sciences, Shanghai, China
YONG XIE
Department of Biology, The Hong Kong University of Science & Technology, Clear Water Bay, Kowloon, Hong Kong, China

INTRODUCTION

With rapid advances in plant molecular biology and biotechnology, plants have been acknowledged as an important system for the expression of many recombinant proteins with industrial or pharmaceutical value. Plants represent an economical and safe alternative to fermentation-based production systems, that is particularly fundamental for developing countries to produce antigens as very safe vaccines at a lower cost. Regarding cost, safety and required equipment, mass production in plants is far easier to achieve than techniques involving other bio-systems. However, the plant expression system requires further development that would greatly increase the expression level of the transgenes. The properties related to some plant viruses, such as tobacco mosaic virus (TMV), their simple and stable structures of the virion and the genome, excellent immunogenecity of CP[1], high yield of the progeny virus in infected hosts and convenient purification of the virus particles have made it a very efficient vector for producing great amounts of foreign proteins. Many strategies based on the TMV genome have been proposed to guide the designing of various expression constructs for the expressions of specific foreign proteins. With the help of one strategy involving the expression of the foreign peptides as the chimerical CP subunits, many peptides for medical purposes, such as vaccines and medicines have been produced in plants [2,3,4,5].

FMD (foot-and-mouth disease) is a highly contagious disease caused by FMDV,

* To whom correspondence should be addressed. E-mail: xuzenkac@online.sh.cn

which impacts the meat and milk producing domestic animals and causes great economic losses. Vaccine is the most successful way to protect animals from FMDV infection. The commonly used vaccines are based on the utilization of an inactivated virus, however, improper use of this type vaccine would result in a dangerous outbreak of this disease. In this report, we have utilized recombinant TMV as a vector to produce a great amount of FMDV peptides in tobacco plants for vaccine purpose. Earlier studies have revealed that the dominant immunogenic site of FMDV is located in a region within the 141-160 aa of VP1[6,7]. The immunogenicity can be enhanced by a neighboring C' terminal sequence of VP1 (200-213 aa)[8,9]. Further studies also revealed that the synthetic peptide (141-160 aa) alone could not elicit sufficient immune response in tested animals unless the peptides were chemically linked or fused to a carrier protein[10]. In addition, the immunogenic activity of short peptide can be greatly enhanced as the peptides presented as multiple linked copies or spontaneously assembled to give high local concentration[11,12]. Therefore, we believe that the peptides expressed in the fused form to TMV CP may topologically present at the surface of the recombinant virus particle to a very high density on the virus surface that would result in high immnogenicity.

RESULTS

Construction of the recombinant TMV expressing FMDV epitopes

The cDNA fragments specifying the epitopes of FMDV, F11 (PNVRGDLQVLA, 142-152 aa of VP1), F14 (RHKQKIVAPVKQTL, 200-213 aa of VP1) and F20 (VPNLRGDLQVLAQKVARTLP, 141-160 aa of VP1), were synthesized using specific primers and individually inserted *in frame* into an infectious TMV cDNA clone, pTMV at the 3' terminal region of the CP gene by oligonucleotide directed mutagenesis. The constructed clones, pTMV11, pTMV14 and pTMV20, contain the corresponding recombinant CP genes that encoded the additional 11, 14 or 20 amino acids between the sites of 154 and 155 aa of the CP.

Expression of the FMDV epitopes in tobacco

The full length infectious RNA of rTMV or wtTMV was synthesized by T7 polymerase driven run-off transcription and directly applied to inoculate tobacco seedlings. Typical systemic mosaic symptoms appeared on the newly formed leaves of rTMV11 or rTMV14 infected plants between 12 to 16 days post-inoculation, which had no apparent difference from those inoculated with wtTMV. Very similar systemic mosaic symptoms appeared, with about 10 days delay, on the young leaves of the tobacco plants inoculated with rTMV20. This may reflect that the long distance transportation of the virus was retarded as the 20 amino acid-long peptide was fused to the C' terminal of TMV CP.

The recombinant viruses in the infected tobacco young leaves were identified at the RNA level by RT-PCR analysis targeting to the inserted FMDV sequence and at the CP level by SDS-PAGE analysis.

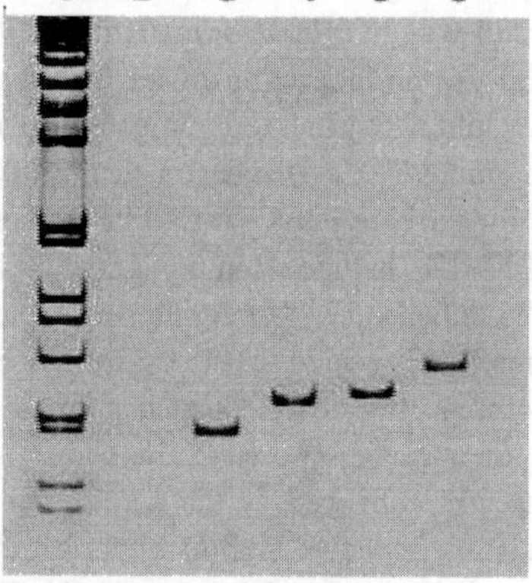

Fig. 1. Analysis of the inserted FMDV fragments in the progeny recombinant TMV genomes in the infected tobacco plants. Tobacco plants were inoculated separately with no virus (lane 2), wtTMV (lane 3), rTMV11 (lane 4), rTMV14 (lane 5) and rTMV20 (lane 6). Fourteen days after inoculation, RNA was extracted from the third leaves above the inoculated leaves of the individual tobacco plants and amplified by RT-PCR using primer SQ(+) and NS(-). The products were separated by 5% PAGE and stained with ethidium bromide. A molecular weight standard of the 1kb DNA ladder (BRL) is shown in Lane 1.

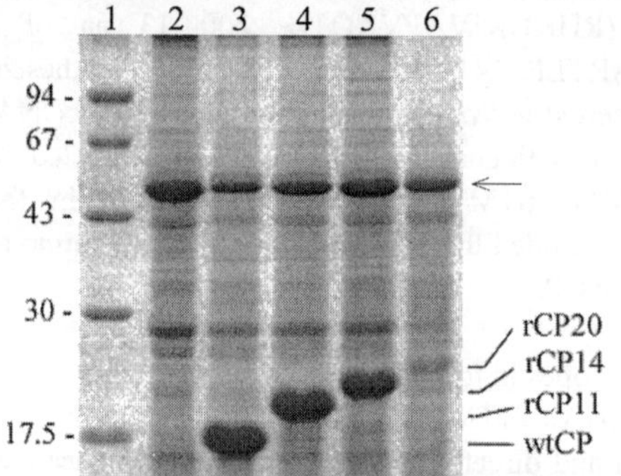

Fig. 2. Detection of the recombinant coat protein subunits in the infected tobacco leaves by SDS-PAGE analysis. Tobacco plants were inoculated separately with no virus (lane 2), wtTMV (lane 3), rTMV11 (lane 4), rTMV14 (lane 5) and rTMV20 (lane 6). Fourteen days after inoculation, total proteins were extracted from the third leave above the inoculated leaves of the individual tobacco plants and separated by 12.5% SDS-PAGE. The gel was stained with Coomassie brilliant blue. The protein molecular weight standard is in indicated in Lane 1. The arrow indicates the large subunit of ribulose-1-5-biphosphate carboxylase.

Primers specific to the TMV sequence flanking the inserted FMDV sequence were used in the RT-PCR analysis. As shown in Figure 1, the size of each RT-PCR product was consistent with the corresponding inserts, 33 bp for rTMV11 (lane 3), 42 bp for rTMV14 (lane 4) or 60 bp for rTMV20 (lane 5).

The recombinant coat protein subunits expressed in infected tobacco were analyzed by 12.5% SDS-PAGE as shown in Figure 2. The distinct protein band of the recombinant CP subunit of rTMV11 (lane 4), rTMV14 (lane 5) or rTMV20 (lane 6) was present. With the subunit of ribulose-1-5-biphosphate carboxylase as an internal reference (indicated by an arrow), the amount of the recombinant CP in the newly formed leaves infected with rTMV11 or rTMV14 was comparable with that of the wild-type CP in the sample infected with wtTMV. The amount of the recombinant CP of rTMV20 was apparently much lower in newly formed leaves. However, after a longer infection period, the amount of rCP20 in the same leaf increased to 1/3 of the wtCP. The slow accumulation of rCP20 in newly formed infected leaves may be due to the 20 aa insertion in CP that resulted in the inefficient long distance transportation of the rTMV20.

TMV-like virus particles were observed in the purified samples by electron microscopy. The recombinant virus particles can be purified by PEG precipitation. The yield of the purified rTMV11 and rTMV14 was estimated to be 1 g per 100 g of infected fresh leaf tissue, similar to that of wtTMV. However, no virus particle of rTMV20 was precipitated after PEG treatment although virus alike virus particles of rTMV20 could be observed clearly by electron microscopy. It seems that the particle of rTMV20 was not stable enough in the conditions during purification.

When mixed with trace amount of the wtTMV, all the recombinant viruses rTMV11, rTMV14 and rTMV20 were unable to cause systematic infection in tobacco, the progeny virus was apparently that of the wtTMV. This suggests that the recombinant viruses were unable to compete with the wild type TMV. Without the competition of wtTMV, the recombinant viruses were also quite stable during the mechanical transmission from plant to plant.

Effects of rTMV11/rTMV14 on the protection of animals from FMDV infection

The guinea pigs (4 individuals) were injected each with 0.6 mg of rTMV11/rTMV14. Forty-two days later, each guinea pig was challenged with FMDV at 100 guinea pig 50% infection doses ($GPID_{50}$). All the guinea pigs were protected completely 10 days after the challenge (Table 1). However, in the control of the same experiment, the guinea pigs who received 0.6 mg of wtTMV were found all infected after the viral challenge showing typical severe lesions on the footpads. Western blot assay (data not shown) identified that the antiserum collected from the protected guinea pigs could react specifically to the VP1 of FMDV.

Table 1. Protection of the guinea pigs from FMDV challenge after the immunization with rTMV11/rTMV14.

Vaccine	No. of Challenged	No. of Protected	Protection rate (%)
wtTMV	4	0	0
rTMV11/rTMV14	4	4	100
BEI-inactivated	4	3	75

Table 2. Effects of the injected amounts of rTMV11/rTMV14 on the protection of guinea pigs from FMDV challenge.

Antigens injected	Dose (ug/guinea pig)	No. guinea pig injected	No. guinea pig infected	Protection rate (%)
rTMV11/rTMV14	500	6	0	100
rTMV11/rTMV14	250	6	1	83
rTMV11/rTMV14	125	6	2	60
wtTMV	500	6	6	0

The guinea pigs were challenged with 100 $GPID_{50}$ of FMDV 42 days after immunized once with 500, 250 or 125 ug of rTMV11/rTMV14 or 500 ug of wtTMV (as a control). The protection rate of each group was calculated according to the numbers of the protected guinea pigs 7 days after the challenge.

Table 3. Protection of the suckling mice from FMDV challenge by pre-injection of the guinea pig antiserum.

The suckling mice pre-injected with guinea pig antiserum against the mixture of rTMV11 and rTMV14			The suckling mice pre-injected with guinea pig antiserum against wtTMV		
Dilution factor of the inoculum FMDV	No. of the mice challenged	No. of the mice survived	Dilution factor of the inoculum FMDV	No. of the mice challenged	No. of the mice survived
$1:10^3$	0	0	$1:10^5$	0	0
$1: 10^4$	4	0	$1:10^6$	4	0
$1: 10^5$	4	4	$1:10^7$	4	4

The suckling mice were pre-injected each with 0.1 ml of the antiserum of guinea pigs raised against wtTMV or the mixture of rTMV11 and rTMV14 one day prior to the challenge with FMDV.

To further determine the protection effects, various amounts of rTMV11/rTMV14 were injected into the guinea pigs. Forty two days after the immunization, the guinea pigs were challenged with FMDV at 100 $GPID_{50}$ and the symptoms were observed 10 days after the challenge. As shown in Table 2, none of the guinea pigs were found infected in the group immunized with 500 ug of rTMV11/rTMV14; two guinea pigs were infected in the group injected with 250 and 125 ug of rTMV11/rTMV14, respectively. However, all of the guinea pigs were severely infected in the group injected with 500 ug of wtTMV. The results clearly demonstrated that the protection of the guinea pigs was increased when more rTMV11/rTMV20 were applied and the protection indeed resulted from the epitopes carried by the recombinant TMV coat protein subunits.

Further experiments (Table 3) using suckling mice also revealed that the suckling mice pre-injected each with 0.1 ml of the antiserum from the protected guinea pigs were also protected from the following challenge of FMDV at 1: 100000 dilution, they all appeared healthy and normal. The results indicated that the antiserum raised against rTMV11/rTMV14 can neutralize specifically the infectivity of FMDV. For the control in the same experiment, all the suckling mice pre-injected with 0.1 ml of the guinea pigs antiserum raised against wtTMV all died after the challenge with FMDV even at 10^6 fold dilution.. Based upon the numbers of the surviving mice and corresponding dilution factors of the FMDV among the tested and control suckling mice, the protection index for rTMV11/rTMV14 was estimated as 2.

The results have demonstrated that the recombinant coat protein of rTMV11/rMTV14 produced in tobacco can indeed induce the specific antibodies that not only react with but also neutralize the infectivity of FMDV. This suggests that tobacco may produce functional vaccine as effectively as the current vaccines for controlling FMDV infection.

DISCUSSION

A similar objective[13,14] has been attempted with a spherical plant virus, CPMV, by inserting the FMDV short peptides into CPMV coat protein. The recombinant CPMV was found to be able to infect plant cells but failed in systemic infection, and no progeny virus could be isolated from the leaf tissue. The authors proposed this might due to the peptide contained in a FMDV-specific amino acid sequence of Arg-Gly-Asp (RGD), an attachment site to the host cell membrane, which caused the recombinant CPMV bind to cell membrane and thus blocked the virus spreading. In our experiments, rTMV11 and rTMV20 both contained the RGD sequence in the fused CP, and rTMV11 infects and spreads systemically in tobacco plants as efficiently as wtTMV. On the other hand, rTMV20, compared to rTMV11, rTMV14 and wtTMV, had about 10 days delay in the virus particle accumulation in the non-inoculated leaves. It seems that the size of the insertion rather than the RGD sequence has affected long distance transportation of the recombinant TMV. Earlier studies have indicated that changes in size, PI and sequence of the CP subunits will influence the interactions among the neighboring CP subunits and viral RNA, and hence the virus particle stability and transportation ability[15,16,17].

According to our experience, the recombinant virus of rTMV20 is a very good example. Although it apparently had a lower efficiency in spreading in the infected plant, the electron microscopy, protein PAGE, immuno blot and RT/PCR all indicated the virus was as normal as the wild type virus. However, the failure of purifying the virus particles suggested that the insertion of the 20 aa long peptide would still strongly affect the stability of the virus and even the recombinant coat protein subunits in the regular conditions. Therefore, an overall conformational change due to the individual insertion and accepter protein sequences, should be taken into consideration when using virus vectors for expressing foreign proteins in plants.

Although there are several other systems to produce FMDV VP1 or short peptides for vaccine purposes, such as chemical synthesis, fermentation or transgenic plant, a novel expression system is required to develop for overcoming the difficulties of low protein yield, biosafety or considerable labor requirement. Using TMV as an expression vector to express foreign protein in plants has several advantages. First, it is simple to obtain a great amount of antigen with low cost. By PEG precipitation, as many as 10 g of the purified recombinant TMV particles of rTMV11 or rTMV14 can be obtained from 1 kg of the infected fresh tobacco leaves, which is equivalent to 0.62 g or 0.77 g of FMDV epitope peptides. Secondly, the recombinant TMV genome is stable but not strong enough in infectivity to compete with the wild-type TMV. It is apparently safe in terms of its undesirable spreading in fields although care should be taken to avoid contaminations of wild type TMV in production of, although the molecular and pathology data presented in this report have clearly demonstrated that the genome of the recombinant TMV during the production of the rTMV based vaccine. Thirdly, the recombinant TMV cDNA clone is readily manipulated to produce various cameric epitopes specific to different serotypes of FMDV which can be mixed and used as a "cocktail" to protect the animals from the infections by different serotypes of the virus. In summary, tobacco may produce, via TMV vector, the safe, efficient and cost effective FMD vaccines on a large scale without requireing of elaborate equipment and materials.

REFERENCES

1. Loor, F. (1967) Comparative immunogenicities of tobacco mosaic virus, protein subunits and re-aggregated protein subunits. *Virology* **33**, 215.
2. Hamamoto, H., Y. Sugiyama, N. Nakagawa, E. Hashida, Y. Matsunaga, S. Takemoto, Y. Wantanabe, and Y. Okada. (1993) A new tobacco mosaic vector and its use for the systemic production of angiotensin-I-converting enzyme inhibitor in transgenic tobacco and tomato. *Bio/Technology* **11**, 930-932.
3. John Fitchen, Roger N. Beachy and Mich B. Hein (1995) Plant virus expressing hybrid coat protein with added murine epitope elicits autoantibody response. *Vaccine* **13**, 1051-1057.
4. Yoshinori Sugiyama, Hiroshi 2. Hamamoto, Shizume Takemoto, Yuichiro Watanabe, Yoshimi Okada. (1995) Systemic production of foreign peptides on the particle surface of tobacco mosaic virus. *FEBS Letter* **359**, 247-250.

5. Thomas H. turpen, Stephen J. Reinl, Yupin Charoenvit, Stephen L. Hoffman, Victoria Fallarme, and Laurence K. Grill (1995) Malarial epitopes expressed on the surface of recombinant tobacco mosaic virus. *Bio/technology* **13**, 53-57.
6. Strohmaier, K., Franze, R. and Adam, K.H. (1982) Localization and characterization of the antigenic portion of the foot-and-mouth disease virus protein. *J. Gen. Virol.* **59**, 295-306.
7. Geysen, H.M., Neloen, R.H. and Barteling, S.J. (1984) Use of peptide synthesis to probe viral antigens for epitopes to a resolution of a single amino acid. *PNAS* **81**, 3998-4002.
8. Parry, N.R., Ouldridge, E.J., Barnett, P.V., Rowlands, D.J., 10. Brown, F., Bittle, J.L. *et al.* (1985) Identification of of neutralizing epitopes of foot-and-mouth disease virus. *Vaccine* **85**, 211-216.
9. DiMarchi, R., G. Brooke, C. Gale, V. Cracknell, T. Doal, and N. Mowat. (1986) Protection in cattle against foot and mouth disease by a synthetic peptide. *Science* **232**, 639-647.
10. Brown, F. (1992) New approaches to vaccination against FMD. *Vaccine* **10**, 1022-1026.
11. Tam, J.P. (1988) Synthetic peptides vaccine design: synthesis and properties of a high-density multiple antigenic peptide system. *PNAS* **85**, 5409-5413.
12. Clarke, B.E., Newton, S.E., Carroll, A.R., Francis, M J., Appleyard, G., Syred, A. D. *et al.* (1987) Improved immunogenicity of a peptide epitope after fusion to hepatitis B core protein. *Nature* **330**, 381-384.
13. Usha, R., Rohll, J.B., Spall. V.E., Shanks, M., Maule, A.J., 3. Johnson, J.E. and Lomonosoff, G.P. (1993) Expression of an animal virus autogenic site on the surface of a plant virus particle. *Virology* **197**, 366-374.
14. Porta, C., Spall, V.E., Loveland, J., Johmson, J.E., Barker, P.J. and Lomonosoff, G.P. (1994) Development of cowpea mosaic virus as a high-yielding system for the presentation of foreign peptides. *Virology* **202**, 949-955.
15. Dawson, W.O., Bubrick, P. and Grantham, G.L. (1988) Modification of the tobacco mosaic virus coat protein gene affecting replication, movement, and symptomatology *Phytopathology* **78**, 783-789.
16. Mohammed 17. Bendahmane, 3. John H. Fitchen, Guangming Zhang, and Roger N. Beachy (1997) Studies of Coat protein-mediated resistance to tobacco mosaic tobamovirus: correlation between assembly of mutant coat protein and resistance. *Journal of virology* **71**, 7942-7950.
17. Bendahmane, B., Koo, M., Karrer, E. and Beachy, R.N. (1999) Display of epitomes on the surface of tobacco mosaic virus: impact of charge and isoelectric point of the epitope on virus-host interactions. *J. Mol. Biol.* **290**, 9-20.

17. RESEARCH RESOURCES WORKSHOP

WORLD FEDERATION OF SCIENTISTS PERMANENT MONITORING PANEL ON CLIMATE, OZONE & GREENHOUSE EFFECT

WILLIAM A. SPRIGG
University of Arizona

The WFS PMP#7 met and held a *Workshop on Research Resources for Addressing Planetary Emergencies* during the 25th Session of the International Seminars on Planetary Emergencies.

On behalf of the WFS, the Climate Panel continues to pursue the means by which equal access to scientific data, information, and the tools of research may be given to scientists of all nations. The 25th Session of the International Seminars on Planetary Emergencies built upon previous Seminars in the WFS series and the WFS statement on data access endorsed during the 24th Session.

The *Workshop on Research Resources for Addressing Planetary Emergencies*, focused on meteorological and related data and information for research and applications in developing countries. Each of the other 14 WFS PMPs were invited to participate.

The workshop began to identify both general and specific research problems and their associated resource requirements. Priority was given to studies designated under the WFS planetary emergencies, with special attention to S&T data and information resources for PMP 7.

Workshop participants attempted to:

1) Broadly characterize the types (e.g., physical characteristics, measurements, geographic distribution, etc) and sources (e.g., types of instruments, institutions, countries) of meteorological data/information that are needed to successfully study, understand, and work toward addressing your specific "planetary emergency."
2) Identify the major gaps and barriers in (a) creating, (b) accessing, and (c) using the databases and other information resources characterized in #1, and the factors that may lead to success.
3) Provide views on what the WFS should do to address research resource problems in this area, particularly in developing countries, and how it might work to achieve those objectives.

The workshop agenda follows.

Workshop on Research Resources for Addressing Planetary Emergencies: Focus on Meteorological and Related Data and Information for Research and Applications in Developing Countries

Agenda

9:30	Opening remarks by the Chair - Introduction of participants - Meeting objectives and schedule	William Sprigg, University of Arizona, USA
9:45	Overview of Barriers to Access and Use of Scientific Data and Information in Developing Countries - *Discussion*	Paul Uhlir, National Research Council, USA
10:20	Policy Issues in the Dissemination and Use of Meteorological and Related Data and Information - *Discussion*	Glenn Tallia, NOAA, USA
11:00	Break	
11:15	Case Studies in Meteorological Data Research and Applications for Addressing Planetary Emergencies in Developing Countries	
	- Examples from Sub-Saharan Africa	Mohamed Boulahya, ACMAD, Niger
	- Examples from Central America	Max Campos CRRH/SICA, Costa Rica
	- *Discussion*	
12:45	Lunch	
14:15	Discussion with PMP Representatives Regarding Meteorological and Related Data and Information Requirements for Addressing the 15 Planetary Emergencies	
16:15	Break	
16:30	Discussion of Conclusions and Possible Next Steps for the WFS	William Sprigg
17:30	End of meeting	

A note of appreciation for organizing the workshop is extended to Glenn Tallia of the U.S. National Oceanic and Atmospheric Administration and Paul Uhlir of the U.S. National Research Council.

WORKSHOP CONCLUSIONS

Following the focused discussions of the PMP #7 workshop, *the principal conclusion of the organizers was that in order for researchers to address the most pressing problems facing humanity, including the 15 "planetary emergencies" identified by the WFS, they must have access to and the ability to use adequate research tools.* All researchers everywhere now require basic infrastructure such as computers and Internet access. In addition, access to many kinds of specialized observational data is needed in order to properly identify, study, and monitor significant environmental problems, while highly specialized information and laboratory tools typically are required to make progress on biomedical research on, for example, infectious diseases.

In the more economically developed OECD countries, the problem of inadequate access to research resources is not usually due to a lack of basic infrastructure, but of tools necessary to perform advanced, cutting-edge research. In many cases, problems may be caused by inappropriate or conflicting statutes and by regulations or policies at institutional, national, and international levels that hinder access to or use of various research resources. In less developed countries, many of these same technical and policy problems exist, but are compounded by more basic research infrastructure inadequacies. Thus, scientists in those areas where planetary emergencies may be most acute frequently have the least sufficient research resources with which to work on solving these emergencies. *Failure to address the inadequacy of research resources in the developing world, therefore, will not only delay creation of appropriate solutions to many of the most pressing planetary emergencies, but will result in a further widening of the gap in economic development between North and South.*

The PMP on Climate, Ozone & Greenhouse Effect proposes to establish a Working Group on Research Resources, organized under the auspices of the World Federation of Scientists and comprised of members from all WFS PMPs, to perform the following tasks:

1) Identify and document both general and specific research resource problems involved in the study of all 15 planetary emergencies.
2) Prepare reports or plans of action for consideration and possible adoption by the WFS.
3) Interact, as appropriate, on behalf of WFS with members of the scientific community, policymakers, national and international organizations, and the media to actively address the problems that are identified and to further the goals of the WFS.
4) Monitor and report back to the WFS on progress made in addressing the identified problems.

Such a Working Group would be composed of 20 members, five appointed from the general body of the WFS and one appointed by each of the 15 Permanent Monitoring Panels on Planetary Emergencies. The working group would meet at least once a year in

August in Erice, Sicily to establish a work agenda and to assess progress. Additional meetings would take place as needed and subject to available funding. Most of the work could be done in a distributed manner through the use of e-mail and other means of communication, including the WFS web site.

INTELLECTUAL PROPERTY RIGHTS IN DIGITAL INFORMATION IN THE DEVELOPING WORLD CONTEXT: A SCIENCE POLICY PERSPECTIVE. SUMMARY OF PRESENTATION

PAUL F. UHLIR
National Research Council, Washington, DC, puhlir@nas.edu

The increasingly pervasive use of electronic networks for disseminating digital data and information has rendered inadequate some of the previous laws for protecting proprietary rights in information in the print paradigm. Major changes to national, regional, and international intellectual property law, therefore, have been proposed and adopted over the past few years. These changes have important implications to the way in which digital information—including scientific and technical data and information—will be created, accessed, and used. The new legal developments, which are being formed with the purpose of protecting primarily large, multinational content providers, are likely to place researchers and educators, especially in developing countries, at a disadvantage and thwart some of the great promises of the Internet.

Under traditional intellectual property law, developed over several centuries of national and international law, information under the print paradigm has been protected by several well-established legal approaches, primarily copyright and trade secret. Copyright protects the creative works of authors, but also provides a variety of public-interest exceptions to the users of those creative works. For databases of factual material, copyright protects only the original selection and arrangement of the contents, but not the underlying compilation of data and facts, which have been considered to be in the public domain. Trade secret protection has been used primarily by businesses to protect confidential and proprietary information that they may contractually license on an individual basis to select end users. In some jurisdictions, commercial unfair competition law has been used to protect businesses that have had their noncopyrightable databases misappropriated by other businesses.

A number of key characteristics of the revolution in computerized information and electronic networks have greatly facilitated the creation, dissemination, and use of information on a worldwide basis. These characteristics include the potential for universal access; direct contact between the vendor and user; and easy copying, transforming, and redisseminating of information. Together, these developments have resulted in unlimited potential for knowledge creation and economic exploitation of information, leading to what is now commonly referred to as the Information Age and the Knowledge Economy. Nevertheless, these same developments have led to some loss of control over proprietary rights in information and to the call for significant changes to the national and international legal regime that governs rights in digital data and information.

The recent revisions to the law have substantially strengthened the legal rights of the owners of digital content, upsetting the balance of rights developed over many decades and tipping the scales towards the creators or vendors of information, at the expense of all downstream users of that information. For example, in March 1996, the European Union issued a Directive on the Legal Protection of Databases that established an unprecedented strong exclusive property right in noncopyrightable databases and collections of factual material. In previous law, such databases had been considered the least worthy of strong intellectual property protection, and overprotection of factual databases is particularly harmful to scientists and educators. A similar law has been vigorously advocated by large database publishers in the U.S. Congress.

In December 1996, the treaty on digital copyright was opened for signature by the World Intellectual Property Organization to its member States. This new copyright law, when combined with technical protection measures such as encryption, has been implemented in both the European Union and in the United States in a way that greatly limits the traditional "fair use"-type exceptions for users, particularly those in not-for-profit research and education

Perhaps most important during recent years, the use of contractual licensing has become the dominant legal mechanism for providing access to and use of proprietary digital information, whether through CD-ROMs or online. Whereas previously the licensing of information was used almost exclusively in large commercial transactions involving trade secrets, such licenses are now used almost universally in individual consumer transactions, as well as for institutional site licenses, such as at universities and libraries. These licenses typically do not include any of the public-interest exceptions for users that were developed under traditional copyright law, and contain restrictions on use and redissemination of material, such as personal copies, loans or resale of single copies, and other uses not specifically authorized by the vendor, that previously were not able to be limited under copyright law. Moreover, in the United States and in other countries there are efforts underway to greatly increase the rights of vendors in such licensing arrangements at the further expense of all downstream users. The growing use of contracts for disseminating proprietary information thus poses one of the greatest limitations on scientists and educators—as well as on legitimate commercial competitors and general consumers, particularly in developing nations.

Finally, the 1994 Agreement on Trade-Related aspects of Intellectual Property (TRIPS), which is implemented under the auspices of the World Trade Organization, requires all signatories, including many of the world's developing and least developed countries, to adopt the intellectual property norms in existence in Europe and the United States. The TRIPS agreement has both positive and negative impacts on access to and use of information in developing countries. From a positive perspective, the TRIPS Agreement and new IP laws are expected to stimulate: trade; foreign investment; availablity of proprietary information (with some possible concessions to less developed countries); private-sector activity within the developing world; and overall economic growth, worldwide.

However, a number of negative effects also are occurring, including, among others: a lack of balance of interests in the new regime, as noted above; strong protection that favors existing, large businesses; legal support of monopoly or strong market positions; anti-competitive effects; increased costs to consumers; unprecedented restrictions on access AND use of scientific and technical information; insufficient public-interest exceptions; and exceptions for developing countries that are unclear and weak.

Although these laws affect only proprietary information, primarily in the private sector, there is a large spillover effect on public-sector information sources as well, further diminishing full and open access to government and academic scientific data and decreasing the information available in the public domain. Strong protection of digital information encourages the protection of new databases and other information produced in government and academia that were previously freely and openly available, either though their direct commercialization or by transferring the rights in the information to private-sector entities. Thus, many research and educational information resources are becoming inaccessible, particularly to scientists and educators in developing countries who need them the most, but who cannot afford even relatively low-cost prices.

As the debate on these issues of global significance continues, it is essential for scientists in both the developed and the developing world to make their voices heard and to publicize the social and economic importance of scientific values and the role of scientists. Scientific values and norms are based on the search for truth, based on facts; openness; cooperation and sharing of data and information; an international perspective; and a reward structure not dependent on commercial success or on making a profit. Strong IP laws and restrictive government information policies tend to hinder, not promote, basic research, education, and other public interests, resulting in direct and collateral social and economic costs. As major stakeholders in the outcome of the new legal regime in digital information, scientists must be involved at both the national and international levels through organizations such as the World Federation of Scientists.

ONLINE BIBLIOGRAPHY

www.nap.edu (U.S. National Academy Press)
See the following reports, all free online:

- *Bits of Power: Issues in Global Access to Scientific Data* (1997)
- *A Question of Balance: Private Rights and the Public Interest in Scientific and Technical Databases* (1999)
- *The Digital Dilemma: Intellectual Property Rights in the Information Age* (2000)

www.codata.org/codata/data_access/index.html (against strong protection)
www.dfc.org (against strong protection)
www.siia.net (for strong protection)

OFFICIAL INTERGOVERNMENTAL AND U.S. GOVERNMENT WEB SITES

www.wipo.org (World Intellectual Property Organization)
www.uspto.gov (U.S. Patent and Trademark Office)
lcweb.loc/copyright (U.S. Copyright Office, Library of Congress)

POLICY ISSUES IN THE DISSEMINATION AND USE OF METEOROLOGICAL DATA AND RELATED INFORMATION SUMMARY OF PRESENTATION

GLENN E. TALLIA, ESQ.
National Oceanic and Atmospheric Administration, Silver Spring, Maryland, USA. Glenn.E.Tallia@noaa.gov

Historically, data produced by public sector entities, such as meteorological data, was made available for open and unrestricted international exchange. Typical data exchanged included observations of temperature, wind, pressure and precipitation. This policy and practice was based on the understanding that weather knows no boundaries, no one nation can generate all the data it requires and the benefits to society, in general, are maximized by data sharing. For its part, the United States traditionally has made all of meteorological data available at no more than the marginal cost of dissemination. It also has placed few restrictions on its use and further dissemination.

Many nations, in addition to the United States are embracing the concept of "open and unrestricted" access to publicly generated data including meteorological data. Other nations, however, do not share these views and are treating their data and information as a government-owned commodity to be "commercialized," and assert essentially a monopoly on certain types of information in order to maximize revenues. This tends to preclude other entities from developing markets for the information or otherwise disseminating the information in the public interest. The scientific and research communities are particularly concerned that such practices are decreasing the availability of critical data and information

In the case of meteorological data, experience with the implementation of World Meteorological Organization (WMO) Resolution 40 has raised concerns as some European meteorological services have moved aggressively to commercialize their information. It appears that European meteorological services are adopting an increasingly narrower view of what constitutes "essential" weather data which they are required to provide freely and promptly to their counterparts under WMO 40. As a result, data are being withheld from international data archives, governments are increasing the prices of data they do make available; scientists and researchers are charged fees for the basic data needed to improve forecasts and are forced to sign restrictive agreements on how data is used; technical assistance to developing meteorological services is decreasing; and cooperation among meteorological services, in general, is decreasing.

The free flow of meteorological data, however, is critical to scientific advancement. It has played a critical role in past advances such as the discovery of the ozone hole and the increasing amount of CO_2 in the atmosphere. Presently, it is critical for forecasting and detecting climate variability and change, natural disaster forecasting and severe weiher forecasting. Moreover, future scientific advancement in such areas as ensemble forecasting are dependent upon the free flow of meteorological data.

There are alternatives to commercialization available to the developing world to help leverage its limited resources. These include the formation of public and private partnerships in furtherance of the delivery of meteorological data and services. In addition, taking a regional approach to the delivery of meteorological data and services offers great promise.

Recently, the Directorate General for the Information Society (the DG XIII) issued a Green Paper titled "Public Sector Information a Key Resource for Europe." That paper suggested a new policy direction for Europe, i.e., increased openness and availability of European public sector data. The United States, at the European Commission Presidency, has supported the policy shift discussed in the Green Paper.

In summary, the record establishes:

- Open and unrestricted access to meteorological data offers the greatest benefit to science and society in general.
- Data commercialization practices are incompatible with advances in science.
- Commercialization practices are detrimental to international cooperation.
- Cooperation not competition is needed between national meteorological services, especially in light of new technologies and the free flow of data over the Internet.
- Developing national, regional and international partnerships to meet the needs of science and the general public can provide adequate, long term support for developing country meteorological services.

18. MOTHER–INFANT HIV TRANSMISSION WORKSHOP

SUCCESSFUL INTERVENTIONS TO REDUCE PERINATAL TRANSMISSION OF HIV

CATHERINE M. WILFERT, M.D.
Professor Emerita, Duke University Medical Center
Scientific Director, Elizabeth Glaser Pediatric AIDS Foundation

Cumulatively, there are eight perinatal trials establishing that antiretrovirals administered to HIV infected pregnant women can decrease transmission of the virus from mother to child. Zidovudine, Zidovudine/Lamivudine, ddI, D4T, and Nevirapine have been used in various regimens. A brief summary of the reduction in transmission rates (37%-67%) and the hypothesized reasons these interventions are successful will be presented. In the developed world optimal suppression of the maternal virus with available combination antiretroviral therapy can occur. For example, in the U.S., an 80% reduction in reported pediatric AIDS has occurred since 1993.

Unfortunately, the knowledge has not been implemented in the developing world where 90% of perinatal transmission of HIV occurs. Thailand will be the first "developing" nation to institute a nationwide program to prevent mother to child transmission of HIV. The regimen is based on clinical trials done in Thailand and utilizes ZDV administered antepartum, intrapartum and to the newborn. Thailand is a nation capable of providing formula to newborns and thus prior to the end of 2000 their Ministry of Health will have a program in place throughout the country.

The Elizabeth Glaser Pediatric AIDS Foundation and Global Strategies to Prevent HIV Transmission announced a Call to Action in September of 1999. This program recognizes the Ugandan trial (HIVNET 012) which demonstrated that Nevirapine could be administered in a single dose to pregnant women at the time of onset of labor and in a single dose to the newborn within the first 72 hours of life and decreased transmission by 50%. Such a regimen costs an estimated $4.00, is easy to administer, and in theory could be administered by traditional birth attendants as well as by physicians and nurses. The program and its progress will be described briefly with 8 sites receiving approximately $l,000,000 and initiating the interventions by June 2000. These first eight programs will train 1200 people, counsel and test 50,000 women, treat about 10,000 mothers and 20,000 infants, and prevent about 2500 infections.

Implementation of advances in technology is difficult and have always been delayed. The ethical dilemmas associated with HIV are numerous and enter into the establishment of Health Policy. For example, what if a country has resources and the political will to plan a program to prevent mother to child transmission and lacks resources for counseling and testing? What if uptake of counseling is not optimal? The U.S. began HIV interventions with an emphasis on counseling. The recent Institute of Medicine report urged routine voluntary counseling and testing with right of refusal as an acceptable option. That is, informing mothers and providing them with the opportunity to refuse testing is a reasonable approach to improve the implementation of the intervention and treatment of mothers with HIV infection. The counseling requirements in the developing world are currently in excess of this, and this may contribute to poor uptake.

Perinatal transmission can be reduced and the means to do so in the developing world currently exists. How can the Scientific Community work together to achieve a significant reduction in the number of infected children?

Catherine M. Wilfert, MD
Professor Emerita
Duke University Medical Center
Scientific Director
Elizabeth Glaser Pediatric AIDS Foundation

1917 Wildcat Creek Road, Chapel Hill, NC, 27516
(919) 968 0008 phone
(919) 968 0447 fax
(919) 684 8772 Office assistant
(919) 684 8514 fax at Duke

READINESS OF PERINATAL HEALTH CARE PROVIDERS IN DEALING WITH MOTHER-INFANT AIDS TRANSMISSION: A CASE STUDY IN INDONESIA

HADI PRATOMO
Faculty of Public Health, the University of Indonesia, Depok Campus, West Java, Indonesia

Currently Indonesia is the fourth largest country in the world. With a total population (1998) of 240,392 million, it has a GNP of U.S. $980 (1997). With an ongoing economic crisis since the mid-year 1997, it became one of the more economically disadvantaged countries.

The first individual case of AIDS was reported in 1987. Over 13 years, as of 31 May 2000, 23 out of 27 provinces in Indonesia reported a total of 934 HIV-positive individuals (of which 40% were women), plus another 323 with AIDS (of which 18% were women). Based on the risk factor, it was reported that the majority were heterosexual, then homo/bisexual, drug users and lastly due to perinatal transmission. The corresponding figures were the following: 67%, 10%, 4%, and 0.6% respectively. Currently there is no good nationwide surveillance of HIV among pregnant women and MCH services. In 1992, results of a KABP study among perinatal health care providers in the country indicated that although their knowledge and attitude towards HIV/AIDS was sufficient, their practice of universal precautions was below standard. To increase their knowledge and awareness of the issues, regular seminars for the topic are given during Perinatal conferences.

Up to June 2000, the Pelita Ilmu Foundation, a non-government organization from Jakarta reported caring for 225 people living with AIDS. About 27.5% of them were drug users and about the same proportion were from the low socioeconomic level and 13% were married. There is an observation that substance abuse is becoming more common and there are about 4-10 new cases of HIV/AIDS every week.

Currently, 8 of them have delivered their babies and three of them are currently pregnant. AZT is given for treatment, costing about U.S. $200.

It was also reported that many perinatal health care providers in the hospitals are still not prepared to assist delivery of HIV-positive women.

CONCLUSIONS AND LESSONS LEARNED

In a big country with a relatively low level of HIV prevalence, it seems that this condition is not conducive for encouraging the readiness of perinatal health care providers. Continuous effort for creating awareness and skills in complying with universal precautions among them is a must activity. The unavailability of good and reliable surveillance among pregnant women could not identify the exact magnitude of the problem. The government has limited funds and HIV/AIDS is not one of its priorities. At the same time, all donors are not interested in providing funds for curing the cases. The high cost of therapy is still a problem. Therefore, an alternative vaccine for HIV which is effective, inexpensive, acceptable and widely available may benefit the women at risk in the country.

BREASTFEEDING AND TRANSMISSION OF HIV

ROLF ZETTERSTRÖM
Acta Paediatricia, Karolonska Hospital, 171 76 Stockholm, Sweden

In many developing countries, breastfeeding during the first 5-6 months after birth is essential for survival, normal growth, and development as has repeatedly been demonstrated[1]. Respiratory infections and diarreahal disease are the main mortal diseases when other means of feeding are practiced. A WHO collaborative study team has recently assessed the beneficial effect of breastfeeding by the use of meta-analytical techniques[2]. In 8 separate studies which were performed 1980-1998 in Brazil, the Gambia, Ghana, Pakistan, the Philippines, and Senegal, the cause of death, the age and sex of the infant, and the educational level of the mother were taken into account. In the African studies all babies were breast-fed into the second year of life but no mention is being made if and for how long they were exclusively breast-fed. In many developing countries, most infants receive various kinds of fluid during the first days after birth in an almost ritual way[3], a habit that may seriously interfere with the protective effect of fresh mother's milk as it often exposes the newborn infant to microbiological pathogens. Instead, the infants are deprived of colostrum, which is particularly rich in immunologically-active compounds[4]. In the collaborative WHO study, it was demonstrated that the protective effect of breast milk declines as the infants grow older. Pooled odds ratio were found to be 5.8 for infants below 2 months of age, 4.1 for 2-3 months olds, 2.6 for 4-5 months olds, and 1.4 for 9-11 months olds[2]. In the first 6 months after birth, the odds ratio of protection against diarrhea was substantially higher than for respiratory infections, i.e., 6.1 and 2.4, respectively.

In addition, mother-infant bonding and possibly also infant development are enhanced by early suckling and breastfeeding[5]. Keeping these aspects of breastfeeding in mind, the risk of acquisition of HIV from breastfeeding[6,7] constitutes a severe dilemma. As has been shown in a recent study from in Durban, South Africa, the risk of transmission is not higher than in formula-fed infants up to an age of 6 months but then increases markedly if breastfeeding continues[8]. On the other hand, the transmission rate was found to be higher in breast-fed infants receiving complementary and/or supplementary food than in those being formula-fed. Provided that safe alternatives are available and the hygienic standard is high, infants of HIV-infected mothers should be formula-fed from birth. This recommendation may, however be altered by a better knowledge about the pathology of the transmission of HIV by human milk. It should also be kept in mind that pasteurized milk from an infected mother is a safe alternative if appropriate collection and transport can be arranged.

In developing countries, the risk of transmission of HIV by mother's milk has to be weighed against the beneficial effect of such feeding. The findings that the risk of transmission is rather low during the first 6 months in exclusively breast-fed infants urgently needs to be confirmed in different settings and in other countries than South Africa. If the same results are obtained, breastfeeding with weaning at an age of 4-6 months may be recommended. In such an instance there is, however, need of a better knowledge about the definition of exclusive breastfeeding if this is a requirement for a low risk of transmission. It also has to be elucidated whether weaning should occur suddenly or as a continuous process during 1-2 months.

REFERENCES

1. Khan, S.R., Jahil, F., Zaman, S., Linblad, B.S., Karlberg, J. Early child health in Lahore, Pakistan: X. Mortality. 1993; Suppl 390: 109-182.
2. WHO Collaborative Study Team on the Role of Breastfeeding on the Prevention of Infant Mortality. Effect of Breastfeeding on infant and child mortality due to infectious diseases in less developed countries: a pooled analysis. Lancet 2000: 355: 451-5.
3. Ashraf, R.N., Jalil, F., Aperia, A., Linblad, B.S. Additional water is not needed for healthy breast-fed babies in a hot climate. Acta Paediatr 1993; 82: 1007-11.
4. Gunnlaugsson, G., da Silva, M.C., Smedman, L. Does age at the start of breast-feeding influence infantile diarrhea morbidity? A case-control study in per urban Guinea-Bissau. Acta Paediatr 1995; 84: 398-401.
5. Zetterström, R. Breastfeeding and infant-mother interaction. Acta Paediatr 1999; Suppl 430: 1-6.
6. Dunn, D.T., Newell, M.L., Ades A.E., Peckham, C.S. Risk if human immunodeficiency virus type 1 transmission through breastfeeding. Lancet 1992: 340: 585-8.
7. Leroy, V., Newel, M.L., Dabis, F., Peckham, C.S., Van de Pierre, P., Bulterys, M., *et al.* for the Ghent International Working Group on Mother-to-Child transmission of HIV-1 infection. Lancet 1998; 352: 597-600.
8. Coutsoudis, A., Pillary, K., Spooner, E., Kuhn, L., Coovadia, H.M., for the South African Vitamin. A Study Group. Influence of infant-feeding patterns on early mother-to-child transmission of HIV-1 in Durban, South Africa: a prospective cohort study. Lancet 1999; 354: 471-6.

UTILIZING THE CLIMATE, WATER, DEVELOPMENT, AND INFECTIOUS DISEASES PERMANENT MONITORING PANEL TO EVALUATE THE COFACTORS FUELING THE HIV/AIDS EPIDEMIC IN SUB-SAHARAN AFRICA

DEBORAH BIRX
dbirx@hivresearch.org

ABSTRACT

HIV/AIDS alone is reversing decades of hard fought progress in Africa in both improved infant mortality and life expectancy. Life expectancy is decreasing for the first time in sub-Saharan Africa from 55/60 years to 30/35 years. This disease is creating the largest concentration of orphaned children ever experienced in the history of the world. This highly vulnerable group over the next decade will fall prey to infectious diseases, malnutrition, and potentially provide the fertile soil for unrest and political instability. This HIV/AIDS crisis does not exist in isolation but is compounded by substantial local environmental issues accelerating the epidemic through infectious and noninfectious cofactors. Understanding the impact of these defined cofactors on the HIV viral load and thus transmissibility of HIV will not only lead to fundamental breakthroughs in the approach to the control of HIV/AIDS but the lessons learned will be applicable to new emerging infectious diseases and their control.

By utilizing a multidisciplinary approach to evaluation of the HIV/AIDS cofactors, moving beyond the classical infectious diseases evaluation of only the pathogen and host, will hopefully lead to new insights and answers to the complex issue of the HIV/AIDS epidemic. The proposed evaluation seeks to explore and dissect the role of physical factors as a surrogate of infectious cofactors and to evaluate the effect of human migration and deforestation on the spread of the current HIV epidemic. Specifically, the impact of climate, development and geography on the HIV pandemic in sub-Saharan Africa will be explored using the marked differences in the HIV prevalence/incidence in West and East Africa as a model.

BACKGROUND

There are no accurate data on the prevalence of enteric and malaria infection, however, water purity and climate data from the regions of interest exist and can be utilized as surrogates for the diseases of interest. Data on deforestation of the rainforest and road construction (increased exposure of humans to diverse primates that may have resulted in the initial transmissions) and human migratory patterns in the areas of interest are available. Complementary epidemiologic and molecular epidemiologic data of HIV prevalence and incidence are available for the regions of interest in East and West Africa. These data demonstrate a clear difference in the incidence and prevalence of HIV/AIDS between West and East Africa that cannot be explained solely on the basis of the different HIV circulating subtypes. West Africa has less than half the HIV/AIDS incidence and prevalence compared to East Africa despite similar onset to the epidemic.

HYPOTHESIS

Differential physical and developmental factors through their influence on known HIV cofactors influenced the differential HIV epidemics in East and West Africa.

APPROACH

Use a multivariate analysis of specific climate data (supportive of malarial infections), water/sanitation data (likelihood of enteric infections), infrastructure of development (road and mobility), and physical geography relative to the regional (East vs. West) HIV incidence over the past 15 years (1985-2000) and the linear part of the HIV epidemic curve for sub-Saharan Africa.

The demonstrations of these linkages will allow leveraging of the HIV/AIDS funds to address other essential cofactors impacting the HIV/AIDS spread as well as address other substantial causes on morbidity and mortality in the region. The investigation of HIV or any other infectious disease should not be done in isolation but as one component in the critical issues confronting sub-Saharan Africa.

KNOWNS

1. Malaria, enteric diseases, and t.b. increase HIV viral load in HIV infected individuals.
2. Increased viral load is associated with increased HIV transmission and increased disease progression.
3. HIV incidence and prevalence data is available for East and West Africa.
4. Climate data and some water sanitation data in East and West Africa.
5. Clear association of climate parameters with malaria incidence.
6. Developmental, human migration, deforestation data is available.
7. HIV incidence and prevalence is substantially different in East and West Africa.

UNKNOWNS

1. Malaria, t.b., and enteric diseases incidence.

MOTHER TO CHILD TRANSMISSION—PERSPECTIVES FROM SOUTH AFRICA

ANNA COUTSOUDIS, HOOSEN COOVADIA
Dept. Paediatrics and Child Health, University of Natal, Private Bag 7, Congella 4013, South Africa

The issues around mother-to-child transmission (MTCT) of HIV in South Africa have been bedeviled by controversy on the one hand, but on the other hand, have added new knowledge to the management of the global problem. The stark fact is that despite 60,000 new HIV-infected births year in and year out, there is as yet no comprehensive state strategy for addressing this aspect of the epidemic. In fact, South Africa, given the magnitude of the disease and the dearth of a matching response for controlling the epidemic in infants, can be considered a case study in the discordance between government responsibility and social need. The various shortcomings in the national AIDS program and the politicization of science are universally known and can be profitably discussed at the think tank. The lessons, as shown by the drawing up of the Durban Declaration, are global in reach and profound in character.

Major studies on reducing MTCT of HIV using antiretroviral drugs have been undertaken in South Africa: these include PETRA and SAINT. The cost-effectiveness of these regimens for South Africa has been presented in numerous scientific and lay fora. The SAINT trial confirmed the effectiveness of nevirapine (one dose to the mother during labor and one dose to the infant after delivery) in reducing MTCT of HIV. The relative inexpensiveness of nevirapine makes it an attractive option for use in South Africa. However, issues around the development of resistance to nevirapine have hampered efforts in promoting nevirapine as a strategy to reduce MTCT of HIV. The role of breastfeeding transmission has been conceptualized in an entirely different and novel way by the work of the Durban research team (headed by Coutsoudis); they have highlighted the safety of exclusive breastfeeding. The ripples in the scientific community from this study have disturbed the bland smooth surface of existing dogma. The contrasts between this Durban study and Nduati's Kenyan randomized controlled trial provide the best basis for an in-depth discussion on breastfeeding in transmitting HIV, and added a new insight in that it highlighted the dangers of artificially feeding infants in disadvantaged environments. The formula-fed infants, despite having a lower risk for HIV transmission, had high infant morbidity and mortality. Other important projects to come out of South Africa have been some preliminary data of destroying the HIV virus using a simple home method to pasteurize expressed breast milk.

Future projects planned in South Africa will be a large study in a rural and urban setting which will set out to elucidate more conclusively the extremely complex issues of exclusive vs mixed breastfeeding. The option of providing breast-fed infants with nevirapine during the breastfeeding period is being explored in the HIVNET 023 study. Finally, a novel idea of using HIV vaccines to provide protection to infants against breastfeeding transmission is being considered. Preliminary work on the development of a clade C vaccine suited for local conditions is being undertaken and involves identification of the epitopes presented by the major HLA molecules which may be required for cellular immunity.

19. LINKING THE CONVENTIONS: SOIL CARBON SEQUESTRATION AND DESERTIFICATION CONTROL WORKSHOP

CARBON SEQUESTRATION TO COMBAT DESERTIFICATION–POTENTIALS, PERILS AND RESEARCH NEEDS

LENNART OLSSON
Centre for Environmental Studies, Lund University, Sweden

BACKGROUND TO CARBON SEQUESTRATION

Carbon in CO_2 is accumulating in the atmosphere at a rate of about 3.5 gigatons (Gt) a^{-1} as a result of combustion of fossil fuel, tropical deforestation and other land use changes. Measures taken so far to reduce future emissions of CO_2 will not be sufficient if we are to fulfil the UN Framework Convention on Climate Change (UNFCCC) goal of stabilising greenhouse gas levels "*at a level that would prevent dangerous anthropogenic interference with the climate system.*" Other options for reducing the amount of CO_2 already emitted to the atmosphere must be considered.

Because reducing emissions of CO_2 is costly, it is important to find many different ways to achieve them. The abatement cost in Europe is estimated by the World Bank at 70-80 \$U.S. t^{-1} compared to 30-40 \$U.S. t^{-1} for the USA. Recent studies have shown that reducing atmospheric concentrations by increasing sequestration might be a cost-effective strategy that conforms to principles of sustainable development. The sequestration of carbon in the biosphere will be made possible by the trading of emission rights under the Kyoto Protocol. It is, however, important to stress that carbon sequestration should be seen as:

- no more than a temporary activity that can help in stabilising atmospheric CO_2 at a lower level than without sequestration in the next few decades;
- an activity that must not be seen as an alternative to other means for reducing CO_2, i.e. changes in the energy system and reductions of energy demand;
- but, a no-regret policy with several advantageous side effects.

The carbon trade is a very controversial issue, to which present official policies of USA and EU are in opposition. However, it is important to acknowledge that carbon trading is already in operation–whether we like it or not[1]. The important issue now is to use the carbon trade in a way that is environmentally, socially and economically benign

[1] By August 2000, over 100 contracts of at least 100,000 tons of carbon each, at a price of 10-11 \$US have already been signed. Personal communication: Dr. Larry Tieszen, Deputy Manager, USGS International Programme.

and where the trade can be used to mitigate climate change *and* to support sustainable development in the poorest countries.

Sequestration is possible, even desirable. The best known method is to sequester carbon in forests. Another major sink could be in soil, and although the Kyoto protocol did not explicitly mention agricultural soil and land use change as a potential carbon sink, there is now mounting pressure to accept them as credible sinks. There is even pressure to divert some of the emphasis on forestry in the Kyoto Protocol to the agricultural sector, recognising that the environmental benefits might be easier to achieve in agricultural soils than in forests[2]. The Kyoto Protocol is still in process of taking final shape and new research is needed to demonstrate how the carbon trade can be used to promote sustainable development around the world.

DRY LAND AGRICULTURE

Dry lands cover some 70% of the total land area and suffer from some of the most serious environmental problems, in many cases affecting the poorest of people. Many millions of people are involved. One way to inject funds into dry-land rural economies would be to invoke the mechanisms opened by the Kyoto protocol, i.e. emissions trading (Article 17) and the Clean Development Mechanism (Article 12).

Land use systems in dry-land areas, such as those round Bara in Sudan, and across the Sahelian belt between Senegal and the Red Sea, must be highly dynamic and responsive to a very variable environment if they are to survive. The principal requirement is rainfall: in good years crops can flourish, some surplus may even be sold, and household granaries are filled. In poor years, men must migrate to find labour, women resort to other, local sources of income, and the granaries quickly empty. Soils, which are for the most part poor, acid and sandy are a further constraint, and, because fertilizer prices are higher in Africa than in Europe, and thus beyond the means of all but a very few farmers, they must use other means to restore the nutrients removed by crops. There are three main sources: manure (for which they must have livestock); "green manure" from household waste, cut branches or crop stubble or stover; or fallow, which allows nitrogen fixation by leguminous trees and weeds, and soil crusts and the accumulation nutrient-rich dust (generally bringing enough calcium and potassium each year to replace that removed).

In Sudan, unlike most of the rest of the Sahel, there has been, for centuries a source of income other than livestock and crops. Gum arabic, for which Sudan is still the main global source, is exuded by the branches of *Acacia Senegal* (locally *hashab*), if careful scarring is applied. Gum arabic can provide a good income to farmers, but only when the price is right, and when the institutional framework for getting it to market works properly. There is presently a very big demand for gum arabic on the world market, in fact bigger than the suppliers can produce.

---

[2] Opinion expressed by several participants at the Expert Panel: *Carbon sequestration, sustainable agriculture and poverty alleviation*, Geneva 30/8 – 1/9 2000, organised by WMO, FAO, IFAD and USAID

ENCOURAGING CARBON SEQUESTRATION IN DRY-LAND AGRICULTURE

There are many technically feasible means for encouraging more carbon sequestration in dry-land agricultural systems. These include increased use of fallow periods, reduced tillage, increased use of rotational crops and the application of agro-forestry. In addition to mitigating climate change, these technologies could provide ancillary benefits, including improvements in soil fertility, water holding capacity and reduction in wind and water erosion. Thus, enhanced soil carbon sequestration in agro-ecosystems could also help to meet the growing demand for food. This option for reduction of atmospheric CO_2, commonly named carbon management, is more and more recognised as an important complement to emission reductions.

A recent expert panel (*Carbon sequestration, sustainable agriculture and poverty alleviation*, Geneva 30/8 – 1/9 2000, organised by WMO, FAO, IFAD and USAID) concluded that carbon sequestration in degraded agro-ecosystems in developing countries might well be a viable means to reduce the net emission of GHG through mechanisms in the Kyoto protocol. However, research on several issues related to carbon sequestration in these regions is necessary prior to an operational phase of emissions trading and the CDM for this purpose.

The highest potential for increasing soil carbon content will most likely be found in the most severely degraded ecosystems around the world. There are vast areas of degraded lands throughout the world, many in developing countries, including semi-arid Sudan, where improvements in rangeland management and rain-fed agriculture can increase the sequestration of carbon in the soil. Preliminary studies have estimated that between 0.6 and 2 Gt of carbon per year could be sequestered by large-scale application of appropriate land management in degraded lands of the world[1]. Results from research at Lund University suggest that it is reasonable to assume that carbon sequestration can be increased by 5-10 grams per m^2 and year by increasing fallow periods in the savannah regions of Africa[2]. Figure 1 shows the effect of intensifying agriculture on soil carbon content during the last five decades (1950–2000). It also shows the recovery of soil carbon contents if improved land management practices are introduced (2000–2200).

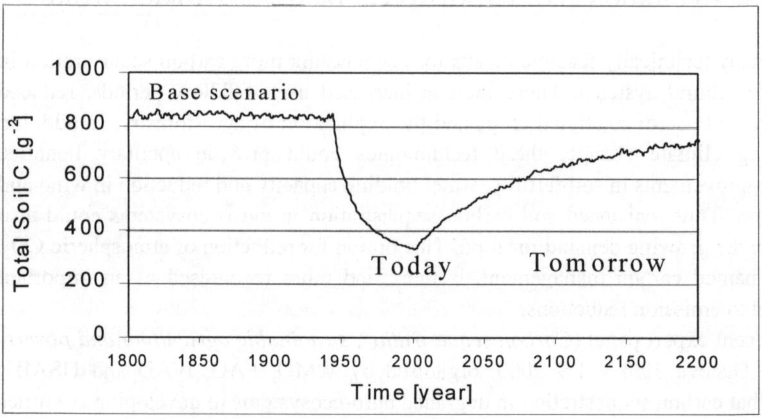

Fig. 1. Total soil carbon (g m^{-2}, upper 20 cm) from year 1800 to year 2200 predicted using the CENTURY model. Up to 1950, long periods of fallow interspersed with short periods of cultivation prevailed. From 1950, the fallow periods were gradually reduced to become permanent cultivation in the 1990s. From 2000 to 2050, the fallow periods were gradually increased to the original state. Source[2].

The technology for this and other improvements has been tested and applied before, but it is very difficult to persuade farmers to adopt it. There are many reasons, but there are two principal ones: the labour required to implement them, which is limiting in many semi-arid agricultural systems (especially those, the majority, where migration is a major source of income); and the fear of adopting new and risky technologies, in situations where risk is already high, where land use systems have been carefully evolved over centuries to deal with these risks, and where farmers have very little economic room for manoeuvre. Hence the need for carefully designed pilot projects, as in the case of the GEF project at Bara described below, which has made an excellent start in this direction, and which can be the basis for testing a range of approaches.

RESEARCH PRIORITIES

Research is needed particularly in two major fields:

- Monitoring and verification–without a reliable and cost-effective monitoring and verification system, carbon trade will not be possible.
- Systems analysis to answer the crucial questions:

    - where is carbon best sequestered?
    - what drives the land use system?
    - what are the possible interventions in terms of carbon sequestration?
    - how do we avoid undesirable side effects?

- what kind of sequestration measures should be implemented (fallow periods, green manure crops, reduced tillage, agro-forestry, controlled grazing, alternative energy sources etc.)?
- how should the economic benefits be transferred to the people?
- how to avoid undesired effects, for example unequal opportunities, deterioration of social coping strategies.

One way to start addressing these issues is to use a systems approach in order to identify all the components of the farming system, identify the driving forces and to elucidate the linkages between the components.

MONITORING AND VERIFICATION

The main obstacle for an operational trade of carbon emission rights including agricultural soil carbon is verification. There is still some doubt whether we can verify accurately enough the amount of carbon being sequestered in the soil at a cost that is reasonable in relation to the amount of carbon being sequestered. If we want to make agricultural soil carbon tradable we have to develop timely verification and monitoring methods. In terms of degraded ecosystems in the tropics, and in particular semi arid regions, it is important to recognise the following facts:

- carbon sequestration in semi arid lands will be low per area unit, implying that large areas must be considered.
- carbon sequestration among smallholding farmers will be low per farmer, implying that a large number of people must be considered.

Monitoring and verification will necessarily consist of a set of techniques combining remote sensing, modelling and field sampling

THE PROJECT AT BARA IN WESTERN SUDAN

The project area covers a total of 24,000 ha comprising five village councils, and is inhabited by 6000 people. The area has an annual rainfall of about 270 mm and land use characterised by a risk-aversion strategy, in which many different kinds of income sources are employed at the same time. The main sources of income are cultivation (millet, sorghum, sesame and ground nuts), animal husbandry (goats, sheep) and production of gum arabic. The project is managed through UNDP and wholly administered by Sudanese staff. The project has successfully implemented a range of carbon sequestration measures, for example:

- Introduction of improved household stoves–stoves are made locally from clay; 97% of the households have adopted these stoves and this has brought about a sharp reduction in fuelwood consumption.

- Change in the composition of herds–in order to better manage the rangeland, it is planned to change the composition of the livestock from a majority of goats to 80% sheep and 20% goats. This is being achieved by economic incentives that favour sheep.
- Controlled grazing–communal grazing reserves have been established and grazing by domestic livestock is controlled by the local people.
- Reforestation–seeds of *Acacia senegal* are distributed and local nurseries have been established in order for people to plant and manage these trees, for subsequent production of gum arabic.
- Improved house construction–in order to reduce the cutting of wood for construction purposes, new ways of house construction using clay have been introduced.

All these activities are explicitly aimed at increasing the amount of carbon stored in the ecosystem, both above and below ground (as specified by the GEF). If we can effectively verify the amount of carbon being stored and monitor the permanence of the carbon storage, for which more years are needed in such a highly variable climate, this kind of project might be a pilot for many more that could contribute to climate change mitigation and at the same time bring economic and environmental benefits to rural people in developing countries–a true win-win situation.

REFERENCES

1. Batjes, N.H., *Management options for reducing CO2-concentrations in the atmosphere by increasing carbon sequestration in the soil.* 1999, International Soil Reference and Information Centre.: Wageningen.
2. Olsson, L. and J. Ardö, *Is there a large potential carbon sink in semi arid rangelands?* Manuscript, in prep.

SOIL CARBON SEQUESTRATION IN AFRICA

PAUL BARTEL, MIKE MCGAHUEY
USAID Africa Bureau, Office of Sustainable Development, Washington, USA

CONTEXT OF THE ISSUE

Soil carbon sequestration forms a nexus between the global process of the carbon cycle and local processes of soil fertility. It also forms a nexus between broad biophysical processes and socio-economic processes. To frame the discussion it is necessary to offer some contextual remarks starting with the linkages of natural resource management to other subsectors of social behavior. In Figure 1, natural resource management links to key global issues on the left and national issues on the right.

First, people invest in sound natural resources management (NRM) when it leads to more secure and prosperous livelihoods. Second, resources are often shared. Consequently, increasing population pressure increases the potential for conflict. Third, maintaining biodiversity and mitigating the effects of climate change requires good natural resources stewardship. In turn, quality of life is linked to healthy biodiversity and stable climates. In order to understand the potential of carbon sequestration and constraints that temper this potential, it is necessary to illustrate some key trends that are affecting the African landscape. Many of the most salient trends are anthropogenic in nature. Others are the result of long-term climatic change that, although potentially independent from human activity, are nevertheless exacerbated by trends and the behavior of the human population.

Land Degradation

The richness of African biological diversity is precariously supported on ancient and highly fragile soils[1]. Traditionally, grazing and cropping systems used an extensive area of land, shifting to new lands as the quality of forage and fertility declined to low levels of productivity. As the ability to exploit large areas of land has become constrained by population growth and political uncertainties, production has become increasingly concentrated on smaller areas of land and the land resource has become degraded. Zimbabwe provides an excellent example of this, showing the effects of colonial and post-colonial land tenure structures with the resulting effect of concentrating a previously extensive production system on small parcels of land. Figure 2 demonstrates this, where the zones of darker color represent relatively undisturbed land while the lighter zones show high levels of degradation.

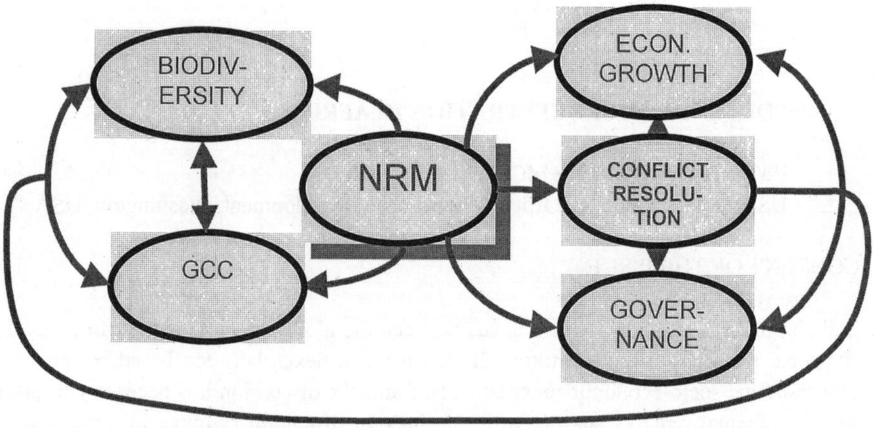

Fig. 1. Linkages Between Natural Resources Management (NRM) and Other Sectors.

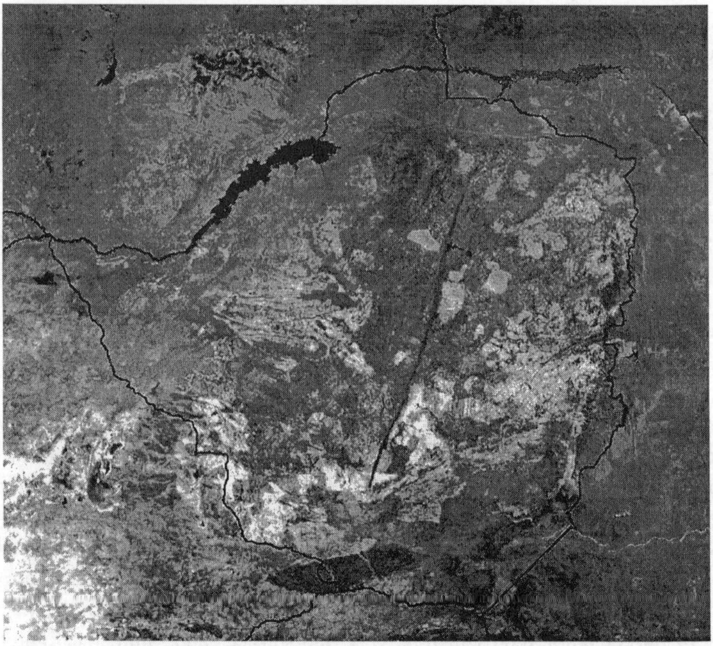

Fig. 2. Satellite Image of Land Degradation in Zimbabwe (EOSAT, 1999).

Loss of Forest Cover

Mainland Africa has suffered considerably lower levels of deforestation in its dense forests than other areas, particularly Asia, Europe and the Americas. This is changing however as commercial interests increasingly look to Africa as a source of hardwoods as these other areas become increasingly degraded. This initiates a process of fragmentation of forests, as is shown in Figure 3. The establishment of large timber concessions, the construction of roads and encroachment of human populations in previously unsettled areas will result in a highly fragmented forest surrounded by savannah if left unchecked. The ultimate result will be loss of habitat for flora and fauna of local and international significance, an expansion of the continuing trend towards degradation of soils and no significant contribution of the forest or soil resource to long-term economic growth in the countries in which this occurs.

Climate Change

The anthropogenic effects of land degradation and deforestation are exacerbated by long-term climatic changes. In the case of the Sahel, humid and sub-humid isohyets (rainfall greater than 500 mm annually) have shifted south by 20 degrees of latitude since 1940 (Fig. 4). Current climate models predict that Africa will become generally hotter and drier. This implies that agricultural production will shift increasingly to unexploited lands and that the forty percent of the African population who inhabit arid and semi-arid zones will be forced to move or live in increasingly inhospitable conditions with even greater levels of vulnerability to poverty and famine.

Population Pressure

Population growth rates in Africa have been between 2.5 to 3.5 percent annually[2]. This growth has taken place most notably in areas which are highly vulnerable to land degradation. They have also taken place in close proximity to highly unique and exceedingly fragile ecological zones (see Fig. 5). Thus, an already overburdened landmass is expected to sustain a larger and younger population.

The Effect of the HIV/AIDS Pandemic

The HIV/AIDS pandemic is tempering population growth trends. Though AIDS-induced reductions in the growth rates in many countries will not drop below 2.5 percent, others will experience negative growth, particularly in Southern Africa. This is in no way a favorable trend. AIDS death rates are greatest among the economically-active population; those men and women between 15 and 65 years of age (see Fig. 6). The result will be that families will become dependent upon an increasingly younger demographic stratum for their livelihoods. An increase in the number of households headed by young women and girls implies increased vulnerability. Traditionally, young women are the least enfranchised portion of the population; lacking access to education, credit and land tenure.

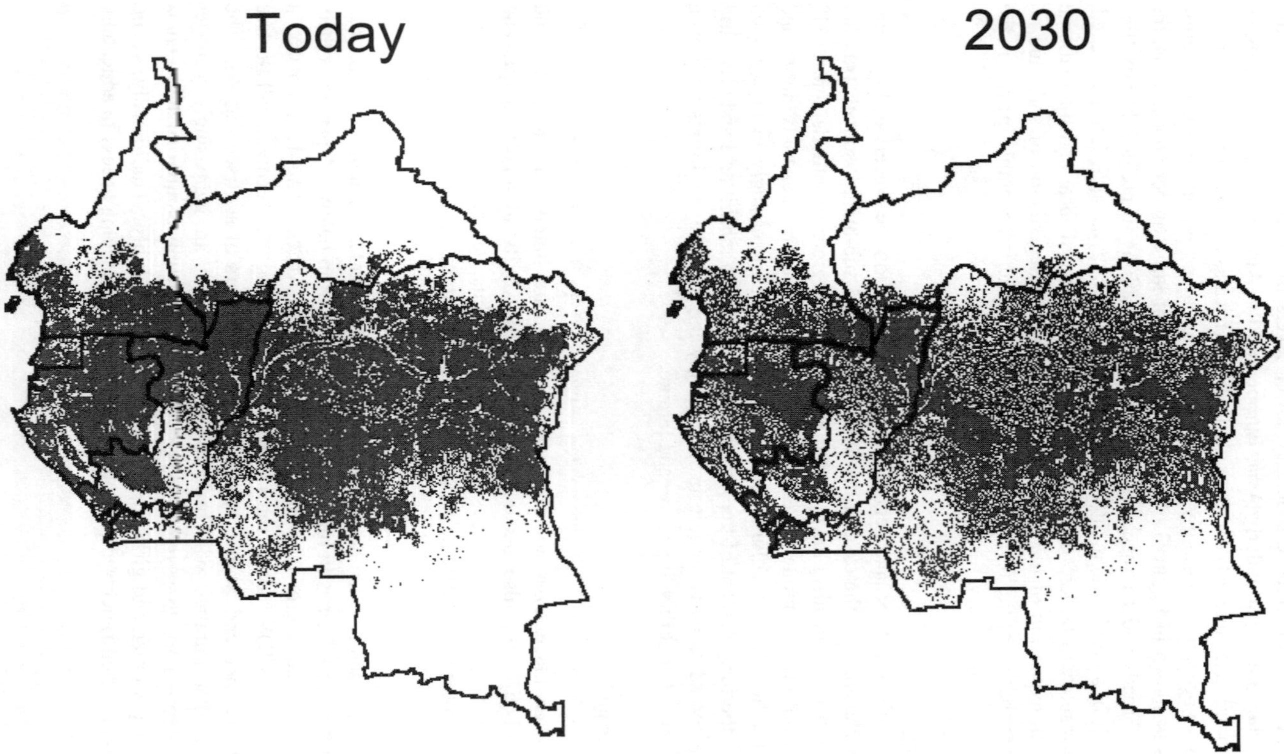

Fig. 3. Forest Degradation in Central Africa (NASA, 1997).

Source: NASA

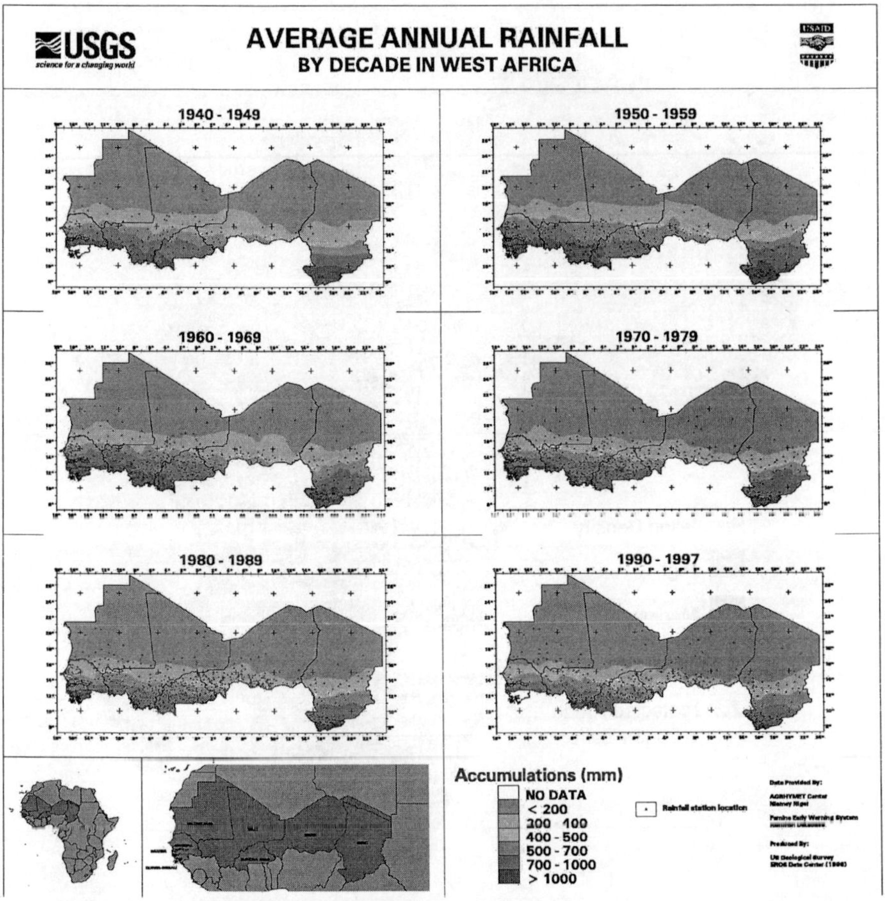

Fig. 4: Historical Changes in Sahelian Rainfall Isohyets (USGS, 2000).

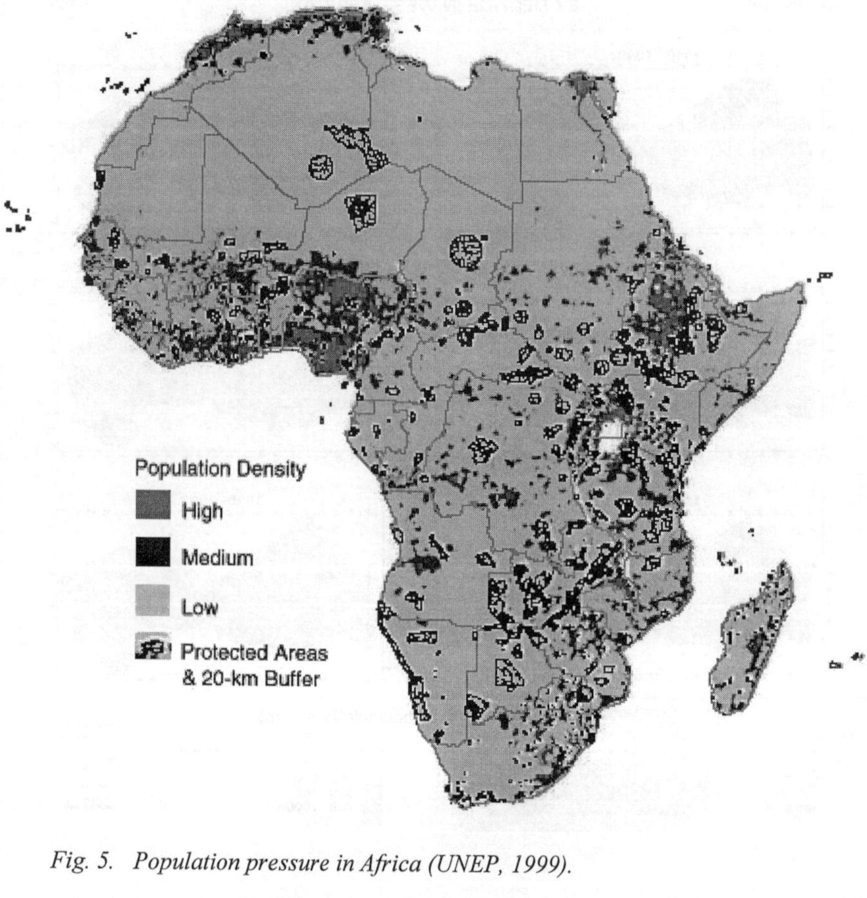

Fig. 5. Population pressure in Africa (UNEP, 1999).

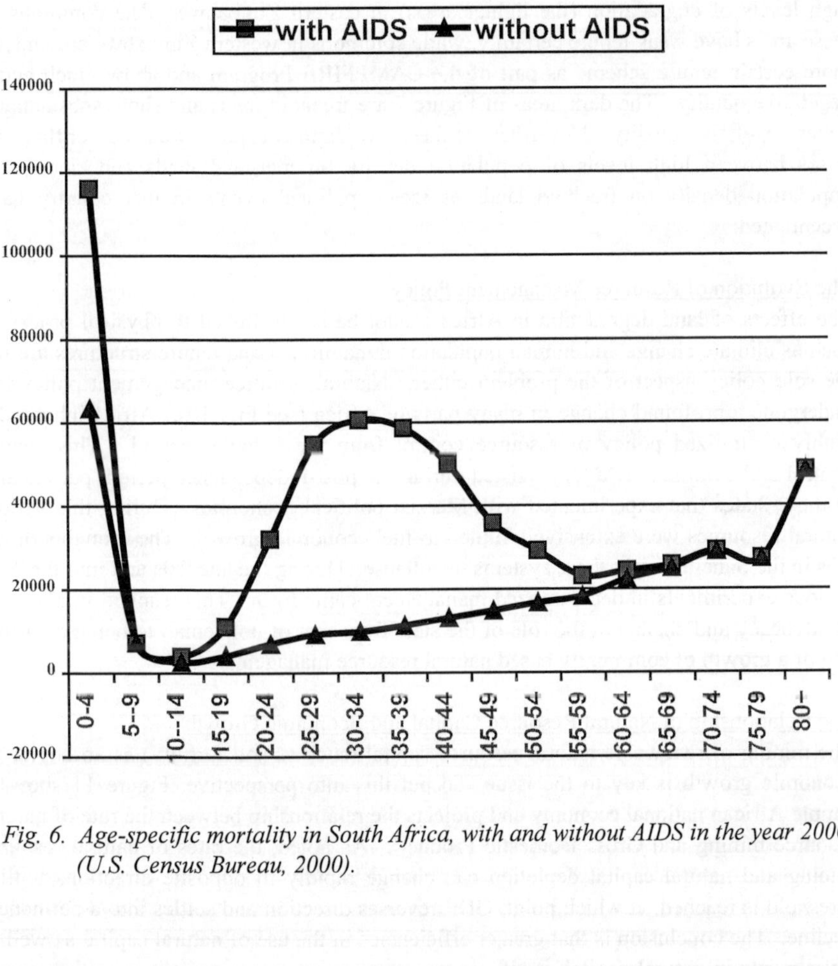

Fig. 6. Age-specific mortality in South Africa, with and without AIDS in the year 2000 (U.S. Census Bureau, 2000).

Juxtaposition of Land Tenure and Land Degradation

Land tenure is an essential element in the implementation of natural resource management upon which effective soil carbon sequestration depends. Common property tenure structures have been viewed as highly vulnerable to land degradation, as is shown in a comparison between Figures 7 and 8 for Zimbabwe. The white areas of Figure 7 show common property tenure zones. Figure 8 (as may be recalled from Fig. 2) shows high levels of degradation (the lighter areas) in eastern Zimbabwe. The commons of these areas have weak tenure certainty, while commons in western Zimbabwe are under a more certain tenure scheme as part of the CAMPFIRE Program and show much better vegetative quality. The dark areas of Figure 7 are freehold lands and show substantially better vegetative quality. Nevertheless, these two figures capture well the conflict that arises between high levels of population density on marginal lands and richer low population density on freehold lands as recent political events in that country have accentuated.

The Evolution of Resource Management Policy

The effects of land degradation in Africa cannot be solely linked to physical processes such as climate change and human population dynamics. Land tenure structures are not the sole policy aspect of the problem either. Natural resource management policy has undergone a profound change in many parts of Africa (see Fig. 10). Africa inherited a highly centralized policy of resource control from the colonial period. This central control and command (C&C) persisted during the post-independence period, particularly in those states that experimented with Marxist political economies. During this period, natural resources were extensively mined to fuel economic growth. The droughts of the '70s in the Sahel brought these systems to collapse. During the late '70s and into the '80s, various experiments in decentralized management came forth. This resulted in a growth of advocacy and change in the role of the state from that of policeman to being a partner and of a growth of community based natural resource management[3].

The Relationship of Natural Resource Capital and Economic Growth

The mining of renewable natural resource capital (such as soil fertility) as an engine of economic growth is key to the issue. To put this into perspective, Figure 11 shows a simple African national economy and projects the relationship between the rate of natural resource mining and Gross Domestic Product[4]. As noted, the rates of natural resource mining and natural capital depletion can change rapidly in opposite directions until a threshold is reached, at which point, GDP reverses direction and settles into a permanent decline. The conclusion is that greater efficiencies in the use of natural capital as well as investments in natural capital, itself, are necessary to sustain economic growth between generations. This is the national context in which soil carbon sequestration exists.

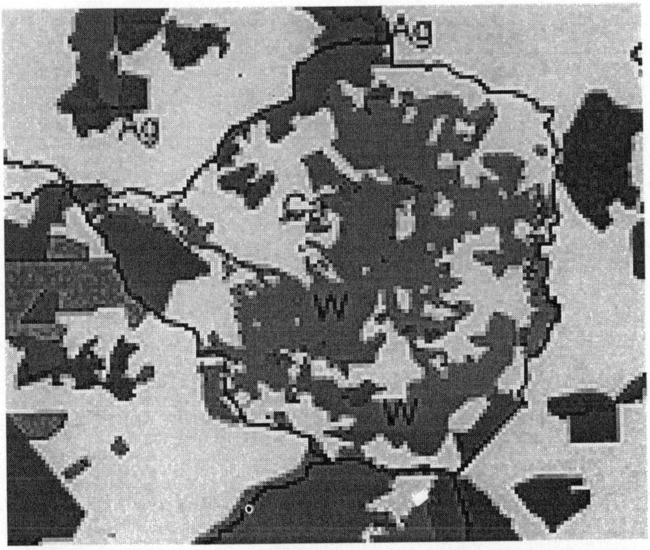

Fig. 7. Land Tenure Zones in Zimbabwe (World Wildlife Fund, 1999).

Fig. 8. Land Degradation in Zimbabwe (EOSAT, 1999).

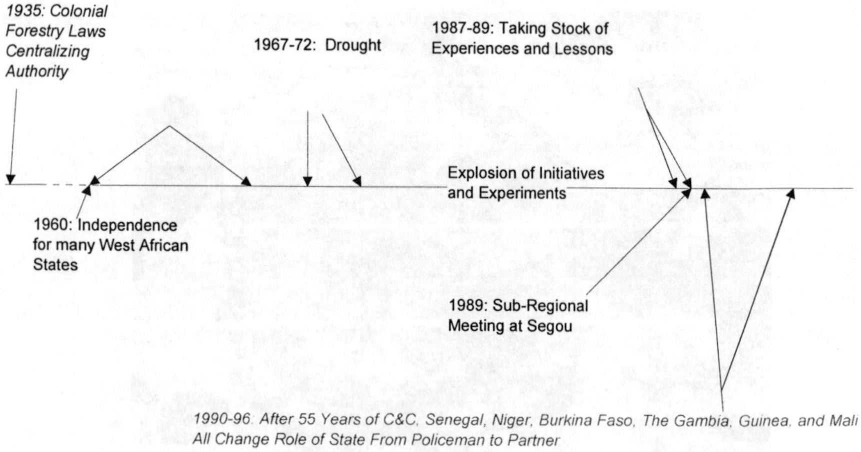

Fig. 10. The Evolution of Resource Management Policy in the Sahel (IRG, 2000).

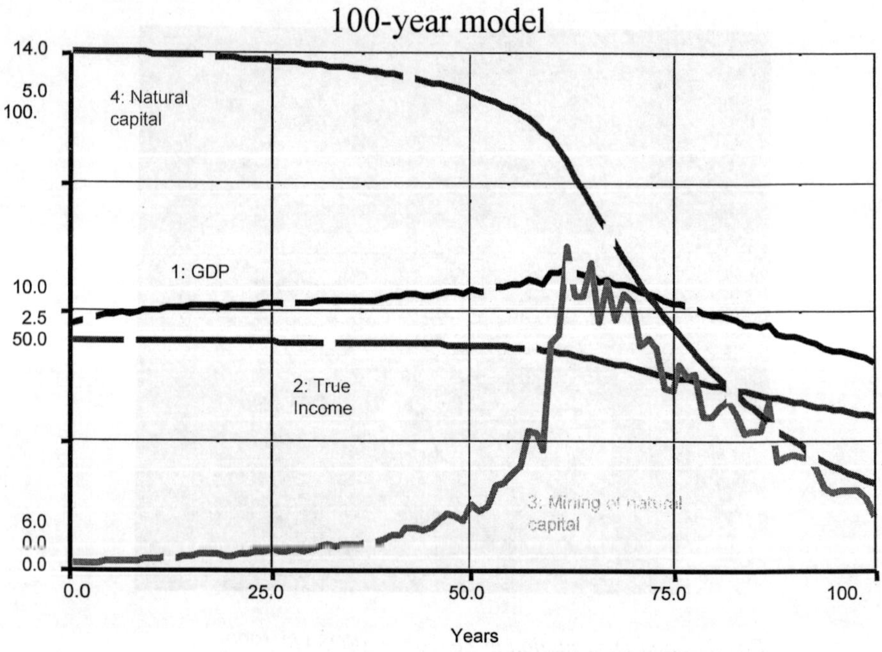

Fig. 11. A Simple Economy Maximizing Mining of Natural Capital (Woodwell, 2000).

Results of 30 Years of NRM Experience: Socio-economic impacts

African natural resource management, with the support of international donors such as USAID, has benefited from thirty years of experience during and following the disastrous droughts experienced in various parts of the continent. The key lesson learned is that decentralized authority, placing authority for natural resource management in the hands of communities and linking benefits from natural resource management directly to those upon whom the burden of conservation rests, have positive effects. Figure 11 shows how decentralization policy has had an effect on community incomes, and in some cases raising entire communities above the poverty datum line of $1,200 annual income as determined by UNDP.

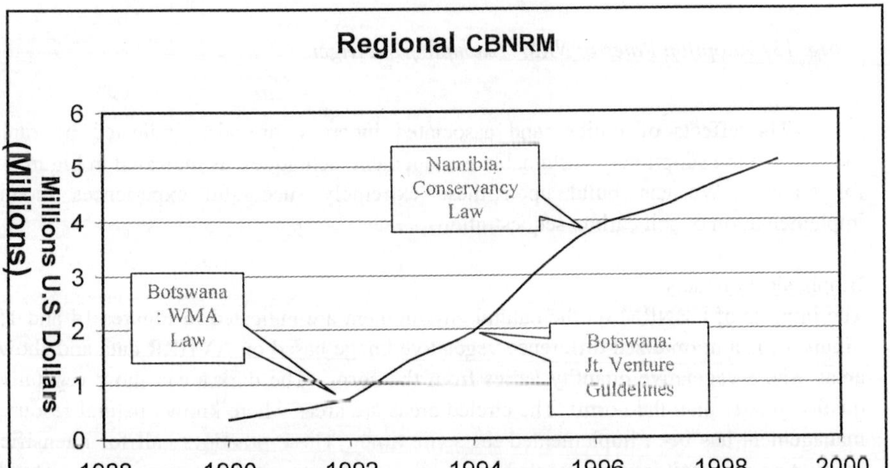

Fig. 12. NRM and Revenue Generation: Southern Africa Example.

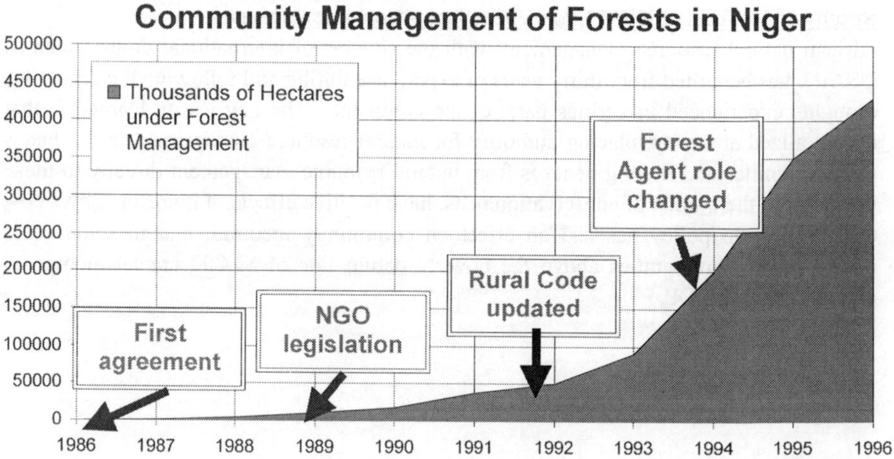

Fig. 13. Adoption Rates of NRM: Example from Niger.

The effects of policy, and associated incomes, are also indicated by rapid increases in the adoption of sustainable management techniques, as indicated in Figure 13 for Niger. We can build upon these extremely successful experiences in the implementation of soil carbon sequestration.

<u>Biophysical impacts</u>

The impacts of CBNRM on the natural environment are indicated in Figures 14 and 15. Figure 14 is a normalized difference vegetative image based on AVHRR data and shows areas where vegetative quantity varies from the norm. The dark areas show vegetative quality greater than the norm. The circled areas are areas where known natural resource management has been implemented for some time. These practices call for intensified production with soil improvements and secure tenure authority by communities. This is shown in the encircled areas of Figure 15 that correspond to the same areas circled in Figure 14. Though the biophysical effects of CBNRM occur on a broader time scale than the social effects, we are now seeing results on the ground, at least in a preliminary form.

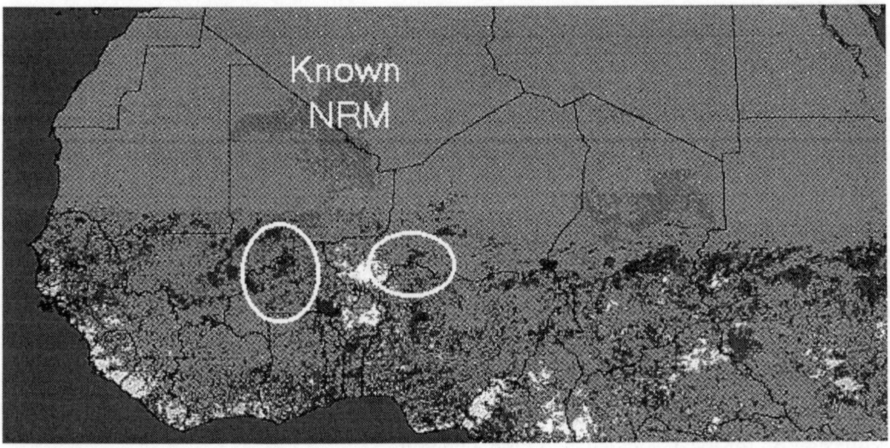

Fig. 14. NRM Impacts: Biophysical Evidence of Vegetative Quality (USGS, 1998).

Fig. 15. NRM Evidence: Biophysical phenomena correlated to Socio-Economic Data (World Resources Institute, 1999).

OPPORTUNITIES

Some scientists suggest that the highest potential for soil carbon sequestration can be found in degraded lands including the semi-arid and sub-humid regions of Africa[5]. Due to a combination of increasing population and animal densities, over-cultivation, extensive fuel-wood gathering and overgrazing, as well as unfavorable economic and agricultural policies, these regions have been experiencing critical losses of biomass and a decline in biological diversity and productivity. Degraded lands not only have exhausted most of their capacity to sequester carbon but also have emitted a substantial amount of CO_2 from the soil to the atmosphere, thereby contributing to the greenhouse effect and global warming.

Improved dryland farming, range management, and irrigation could reverse this trend by replenishing depleted soil carbon stocks. It is the rehabilitation of degraded areas through successful carbon sink management that is at the heart of soil carbon sequestration for local, societal, and global benefits.

Most African countries rank very low among the global emitters of CO_2 from fossil fuel burning and, therefore, have played only minor roles in international projects that address global warming. However, many of these countries experience alarming rates of declining soil carbon, soil quality and fertility due to land degradation and desertification. Soil carbon sequestration would permit African countries to proactively participate in and benefit from global change mitigation, simultaneously addressing three important international conventions: The Convention on Climate Change, The Convention to Combat Desertification, and The Convention on Biological Diversity.

ELEMENTS OF THE DIALOGUE

Framing Global Issues

International interest in soil carbon sequestration stems from three international conventions in negotiation since the Rio Summit on the Global Environment. The biodiversity convention calls for efforts aimed at maintaining biological diversity in key areas of the world. Soil carbon sequestration offers a landscape management approach that mitigates biodiversity loss. The Global Convention on Climate Change offers opportunities for promotion of carbon sequestration, though major actors in this debate question the use of carbon sinks as an approach to mitigating climate change and the U.S. Congress has not ratified the convention, prohibiting the U.S. Agency for International Development (USAID) from actively promoting the convention itself. The Convention to Combat Desertification offers an opportunity to consider carbon supplementation as an approach to mitigate desertification trends and the U.S. Congress recently held hearings on ratification.

Workshops and Pilots

St. Michaels: A workshop to explore issues central to achieving the 40 to 80 billion metric ton potential for carbon sequestration through soils was organized by the Pacific

Northwest National Laboratory, the Oak Ridge National Laboratory and the Council for Agricultural Science and Technology in St. Michaels, MD in December 1998. Nearly 100 Canadian and U.S. scientists, practitioners, and policy-makers representing agricultural commodity groups and industries, Congress, governmental agencies, national laboratories, universities and the World Bank attended the workshop. The U.S. Environmental Protection Agency, the Monsanto Company, and the National Aeronautics and Space Administration provided support for the workshop[5].

The St. Michaels workshop found that reductions in atmospheric content can be achieved by large-scale application of tried-and-true land management practices such as reduced tillage; increased use of rotational corps such as alfalfa, clover and soybeans; and by an efficient return of animal wastes to the soil. Forests and grasslands afford additional capacity for carbon sequestration when established on former croplands. Programs to further soil carbon sequestration will provide ancillary benefits, including improvements in soil fertility, water holding capacity, and tilth; and reductions in wind and water erosion. However, although practical and economically viable farming methods that increase carbon storage in soil already exist, research is needed to develop new methods that increase the amount stored and the length of time carbon remains in the soil. Promising research leads involving applications of molecular science, colloidal chemistry, bioengineering and traditional plant breeding were discussed.

There is opposition to using soil carbon sequestration in the Kyoto Protocol calculations. One cause of the opposition is the perception that it will be difficult, if not impossible, to verify claims from around the world that carbon is actually being sequestered in soild. The workshop concluded that it *is* currently possible to monitor changes in soil carbon content, but current methods are expensive and not well suited to global monitoring. Technology, it was concluded, can provide new and widely applicable methods at a reasonable cost. These methods can be based on applications of remote sensing, direct non-destructive sampling, carbon flux monitoring in the field and the use of simulation models validated against observations.

One special opportunity to link the objectives of two United Nations Conventions—the Framework Convention on Climate Change and the Convention to Combat Desertification—was discussed in considerable detail. The workshop concluded that there are vast areas of degraded and desertified lands throughout the world, many in developing countries, where improvements in rangeland management, dryland farming and irrigation can add carbon to the soil. Soil carbon sequestration can provide the impetus for changes in land management practices that will begin the essential process of stabilizing the soil against further erosion and degradation with concomitant improvements in fertility and productivity. Specific management practices, useful species and other appropriate steps in combating desertification were discussed.

Uncertainty about the costs, benefits and risks of new practices could impede the adoption of new technologies to increase carbon sequestration. The workshop concluded that financial incentives could increase the adoption of such practices and potentially provide an addition to farmer income. Government payments, tax credits, and/or emissions trading within the private sector are mechanisms that could be employed to overcome farmer reluctance. Impediment and adoption issues in need of clarification and research were discussed.

A number of recommendations emerging from drafts of the workshop papers and working group reports presented in the St. Michaels final report have already affected agency planning with regard to soil carbon research. First among these recommendations is that targeted research in the basic science of soil and plant relationships will yield understanding of how to enhance carbon sequestration in soils and reap the ancillary benefits of improved soil productivity, perhaps especially in desertified lands. Second, are sampling technologies, computer modeling, eddy flux approaches, scaling up of data and other methods such as remote sensing techniques. Third, the joint global goals of sequestering carbon and combating desertification could be facilitated through the use of mechanisms such as carbon trading, the Clean Development Mechanism, and an international carbon fund, enabling large-scale international projects and establishment of baseline data. Multiple environmental and productivity benefits could result from congruent efforts in basic science, monitoring and verification methods, and coordinated policy programs.

Sioux Falls: The U.S. Geological Survey, EROS Data Center hosted a follow-up workshop in Sioux Falls, SD in May 1999. USAID, the Sand Country Foundation and the International Program at the EROS Data Center sponsored the workshop. The main purpose of the conference was to: 1) extend the understanding of the potential roles of land use and land management in the sequestration of carbon in soil and 2) identify mechanisms for developing country implementation[7]. The focus was on:

- Semi-arid and sub-humid areas, grasslands, savannas and agricultural lands with an emphasis in developing countries in Africa;
- Defining the potential for carbon sequestration, the economic value, its importance for sustainability, and possible implementation mechanisms; and
- Ensuring appropriate participation in carbon crediting opportunities stemming from the Kyoto Protocol for Climate Change and subsequent conferences.

The workshop effectively built upon the St. Michaels seminar which established that substantial carbon sequestration could be achieved with some modifications in land use or land management. The two-day Sioux Falls workshop was designed with key presentations and working group sessions to achieve a robust set of goals:

1. Review the new agreements, policy implications and developing opportunities for joint U.S. and African participation in programs arising form the KPCC.
2. Support participation by African specialists and ensure that the potential contributions by developing countries and small holders are realized.
3. Describe the potential for carbon sequestration in parts of Africa; confirm measurement, monitoring and verification procedures; and review financial instruments to assure transfer of accrued funds to individuals.

4. Help equip industrial representatives, landowners and other stakeholders to understand both the science and the opportunities.
5. Review current carbon sequestration projects and industry participation, encourage networking, and consider a well-monitored demonstration project.

Senegal Pilot: The Sequestration of Carbon in Soil Organic Matter (SOCSOM) is a pilot feasibility study project. The project identifies representative project areas in important agro-ecological zones; quantifies the biophysical potentials for carbon sequestration as a function of management strategy on a geo-spatially explicit basis; analyzes the potential economic value to the smallholder of selling "certified carbon reduction credits"; evaluates the total benefits associated with improved management; estimates the total economic potential for Senegal; and provides opportunities for national capacity building and regional training. An important component of the project identifies current practices that are conducive to the enhancement of soil carbon and fertility, other socioeconomic incentives required for smallholder cooperation and ownership, and existing policy constraints. Although there is currently interest in "buying carbon" on the part of industry and power companies, USGS will not serve as brokers. USGS hopes to define the realities of this potential and means by which developing countries can become active contributors to climate mitigation, as well as economic beneficiaries of carbon trading, stemming from the Kyoto Protocol.

THE UNIVERSE OF ACTORS

A number of actors have maintained a dialogue on soil carbon sequestration within the framework of workshops, activities and negotiations on the various conventions discussed. The following table shows key actors that USAID is collaborating with as well as others who, though not funded by USAID, are important in the ongoing dialogue. Though not a complete list, it serves to provide an institutional context in which the discussions are taking place.

TOWARDS AN UNDERSTANDING OF SOIL CARBON SEQUESTRATION

Soil carbon sequestration is a dynamic issue, linking biophysical and socio-economic processes. Certainly, there is a key role for science to develop relevant hypotheses, organize data and test these hypotheses in an effort to prove the theoretical validity of this proposed means of mitigating carbon levels at the global level, as well as addressing the growing crisis of soil fertility in Africa. We should not lose sight, however, of the large store of knowledge already existent of valid means for addressing soil fertility that has been developed over the decades of development experience.

Linking Global Objectives With Local and Household Objectives

To adequately understand the dynamics, potentials and constraints in which soil carbon sequestration takes place in the biophysical and socio-economic realms, modeling approaches may be used. Figure 16 shows a highly simplified representation of the processes and flows associated with soil carbon sequestration. Here, we see three general processes; carbon storage, the human ecology or production system, combined with income and benefits at the level of the political economy. We measure two key flows: carbon and money through the system in order to measure progress toward the two objectives of increasing carbon stocks, mitigating climate change; and sustained economic growth and accrual of household benefits to the human population.

AGENCY/ORGANIZATION	AFRICA EXP?	AREA OF INTEREST/EXPERTISE	KEY COLLABORATIVE LINKS
USAID/G/EGAD/AFS/AEMD	X	Policy and international conventions, Poverty alleviation, Analysis and Land Grant and IARC collaboration	NASA, USGS, Winrock, IFDC, ICRAF, U Mont
USAID/AFR/SD/ANRE/ENRM	X	Promotion of NRM at farm and community level, economic linkages of Environment and Economic Growth, analyses and networking	
NASA/APS	X	Application of remote sensing for baseline development and verification methodologies	USAID/G & AFR
NASA/MPE		Analysis and Science applications of remote sensing	USAID/AFR
USRA			
USGS/EDC	X	Analysis and remote sensing, data archiving	USAID/AFR
USDA/NRCS	X	Soil mapping and GIS applications	
USDA/ARS		Carbon estimation, tool development	
USDA/FAS	X	Provision of technical assistance (personnel)	USAID/G & AFR
Pacific NW Laboratory		Analysis (economic) and technology development	DOE
Oak Ridge National Laboratory		Analysis	DOE
Los Alamos Larwence Livermore Lab		Tool Development	DOE
ICRAF	X	Forest and agro-forestry development	USAID/G
IFDC	X	Agricultural Development and soil fertility	USAID/G & AFR
IFAD	X	Agricultural Development, Soil Carbon Sequestration and Desertification projects	USAID/G
Univ of Montana		Soil carbon estimation and soil fertility	
U Arizona Arid Lands Center	X	Research in arid lands development, SoCSeq pilot, remote sensing	NASA/APS, USRA
Winrock International	X	Agricultural development, natural resource management project implementation, soil carbon sequestration verification methodologies	USAID/G
Technoserv	X	Agricultural Development, natural resource management	
The Nature Conservancy	X	Conservation of natural resources, project implementation	USAID/G
World Resources Institute	X	Environmental policy, environmental economics, carbon markets, analysis	USAID/AFR
Sand County Foundation		Conservation, business linkages	USGS

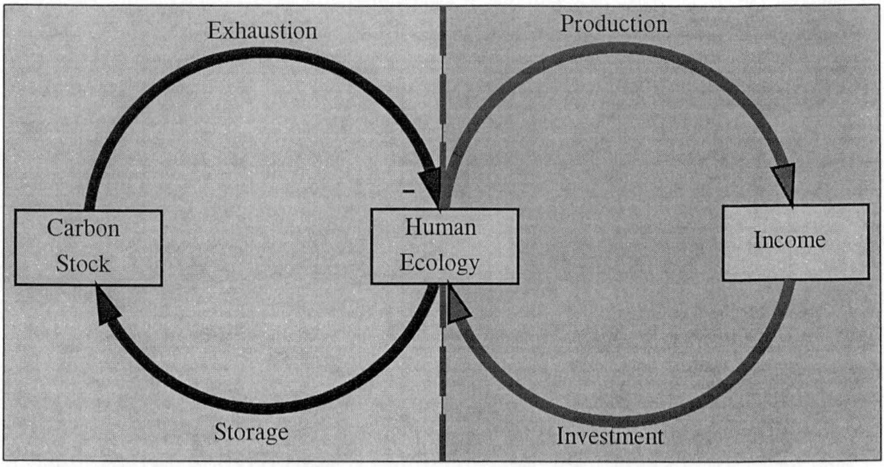

Fig. 16. A conceptual model for soil carbon sequestration.

<u>The Household Economy and the Carbon Cycle</u>

The household economy serves as a surrogate for the human processes which form the central process in the model described above. In Figure 17, we see four key processes: the household unit itself, cropland, common land and livestock production. Flows of carbon circulate between livestock, the commons, and cropland. Goods and services flow from livestock and cropland to the household. In this depiction, carbon stock is storage of carbon, which would have certain market values under a carbon sequestration scheme as well as serving as a reservoir for fertility for the three agro and biophysical processes.

In this depiction, natural processes on common land and cropland store carbon. This storage occurs by vegetative processes stemming from the growth of natural vegetation on common land and cultivars on cropland. Livestock also serve as a source of carbon indirectly in so far as livestock wastes are used to develop soil fertility on cropland. Carbon storage is depleted as well through utilization of the two land production processes. The objective of sustainable use is to ensure that carbon extraction is less than or equal to the storage accomplished by vegetative processes. In a carbon sequestration scheme, extraction should be less than the storage of carbon.

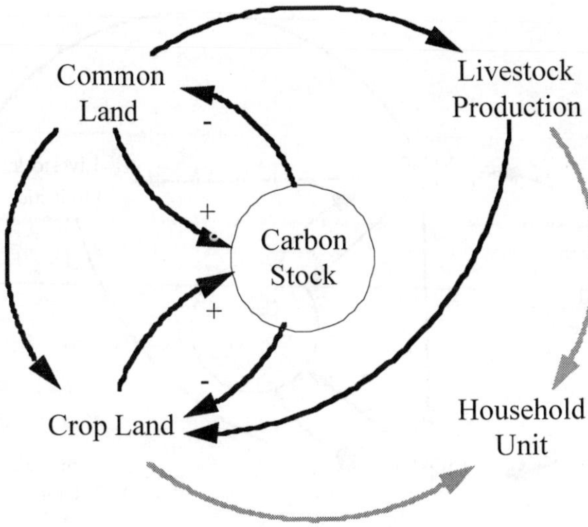

Fig. 17. Generalized Model of Key Components.

The Household Economy and Outside Factors

The household production system is not closed. Various exogenous factors have critical impacts on the performance of the system. In this case, we direct our attention to outside natural processes and outside social processes. At the same time, we take the opportunity to include services provided to the household from the commons; particularly provision of fuel, fiber, building materials, and non-farm products (food and medicines). In Figure 18, we show natural processes having a primary impact on the natural processes of common lands and croplands. Social processes have a primary impact on the household unit. This is not a definitive presentation of such models. Excellent models of household dynamics in relation to the key production factors have been constructed, notably at the University of Lund, in Sweden. Rather, this serves to identify the primary flows for consideration in assessing potential production and soil sequestration potential of various soil conservation approaches that may be applied in this context.

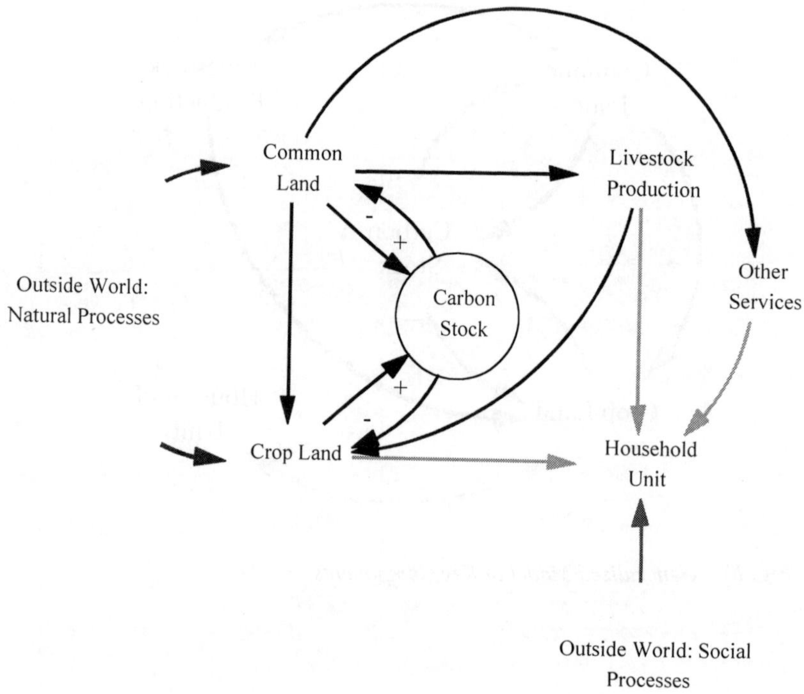

Fig. 19. Local Components and Outside Factors.

<u>The Social Environment</u>

The rural African household, even with the simplest of production systems, exists in a rather complex social and economic milieu. This milieu impacts the household through outside influences including culture and norms, a political and policy structure, market structures, and demographic trends that are often tempered by disease. In order to assess the potential of carbon sequestration and, more importantly, to determine key implementation factors, we must show the dynamics and interactions among these key factors. This representation serves to develop a context in which the broad trends discussed earlier fit the consideration of soil carbon sequestration approaches that depend upon numerous household production systems to sequester sufficient carbon to be of interest.

Of particular interest are structures already existent that connect the household production system to markets. In many African countries, agricultural and natural resource marketing structures have been developed in order to more effectively integrate rural production to national and international markets. These may be built upon to promote marketing of carbon stocks sequestered in soils. Further, an analysis of policies at the national level will indicate constraints that may exist to such schemes. Finally, all analyses should consider culture and norms, which truly frame the social context of the household economy.

The Natural Environment

Equally complex, are the natural processes which affect household production. The dynamics presented in Figure 20 are very broad and do not attempt to be exhaustive. However, there has been considerable progress in modeling these processes in an agro-climatologic context. The challenge is to integrate disparate models in a cohesive manner in order to assess the viability of approaches and to describe the potential that carbon sequestration has in this African context.

There has been considerable analysis of the effects of climate and rainfall on African production systems. This is one of the key factors that determine risk and variability of the productivity (and ultimately, the carbon sequestration potential) of household systems. Inclusion of this analysis requires going beyond the seasonal variability of rainfall, but includes an assessment of the impact of inter-seasonal dynamics, drought and flood, on the production system. Considerable data exist for certain areas. Most likely, recorded data may be characterized for these models. The dynamics of crops is well understood. Less understood are the dynamics of natural vegetation. Considerable benefit will be achieved through analysis of this, as well as that of soil dynamics, which is the least understood component of the general system.

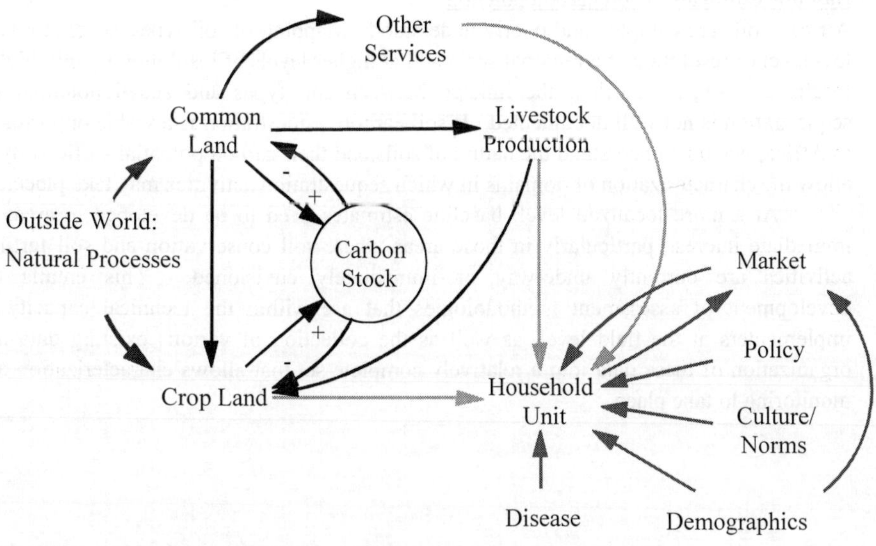

Fig. 20. Outside Natural Factors Affecting the Household Production

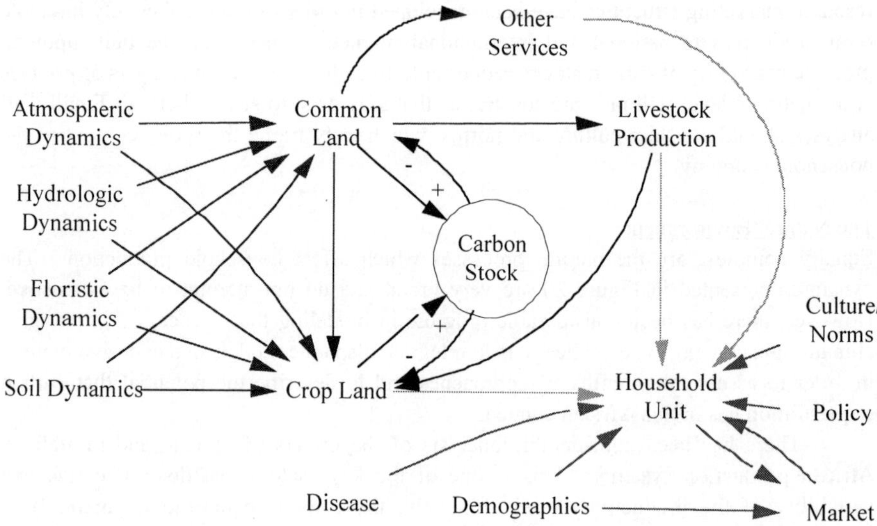

Fig. 21. Outside Social Factors Affecting the Production System.

ISSUES

Baseline estimates of soil carbon potential

Africa's soils are complex and poorly understood. Mapping of soil types exists at a very low level of resolution at a regional scale and at higher levels of resolution in only highly localized settings. Further, the linkage between soil types and actual potential for sequestration is not well documented. If soil carbon sequestration is a viable opportunity in Africa, we must understand the nature of soils and their carbon potential sufficiently to allow the characterization of domains in which sequestration activities may take place.

At a more localized level, baseline estimates need to be developed in areas of immediate interest, particularly in those areas where soil conservation and soil fertility activities are currently underway or immediately envisioned. This entails the development of assessment methodologies that are within the technical capacity of implementers at the field level, as well as the collection of various existing data and organization of these data into a relatively complete set that allows characterization and monitoring to take place.

Evaluation of carbon sequestration potential from common soil fertility and land management approaches

USAID and other donors have decades of experience in applying a wide variety of approaches that promote soil fertility. Further, USAID and donors have been engaged in the management of common property resources, which forms a second area for potential carbon sequestration. The actual levels attained or potential of these activities for carbon sequestration are not well documented. In order to determine the overall potential for carbon sequestration in Africa, the carbon sequestration value of a discrete set of approaches needs to be documented in order to develop a program which achieves the joint aims of carbon sequestration—carbon for trading (or offset) and improved productivity or income from the resources themselves.

This may require modeling of production systems to capture social, economic, and biophysical processes involved in soil fertility management. Numerous studies of on-farm soil fertility and economic production have been conducted. Management of common property systems, including range management and wildlife management, have been less fully analyzed.

Verification techniques for soil carbon sequestration from on-farm and landscape management applications

Verification of carbon stocks is a key element linking soil fertility-driven carbon sequestration efforts to global sequestration objectives. Verification methodologies in Third World settings have been developed and applied since 1993. Most of these have dealt with biomass sequestration approaches and not in soils. Considerable agronomic knowledge of carbon sequestration potential and tools for measurements in soils are available. However, methods that are cost-effective and within the technical capabilities of field personnel in Africa must be developed. Such methodologies will also need to be at sufficient levels of confidence to meet global demand.

Challenges for local organization

As envisioned, soil carbon sequestration may be adapted to existing agricultural productivity, rural commodity marketing, and community based natural resource management activities. Existing activities benefit from the organizational experience necessary to launch an enterprise-based marketing approach for supplying carbon as a commodity. However, carbon, as opposed to most agricultural commodities is "lumpy" in nature; i.e., large quantities must be supplied to attract demand. It will be necessary to understand the dynamics of scaling up farmer and community organizations to provide large quantities of verifiable carbon stocks as well as develop efficient and equitable distribution systems for revenues.

The development of a carbon stock market in Africa and links to global demand

The global carbon stock market is in its infancy. Currently, most stock trades occur within the Americas. Africa presents a new frontier for market development. Analysis and field experience will be necessary to foster an African-based market. This includes the development of information systems to link producers with clients, a pricing and bidding structure to define trades, as well as systems of verification and audit. Though not of initial concern (initial trades will most likely be brokered between previously identified producer entities and clients), some effort should be made to develop a future vision and facilitate market formation.

The national and international policy frameworks necessary to foster soil carbon sequestration

A primary factor affecting the likelihood of the success of carbon sequestration is the policy framework necessary to authorize such trades. At the international level, negotiations on the Global Climate Change Convention, particularly the acceptance of carbon sequestration as offsets to emissions will determine the future of the concept. This is a primary concern during the next few months as the GCCC negotiations enter a new phase. At the same time, official U.S. policy and Congressional directives circumscribe the level to which USAID may be an active partner in this.

Since carbon stocks represent a new commodity and in particular, one that is created by international policy, national policy structures will be necessary in order to facilitate this. Though policies will vary among states, a set of essential policy elements should be developed to guide this process.

Risk

Since the concept of soil carbon sequestration is new and not widely tested, it carries a high level of risk. First, there is a considerable element of uncertainty regarding the likelihood that carbon markets will form to enable this African experiment to bear fruit. Though this should be clarified in the next twelve to eighteen months, a widespread application will not likely occur until this decision is made.

Second, since there are few models for organizing producers of carbon stocks into sufficiently large groups to attract clients, there is risk associated with the formation of supply markets or entities that can guarantee delivery of sufficient quantities of carbon stocks to meet world demand.

Third, without robust and widely accepted verification methodologies, auditing standards, and an understanding of the dynamics of soil carbon sequestration, there is considerable risk that producer groups may not be able to deliver the actual stock of carbon contracted for, even if organizational hurdles may be overcome.

Finally, the rural Africans themselves, who will ultimately supply carbon stocks, have their own calculation of risk in terms of investment. A clear understanding of the elements of their perceived risk will be necessary if the idea is to spread sufficiently to allow an African carbon market to form.

THE WAY FORWARD

The way forward should be viewed as a process that first seeks to engage African decision makers and resource managers in a dialogue regarding global climate change. In addition to the objective of mitigating current CO_2 levels from a global perspective, they should be engaged in such a way that Africa becomes part of the solution, acting in their own best interests, and not removed from the global issue as has largely been the case thus far.

Taking the dialogue to Africa

African states and stakeholders have not been centrally involved in the discussion of carbon sequestration, partly because of the prevailing view that global climate change is a "Northern" problem which is being imposed upon them and partly because of a lack of understanding of the dynamics, vulnerabilities and opportunities presented by climate change. This is tempered somewhat by the dialogue focusing on desertification; a phenomena that has real and immediate implications to the African landscape.

Three areas of dialogue should be initiated with African stakeholders: policy, analysis, and approaches. Dialogue on all three themes may be initiated simultaneously. A series of meetings and workshops are already envisioned, starting with the meeting in Geneva to discuss inclusion of carbon credits in the GCCC. The Geneva meeting will be followed upon by a meeting tentatively scheduled for Iceland to engage African decision-makers more fully in the dialogue. A second series of meetings building upon the US-based workshops on soil carbon sequestration in St. Michaels and Sioux Falls, will take place, first in Erice, Sicily and later in Dakar, Senegal.

These meetings should result in a general framework for approaches, analyses, and policies to facilitate carbon sequestration. A series of follow-on meetings, perhaps in western, southern, and eastern Africa should be undertaken. These workshops would build upon the previous workshop in Senegal and aim toward the identification of further pilot activities. Alternatively, a series of workshops addressing particular themes, such as verification, markets, policy structures, and approaches may be held. These workshops should be started during FY 2001.

In addition, USAID missions should be engaged in the dialogue. The concept has been discussed informally with mission environmental and agricultural development officers from Zambia, Madagascar and Malawi. We may be in a position to introduce the concept in more formal terms during the upcoming Agriculture, Environment and Private Sector Conference, scheduled for the second week of November in Nairobi. Field visits to present this to particular missions may follow from this workshop.

Development of tools

A series of tools will need to be developed to prove and verify the value of carbon sequestration in Africa. The first category of tools involves the use of remote sensing to establish baseline carbon values in Africa, to verify carbon content at a level of confidence acceptable and accountable for carbon trades or offsets. These same tools may be used to establish baselines from which carbon sequestration may be measured. This is a primary area of interest to NASA and USGS.

At the same time, ground-level tools should be developed to validate remotely sensed data as well as provide tools for local applications of carbon verification. The key issue in this is that tools be cost-effective and within the technical capacity of field personnel. Several Land Grant institutions as well as Winrock International are developing verification technologies and approaches.

Finally, a robust information system should be developed that enables characterization of areas which possess the appropriate conditions (both socio-economic and biophysical) for carbon sequestration activities. This tool would help to identify likely sites, serve as an information point for potential investors or purchasers of carbon stocks, as well as a means for monitoring progress and spread of carbon sequestration efforts.

Economic, social and biophysical analyses

Carbon sequestration is essentially a union of economic and biophysical phenomena. In order to effectively understand these, a series of analyses should be carried out. First, an understanding of the micro-economic aspects of the supply of carbon stocks should be carried out. This analysis should look at the production system, identifying key decision factors that lead a producer or group of producers to offer carbon stocks on the market. Considerable work has been done on economic decision making at the farm level and less has been done concerning common property management. However, this is essential if programs are to be effectively designed. The Land Grant institutions have considerable experience in this field.

Second, a study of market development, linking supply and demand for carbon stocks should be conducted. This should lead to an understanding of the types of institutions necessary to facilitate trading or off-sets, key areas of market risk and the means to mitigate them, as well as an understanding of the pricing structure necessary to motivate trades to take place. World Resources Institute may have the technical capability for this. Land Grant institutions may also contribute, as may institutions such as IFPRI and ICRAF.

Simultaneous to these, analyses of the biophysical dynamics of carbon sequestration among key production systems should take place. These should lead to an understanding of the process and temporal aspects of carbon stock accumulation. They can be linked to key production systems that may be exploited in field tests. These analyses should also contribute to the establishment of verification tools for specific site applications.

A feature of these studies should be the development of analytical and heuristic models that allow us to predict in broad terms, the scope of carbon sequestration potential, the dynamics involved, and understanding of risk. These models should have the ability to communicate one to the other and link to the information systems developed. At another level, the models should be interactive in so far as they facilitate dialogue among stakeholders and decision makers.

An essential feature of the analytic process is the inclusion of African expertise in the design and execution of the analyses. This serves several purposes. First, it captures the rich level of knowledge already existent in Africa. Second, it serves to increase the technical capacity among those who will ultimately determine the spread of the concept.

Finally, it offers a legitimization of the process, making it increasingly African based in nature.

Establishment of pilot projects

Currently, there is one pilot activity based in Senegal. Carbon sequestration may be carried out in two broad categories of production systems; on-farm crop production systems and common property systems. A small set of pilot projects looking at the feasibility of carbon sequestration in crop systems, wildlife or landscape management systems (such as those in southern Africa) and range management systems should be tested. Ideally, these would have some geographic spread so that a continent-wide generalization could be obtained. During the first two years, three or four such projects could be established. Potential implementers include Technoserv, CLUSA, CARE, Africare, WWF, and WCS. These should include collaboration with Winrock International for the implementation of verification methodologies, though Land Grant institutions and various USDA agencies or services could participate.

USAID Involvement

USAID supports the concept of soil carbon sequestration because it promotes effective natural resource management and contributes to poverty alleviation. The Africa Bureau of USAID will continue to play a facilitative role engaging partners in a dialogue of the issues and development of a path forward. USAID has nearly forty years' experience in agriculture and natural resource development. The Africa Bureau also has excellent means of disseminating soil carbon sequestration concepts and experiences throughout its network of partners, particularly in the 22 countries in which it maintains a field presence.

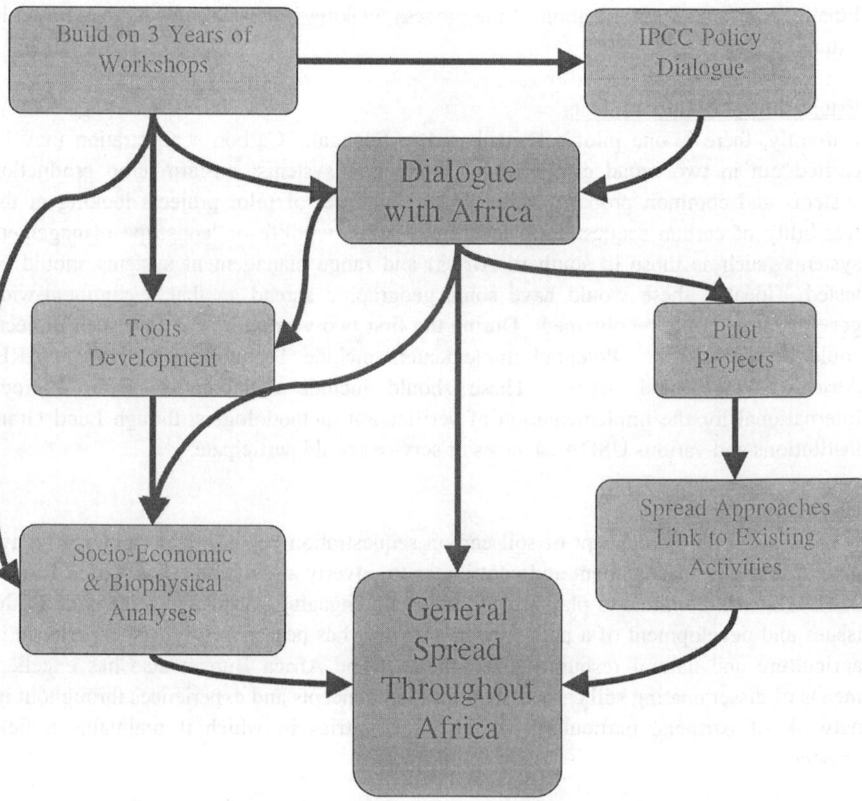

Fig. 22. The Way Forward.

During the past fifteen years, USAID has made great strides in developing approaches which organize communities into representative groups and link these to markets. USAID therefore, provides an excellent set of potential cases in which pilot projects may be initiated without starting from scratch. Additionally, the Africa Bureau will continue to support regional monitoring and data collection in support of this effort.

USAID maintains linkages to a wide variety of international donors and non-governmental organizations. Soil carbon sequestration is a complex concept which will be applied over a wide area. USAID will use its partnership network to leverage funding and coordinate the spread of the concept.

REFERENCES

1. Almaraz, Russal et al.: **Carbon Stocks in Soils of Africa, The 1996 *Annual Report***, United States Department of Agriculture Natural Resources Conservation Service.
2. Singh, Asbindu, Amadou M. Dieye, et al: **Early Warning of Selected Emerging environmental Issues in Africa: Change and Correlation from a Geographic Perspective**. UNEP, 1999.
3. International Resources Group: **Rapport de synthèse de la consultation technique régional sur les expériences de la gestion des ressources naturelles: Evolution et perspectives, (version provisionnelle)** Koudougou, Burkina Faso 1999.
4. Woodwell, John C.: *Dynamic Modeling of Ecological-Economic Systems: An Introduction for International Resources Group and the United States Agency for International Development.* IRG, 2000.
5. USGS: *Soil Carbon Sequestration in Semi-Arid and Sub-Humid Africa.* USGS, Eros Data Center. 2000.
6. Rosenberg, Norman J., et al. **Carbon Sequestration in Soils: Science, Monitoring and Beyond**. Battelle Press, 1998.
7. USGS: **Carbon Sequestration in Soils and Carbon Credits: Review and Development of Options for Semi-Arid and Sub-Humid Africa**, United States Geological Survey, 1999.

20. LIMITS OF DEVELOPMENT: FOCUS AFRICA

FOOD INSECURITY IN SUB-SAHARAN AFRICA DUE TO HIV/AIDS

CURT A. REYNOLDS
U.S. Dept. of Agriculture (USDA), Foreign Agricultural Service (FAS)
USDA South Building, Rm 6053, 1400 Independence Ave, SW, Washington DC 20250, USA. reynoldsc@fas.usda.gov

ABSTRACT

Sub-Saharan Africa, a region of 600 million people, is experiencing a tragedy of unprecedented proportions due to loss of lives from HIV/AIDS. With only 10 per cent of the world's population, sub-Sahara Africa has 70 per cent of global HIV/AIDS cases. AIDS in sub-Sahara Africa now is claiming more lives than the sum total of all wars, famines, and floods on the continent. The erosion of national human resource bases, due to HIV/AIDS deaths, is turning back decades of development and reversing economic growth in various countries. The result is reduced growth in agricultural productivity, capital generation, and labor industries. It is imperative to arrest the AIDS pandemic in sub-Saharan Africa to improve the food security and economic growth for the region. Food insecurity, poverty, migration, gender inequality, and youth unemployment are major contributors towards the spread of HIV. This makes the HIV/AIDS epidemic not only a health issue, but a development issue as well. Development projects and policies that help to alleviate food insecurity and poverty can assist to reduce the transmission of HIV. This work will require the efforts of not just health organizations, but the entire development community. Virtually no sector or sub-sector involved in planning, design, and implementation of agricultural projects can be regarded as beyond the reach of HIV/AIDS. It is essential to integrate HIV prevention strategies within all development activities, including agriculture, in order to maximize agriculture production and reduce family farm costs from loss of lives, medical and funeral expenses.

INTRODUCTION

The two World Wars from last century were the two major political challenges that faced mankind where over 60 million lives were claimed. As the twenty-first century begins, the morbidity and mortality statistics from HIV/AIDS already indicate this plague can easily exceed the death toll from the past two World Wars. An estimated 18 million people have already died from AIDS (with almost 15 million deaths in Africa) and 34.3 million people are currently infected with HIV.[1] It does stretch the imagination to foresee that the HIV/AIDS plague will soon surpass the combined death toll from last century's two World Wars. Correspondingly, the first major political challenge facing this century will be how national governments and international organizations cooperate to help prevent the spread of HIV.

Statistics further indicate the frontline of the HIV/AIDS battle will be in the sub-Sahara where the UNAIDS organization estimates 24.5 million people are carrying HIV, the catastrophic virus that causes AIDS (Table 1). As in past wars, many national governments and international organizations hesitated to take action in reducing the loss of human lives. The same can largely be said for the HIV/AIDS plague in sub-Saharan Africa. Delayed HIV mitigation and prevention programs in sub-Saharan Africa have made HIV/AIDS not only a medical problem, but one with immense social and economic dimensions as well.

Continued delays by national governments to not recognize the severity of the HIV/AIDS epidemic will not only cost lives, but will decrease agriculture production and the gross domestic product for most nations in sub-Saharan Africa. A strong political commitment is necessary to contain the epidemic, which often is perpetuated through poverty, lack of information, food insecurity, unemployment, and population movements within rural and urban African communities.

HIV/AIDS IMPACTS ON NATIONAL GOVERMENTS

The erosion of national human resource bases, due to HIV/AIDS deaths, is turning back decades of development and reversing economic growth in various countries. The result is reduced growth in productivity, capital generation, and labor industries. With government funds and household savings being diverted to purchase health and health-related goods and services, less capital is available for investment and ultimately resulting in a significantly stunted growth in the economy (GDP, GNP and employment).

It is difficult to forecast the effect of AIDS on future population growth, because of the uncertainty in how long it takes HIV infections to develop into AIDS. Some demographers project drastic declines in life expectancy in the most affected countries over the next 15 years, which will slow or even stop population growth. Other experts foresee less extreme demographic consequences from the epidemic.[2]

In either case, the region's basic infrastructure, financial and managerial resources, and health care personnel are insufficient to stem the epidemic. Some experts predict a major deterioration of industry and services Already AIDS is killing millions of people in their most productive years and destabilizing all walks of African life: health, education, industry, transport and agriculture. For example, the human-power shortage is already beginning to affect agriculture, mining and crude oil extraction industries, which account for a significant proportion of the GDP in most African countries.

AIDS has also hit the educated populations in sub-Saharan Africa during their most productive years. This trend could have devastating results for economic growth in developing countries. For example, the ability for governments to run hospitals, banks,

Table 1. HIV/AIDS Statistics for sub-Sahara Africa[3].

	Estimated number of people living with HIV/AIDS, end 1999					Orphans	Estimated AIDS deaths		Population 1999		Adults and children living with HIV/AIDS, end 1999	
Country	Adults and children	Adults (15-49)	Adult rate (%)	Women (15-49)	Children (0-14)	Orphans cumulative	Adults and children, 1999	Adults and children, cumulative	Total (thousands)	Adults 15-49 (thousands)	Low Estimate	High Estimate
Global Total	**34,300,000**	**33,000,000**	**1.07**	**15,700,000**	**1,300,000**	**13,200,000**	**2,800,000**	**18,800,000**	**5,958,849**	**3,082,548**	**25,800,000**	**41,800,000**
sub-Saharan Africa	**24,500,000**	**23,400,000**	**8.57**	**12,900,000**	**1,000,000**	**12,100,000**	**2,200,000**	**14,800,000**	**596,272**	**273,488**	**19,000,000**	**29,900,000**
Angola	160,000	150,000	2.78	82,000	7,900	98,000	15,000	110,000	12,497	5,389	110,000	200,000
Benin	70,000	67,000	2.45	37,000	3,000	22,000	5,600	27,000	5,945	2,730	56,000	84,000
Botswana	290,000	280,000	35.80	150,000	10,000	66,000	24,000	95,000	1,592	775	230,000	350,000
Burkina Faso	350,000	330,000	6.44	180,000	20,000	320,000	43,000	370,000	11,633	5,116	280,000	420,000
Burundi	360,000	340,000	11.32	190,000	19,000	230,000	39,000	290,000	6,587	3,006	260,000	460,000
Cameroon	540,000	520,000	7.73	290,000	22,000	270,000	52,000	340,000	14,704	6,723	430,000	650,000
Central African Republic	240,000	230,000	13.84	130,000	8,900	99,000	23,000	140,000	3,550	1,649	190,000	280,000
Chad	92,000	88,000	2.69	49,000	4,000	68,000	10,000	86,000	7,462	3,283	60,000	120,000
Comoros	...	400*	0.12*	...	...	...	...	...	676	327	...	...
Congo	86,000	82,000	6.43	45,000	4,000	53,000	8,600	66,000	2,867	1,277	56,000	120,000
Cote d'Ivoire	760,000	730,000	10.76	400,000	32,000	420,000	72,000	530,000	14,534	6,785	610,000	910,000
Dem. Republic of Congo	1,100,000	1,100,000	5.07	600,000	53,000	680,000	95,000	750,000	50,407	21,564	830,000	1,500,000
Djibouti	37,000	35,000	11.75	19,000	1,500	7,200	3,100	12,000	631	299	24,000	50,000
Equatorial Guinea	1,100	1,000	0.51	560	<100	860	120	1,000	442	200	690	1,400
Eritrea	...	49,000*	2.87*	...	...	...	...	...	3,717	1,705	...	...
Ethiopia	3,000,000	2,900,000	10.63	1,600,000	150,000	1,200,000	280,000	1,400,000	61,123	26,931	2,200,000	3,800,000
Gabon	23,000	22,000	4.16	12,000	780	8,600	2,000	12,000	1,196	527	15,000	31,000
Gambia	13,000	12,000	1.95	6,600	520	9,600	1,400	12,000	1,266	616	8,200	17,000
Ghana	340,000	330,000	3.60	180,000	14,000	170,000	33,000	230,000	19,699	9,164	280,000	410,000
Guinea	55,000	52,000	1.54	29,000	2,700	30,000	5,600	35,000	7,375	3,414	40,000	70,000
Guinea-Bissau	14,000	13,000	2.50	7,300	560	6,100	1,300	7,800	1,188	532	9,000	19,000
Kenya	2,100,000	2,000,000	13.95	1,100,000	78,000	730,000	180,000	960,000	29,507	14,217	1,600,000	2,500,000
Lesotho	240,000	240,000	23.57	130,000	8,200	35,000	16,000	52,000	2,108	998	160,000	330,000
Liberia	39,000	37,000	2.80	21,000	2,000	31,000	4,500	34,000	2,941	1,335	26,000	53,000
Madagascar	11,000	10,000	0.15	5,800	450	2,600	870	3,400	15,502	7,199	8,700	13,000
Malawi	800,000	760,000	15.96	420,000	40,000	390,000	70,000	470,000	10,674	4,733	640,000	960,000
Mali	100,000	97,000	2.03	53,000	5,000	45,000	9,900	56,000	10,976	4,770	66,000	140,000
Mauritania	6,600	6,300	0.52	3,500	260	...	610	2,900	2,602	1,209	4,300	8,900
Mauritius	...	500*	0.08*	...	...	...	...	...	1,149	647	330	690
Mozambique	1,200,000	1,100,000	13.22	630,000	52,000	310,000	98,000	430,000	19,222	8,607	860,000	1,500,000
Namibia	160,000	150,000	19.54	85,000	6,600	67,000	18,000	89,000	1,689	790	130,000	190,000
Niger	64,000	61,000	1.35	34,000	3,300	31,000	6,500	38,000	10,414	4,524	42,000	87,000
Nigeria	2,700,000	2,600,000	5.06	1,400,000	120,000	1,400,000	250,000	1,700,000	108,995	50,705	2,100,000	3,200,000
Reunion	...	...	...	...	...	...	...	...	690	370	...	...
Rwanda	400,000	370,000	11.21	210,000	22,000	270,000	40,000	370,000	7,238	3,338	320,000	480,000

Table 1. HIV/AIDS Statistics for sub-Sahara Africa[3] *(Continued).*

	Estimated number of people living with HIV/AIDS, end 1999					Orphans	Estimated AIDS deaths		Population 1999		Adults and children living with HIV/AIDS, end 1999	
Country	Adults and children	Adults (15-49)	Adult rate (%)	Women (15-49)	Children (0-14)	Orphans cumulative	Adults and children, 1999	Adults and children, cumulative	Total (thousands)	Adults 15-49 (thousands)	Low Estimate	High Estimate
Senegal	79,000	76,000	1.77	40,000	3,300	42,000	7,800	53,000	9,251	4,268	63,000	95,000
Togo	130,000	120,000	5.98	66,000	6,300	95,000	14,000	110,000	4,515	2,013	100,000	150,000
Sierra Leone	68,000	65,000	2.99	36,000	3,300	56,000	8,200	67,000	4,721	2,164	44,000	92,000
Somalia	...	...	...	...	...	...	..	...	9,718	4,280	...	...
South Africa	4,200,000	4,100,000	19.94	2,300,000	95,000	420,000	250,000	710,000	39,796	20,630	3,400,000	5,100,000
Swaziland	130,000	120,000	25.25	67,000	3,800	12,000	7,100	20,000	981	480	100,000	150,000
Togo	130,000	120,000	5.98	66,000	6,300	95,000	14,000	110,000	4,515	2,013	100,000	150,000
Uganda	820,000	770,000	8.30	420,000	53,000	1,700,000	110,000	1,800,000	21,209	9,222	650,000	980,000
United Rep. of Tanzania	1,300,000	1,200,000	8.09	670,000	59,000	1,100,000	140,000	1,300,000	32,799	15,068	1,000,000	1,500,000
Zambia	870,000	830,000	19.95	450,000	40,000	650,000	99,000	800,000	8,974	4,137	690,000	1,000,000
Zimbabwe	1,500,000	1,400,000	25.06	800,000	56,000	900,000	160,000	1,100,000	11,509	5,771	1,200,000	1,800,000

schools, and court/legal systems is becoming seriously affected, because they cannot find enough skilled and trained people to replace employees who were victims of AIDS. Hospitals are also reaching threshold limits, where HIV/AIDS in 1999 accounted for the greatest number of deaths and hospital admissions, and HIV/AIDS patients occupied up to 50 per cent of the beds in some hospitals.

It is clear that HIV/AIDS represents a multi-faceted complex set of problems that crosses all social and economic boundaries in the sub-Saharan region. The scope and the gravity of this disease on the social and economic infrastructure have been overwhelming, and it has impoverished governments, as well as individuals, families, and communities.

HIV/AIDS IMPACTS ON RURAL COMMUNITIES

The Food Agriculture Organization (FAO) was one of the first of the United Nations agencies to recognize the socio-economic impact of HIV/AIDS on rural economies. They found HIV/AIDS tends to primarily reduce labor on the farm and increases family medical expenses.[4]

Besides identifying the impacts on communities, the FAO also attempted to identify linkages and patterns between agricultural zones and social settings. Their studies revealed that geographic and ethnic factors, religion, gender, age, marriage, customs and agro-ecological conditions all play different roles in different regions, implying that many HIV prevention programs will need to be specially tailored for specific regions. However, in general terms, gender, age, and marital family status played a decisive role in determining susceptible and vulnerable groups to HIV/AIDS, with such factors as food insecurity, poverty, migration, gender inequality, and unemployment (especially among the youth) contributing greatly to the spread of HIV.

Some common patterns and impacts on these vulnerable groups are briefly described below.

Food Insecure and Poverty Stricken Populations

One of the most important points from the FAO studies showed that poverty and food insecurity create a high-risk environment fertile for HIV transmission, and a cycle is present similar to a poverty trap.[5] For example, parents are dying of AIDS everyday, and children are fast becoming orphans with no family or government support system in place to help them. These orphaned children, while seeking survival, then become vulnerable to abuse such as girls turning to prostitution and becoming infected just like their parents – thus perpetuating a vicious cycle.

Poverty alleviation programs are required which must be gender sensitive and address broad issues pertaining to rural development such as agriculture, health, education, rural infrastructure, and income generation. Understandably, the challenge of rural development and poverty alleviation is an area where national governments need to work in close collaboration with the private sector and non-governmental organizations.[6]

But Topuzis and du Guerny (2000) note that development is a doubled-edge sword, where intentions to reduce poverty may also increase HIV vulnerability, if HIV prevention strategies are not implemented simultaneously.[7] These results indicate that all development projects must be implemented in conjunction with HIV prevention strategies to reduce both poverty and HIV vulnerability in the long-term.

Migration

In the early 1980s, the disease was found mainly in the swathe of territory stretching from west Africa to eastern Africa, with countries in southern Africa hardly affected. However, today, no part of the African continent is unaffected, and as many as one person in five is estimated to carry HIV in some of Africa's most affected regions (Fig. 1).

A decade ago HIV was perceived as being largely urban, especially in high population density areas. But now HIV/AIDS is no longer restricted to cities, as the disease has spread into rural areas and affected farming populations, especially people in their most productive years (ages 15-45). Until recently, the spread of HIV into rural areas was largely overlooked because of poor data, irregular patterns of spread, and lower prevalence than in urban areas.

Increased mobility has contributed to the transmission of HIV into rural areas, as mobility is a prerequisite for the spread of most diseases. Of course, roads play a major factor in permitting HIV transmission, as well as rural-to-urban migration that accounts for at least half of the population growth of African cities.[8]

Rural-to-urban migration is not unique to Africa, where global rural-to-urban migration during the twentieth was the greatest human migration ever recorded in time. Some of the reasons for the rural-to-urban migration include: population pressure on agricultural land; pursuit of employment; pursuit of higher education; and expectation of better social amenities including housing, water and health care.

Migration can be seasonal or permanent, but both cause the major burden of farming to fall on women, children and older men, and it may cause spousal separation for extended periods of time. The development process also allows migrant workers and other mobile populations such as long-distance truck drivers to carry HIV to new destinations and back to their home areas.

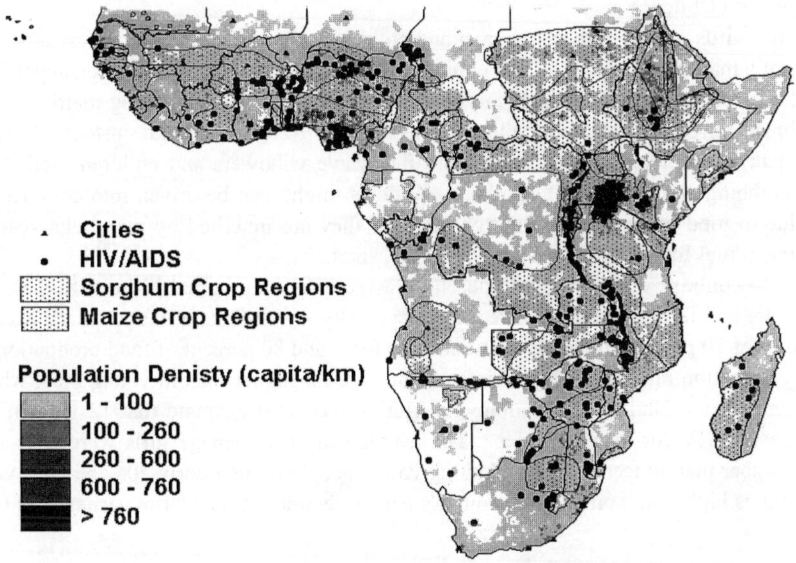

Fig. 1. Location of HIV/AIDS cases (1994-1999) and crop areas.

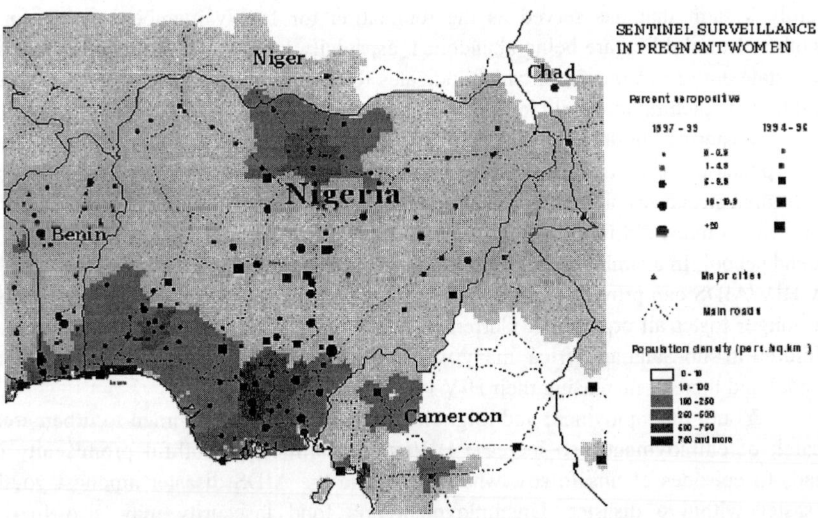

Fig. 2. HIV cases (1994-1999) in both urban and rural areas where roads serve as transmission routes[9].

Women and Children

The HIV virus greatly impacts women and children, who may be left homeless after the death of a male householder. Their coping mechanisms vary greatly and often depend on tribal inheritance systems, which can be highly complex, organized along matrilineal or patrilineal lines, and can divide the deceased's estate based on special customs. Results from patrilineal inheritance systems can often leave widowers and children victims of land grabbing and subjected to poverty. Widows might also be driven into commercial sex due to food insecurity for their children, or they are inherited by male relatives--all which can fuel further transmission of the HIV virus.

In contrast, distribution of household assets may be minimal if the wife dies, but child care and farm labor often is reduced, especially in sub-Sahara Africa where women account for 70 percent of the agricultural labor force and 80 percent of food production.

Infection rates in young African women are far higher than in young men, where the Joint United Nations Programme on AIDS (UNAIDS) estimate that 12 women are living with HIV for every 10 men. The average rates in teenage girls were over five times higher than in teenage boys. Among young people in their early 20s, the rates were three times higher in women, and women's peak infection rates occur at earlier ages than men.[10]

Youth (Unemployed and Orphans)

HIV/AIDS continues to disintegrate and destabilize the traditional African extended family system that has served as the foundation for family members caring for one another. But children are being abandoned, especially when AIDS kills both parents and devastate entire communities, which indicates the community's capacity to care for extra children is reaching threshold levels.

Fostering children is a common practice in many regions as it is common practice for orphans to be redistributed within the extended family. Before the AIDS epidemic, traditional structures only had to host several orphans, but hosting more than four or five can lead to threshold limits where some children will go hungry or not all children will attend school. In a similar way, a community in which only a few households are affected by HIV/AIDS can provide support, but beyond a certain point, the community itself can no longer foster all orphans.[11] Currently, AIDS has orphaned more than twelve million children in sub-Saharan Africa, many whose future may be jeopardized by not attending school and thereby increasing their HIV vulnerability (Table 1).

Youth unemployment and migration of young men and women to urban areas in search of employment also increase HIV vulnerabilities. Youthful promiscuity often leads to episodes of unsafe sex, which has made the AIDS disaster amongst youths--a disaster within a disaster. Unemployment and food insecurity may also fuel HIV transmission in both urban and rural areas, where young girls and young women may be driven to casual sex to buy food for their families.

HIV/AIDS IMPACTS ON FARMS AND PASTORALISTS

Small Farms

Subsistence and labor-intensive farming systems with low levels of mechanization and agricultural inputs are particularly affected by AIDS related deaths. For example, food supplies drop precipitously when the first adult develops full-blown AIDS. This deprives the family not only of this worker in the fields, but also of the work time of another adult caring for the AIDS victim. All result in a loss of potential income, and the situation is aggravated in farming systems that are marked by gender division of labor.

As the illness progresses, families are constantly faced with an exorbitant medical cost, depleting all their savings and even forcing them to dispose of their assets such as selling livestock, land, etc. Cash income and labor are partly diverted to cope with or compensate for the effect of HIV/AIDS, leaving less labor for farm and off-farm activities as well as reducing the amount of money available to the household.[12]

For example, where households own livestock, and there is no cash income, cattle may be sold to pay for medical and funeral expenses. But the sale of livestock adversely affects household crop production capacity, and draught power is lost. Once the limited stock owned is exhausted, households face serious food insecurity and malnutrition. Less draught power results in reduced cultivated areas. Therefore, the sacrifice or sale of cattle might be regarded as one of the most destructive process related to HIV/AIDS.[13]

Some other effects from labor shortage at the farm-level are:[14]

- Reduction in the acreage of land under cultivation
- Delay in farming operations such as tillage, planting and weeding
- Reduction in the ability to control crop pests
- Decline in crop yields
- Loss of soil fertility
- Shift from labor-intensive crops (e.g. banana) to less labor-intensive crops (such as cassava and sweet potatoes)
- Shift from cash-oriented production to subsistence production
- Reduction in the range of crops per household
- Decline in livestock production
- Loss of agricultural knowledge and management skills.

Commercial Farms

FAO/UNDP found that several commercial farms in Kenya were particularly susceptible to the AIDS epidemic, and morbidity and mortality are costing the industry direct losses (medical and funeral expenses) and indirect costs through the loss of valuable skills and experience.[15] The epidemic thus adversely affects the commercial farms' efficiency and productivity (Fig. 3).

Rugalema (et al, 1999) found commercial farms are highly susceptible to the transmission of HIV because their social and economic environments constitute a risk due to:[16]

- Overcrowding and lack of privacy.
- Casual and commercial sex is common.
- Poverty encourages commercial sex.
- High incidence of sexually transmitted diseases (STDs).
- Recreation facilities are seriously lacking.
- Resistance to use condoms remains strong and misinformation widespread.
- Cultural beliefs still dominate HIV/AIDS discourse.
- Knowledge about HIV prevention is not always practiced.

Rugalema (et al, 1999)[17], Baier (1997)[18], and Topouzis (1998)[19] also proposed many useful HIV prevention strategies for governments and agencies to integrate with development projects and policies.

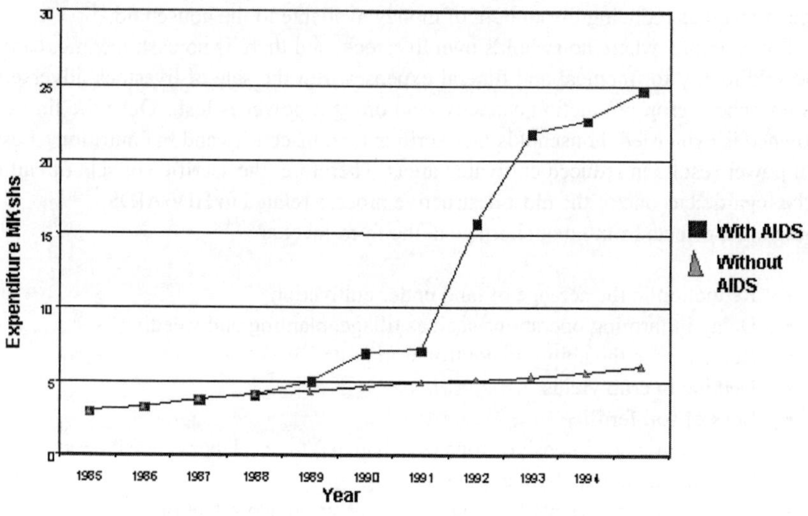

Fig. 3. Medical expenditure of an agro-estate in Kenya (with and without AIDS)[20].

<u>Pastoralists</u>

Pastoralists and transhumant societies traditionally inhabit arid and semi-arid lands (ASAL) regions, which are grasslands typically only suitable for grazing livestock or marginal agricultural environments which are used as drought reserves. Pastoralists are typically food insecure due to their vulnerability to famine from high rainfall variability.

As a general rule, pastoralists have diversified their livestock economies by sending young men, as well as older men, to seek employment in urban areas as a mechanism to cope with drought. These younger men are expected to send portions of their income home, especially during periods of drought for drought recovery purposes such as livestock restocking. However, these urban employees often return to their homelands during annual leave, only to unknowingly transmit HIV to their community.

CONCLUSIONS

About 90 percent of all HIV transmission in Africa occurs via heterosexual sex. This is 100 percent preventable with implementation of HIV prevention and AIDS mitigation strategies. However, economic intervention must be part of the overall strategy as some 95 per cent of Africans infected with HIV/AIDS live in abject poverty.

In addition, the spread of the disease in rural areas is not properly understood because of poor data and irregular patterns of spread. In view of the rapid spread of the HIV/AIDS epidemic in rural areas, socio-economic and cultural research needs to be conducted on the impact of the disease on agricultural production systems, household food security, traditional coping mechanisms, etc. to enable the development of appropriate prevention and mitigation strategies.

During the past decade, the linkages of HIV/AIDS to agriculture were largely ignored in rural areas because the epidemic was perceived as being urban, especially in high-density areas. Information of the prevalence and incidence rates by sex and age for rural populations is still not available due to the lack of data from rural areas. Therefore, more surveillance data as well as cultural practice information is required to design effective education programs to reduce the spread of the disease.

The results and findings from additional rural HIV/AIDS data can then be used as inputs for national planning and program formulation purposes, as well as for regional and local mitigation strategies targeting individual farmers, families and communities. In addition, agricultural extensions officers and other employees from relevant agencies should be sensitized to the socio-economic impact of HIV/AIDS on agricultural production, food security and rural development.

It is imperative to mitigate the AIDS pandemic to improve food security, reduce poverty, and promote economic growth in sub-Saharan Africa. This will require the efforts not only of health organizations, but the entire development community. Modern (print and digital) as well as traditional (folk dance, drama, etc.) media have a role to play in the prevention and mitigation of HIV/AIDS. This will require national and international agencies involved in agriculture, planning and national development, health, labor, rural development, and women/child welfare programs to create focal points for implementing HIV/AIDS mitigation and preventions strategies.

REFERENCES

1. UNAIDS, 2000. AIDS and Population Fact Sheet. Joint United Nations Programme on AIDS (UNAIDS), http://www.unaids.org/fact_sheets/files/Demographic_Eng.html
2. Brown, L. 1996. The Potential Impact of AIDS on Population and Economic Growth Rates, International Food Policy Research Institute.

3. UNAIDS, 2000. Excel Table of country-specific HIV/AIDS estimates and data, Joint United Nations Programme on AIDS (UNAIDS), http://www.unaids.org/epidemic_update/report/index.html
4. FAO. 1994. What has AIDS to do with agriculture? FAO, Rome.
5. FAO. 1997. The rural people of Africa confronted with AIDS: a challenge to development, Summary of FAO studies on AIDS. Rome, December.
6. Rugalema, G, S. Weigang and J. Mbwika. 1999. HIV/AIDS and the commercial agricultural sector of Kenya - impact, vulnerability, susceptibility and coping strategies, FAO/UNDP.
7. Topouzis, D. and J. du Guerny, 1999. Sustainable agricultural/rural development and vulnerability to HIV/AIDS, FAO/UNAIDS.
8. Hsu, Lee-Nah, and J. du Guerny, 2000. Population Movement, Development and HIV/AIDS: Looking Towrads the Future, FAO and UNDP.
9. UNAIDS, 2000. Epidemiological Fact Sheets by Country, http://www.unaids.org/hivaidsinfo/statistics/june98/fact_sheets/index.html
10. UNAIDS, 2000. HIV/AIDS in Africa Fact Sheet, Joint United Nations Programme on AIDS (UNAIDS), http://www.unaids.org/fact_sheets/files/Africa_Eng.html
11. du Guerny, J. 1998. Rural children living in farm systems affected by HIV/AIDS: some issues for the rights of the child on the basis of FAO HIV/AIDS studies in Africa. Paper presented by at the UNHCHR Committee on the Rights of the Child: day of discussion on Children living in a world with AIDS, Geneva, October 5.
12. FAO. 1994. What has AIDS to do with agriculture? FAO, Rome.
13. Engh, Ida-Eline, L. Stloukal, and J. du Guerny. 1999. HIV/AIDS in Namibia: The impact on the livestock sector, Population Programme Service FAO, Rome.
14. FAO. 1995. The effects of HIV/AIDS on farming systems in Eastern Africa. FAO, Rome.
15. Rugalema, et al, 1999. Executive Summary, p. ix.
16. *Ibid.* Executive Summary, p.x.
17. *Ibid.* p.59-59.
18. Baier. 1997. Conclusions.
19. Topouzis, D. 1998. The implications of HIV/AIDS for rural development policy and programming: focus on sub-Saharan Africa, Rome, June 1998. Also published with same title as HIV and Development Programme Study Paper No. 6, New York, UNDP.
20. *Ibid.* p.11.

MIGRATION IN UGANDA: MEASURES GOVERNMENT IS TAKING TO ADDRESS RURAL-URBAN MIGRATION

HON. JANE FRANCES KUKA
Minister of State, Office of the Prime Minister (Disaster Preparedness and Refugees), P.O. Box 341, Kampala, Uganda

INTRODUCTION

Migration is defined as any permanent change in residence (Weeks 1994). It involves detachment from the organisation of activities in one place and movement of the total round of activities to another. In general terms, migration is a phenomenon associated with industrialisation and urbanisation in both first world and third world countries. It relates particularly to special differences in employment opportunities, as most migration may be characterised as "labour migration." Further, migration also can be defined as the movement from one place to another or regular travel from one region to another.

Migration has received much less attention at the beginning of this millennium than other demographic changes, such as fertility and mortality, yet migration is closely related to many broad social and economic world developments.

It is important, therefore, to note that when people move from traditional rural communities to large urban centres, where they have greater freedom to participate in educational opportunities and economic activities, these new social and economic conditions alter gender relations and family structure. In some contexts, also, the migration of women is subject to greater constraints than migration of men because of the dependent position of women within the family and in society at large. Yet, even in such contexts, autonomous female migration may increase if households are in need of income and there are employment opportunities and housing for women in the place of destination.

However, people move largely because of complex reasons, many of which are non-economic, and such reasons can be explained by the "push-pull" theory. That is, some people move because they are pushed out of their former location as a result of land crisis or circumcision, for instance; whereas others move because they have been pulled or attracted to some other place. Thus the main reasons for migration may include economic, social, cultural, and political factors.

There are basically two types of migration: external migration of "refugees" who cross national borders, and internal migration of "internally displaced persons" within a country. Migration can also be classified as being either forced or voluntary.

REASONS FOR MIGRATION

Reasons for migration include the following:

- Economic factors mainly related to the search for wealth and employment.
- Social factors such as marriage. It is almost inevitable that women migrate on marriage when the spouse lives or works in an urban area.
- Cultural practices and roles also shape migration. Women and girls sometimes move to escape cultural pressures from, for instance, female circumcision. However, some cultures deny girls the right to move to protect the family and the girls' reputations.
- Political persecution causes men and women to migrate to other countries, and some men and women also migrate during periods of war, both internal and external.

It is worth noting that the differences between migration of women and men have been studied very little. There is some indication that female migration, especially of women working in domestic services or engaged in illegal or socially unacceptable activities such as prostitution, is rarely reported.

MIGRATION IN UGANDA

Uganda has endured some of the worst tragedies that could befall a country. Uganda gained independence in 1962, but suffered severely under two successive military regimes through the 1970s, and then experienced coup attempts which culminated in a civil war in 1986 that brought the Movement Government (National Resistance Movement) to power. During this time hundreds of thousands of people were killed, and millions migrated to other countries as refugees. This unrest resulted in a deep decline of all indices of human development in Uganda, with increased insecurity, a collapsed economy, and corruption and inefficiency in the public sector. With thousands of refugees and internally displaced persons (IDPs) there was increased poverty and increased rural-to-urban migration.

Major urban centres and roads in Uganda are shown in Figure 1.

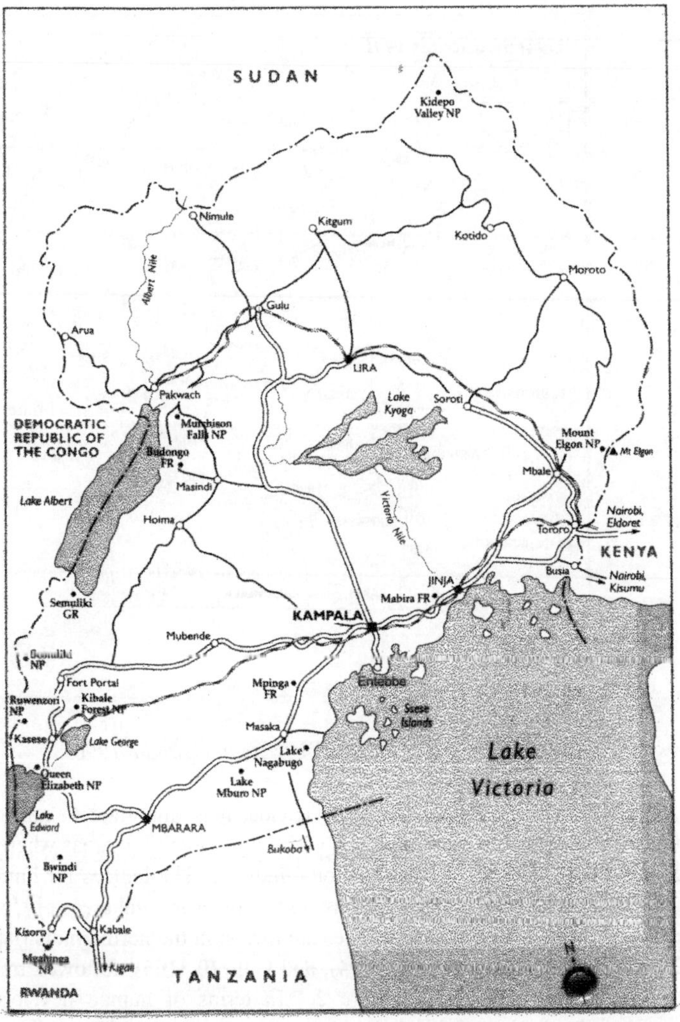

Fig. 1. Map of Uganda.

Uganda is divided into ten provinces and, at this time, 45 districts, including six recently formed in addition to the 39 districts established earlier and shown on Figure 2. Local administration and planning boundaries are at district, county, and sub-county levels.

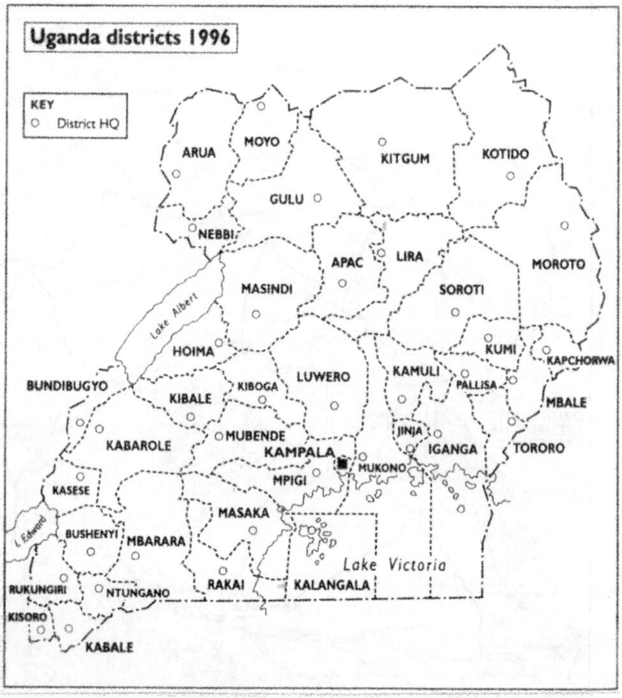

Fig. 2. Political Districts in Uganda, 1996.

To understand migration trends in Uganda, it is important to note the regional disparities in the country. Poverty is much worse in remote rural areas where it is largely masked by national averages and international indices. The indices for human survival, knowledge and literacy, and standards of living are lower in rural areas of Uganda than in urban centres. Human development indices are lowest in the northern districts of Moroto (0.1652), Kotido (0.178), Kitgum (0.3094), and Gulu (0.3165), followed by the western district of Bundibugyo (0.3105), Figure 2. In terms of human development, these districts have fallen far below the already low national average. These disparities reflect the "push" factor to migration and are largely attributable to the remoteness of these areas; to ethnic, cultural, and religious diversity; and to marginalisation over the years.

It should be noted that migration tends to follow the trend of public sector spending and private investment, both of which are heavily skewed towards the central and southern regions in Uganda that are more accessible, more developed, and more stable. The variation in population density by district, based on 1991 census data, is shown in Figure 3.

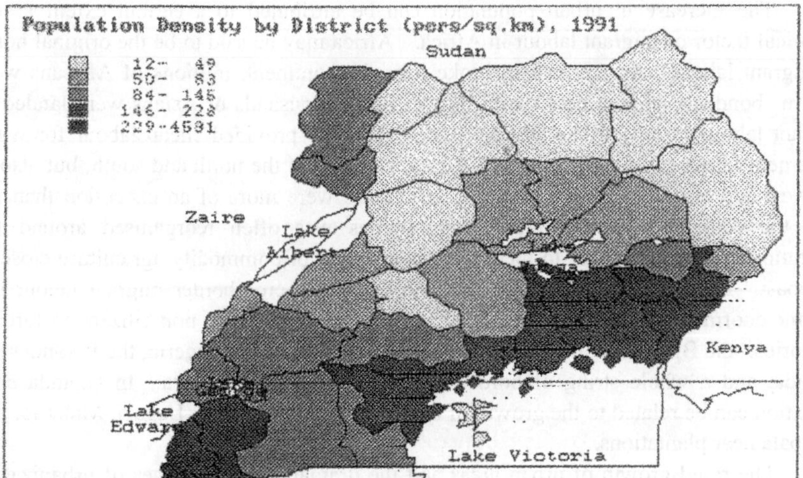

Fig. 3. Uganda Population Density by District, 1991. Population and Housing Census, Uganda Census Office.

Presently, only about 20 percent of the Ugandan population lives in urban areas; most people live in rural areas, and their livelihood is agriculture, which is the backbone of the Ugandan economy. At district levels, population densities are highest in eastern, central, and western regions where rainfall is regular and soils are fertile, but the northern region is sparsely populated, as shown in Figure 2. Kabale and Mbale Districts have the highest population densities. About 40 percent of the people in urban areas have come from somewhere else.

Population data in the table below indicate that in the 1960s over 90 percent of the population lived in rural areas, but the urban population in Uganda is growing, and today about 20 percent of the population is urban.

Table 1. Uganda Population Data.

Year	Total population	Percent urban	Percent annual growth rate	
			National	Urban
1950	5,522,000			
1960	7,262,000			
1970	9,728,000	7.9		
1980	12,298,000		2.8 (1969-80)	4.7 (1965-80)
1990	17,186,000			
2000	23,318,000	20	2.5 (1980-91)	5.1 (1980-88)

The increase in urban population can be attributed to a certain extent to the historical factor of migrant labour in Africa. Africa may be said to be the original home of migrant labour. As capitalism broke over the continent, millions of Africans were sent in bondage to the new world, just as many thousands of Asians were landed as migrant labourers on African shores. Later, migrants provided cheap labour for white settlement farms and for the mines in Africa, not only in the north and south, but also in the west and east. Whereas settlers in agriculture were more of an exception than the rule, the economic life of inland communities was often reorganised around the institution of migrant labour to meet the labour needs of commodity agriculture close to the coast. Following independence the significance of crossborder migrant labour has become enormous. Entire communities now migrate to labour as non-citizens in foreign territories: the Burkinabe in the Ivory Coast, the Ghananians in Nigeria, the Rwandese in Uganda, and a whole string of border nationalities in South Africa. In Uganda such migration can be related to the growth of the urban centres of Jinja, Lugazi, Mukono, and Kampala near plantations.

The rapid growth of urban areas and the dramatic consequences of urbanization have drawn attention to rural-urban migration in recent years. In many countries rural-to-rural and urban-to-urban are common forms of migration, but in Uganda the common form of migration is rural-to-urban.

INTERNAL MIGRATION

Rural-urban migration

Rural-urban migration occurs when a person or family moves from a rural area to an urban area, to a new place but within the same country of origin. Rural-urban migration tends to move the most youthful, strong, and energetic men and women from rural areas, and those left in the rural areas are mainly the elderly and children who are not strong enough to do agricultural work and who are less productive. The resulting increase in urban population puts pressure on social services in urban areas, as seen in sprouting areas of slums and shacks, inadequate schools and health services. Social problems, such as the number of street children, lack of jobs, thefts, and burglaries also increase.

Internally displaced persons

Recognition must also be made of forced internal migration which results in internal displacement. There is no universally agreed upon definition for those internally displaced, or of whom should be considered to be in need of assistance and protection by governments and the international community. However, the United Nations Commission on Human Rights defines internally displaced persons (IDPs) as those who have been forced to flee their homes suddenly or unexpectedly, and in large numbers, as a result of armed conflict, internal strife, systematic violation of women's rights, or natural or man-made disasters within the territory of their own country.

Internally displaced persons fall under the sovereign authority of their government which, if not actually the persecutor, may be unwilling or unable to help them. No organisation or group can be counted to come automatically to their assistance, not the United Nations High Commission for Refugees (UNHCR) or even the International Committee of the Red Cross. Yet IDPs are the single largest at-risk population in the world. They are beset by hunger and disease and lack adequate shelter. In Uganda there are currently about 700,000 people who are internally displaced, especially in the districts of the north and southwest.

Assistance for IDPs has remained a problem, with serious situations still left unattended. Assistance is still ad hoc, carried out on a case by case basis with each agency doing what it believes it can and should do.

Beyond the obvious humanitarian and human rights aspects, IDPs also raise serious problems for the international economic and political order. Instances of large numbers of IDPs almost always arise from complex causes, and it is rare that the crisis that generated the displacement remains confined to a single country. More commonly, massive internal displacement becomes the spark that ignites the flow of refugees across national borders. Assisting and protecting IDPs within their own countries keeps them from becoming refugees. The violence and instability that sparks internal displacement often infests neighbouring states and spreads through an entire region. The worst situations can require international armed intervention. Africa has about 20 to 25 million IDPs in addition to more than 13 million refugees at this time.

All these factors can be said to reflect a breakdown in the basic mechanisms of society to one extent or another, to be crises of national identity. Recently national identity has become the main cause of internal displacement. However, it is not only race, language, religion, and culture that cause conflict and displacement, but the consequences of distribution of resources and opportunities are also important. These have caused massive violations of fundamental human rights and freedom, gravely compromising economic and social development, and leading to breakdown of civil order, attempts at ethnic cleansing, and even genocide. For example, the situations in Rwanda and Burundi between the Tutsi and Hutu and in Sudan between Moslems in the north and indigenous animist people in the south.

EXTERNAL MIGRATION

Refugees

Refugees are migrants who move from one country to another permanently; and their new place or destination is outside the boundaries of their country of origin. Refugees who cross national borders are under the protection of the Refugee Convention of 1951, supplemented and strengthened by the 1968 Refugee Protocol, that gives the international community legal authority, through the intermediacy of the UNHCR, to protect and assist them.

Uganda has hosted refugees since 1956 when the first refugees arrived as a result of the Anyanya war in the Sudan. They were followed by the Rwandese in 1959, and since then Uganda has had a continuous flow of refugees into the country. Uganda now hosts about 200,000 refugees from eight neighbouring countries, including about 180,000 from Sudan; the rest are largely from the Democratic Republic of Congo and Somalia.

The government of Uganda has committed itself to accepting refugees and has established special settlements with access to farmland for them, although in some cases they are also permitted to reside in urban areas. Uganda allocates land to refugees so that ultimately they can become self-sufficient and has started a new approach, the "self-reliance strategy," to streamline delivery of refugee services with national and district development plans. It is hoped this will support development activities in those districts that host refugees and also will promote self-sufficiency of the refugees.

MEASURES TO REDUCE RURAL-URBAN MIGRATION IN UGANDA

Uganda has put in place several measures to reduce rural-urban migration, as discussed below.

Decentralization

Decentralization involves shifting government decision making and the workload from central officials and offices to local administrative units to take services closer to the people and to transfer real power to the districts and thus reduce the workload of remote, under-resourced central government offices. In Uganda the objectives of decentralisation include:

- Bringing political and administrative control over resources to the point where they are actually delivered, thus reducing competition for power at the centre and improving accountability and effectiveness at the local level.
- Freeing local managers from central constraints, thus allowing them to develop organisational structures tailored to local circumstances.
- Improving financial accountability and responsibility by establishing a clear link between payment of taxes and provision of services, financing at the local level.
- Restructuring government machinery to make administration of the country more effective.
- Creating a democracy that will bring about more efficiency and productivity in state machinery through involvement of the people.

Decentralisation in Uganda was initiated in 1993 when the Local Government (Resistance Council) Statute was passed by the National Resistance Council (Parliament). In that year, the first 13 districts began the two-phase decentralisation process, beginning with the transfer of votes and followed by block grants for recurrent district budgets. By fiscal year 1996/97, all 39 districts were fully decentralised in accordance with the 1993 Local Government Statute and the National Constitution. However, decentralisation of the development budget is still being worked out.

Decentralisation in Uganda has been the result of the central government's commitment to devolve power to the lowest levels of government. This commitment was further strengthened by the Constitutional Commitment of 1995 and the new Local Government Act of 1997, both of which emphasize government policy of devolving powers to district and lower levels and decentralisation of decisions on district manpower development and deployment.

Problems associated with political, administrative, and financial powerlessness at the district level have been constitutionally addressed. For example, under articles 188 and 200 of the Constitution of Uganda, District Service Commissions are empowered to hire and fire employees of the district. Local authorities, which previously were only in charge of primary education and health centres under the Local Government Statute of 1993, are now also responsible for secondary education, trade, special education, and hospitals (other than hospitals providing referral and medical training).

Local authorities have large budgets that reach to U.S. $3 million dollars. A sub-county retains 65 percent of the amount of revenue collected and distributes five percent to county councils, five percent to the parishes, and 25 percent among the village councils.. This retention of money at the lower levels enables them to plan, in a predictable manner, credible methods of eradicating poverty in those areas where the "push" factor is a major contributor to rural-urban migration.

Decentralisation of government also can address various factors underlying poverty, including lack of self determination, lack of planning skills, lack of organising competence, and lack of efficient service delivery systems. The legal and administrative framework of the decentralisation programme addresses all these issues, but will be successful only if the people eventually acquire the ability to make the process work for themselves. Training programmes are being implemented as a part of the decentralisation process to empower both leaders and stakeholders to make full use of the benefits of decentralisation. The Ugandan Constitution further provides for special equalisation grants which are given to local governments lagging behind national averages.

A recent study by the Makerere University Institute of Social Studies (MISR) identifies a number of favourable effects of decentralisation. The most important effect observed is that the programme has tremendously improved planning, coordination of field activities, accountability, and reporting systems. Resources have become more readily available, and delivery of services has been greatly improved. For example, before decentralisation in 1991/92, the local government in Rakai District had a total of over 26 departments, each working independently of the others, and administrative costs were 75 percent of the total district revenue. Following decentralisation, the number of departments was reduced to eight, and administrative costs fell from 75 percent of the total budget in 1991/92 to 30 percent in 1994/95 and to 18 percent in the 1995/96 fiscal year. There are also tremendous savings in other areas of resource use.

In short, decentralisation has given power to the districts and on down to sub-counties, parishes, and villages. Government officials at all levels are accountable to the people as the Constitution confers powers upon the people to recall non-performing elected officials. This works to address the feeling of political powerlessness, which was a serious disadvantage before decentralisation, and there are now reasonable governmental financial resources, even at sub-county levels, that make planning based on predictable financial resources possible. Decentralisation can, therefore, be viewed as a constitutional and integrated package of enabling interactions.

Limitations of Decentralisation. The constraints of Uganda's programme of decentralisation are basically of two categories, namely those associated with the leadership and those associated with weaknesses of ordinary citizens.

Constraints of leadership. There is generally a shortage of adequately trained manpower at all levels in Uganda. There is a shortage of planning skills, a lack of knowledge about provisions of the Constitution and the Local Government Statute concerning the rights and duties of the leaders, and there are numerous reports of corruption and misappropriations of funds. For example, in 1995/96 the Mbarara District administration lost about 400 million shillings (about 25 percent of the total revenue collected for that year) because false tickets were printed and money budgeted for teachers' salaries was diverted to pay tenders. In the Tororo District about 1.5 billion shillings was spent in four months in 1996 without the authority of the council or any of its committees. In 1997, Mpigi District councillors proposed to give each other gratuities of one million shillings for having successfully served their term of office, but the Minister of Local Government rightly disallowed this expenditure.

Recent surveys of health services delivery in the field (Cock Croft/CIET International, 1995) revealed important information about the performance of public servants and perceptions of the recipient beneficiaries (the general public). For example, the survey found that drugs distributed from the centre sometimes do not reach targeted groups at the district level and that in some case users now pay more for health services than before decentralisation. Beneficiaries sometimes pay cost-sharing fees and informal fees demanded by service providers. But the surveys also revealed cases of honest and competent service.

Constraints of ordinary citizens. Ordinary citizens have weaknesses arising from inadequate information about their rights under the Constitution and the Local Government Act. Thus, they are not in a position to police activities of civil servants or to demand accountability and transparency from public servants who are supposed to deliver enabling services to them. However, at this time, all of such shortcomings ought to be regarded as temporary. Costs associated with weaknesses of the technical staff and ordinary citizens (who are supposed to monitor and evaluate staff performance) must be regarded as part of the training cost related to implementing decentralisation.

Effectiveness of Decentralisation. Decentralisation can be made a more effective instrument for reducing rural-urban migration in Uganda if:

- Training programmes being implemented for the various stakeholders and other institutions are continuously updated, using follow-up data that should be collected continuously in the field.
- Sensitisation concerning the national purpose of the decentralisation programme is promoted at all levels.
- There are more frequent inspections of books of accounts by independent auditing institutions, such as the Auditor General, and corrupt people are promptly punished when courts of law have found them guilty.
- Surveys of service delivery should be conducted at an expanded level, and all levels of local government should be trained to conduct and use such surveys for themselves.

Economic Empowerment

Full development, both local and national, cannot be achieved without both men and women being involved. Economic empowerment is the key to women's emancipation, without which full empowerment cannot be attained. The Ministry of Gender, Labour and Social Development was established by the government of Uganda in 1988 as a Ministry of Women in Development, and now has a Minister of State in charge of "Entandikwa" a micro finance department that gives loans mainly to poor rural men, women, and youths.

The poor in Uganda are not a homogenous group in need of homogenous packages of intervention. The Uganda Country Report, presented to the World Summit on Social Development in Denmark in 1995, identified the following categories of poor people:

- Disadvantaged peasants, including the landless, squatters, and pastoralists with inadequate livestock.
- People who are too handicapped to work by reason of physical or mental disabilities or by reason of old age or tender age.
- One-parent families, particularly those headed by females, such as divorcees, widows, and unmarried mothers.
- Children in need of care and protection, including orphans, displaced children, and street children.
- Disadvantaged urban dwellers (especially the unemployed), informal sector workers, slum dwellers, people in very remote areas who lack access to services and profitable markets, people who live in insecure areas, and those frequently affected by natural disasters such as droughts and earthquakes.

In addition, there are fundamental imbalances in the respective rights and obligations of men and women, which lead to a situation where men and women have highly differentiated economic opportunities. There is poverty associated with gender having to do with the following:

- Women have less lucrative economic roles than men.
- Women have restricted access to economically productive assets such as land, compared to men in general.
- Women lack meaningful control over productive resources and even over the crops they produce.
- Women are poorly represented in key positions of decision making.
- Women are economically, legally, and culturally disadvantaged.

There are also non-government organisations, such as Action for Development (ACFODE), that target rural areas. One of ACFODE's programmes, called Economic Empowerment, targets women's groups and out-of-school youths in rural areas and gives them a full package on loan management. This assistance has tended to make rural areas more attractive and to keep women and young people from migrating to urban areas. There are also other micro-finance institutions that give loans to rural women and men in Uganda. Micro Credit Finance has played a great role in improving the quality of life in rural areas.

However, it should be noted that the 1995 Uganda Constitution addresses most of these problems. For example, Article 33(1) of the Constitution states: "Women shall be accorded full and equal dignity of the person with men," and Article 33(6) provides that laws, cultures, customs, or traditions that adversely affect the dignity, welfare, or interest of women, or which undermine the status of women, are prohibited.

Given that Uganda is primarily an agricultural country generously endowed with fertile soil and reasonable amounts of rainfall, the central government and local governments have adopted the following measures designed to improve agricultural productivity, food security, and other outcomes that would reduce rural-urban migration:

- The National Budget of the Ministry of Agriculture has been increased to deal with agricultural policy and to address problems of appropriate technology (e.g animal traction and improved seeds).
- Government has introduced appropriate land reforms to make land more accessible to the people, and especially to women.
- Government is strengthening the food supply system by improving storage facilities, food processing, and the transportation and marketing of food locally, regionally, nationally, and internationally.
- Government has implemented the Universal Primary Education Programme which provides for free universal education for four children per family.

The poverty of knowledge is being at least addressed at this critical stage. Illiterate people are too powerless to tackle poverty because they must depend on what they hear from other people. It is very easy to mislead such people, and they tend to move to urban areas in search of a better life.

GOOD GOVERNANCE

The current government is operating under the "Movement System" (National Resistance Movement) of governance which is a broad-based, inclusive, and non-partisan political system based on the following principles:

- Participatory democracy.
- Accountability and transparency.
- Accessibility to all positions of leadership by all citizens.
- Individual merit as the basis for election to political office.

The Movement System of governance has brought about not only political stability, but also has created an environment conducive to economic empowerment.

CONCLUSIONS

It should be clear that rural-urban migration in Uganda is a multi-faceted problem which includes various "push" and "pull" factors. Further, despite measures the Government has taken to reduce rural-urban migration, the problem still persists.

Democratic decentralisation, which gives money, decision making power, and planning authority to local governments and provides participation opportunities to ordinary people, should be able to develop in rural areas similar opportunities and quality of life that up to now have pulled people to urban centres.

The empowered local authorities are advised to work closely with NGOs, both local and international, because the NGOs possess the organisational capacity and financial resources from which the local authorities can benefit.

More research needs to be done to establish more facts in order to put other measures in place to reduce rural-urban migration. Research is needed on measures to eradicate poverty, to strengthen communities in rural areas, to develop private sector initiatives as a way of increasing job opportunities, and last, but not least, research is needed on oppressive cultural practices, such as female genital mutilation, that have to a great extent pushed people to urban areas.

REFERENCES

1. Cock Croft/CIET International, 1995.
2. Mbilinyi and Omari.
3. Makerere Institute of Social Research. Study of the effects of decentralisation, 1997.

4. Obbo, C. African Women: Their Struggle for Economic Independence, Zed, London, 1980.
5. United Nations Under Secretary General Deing. Exodus from Within, 1999.
6. Weeks, John R. Population: An Introduction to Concepts and Issues, 5th Edition, Wadsworth Publishing Company, California, 1994.

THE IMPACT ON AFRICAN ECONOMIC DEVELOPMENT OF ORPHANS BY AIDS IN AFRICA: A CASE STUDY OF UGANDA

MARGARET FARAH,
Executive Director, Uganda Centre for Disasters, Kampala, Uganda

INTRODUCTION

It is estimated that there are about 1 million children under age 15 orphaned by AIDS in Uganda, and the problem of orphans is enormous. Because of the increasing number of AIDS cases, the impacts of AIDS orphans on social and economic conditions is becoming more critical each year. Most orphans do not have enough support, and yet they are of a tender age when parental care is most needed for proper growth. A lot of Government expenditures have been switched to address the problem. However, due to economic constraints, Government support is limited.

Almost ten million women in Africa are HIV positive, and four out of every five HIV-infected women in the world live on this continent, according to Dr. Awa Marie Coll-Seck of UNAIDS Department of Policy, Strategy and Research. These numbers are high because most African women are not in charge of their own sexuality, even though they could be in a position to be informed about AIDS. Cultural and social customs and norms rule against them, and without economic power, women cannot decide for themselves. It is their sexual availability, therefore, that brings them men with money for clothing, food, education, and companionship. AIDS among women creates many orphans of AIDS the world over, but especially in Africa.

In 1992, the Uganda Ministry of Health estimated that 50-70 percent of adult patients in Uganda's major hospitals had HIV-related illnesses. Their predictions were that both the cost of and demand for health care for HIV-AIDS patients would increase.

The extent of illness and deaths caused by AIDS has depleted critical sectors of the Ugandan labour force, and the impact of AIDS on individuals, families, and the nation is increasing drastically. AIDS has affected those who are responsible for the support and care of the children and the elderly. Children have been traumatized by the loss of their parents. They are left destitute; families have scattered; and the orphans are left homeless.

At the household level, AIDS reduces the earning capacity and increases expenditures for medical expenses. It is common for families to sell off assets such as land and reared animals in order to care for terminally ill members, pay burial costs, or support the household after a death. Children are pulled out of schools, adding to the country's number of illiterates. In Uganda today it is common to find grandparents surrounded by a fleet of grandchildren, families headed by adolescents, and children being scattered among relatives as a result of deaths of parents due to AIDS. Life expectancy at birth in Uganda has dropped from 46 years in 1991 to 40 years in 1998 World Bank[1].

ECONOMIC SITUATION OF UGANDA

Uganda is a predominantly rural country, with a population of about 22,000,000 (2000 estimate), located at the equator at a relatively high elevation on the East African plateau. It encompasses 200,000 square kilometres, including parts of Lake Victoria, Lake Edward, Lake Albert, and the Nile River from Lake Victoria to the Sudan border, Figure 1. Population density averages about 104 persons per square kilometer, but varies greatly across the country, ranging from fewer than 50 persons per square kilometer in the north to more than 4000 persons per square kilometer in some urban areas in the southeast near Lake Victoria, Uganda Census Office[2], Figure 2.

At independence in October 1962, Uganda had one of the strongest and most promising economies in sub-Saharan Africa. With good climate and fertile soils, the agricultural sector was self-sufficient and also generated adequate foreign exchange. The country had a reputation for quality education at all levels. In the 1990's the gross domestic product (GDP) grew by 4.3 percent. The gross national product (GNP) per capita was U.S. $180 in 1990 and U.S. $330 in 1997, World Bank[3].

The proportion of people who depend on agriculture for their livelihood is very high, 80 percent, including 81 percent of men and 88 percent of women in 1994, World Bank[1]. Agriculture contributes 95 percent of the country's exports, and it is the economic base for many Ugandan manufacturing and service industries. The real GNP data for the 1990's indicate that increase in agricultural production, for both monetary and non-monetary sectors, is around 3.3 percent.

Social indicators of development for Uganda in 1989 indicate that there were 200,000 people per doctor, 15,000 people per health unit, 2,332 people per nurse, and 800 people per health facility bed.

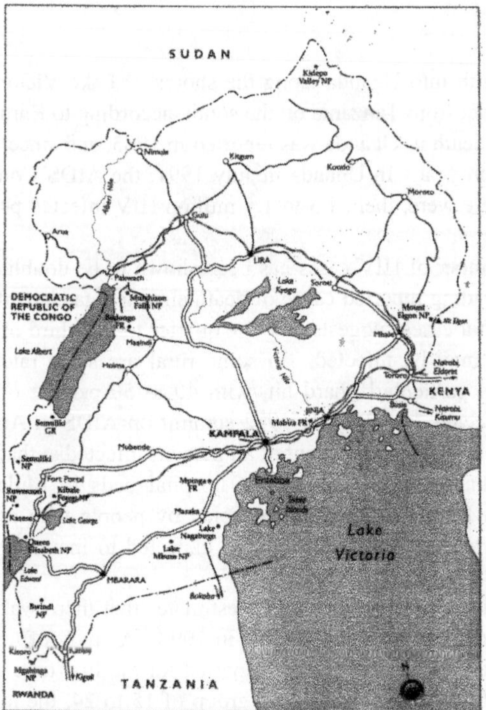

Fig. 1. Map of Uganda.

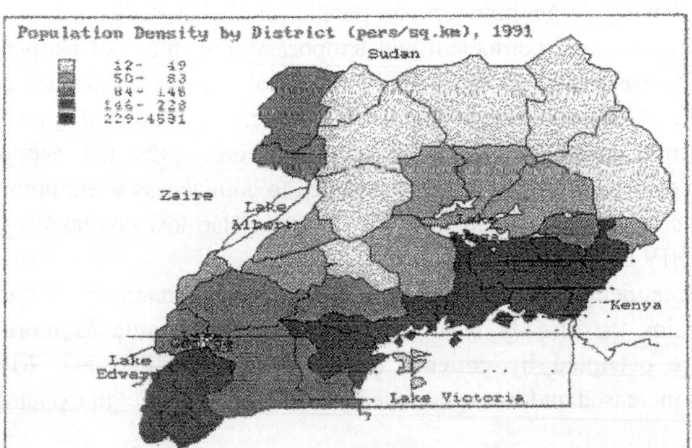

Fig. 2. Uganda population density by District, 1991, Uganda Census Office.

AIDS IN UGANDA

AIDS made its way north into Uganda along the shores of Lake Victoria in the 1970s, through the Rakai District from Tanzania on the south, according to Barnett and Blaikie[4]. The first official AIDS death in Uganda was reported in 1983, and since then the number has been increasing each year. In Uganda in July 1991, the AIDS Control Programme (ACP) reported that there were, then, 1.3 to 1.5 million HIV infected persons - very big figure at that time.

Recently the number of HIV cases has been shown to be doubling every 8 to 12 months. AIDS is the leading reported cause of death among Ugandan adults. Data from blood donors and hospital cases suggests that one quarter to one third of the residents of Kampala (capital city) may be infected. In some rural areas the rates are also high. Rakai District has been particularly hard hit, with 40 to 50 percent of those of child-bearing age already infected. At the July 1992 summit on AIDS in Amsterdam, Rakai District was reported to have had 70 percent of all persons infected at that time.

It is reported that 60 to 70 percent of hospital beds in Mulago, the biggest government hospital in Kampala City, are occupied by people who are HIV positive, Ministry of Health[5]. The infant mortality rate is expected to increase beyond 101 per thousand due to HIV.

The World Health Organization (WHO) estimates that there was little difference in adult HIV infection rates in men and women in 1992. A study of new clinical cases has shown that 91 percent of cases with HIV reported for the first time are adults, and 9 percent are children below eleven. In the age group of 15 to 24, the number of female cases of HIV is five times that of males, while for the age group 25 to 39, both males and females have almost the same infection level.

The major variables influencing the spread of AIDS in Uganda include: demographics - having large numbers of people in the sexually active age group; increasing urbanisation; and permanent and temporary movement of people, especially from severely hit areas to less affected areas. Political and economic factors have also led to the spread of the scourge in Uganda. Politically, wars and civil disturbance encourage migration and extramarital sexual behaviour. On the economic front, differences between men and women enable women to gain access to economic resources through offering sex—a form of prostitution. There is also low coverage by the health system for early HIV diagnosis.

Medical treatment costs per AIDS case are high, and treatment of these patients, therefore, constrains the already weak health system in Uganda as more and more hospital beds are occupied by patients with HIV-related illnesses. Medical drug consumption has increased and has aggravated the already bad foreign exchange situation in Uganda.

The majority of the HIV-infected people are in the productive age group of 14 to 49 years, which is the age group that provides labour in productive sectors of the economy, such as agriculture. There is a serious shortage of labour, especially skilled labour, in all productive sectors of the Ugandan economy.

So far the AIDS epidemic has claimed the lives of three million children worldwide, and another one million are now living with HIV. One in ten of the persons who became newly infected in 1998 was a child. While Africa has only 10 percent of total world population, about 90 percent of all HIV-infected babies are born in Africa, World Bank[6].

The number of AIDS orphans in Uganda is estimated to range between 700,000 and 1.5 million, or in the order of 3 to 5 percent of the total population. With the increasing numbers of adult AIDS cases, it is expected that the number of AIDS orphans will continue to increase. The number of AIDS orphans varies widely throughout Uganda, as indicated by Figures 3 and 4, which show the number of orphans per square kilometer by county, as related to maternal deaths and deaths of both parents in 1991, Uganda Census Office [2]. In 1995 about 300,000 children had lost both parents to AIDS, and this has required establishing 78 new child health units in the last two years.

SOCIAL AND ECONOMIC IMPACTS OF AIDS IN UGANDA

The AIDS epidemic has drastically altered many aspects of life in sub-Saharan Africa, including household structure and income; nutrition, health, and mortality; and agriculture, infrastructure, and the private sector.

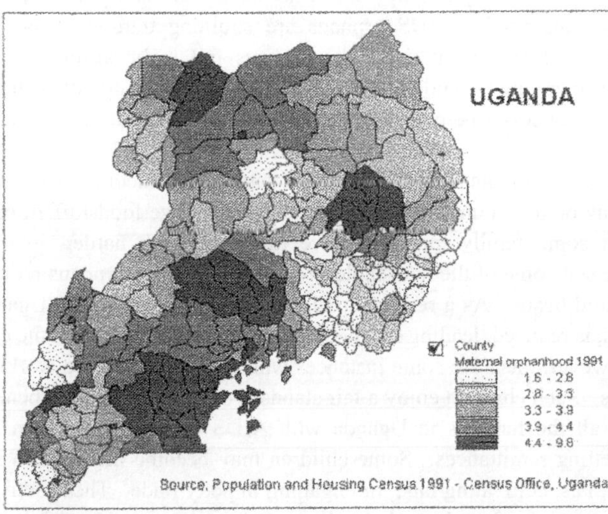

Fig. 3. Maternal orphanhood by county, 1991.

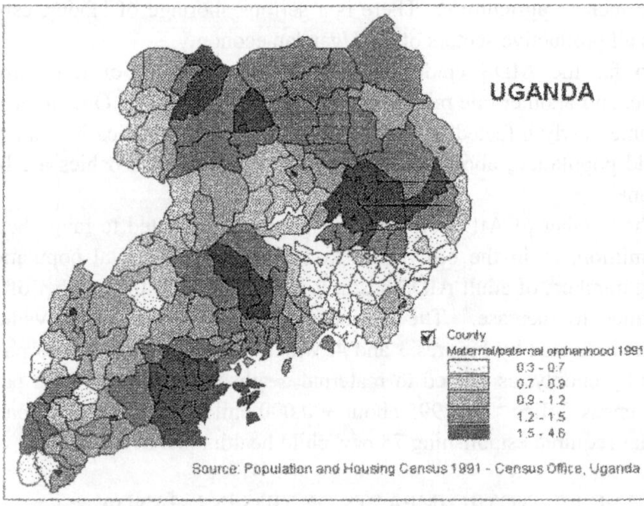

Fig. 4. Maternal and paternal orphanhood by county, 1991.

Orphans and Household Income

The extended family system in Uganda has enabled some orphans to be absorbed by relatives in their homes. However, this is not without problems. Some of these families may already be resource constrained. Some may be already too large, while others may be headed by grandparents who are too old and poor to provide the support that the orphans need. AIDS orphans are straining traditional systems of fostering children in Uganda, and at times the orphans fail to be absorbed into family units, especially if there are no known relatives. These ones end up either in orphanages or on the street. Fifty-seven percent of all the families in Uganda include at least one AIDS orphan.

Peasant earnings mainly constitute subsistence income, and while some of the orphans may be too young to work, they need to receive foodstuff from the home. This means that some family members have to either work harder, in order to feed the orphans, or sell some of the household properties; thus, the orphans reduce the income of the household heads. As a result, the taxes (graduated tax) which Uganda collects from individuals, is reduced, leading to a declining economic situation in the country.

However, there are some instances where children may be working and receiving remittances. Such children enjoy a fair standard of living and home benefits too. Twelve percent of all the families in Uganda with AIDS orphans have been found to include children getting remittances. Some children may become involved in such activities as rearing animals, cultivating land, or engaging in petty trade. These activities improve the economic situation of the households because families are able to buy commodities like soap, sugar, salt, etc. However, there is complaint from some parents that their children do not send them assistance, which is an additional problem of child labour.

Land Ownership and Use

Availability of land determines the economic activities that families engage in, and the system of land ownership determines how land is used. Agricultural policies, as well as cropping patterns, depend on the distribution of land, among other factors. Generally in Uganda, land distribution is uniformly small; 60 percent of all the households own less than 2 acres, and these small plots cannot adequately meet the agricultural needs of the households.

The existence of orphans has led to land fragmentation as each orphan must get a share of the land, resulting in further reduction of land size available for economic activities. Lucky families, with land in small plots, hire extra land from those who have it in order to grow basic foods like potatoes and cassava.

This land ownership system does not enable people to obtain loans from banks and other credit-providing institutions. When landlords "hire" land to poor families, the tenants have to give part of their harvests to the landlords since they wish to be allowed to use the land another time. The problem of land shortage is a complaint largely among those households that have experienced the problems associated with caring for AIDS orphans.

Such land problems have a number of impacts. First of all, the families tend to change from cash to food crops although there may be some who resort to intercropping. The desire to have food forces farmers toward food production and results in a decline in the country's foreign exchange earnings, since agricultural products are the main source of foreign exchange for Uganda. Second, there is also a reduction in farming area, as mentioned earlier, due to partitioning of the land. This change in the area cultivated is a measure of the impact of AIDS on supply of labour and ultimately on incomes of families.

Education

The Government of Uganda has taken action to raise the educational level of children, especially of poor children. The Universal Primary Education (UPE) program, introduced in 1998, provides free primary education for up to four children in every family. Prior to that time, about 43 percent of school-age children were out of school. This was mainly due to lack of funds for school fees, but also some families withdrew children from schools to provide family labour. Of all the children who were out of school, 31.3 percent were AIDS orphans. Some of the orphans lose interest in studying after the loss of their parents, while others stubbornly refuse to go to schools. Given the fact that these orphans stay with relatives and the costs of education are high, the guardians often feel relieved if such children refuse to study.

However this increases the illiteracy level of the population of the country. In Kampala, 15 percent of orphans aged 10 to 15 were illiterate in 1991. With the introduction of UPE in 1998, however, all these children were forced to go to school, especially those who were of primary school age. As a result of this policy, Government is spending as much as U.S. $300 million annually, and this has had a very big impact on the economic situation of a country like Uganda. Primary enrollment increased enormously following introduction of the UPE program, from 2.6 million in 1996 to 6.5 million in 1999, Uganda Ministry of Finance[7]. Enrollment rates in secondary schools have increased somewhat in recent years, but remain low. Total secondary enrollment rose from 336,000 in 1997 to 428,000 in 1999.

Nutrition

Nutrition is a very big problem in Uganda, and the problem is increasing due to the AIDS impact on crop yields. The argument here is that reduction of household size by AIDS leads to poor maintenance of crops and the cutting off of income necessary for the purchase of inputs. Caring for orphans and people devoting much time to treating patients divert household resources from agriculture, both financially and in terms of available labour.

Yields, therefore, are reduced, and hence output drops. This drop in output comes as a result of both lack of labour and financing for inputs like fertilisers, herbicides, pesticides, and mulching materials. Other reasons reported include children migrating to towns, and families being unable to afford hired labour.

As a result, there is scarcity of food variety leading to poor diets and malnutrition. Today in Uganda 54 percent of children aged 5 years and younger are reported to be malnourished. Malnutrition rates have risen due to loss of adults, especially mothers who are the principle food preparers and caretakers of the young, reduced breast milk, and dietary changes.

Health

Most of the orphans acquire AIDS at birth. The Uganda health sector has focused mainly on the essential drug requirements needed to treat the opportunistic infections that strike individuals whose immune systems are particularly weakened. Drugs are a substantial component in Uganda's public health budget (about 33 percent of recurrent expenditures) and consume scarce foreign exchange.

There are a number of signs and symptoms of HIV-related conditions which, however, require treatment, including chronic diarrhea, tuberculosis, body rash, and herpes zoster among others. The average annual cost of treatment per person is estimated at U.S. $13.82. The annual drug costs for AIDS treatment for the years 1991-1995 (in million U.S. dollars) have been 1.4, 1.7, 1.9, 2.1, and 2.3, respectively.

Under the most optimistic scenario, it is estimated that 100,000 children in Uganda will require treatment for AIDS in the year 2000 at a cost of around U.S. $10 million. The drugs for AIDS patients might consume between 8 and 24 percent of a constant public sector drug budget.

Drugs needed to treat AIDS-related illnesses, however, are only one cost imposed on the health system and on families and communities. Other costs involved in caring for AIDS orphans are significant and include direct costs such as medical manpower, hospital overhead, and orphanage costs. The total direct medical costs of treating AIDS patients and orphans, although not quantified here, do consume major shares of the public budget for health care. Note that Tanzania spends around 40 percent of the public health budget on treating AIDS, while Rwanda spends roughly 65 percent for the same.

Agriculture

As noted earlier, agriculture is the engine of growth for the Ugandan economy. The emergence and predominance of food crops as an attractive and profitable cash crop alternative to traditional cash crops such as coffee and tea came about during the recovery period of 1986-91. During that time growth in the agricultural sector was led by growth in food crops, particularly bananas.

Today the government hopes to encourage a shift away from food production for domestic markets toward production of raw materials for processing, direct export, or both, and secondly, to increase the production of more traditional export crops while at the same time diversifying into non-traditional, low-investment intensive agricultural exports. Measures to achieve this focus largely on supply side incentives.

The effects of AIDS, however, restrict the capacity of smallholder producers to respond to changed macroeconomics and microeconomics incentives. That is why some AIDS-fragmented families virtually withdraw into subsistence food production.

For AIDS survivors like orphans, the extent to which they are able to respond to market signals is constrained. Specifically, their access to rural credit is limited. Agricultural extension services and training are predominantly for adults, not for children. Moreover, it is not uncommon for AIDS orphans to lose property rights, posing a threat for their ever attaining food security from subsistence crops.

AIDS orphans, therefore, have major effects on the rural labour market and aggregate agricultural output. Severely affected households need special assistance, and orphan households, with limited or no extended support, are the prime group needing help. Such assistance to AIDS orphans should not be limited to emergency relief, but should also provide technical training, basic implements, and advice on issues like nutrition, marketing, and crop management.

Due to the magnitude of the AIDS epidemic, there is need to develop an early warning system to prevent the most vulnerable households from being thrown more deeply into poverty and for long-term planning purposes such as urban food security and nutrient value of changing crop composition.

Private Sector

AIDS orphans have a very great impact on the private sector. Most firms provide health costs for orphans of their former employees, either through company clinics or by reimbursing expenses from private health care providers. The amount spent on medical care by private firms is always beyond the amounts budgeted. In addition to an average hospitalisation cost of around U.S. $70 per person annually, companies provide social benefits for dependants of their former workers, including health, housing, and at times education assistance.

With regard to salary and termination/pension entitlements, many employers usually give AIDS orphans of their former workers a fixed proportion of the employee's salary, taking into account years of service. These lenient and humanitarian policies have given such firms an economic blow, and the impacts are observable in the economic situation.

There is, therefore, a need to have company policies on HIV/AIDS that enable the firms to reduce the expenses incurred in providing the assistance outlined above.

AIDS Mortality

The AIDS orphans are increasing the mortality rate of Uganda. The target of Uganda is to reduce infant mortality rates from 110 per thousand live births to 80; however, the advent of HIV/AIDS has prevented the success of these efforts. It is estimated that by the year 2005 the Ugandan infant mortality rate (for children under 5), will be 132 per thousand live births if AIDS continues to advance as it is today.

Overall mortality is above normal levels in the country's productive firms in Uganda, and national figures also indicate as many as 30 AIDS deaths in a village per year! This is a very big figure for a village and reduces the economic situation of the area by the costs incurred. Money is spent on coffins and transporting dozens of mourners to home villages for burial. It is estimated that for each individual orphan who stays away from his/her burial home area, a minimum of U.S. $300 is spent by the village and friends, compared with an average monthly salary equivalent to U.S. $80.

CONCLUSIONS

AIDS is a complex disease, and its effects cannot be fully explained in a small paper like this. Its effects cut across all sectors of the Ugandan economy. Also, most often, the effects of AIDS are similar to the effects of other factors.

In Uganda we are very grateful to all the NGO's, both international and indigenous, for the work they extend to AIDS orphans in the country. Without such assistance, the impact of AIDS orphans on our economy would even be worse. Despite the constraints outlined, the Government has continued to mobilise funds so as to ensure that the orphans are looked after.

The problems of orphans is quite noticeable given that their numbers are growing by leaps and bounds. These children have a big role to play in society in the future, and there are many advantages to be gained if these children are educated and join the labour force, either skilled or otherwise. Their health needs to be equally emphasized. In the future, they must become self-sustaining and support their families.

Women who have lost husbands due to AIDS have started to realise the problem of looking after their children, and they are coping with the situation. They are now heading households involved in income-generating enterprises, and in educating, clothing, and feeding their children. As this trend continues, with assistance from Government and the NGOs, the problem can be reduced. I am, therefore, hopeful that the presence of AIDS orphans will, instead of damaging our economy, help to improve it in the long term.

REFERENCES

1. World Bank Development Indicators, 1998.
2. Uganda Census Office, Statistics Department, Kampala, 1991.
3. World Bank Atlas, 1991,1999, and 2000/2001.
4. Barnett T. And Blaikiep. AIDS in Africa, its Present and Future Impact, Belhaven Press, London 1992.
5. Uganda Ministry of Health, AIDS Surveillance Reports, Kampala, 1991, 1999.
6. World Bank. Intensifying Action Against HIV/AIDS in Africa, 2000.
7. Uganda Ministry of Finance. Poverty Reduction Strategy Paper, Uganda's Poverty Eradication Action Plan, Summary and Main Objectives, Kampala, March 2000.

OTHER SOURCES

Berker CM (1990), The Demo - Economic Impact of the AIDS Pandemic in Sub-Saharan Africa, 1990.

Dunn A. Enumeration and Needs Assessment of Orphans in Uganda: Survey Report, Social Works Department, Save the Children Fund, Kampala, (date unavailable).

Sylvester. K (1999),The Linkages Between AIDS and Nutrition/ Food Security in Uganda, World Bank, 1999.

Uganda Ministry of Finance. Background to the Budget, 1998/99, Kampala, 1998.

World Bank Uganda's AIDS Crisis, its Implications for Development,1999.

LIMITS OF DEVELOPMENT – FOCUS ON AFRICA CONSTRAINTS AND TENDENCIES OF RURAL DEVELOPMENT IN SENEGAL

COLONEL MBARECK DIOP
Technical Adviser to The President of Senegal, Dakar, Sengal
AMADOU MOCTAR NIANG
Director of Ecological Monitoring Center of Senegal, Dakar, Senegal

INTRODUCTION

After 40 years of independence Senegal is facing constraints of development particularly where 60% of the population is concerned.

Located in the western part of Africa, the country covers 200,000 km^2 with a population of 9 million growing at a rate of 2.7%.

Dakar, the capital, covers only 3% of the area of the country, but concentrates 25% of the total population of the country, and is facing the same problems of Megacities-like growth of the population (6%): urbanisation; sewage and garbage system; transport; water supply, electricity supply; and unemployment .

This short presentation tries to underline main constraints and tendencies of rural development in Senegal, an example of a Sahelian country.

The limits of development in Senegalese rural area is analysed through five factors:

- climate factors,
- population growth and repartition,
- public investments in agriculture,
- food security,
- potential development of irrigation.

CLIMATE FACTORS

Comparing rainfall between the two periods, 1980-1989, and 1990-1994, we can see a general decrease. (Maps 1 & 2).

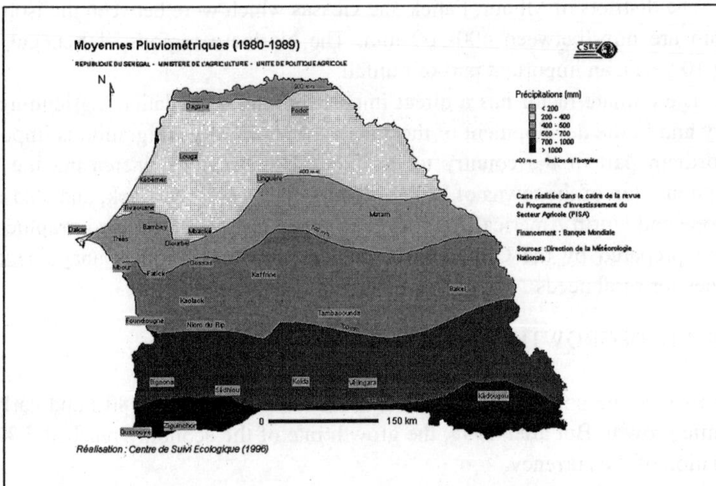

Map 1: Rainfall 1980 - 1989.

For example, isohyete 200 mm is going southward, from Dagana to Louga. Isohyete 400 mm is also going southward from a line Tivavouane-Linguere towards a line Mbour-Diourbel-Matam.

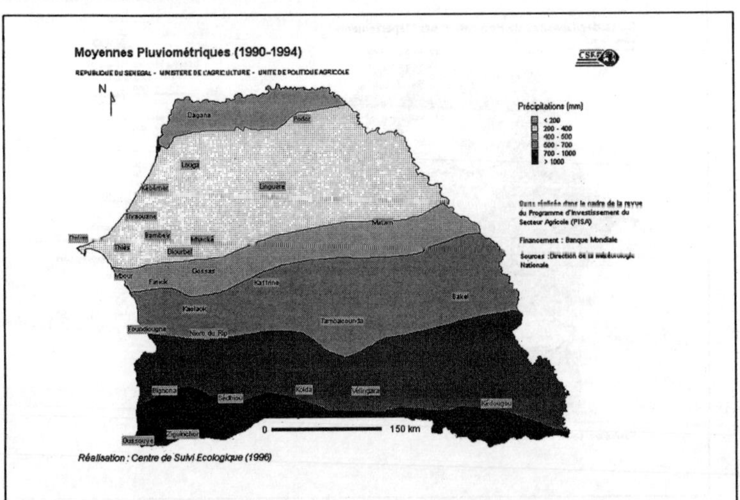

Map 2: Rainfall 1990 - 1994.

The districts of Mbour, Fatick and Gossas which were between the isohyetes 500-700 mm are now between 400-500 mm. The Northern regions of Senegal have lost, during 10 years, an important part of rainfall.

The climate factor has a direct impact on land degradation, agriculture, forestry, poverty and in the development of the country. That is why emigration is important from the Northern part of the country to the South-West regions, increasing the growth of urban population of the towns of Dakar, Saint-Louis, Thies, Kaolack, and also emigration to Europe and North America. An important project of national hydro-graphical network is being prepared by the Government to contribute to a better management of water resources for rural needs .

POPULATION GROWTH AND REPARTITION

The growth of the population (2.7 %) exceeded, during the 70's, 80's and early 90's, the economic growth. But after 1995, the growth rate of the economy reached 5 %, after the devaluation of the currency.

However the growth of urban population remains very high, chiefly for the capital Dakar (6%). The population is concentrated in the Western part of the country, chiefly in Dakar where the density is 400 inhabitants/km^2, while the Central and South-eastern regions have a density of less than 10 inhabitants/ km^2. (Cf. Map 3)

So the quick growth and the bad repartition of the population create other constraints to the development of Senegal.

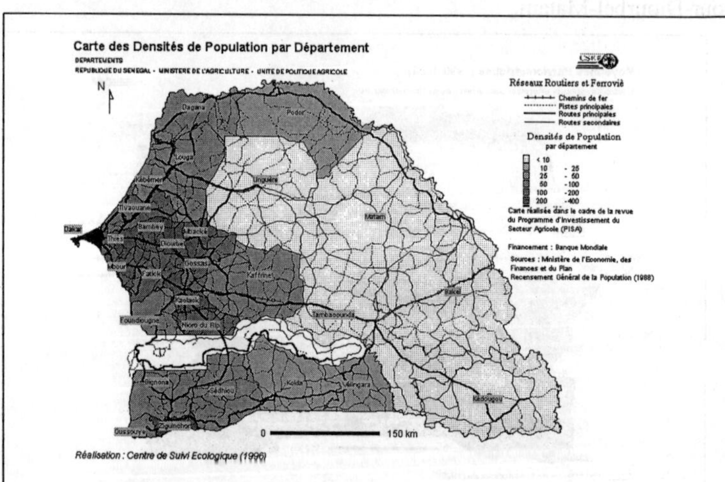

Map 3: Population Density.

PUBLIC AGRICULTURAL INVESTMENTS AND PRODUCTION GROWTH

During the period 1987-1995, a comparison of public investments in agriculture between the regions shows an imbalance in the growth of production and the efficiency of the financing system.

The region of the Valley of the Senegal River has received the most important part of public agricultural investments per capita (300 U.S. $/inhabitant) with a production growth of 40 %, while the Casamance region received 150 U.S. $ per capita, but the rate of production is poor (- 20 %). The Niayes area received only 2 U.S. $ per capita but realised 25 % of production growth (Map 4) .

Region	Invest./hbt.(U.S. $)	Production growth	Observations
Fleuve	300	40%	
Casamance	150	- 20 %	Insecurity
Senegal Oriental	70	- 10 %	Emigration. Low density
Zone sylvo-pastorale	20	- 20 %	Livestock activities
Bassin arachidier	6	- 10 %	Land degradation
Niayes	2	25 %	Private investments

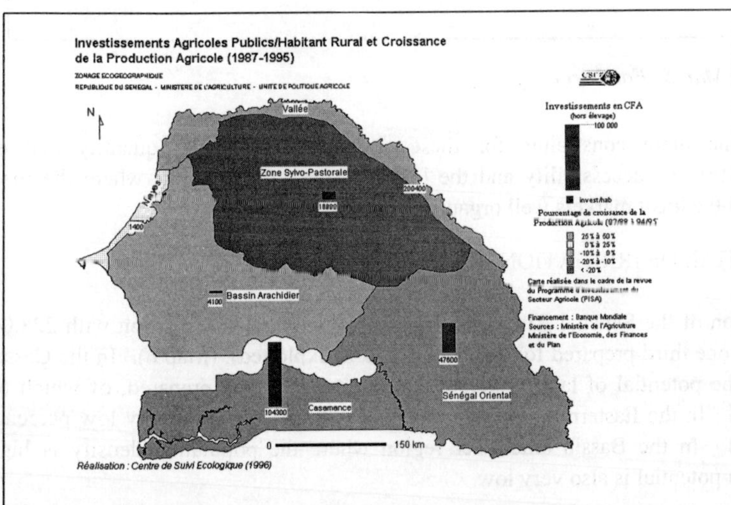

Map 4: Public Agricultural Investments and Production Growth.

FOOD SECURITY

In comparison with FAO standards (185 kg/capita), the following departments do not cover their basic needs, Matam, Fatick, Bambey, Louga, Thiès and Ziguinchor. (Map 5).

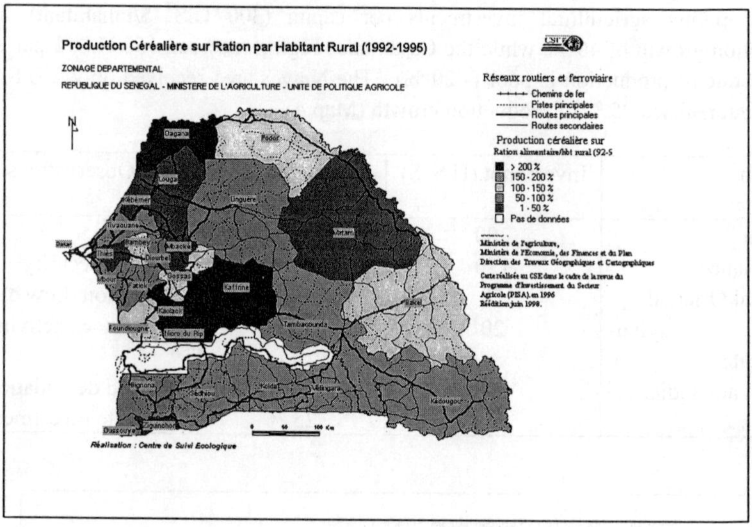

Map 5: Food security.

The main constraints for these areas are the seeds (quantity and quality availability), the accessibility and the loss of crops. The regions where the roads are good and the local markets well organised, the situation is better.

POTENTIAL OF IRRIGATION

The region of the Senegal River has the highest potential of irrigation with 228.000 ha, but only one third prepared for agriculture (54 % exploited). (Map 6). In the Casamance region, the potential of land is 88,000 ha, but only 18% is prepared, of which 60% is exploited. In the Eastern region, the potential is 8.000 ha, but a very low percentage is exploited. In the Bassin arachidier region where the population density is high, the irrigation potential is also very low.

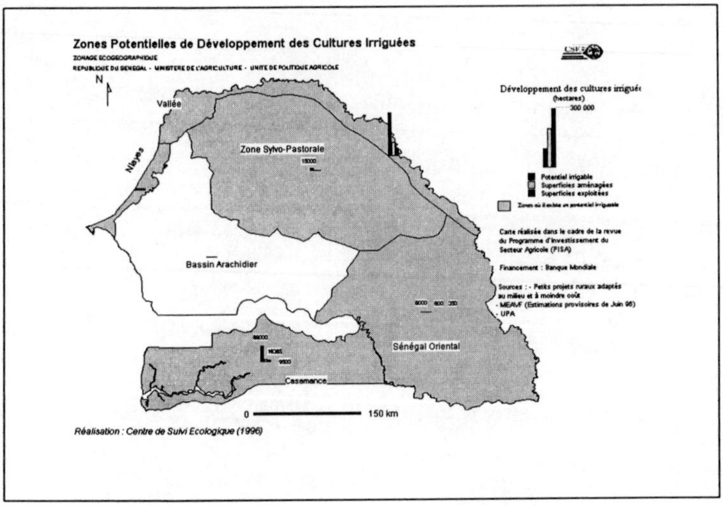

Map 6: Potential of irrigation.

CONCLUSION

The rural development in Senegal and in the Sahelian region is facing many constraints increasing poverty, land degradation, desertification and rural exodus. The challenge will be, for the 21st century, to overcome these environmental, social and economic constraints.

21. SEMINAR PARTICIPANTS

SEMINAR PARTICIPANTS

Mr. Vitor Adefela	WHO Communications Lagos, **Nigeria**
Dr. Jean Pierre Dovie Akue	Polyclinique St. Joseph Lome, **Togo**
H. E. Mario Alessi	Ministry of Foreign Affairs Rome, **Italy**
Professor Ismail Amer	Faculty of Engineering Al-Azhar University Cairo, **Egypt**
Dr. Charles J. Arntzen	Boyce Thompson Institute for Plant Research Inc. Ithaca, **USA**
Dr. Paul Bakai	Johns Hopkins University Research Collaboration Makerere University Kampala, **Uganda**
Professor F. Barre-Sinoussi	Pasteur Institute Paris, France
Dr. Paul Bartel	USGS Washington, **USA**
Dr. Deborah Birx	Walter Reed Army Institute for Research Rockville, **USA**
Dr. David Bodansky	Department of Physics University of Washington Seattle, **USA**
Professor J.M. Borthagaray	Higher Institute of Urbanism University of Buenos Aires Buenos Aires, **Argentina**
Professor Enzo Boschi	National Institute for Geophysics Rome, **Italy**

Dr. A. Farid Boukri	Hydrometeorologic Institute for Training and Research-IHFR Oran, **Algeria**
Dr. Mohamed S. Boulahya	African Centre of Meteorological Applications For Development (ACMAD) Niamey, **Niger**
Dr. Paul Brown	Department of Health and Human Services National Institutes of Health Bethesda, **USA**
Professor Herbert Budka	Institute of Neurology University of Vienna Vienna, **Austria**
Dr. Max Campos	Regional Hydrological Committee San Jose, **Costa Rica**
Dr. Gregory Canavan	Los Alamos National Laboratory Los Alamos, **USA**
Mrs. Tullia Carettoni	UNESCO Italian National Commission Rome, **Italy**
Professor Alberto Cellino	Pino Torinese Astronomical Observatory Torino, **Italy**
Professor Joseph Chahoud	Physics Department Bologna University Bologna, **Italy**
Dr. Nathalie Charpak	Kangaroo Foundation Bogota, **Colombia**
Dr. Andrew F. Cheng	Applied Phyics Laboratory Johns Hopkins University Laurel, **USA**
Professor Robert Clark	Hydrology and Water Resources University of Arizona Tucson, **USA**

Mrs. Deborah Cohen	British Broadcasting Corporation London, **UK**
Dr. William J. Cosgrove	Ecoconsult, Inc. Montreal, **Canada**
Professor Guy de Thé	Pasteur Institute Paris, **France**
Dr. M'Bareck Diop	Technical Advisor to the Presidency Dakar, **Sengal**
Dr. Dumitru Dorogan	National Institute for Marine Research and Development Constanta, **Romania**
Professor Tim Dyson	London School of Economics London, **UK**
Mrs. Viola Egikova	Moscowskaya Pravda Moscow, **Russia**
Dr. Zelig Eshar	Department of Chemical Immunology The Weizmann Institute of Science Rehovot, **Israel**
Professor Lorne Everett	University of California Santa Barbara, **USA**
Progessor Fang Riong-Xiang	Institute of Microbiology Chinese Academy of Sciences Beijing, **China**
Dr. Margaret Farah	Uganda Centre for Disaster Management Kampala, **Uganda**
Dr. Marina Ferreira Rea	Instituto de Saude Sao Paulo, **Brazil**
Professor Anna Ferro-Luzzi	National Institute for Nutrition Rome, **Italy**

Mrs. Lisbeth Fog	Colombian Association of Science Journalism Santafe de Bogota, **Colombia**
Dr. Piero Forcella	Italian Union of Scientific Journalists Rome, **Italy**
Dr. Robert E. Ford	Natural Resources Strategic Planning and Policy USAID Washington, DC., **USA**
Professor Andrei Gagarinski	Kurchatov Institute Moscow, **Russia**
Dr. Carleton Gajdusek	Academic Medical Center University of Amsterdam Amsterdam, **The Netherlands**
Mr. Bertil Galland	24-HEURES Lausanne, **Switzerland**
Professor Robert Gallo	Institute of Human Virology University of Maryland Baltimore, **USA**
Professor Geng Jia-Guo	Institute of Cell Biology Chinese Academy of Sciences Shanghai, **China**
Mr. Wolfgang C. Goede	PM Magazin Munich, **Germany**
Dr. Alberto Gonzales Pozo	Autonomous Metropolitan University Azcapotzalco, **Mexico**
Professor J. Mayo Greenberg	University of Leiden Leiden, **The Netherlands**
Professor L. Hammarstrom	Clinical Immunology Karolinska Institute Huddings, **Sweden**
Professor Majid Hazzanizadeh	Delft University of Technology Delft, **The Netherlands**

Dr. Walter F. Huebner	Southwest Research Institute San Antonio, **USA**
Professor Huo Yu Ping	College of Physics and Engineering Zhengzhou University Zhengzhou, **China**
Dr. Christine Huraux	Mother-Infant HIV Transmission Paris, **France**
Dr. Charles Hutchinson	Office of Arid Lands University of Arizona Tucson, **USA**
Dr. P.K. Iyengar	Atomic Energy Commission Mumbai, **India**
Dr. Cesar Izaurralde	Pacific Northwest National Laboratory Washington, DC., **USA**
Professor Philip T. James	International Obesity Task Force Public Health Policy Group London, **UK**
Professor Oleg Jardetzky	Magnetic Resonance Laboratory Stanford University Palo Alto, **USA**
Professor Douglas Johnson	Graduate School of Geography Clark University Worcester, **USA**
Professor Leonardas Kairiukstis	Laboratory of Ecology and Forestry Vilnius, **Lithuania**
Professor Arturo A. Keller	Bren school of Environmental Science University of California Santa Barbara, **USA**
Mr. Prakash Khanal	Science Writers Association of Nepal Kathmandu, **Nepal**

Professor J-P Kraehenbuhl	Insitute of Biochemistry, ISREC University of Lausanne Lausanne, **Switzerland**
Dr. Andrei Krutskih	Department of Science and Technology Russian Foreign Ministry Moscow, Russia
Dr. Jane-Francis Kuka	State Minister for Disaster Preparedness and Refugees Kampala, **Uganda**
Professor Valery Kukhar	ICSC World Laboratory Branch Ukraine Kiev, **Ukraine**
Ms. Melissa Voss Lapsa	Energy Division Oak Ridge National Laboratory Oak Ridge, **USA**
Professor Tsung-Dao Lee	Department of Physics Columbia University New York, **USA**
Professor Axel Lehmann	Institut fuer Technische Informatik Universitaet der Bunderswehr Muenchen Neubiberg, **Germany**
Dr. Giovanni Levi	Advanced Biotechnology Centre Genova, **Italy**
Dr. Alan D. Lopez	World Health Organization Geneva, **Switzerland**
Dr. Julian Ma	Division of Immunology Guy's Hospital London, **UK**
Mr. Kenji Makino	Department of Liberal Arts University of Tokyo Tokyo, **Japan**
Professor Sergio Martellucci	University of Rome Tor Vergata Rome, **Italy**

Dr. Colin L. Masters	Department of Pathology University of Melbourne Victoria, **Australia**
Mrs. Odile Meuvret	Agence France Presse Paris, **France**
Mrs. Gitte Meyer	Science Journalist Valby, **Denmark**
Professor Valery Mikhailov	Scientific Center of Sea Ecology Odessa, **Ukraine**
Professor Arthur Miller	Department of Science and Technology Studies University College, London London, **UK**
Dr. Vladimir Mirianashvilli	Institute of Geophysics, Academy of Sciences Tbilisi, **Georgia**
Dr. Douglas Morrison	CERN Geneva, **Switzerland**
Dr. P.M. Mullineaux	Sainsbury Laboratory John Innes Institute Norwich, **UK**
Dr. Bineta Ndiaye	Dakar Hospital Dakar, Sengal
Dr. Amagou Moctar Niang	Ecological Followup Centre Dakar, **Senegal**
Mrs. Anna Nolan	Science Journalist Cratloe, **Ireland**
Dr. David Norman	Dept. of Earth and Environmental Sciences New Mexico Institute of Mining and Technology Socorro, **USA**

Dr. Ender Okandan	Dept. of Petroleum and Natural Gas Engineering Middle East Technical University Ankara, **Turkey**
Professor Lennart Olsson	Department of Physical Geography Lund University Lund, **Sweden**
Dr. Jef Ongena	JET Fusion Centre Belsele, **Belgium**
Professor Gennady Palshin	ICSC World Laboratory Branch Ukraine Kiev, **Ukraine**
Professor Donato Palumbo	World Laboratory Fusion Centre Brussels, **Belgium**
Professor Lucio Parenzan	International Heart School Bergamo, **Italy**
Professor Margaret Petersen	Hydrology and Water Resources University of Arizona Tucson, **Arizona**
Professor Mario Pezzotti	University of Verona Verona, **Italy**
Professor Andrei Piontkovsky	Strategic Studies Centre Moscow, **Russia**
Professor Juras Pozela	ICSC World Laboratory Branch Vilnius, **Lithania**
Dr. Hadi Pratomo	Faculty of Public Health University of Indonesia Jakarta, **Indonesia**
Mr. Sergio Prenafeta-Jenkin	Ibero-American Association of Science Journalists Santiago, **Chile**
Professor Richard Ragaini	Department of Environmental Protection University of California Livermore, **USA**

Professor Vittorio Ragaini	University of Milano Milano, **Italy**
Professor Karl Rebane	ICSC World Laboratory Estonian Branch Tallinn, **Estonia**
Dr. Curt Reynolds	USDA-ARS Hydrology Laboratory Beltsville, **USA**
Professor Paolo Ricci	University of San Francisco San Francisco, **USA**
Dr. Maura Ricketts	World Health Organization Geneva, **Switzerland**
Dr. George O. Rogers	Dept. of Landscape Architecture and Urban Planning Texas A&M University College Station, **USA**
Dr. Norman Rosenberg	Pacific Northwest National Laboratory Washington, DC., **USA**
Professor Zenonas Rudzikas	Institute of Theoretical Physics and Astronomy Lithuanian Academy of Sciences Vilnius, **Lithuania**
Professor Francesco Sala	Biology Department University of Milano Milano, **Italy**
Professor Ilkay Salihoglu	Institute of Marine Sciences Middle East Technical University (METU) Erdemli, **Turkey**
Professor N.M. Samuel	Experimental Medicine and AIDS Resource Center The Tamil Nadu Medical University Guindy-Chennai, **India**
Dr. Hiltmar Schubert	Fraunhofer-Institut fur Chemische Technologie Pfinztal, **Germany**

Dr. Beat Schurch	Nestle Foundation Lausanne, **Switzerland**
Professor Geraldo G. Serra	Sao Paulo State University NUTAU Sao Paulo, **Brazil**
Dr. Ed Sheffner	NASA Headquarters Washington, DC., **USA**
Professor Prakash Shetty	London School of Hygiene and Tropical Medicine London, **UK**
Professor Kai M.B. Siegbahn	Institute of Physics University of Uppsala Uppsala, **Sweden**
Professor K. Sivaramakrishnan	Centre for Policy Research New Delhi, **India**
Professor Soroosh Sorooshian	Hydrology and Water Resources University of Arizona Tucson, **USA**
Professor William A. Sprigg	Insitute for the Study of Planet Earth University of Arizona Tucson, **USA**
Dr. Bruce N. Stram	Enron Energy Services Houston, **USA**
Dr. Glenn Tallia	National Oceanic and Atmopspheric Administration Silver Spring, **USA**
Professor Albert Tavkhelidze	National Academy of Sciences Tbilisi, **Georgia**
Ms. Kay Thompson	U.S. Department of Energy Washington, DC., **USA**
Dr. Larry Tieszen	International Programs EROS Data Center, USGS Sioux Falls, **USA**

Professor Vitalii Tsygichko	Institute for System Studies Russian Academy of Sciences Moscow, **Russia**
Mr. Geir Tveit	Science Journalist Valby, **Denmark**
Dr. Paul F. Uhlir	National Academy of Sciences Washington, DC., **USA**
Professor Marcel Vivargent	CERN Geneva, **Switzerland**
Professor Francois Waelbroek	Juelich Fusion Centre St. Amandsberg, **Belgium**
Dr. Robert Walgate	Open Solutions Northwood, **UK**
Dr. Warren M. Washington	National Venter for Atmospheric Research Boulder, **USA**
Dr. Henning Wegener	German Ambassador in Spain (former) Madrid, **Spain**
Dr. Catherine Wilfert	Elizabeth Glaser Pediatric AIDS Foundation Chapel Hill, **USA**
Dr. Robert G. Will	Western General Hospital National CJD Surveillance Unit Edinburgh, **UK**
Professor John Wilson	Department of Earth and Environmental Sciences New Mexico Institute of Mining and Technology Socorro, **USA**
Professor Hans Wolf	Regensburg University Regensburg, **Germany**
Dr. Lowell Wood	Lawrence Livermore National Laboratory Livermore, **USA**

Professor Zheng Kai Xu	Shanghai Institute of Plant Physiology China Academy of Sciences Shanghai, **China**
Dr. Rolf Zetterstrom	Acta Paediatria Stockholm, **Sweden**
Professor Antonino Zichichi	CERN & University of Bologna Geneva, **Switzerland**